Norwegian Petroleum Society (NPF), Special Publication No. 11

Hydrocarbon Seal Quantification

Papers presented at the Norwegian Petroleum Society Conference,
16–18 October 2000, Stavanger, Norway

Further titles in the series:

1. R.M. Larsen, H. Brekke, B.T. Larsen and E. Talleraas (Editors)
STRUCTURAL AND TECTONIC MODELLING AND ITS APPLICATION TO PETROLEUM GEOLOGY – Proceedings of Norwegian Petroleum Society Workshop, 18–20 October 1989, Stavanger, Norway

2. T.O. Vorren, E. Bergsager, Ø.A. Dahl-Stamnes, E. Holter, B. Johansen, E. Lie and T.B. Lund (Editors)
ARCTIC GEOLOGY AND PETROLEUM POTENTIAL – Proceedings of the Norwegian Petroleum Society Conference, 15–17 August 1990, Tromsø, Norway

3. A.G. Doré et al. (Editors)
BASIN MODELLING: ADVANCES AND APPLICATIONS – Proceedings of the Norwegian Petroleum Society Conference, 13–15 March 1991, Stavanger, Norway

4. S. Hanslien (Editor)
PETROLEUM: EXPLORATION AND EXPLOITATION IN NORWAY – Proceedings of the Norwegian Petroleum Society Conference, 9–11 December 1991, Stavanger, Norway

5. R.J. Steel, V.L. Felt, E.P. Johannesson and C. Mathieu (Editors)
SEQUENCE STRATIGRAPHY ON THE NORTHWEST EUROPEAN MARGIN – Proceedings of the Norwegian Petroleum Society Conference, 1–3 February, 1993, Stavanger, Norway

6. A.G. Doré and R. Sinding-Larsen (Editors)
QUANTIFICATION AND PREDICTION OF HYDROCARBON RESOURCES – Proceedings of the Norwegian Petroleum Society Conference, 6–8 December 1993, Stavanger, Norway

7. P. Møller-Pedersen and A.G. Koestler (Editors)
HYDROCARBON SEALS – Importance for Exploration and Production

8. F.M. Gradstein, K.O. Sandvik and N.J. Milton (Editors)
SEQUENCE STRATIGRAPHY – Concepts and Applications – Proceedings of the Norwegian Petroleum Society Conference, 6–8 September 1995, Stavanger, Norway

9. K. Ofstad, J.E. Kittilsen and P. Alexander-Marrack (Editors)
IMPROVING THE EXPLORATION PROCESS BY LEARNING FROM THE PAST – Proceedings of the Norwegian Petroleum Society Conference, September 1998, Haugesund, Norway

10. O.J. Martinsen and T. Dreyer (Editors)
SEDIMENTARY ENVIRONMENTS OFFSHORE NORWAY — PALAEOZOIC TO RECENT – Proceedings of the Norwegian Petroleum Society Conference, 3–5 May 1999, Bergen, Norway

Norwegian Petroleum Society (NPF), Special Publication No. 11

Hydrocarbon Seal Quantification

Papers presented at the Norwegian Petroleum Society Conference,
16–18 October 2000, Stavanger, Norway

Edited by

Andreas G. Koestler
GEO-RECON A/S, Munkedamsveien 67, N-0270 Oslo, Norway

and

Robert Hunsdale
Phillips Petroleum Company, P.O. Box 220, N-4098 Tananger, Norway

2002
ELSEVIER
Amsterdam – Boston – London – New York – Oxford – Paris – San Diego –
San Francisco – Singapore – Sydney – Tokyo

ELSEVIER SCIENCE B.V.
Sara Burgerhartstraat 25
P.O. Box 211, 1000 AE Amsterdam, The Netherlands

First edition 2002

Library of Congress Cataloging-in-Publication Data
Norsk petroleumsforening. Conference (2000: Stavanger, Norway)
Hydrocarbon seal quantification: papers presented at the Norwegian Petroleum Society Conference, 16–18 October 2000, Stavanger, Norway/edited by Andreas G. Koestler and Robert Hunsdale. – 1st ed.
p. cm – (Norwegian Petroleum Society (NPF) Special Publication; no. 11)
Includes bibliographical references and index.
ISBN 0-444-50661-6 (hardcover)
1. Traps (Petroleum geology)–Congresses. I. Koestler, Andreas G, 1953- II. Hunsdale, Robert. III. Title. IV. Series.
TN870.57 .N67 2000
553.2′8–dc21 2001058545

British Library Cataloguing-in-Publication Data
Hydrocarbon seal quantification: papers presented at the Norwegian Petroleum Society conference, 16–18 October 2000, Stavanger, Norway
– (Norwegian Petroleum Society (NPF), Special Publication; no. 11)
1. Petroleum engineering – Congresses. 2. Petroleum – Geology – Congresses
I. Koestler, A.G. II. Hunsdale, Robert III. Norsk Petroleumsforening IV. Norsk Petroleumsforening Conference (2000: Stavanger, Norway)
622.3′38
ISBN: 0-444-50661-6

ISBN: 0-444-50661-6

Series ISSN: 0928-8937

⊗ The paper used in this publication meets the requirements of ANSI/NISO Z39.48-1992 (Permanence of Paper).

Printed in The Netherlands

Preface

In October 2000 the Norwegian Petroleum Society hosted the Hydrocarbon Seals Quantification Conference in Stavanger, Norway. The conference was a follow up to the successful Hydrocarbon Seals (Importance for Exploration and Production) meeting held in Trondheim five years earlier, and as such this conference gave the opportunity for the petroleum community to view how the subject had developed over those years. A total of 181 delegates attended the conference with 50 papers presented in oral and poster form over the three days of the conference. In addition a series of software demonstrations were run throughout the conference exemplifying, at least in the software realm, the advances that have been made in both automating and visualising the quantification of hydrocarbon seals.

Out of the papers presented, 17 are published in this volume. These selected papers reflect the flavour of the conference, falling into three broad categories: methodologies addressing cap-rock integrity, methodologies relating to fault seal and case studies both from the hydrocarbon basins of North-western Europe and in the form of outcrop examples. With the North Sea, Norwegian Sea and Atlantic Margin moving along their respective basin maturity and development curves, exploration is being forced deeper into high pressure/high temperature terrains, while exploitation and development requires greater precision and realism in reservoir simulations to maximise drilling strategies to prolong field life. In all instances the need for predictive tools and methodologies that address the integrity and behaviour of top and lateral (fault) seals to hydrocarbon traps, both in the static and dynamic state, have been identified as key risk factors and this is reflected in this volume.

Five years ago, K.J. Weber provided a historical overview to the subject that set up the thread for discussion through the rest of NPF Special Publication No. 7 (P. Møller-Pedersen and A.G. Koestler, 1997). In the opening paper of this volume, Yielding presents a general overview of fault seal methodology and its limitations, while touching upon techniques and methods more definitively dealt with in the succeeding papers.

In the section pertaining to methodologies on cap-rock integrity a range of papers from the theoretical to basin-scale observations are presented. H.M.N. Bolås and Ch. Hermanrud show how an understanding of rock stress within sedimentary basins can have implications for trap integrity, while H.M. Helset et al. raise the issue of diagenesis and chemical compaction in over-pressure development and cap-rock failure. K. Nakayama and D. Sato present a theoretical model for predicting top seal capacity using the equivalent grain size method. M. Wangen discusses the issue of the effective stress of sedimentary rocks that have failed through hydro-fracturing. H. Lewis et al. take a geomechanical approach to investigating top seal integrity. This section is concluded by a review of top seal capacity in exhumed settings by D.V. Corcoran and A.G. Doré, who utilise the Atlantic Margin and its boarderland basins for illustration.

Empirical observations on fault rock properties from deformation rig experiments are presented by S. Sperrevik et al. and introduce the section on methodologies for addressing fault seal potential. C. Childs et al. describe a method for incorporating the capillary properties of fault rocks into migration models by combining several established techniques in a novel way, while D. Grauls et al. show how pressure data can be used to address field compartmentalisation. K. Hollund et al. show how data and methods such as those described in the three preceding papers can be incorporated, via computer software, into full flow reservoir simulations. In concluding this section J.C. Rivenæs and C. Dart pose the question of whether or not we have the tools to evaluate if two-phase flow can lead to reservoir compartmentalisation.

The final section presents case studies from North and Norwegian Sea fields as well as an outcrop study from the USA. C. Childs et al. show how large pressure differentials can exist across relatively small faults using examples from the Northern North Sea and discuss the implications of such phenomena. D. Wiprut and M.D. Zoback focus on how faults affect the migration of fluids in reservoirs using four oil and gas fields from the northern North Sea to illustrate the significance of geometrical structures and pressure relationships. C. Hermanrud and H.M.N. Bolås describe the relationship between leakage from overpressured reservoirs on the Haltenbanken, suggesting the potential for deeper hydrocarbon traps exists in that area. In another paper highlighting the Haltenbanken area, G.M.G. Teige et al. utilise the seismic signature of the overburden to try and evaluate the cap-rock integrity of potential structures at depth.

The paper of G. Lewis et al. concludes this section, and the volume. Using field examples from Kentucky, USA, the authors show how sand can be injected and flow when strata are faulted at low confining pressures. Such analysis provides a limiting factor when utilising fault seal methods and stresses to focus on the importance of understanding the geohistory. Although last this should not be considered least, as it is field studies that provide the analogues for the concepts that the preceding papers strive to describe. This paper provides a good example of how natural examples can help develop concepts and provide information needed to explain anomalies to analytical results.

The editors feel that the papers included in this volume show significant advances in the understanding and application of hydrocarbon seal methodologies since the initial conference in 1995. With interest in and application of the subject growing further advances will be forthcoming, perhaps to be debated in an NPF forum five years ahead. We would like to thank all the contributors for their efforts and co-operation during the preparation of this volume. The responsibility taken on by the reviewers is also greatly acknowledged, while thanks are extended to the NPF, without whom this conference would not have been possible.

Andreas G. Koestler and Robert Hunsdale

Oslo/Stavanger, October 2001

List of Contributors

L.M. BONNELL	*Geologica A/S, P.O. Box 8034, N-4003 Stavanger, Norway* *Present address: Geocosm, Austin, TX, USA*
C. CHILDS	*Fault Analysis Group, Department of Geology, University College Dublin, Dublin, Ireland*
M.G. CONTURSI	*Statoil's Research Center, N-7005 Trondheim, Norway*
D.V. CORCORAN	*Statoil Exploration (Ireland) Ltd., Statoil House, 6, George's Dock, IFSC, Dublin 1, Ireland*
G.D. COUPLES	*Department of Petroleum Engineering, Heriot-Watt University, Edinburgh EH14 4AS, Scotland, UK*
C. DART	*Norsk Hydro ASA, P.O. Box 7190, N-5020 Bergen, Norway*
A.G. DORÉ	*Statoil (UK) Ltd., 11a, Regent Street, London SW1Y 4ST, UK*
P.E. ELIASSEN	*Statoil ASA, P.O. Box 300, N-4035 Stavanger, Norway*
Q.J. FISHER	*Rock Deformation Research Group, Earth Sciences Department, University of Leeds, Leeds, UK*
I. FRETTE	*Geologica A/S, P.O. Box 8034, N-4003 Stavanger, Norway*
M. GADING	*Statoil's Research Centre, N-7005 Trondheim, Norway*
J. GJERDE	*Norwegian Computing Center, P.O. Box 114, Blindern, N-0314 Oslo, Norway*
P.A. GILLESPIE	*Norsk Hydro Research Centre, P.O. Box 7190, N-5020 Bergen, Norway*
D. GRAULS	*Total Fina Elf e.p., Subsurface and Petrophysics Avenue Larribau, 64018 Pau, France*
T. HALVORSEN	*Department of Geology, University of Bergen, Allegt. 41, N-5007 Bergen, Norway*
A.E. HEATH	*Fault Analysis Group, Department of Earth Sciences, University of Liverpool, Liverpool L69 3BX, UK*
H.M. HELSET	*Geologica A/S, P.O. Box 8034, N-4003 Stavanger, Norway*
C. HERMANRUD	*Statoil's Research Centre, N-7005 Trondheim, Norway*
L. HOLDEN	*Norwegian Computing Center, P.O. Box 114, Blindern, N-0314 Oslo, Norway*
K. HOLLUND	*Norwegian Computing Center, P.O. Box 114, Blindern, N-0314 Oslo, Norway*
O.S. KLØVJAN	*Statoil, P.O. Box 40, N-9401 Harstad, Norway*

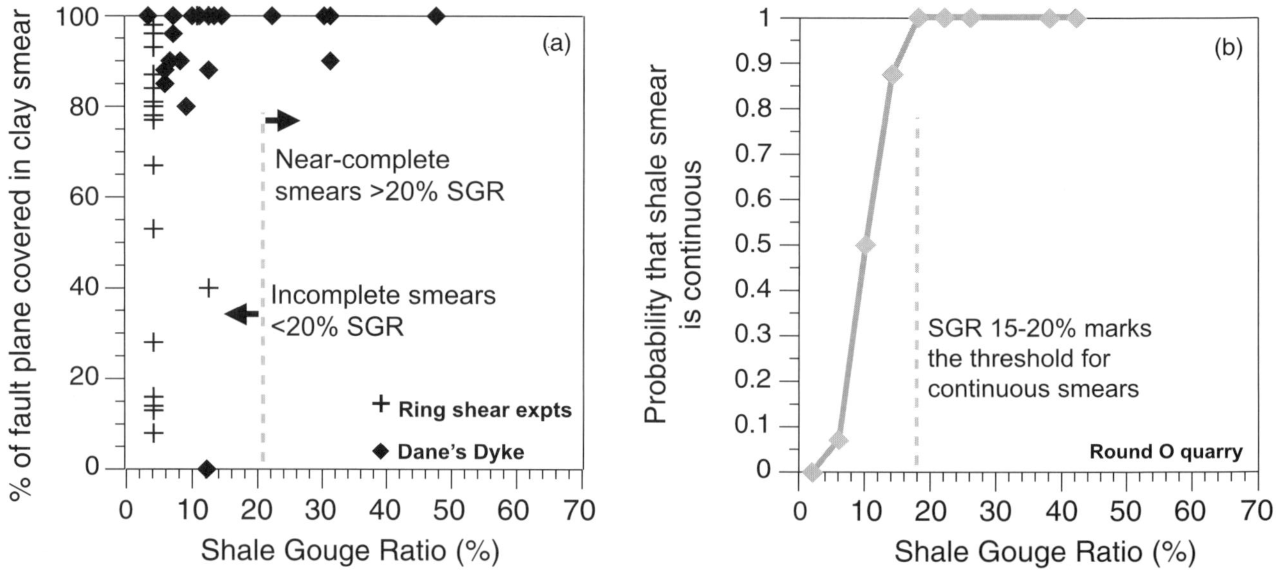

Fig. 3. Outcrop and experimental data relating to clay smear continuity. (a) Individual measurements from Dane's Dyke marls and ring-shear experiments. For each faulted clay layer, the proportion of the separation that is covered in a clay smear is recorded; 100% = smear is continuous from upthrown part of layer to downthrown part. Shale Gouge Ratio was also calculated, as a proportion of clay beds in the 'throw window' (cf. Fig. 1). Diamonds show measurements from marl smears in the Cretaceous Chalk at Dane's Dyke, Yorkshire (Childs, 2000). Maximum burial depth was ca. 1–1.5 km (Hillis, 1995). Crosses show measurements based on ring-shear experiments by Sperrevik et al. (2000), using unconsolidated sand and clay at low confining pressure (equivalent to <50 m burial depth). Both sets of data show many incomplete smears below SGR = 20%, but most smears are continuous at SGR > 20%. (b) Smear probability plot for shale smears in the Carboniferous Coal Measures at Round O Quarry (Lancashire, UK) using data collected by Childs (2000). This plot summarises observations of 80 m of fault traces on vertical quarry faces, with throws up to 5.5 m. Faulting probably occurred after lithification at 2–3 km depth (see Lindsay et al., 1993). As with the Dane's Dyke and ring-shear data in (a), SGR values of ca. 20% correspond to continuous smears.

they often have a very heterogeneous structure with a mixture of deformation products (e.g. Burhannudinnur and Morley, 1997; Foxford et al., 1998; Walsh et al., 1998; Heynekamp et al., 1999; Lewis et al., 2002). Therefore the upscaled composition predicted by the SGR algorithm may mask significant internal variation. For example, a fault in a sand-dominated sequence with only occasional clay layers may give SGR values <10% but still contain clay smears. The critical parameter here (for trap integrity) is probably the continuity of the smears. Fig. 3 summarises smear continuity data from a variety of outcrops and ring-shear experiments. The experimental data (based on work by Sperrevik et al., 2000) is for clay smears resulting from fault slip at low confining pressure (equivalent to <50 m burial depth). The outcrop data (from Childs, 2000) are for marl smears in the Cretaceous Chalk at Dane's Dyke Yorkshire (lithified before being faulted at <1.5 km depth), and for shale smears in Carboniferous Coal Measures (faulted at 2–3 km depth). All these data sets show that, although shale smears may occur when shale beds form only a small fraction of the faulted sequence, the smears tend to remain discontinuous until the proportion of shale beds reaches 15–20%. Similar observations were made at the Moab Fault by Foxford et al. (1998). On this basis a Shale Gouge Ratio of ca. 20% should be a threshold above which continuous shale smears can offer the prospect of an intact seal, and this is borne out by subsurface studies (see below). The precise value for the threshold will be affected by local conditions of the faulting. For example, in superficial near-surface slumping, sand injection processes can enhance the sand contribution into the fault zone relative to the SGR calculation, and would give a higher value for the threshold (Lewis et al., 2002).

An important point relevant to outcrop studies is that SGR is a predictor of upscaled fault-zone *composition*, not shale smear *thickness*. A number of studies have demonstrated that calculated SGR does *not* correlate with the thickness of shale smear or shaly gouge (e.g. Childs, 2000; van der Zee and Urai, 2000).

Seal strength of fault-gouge samples

For thorough trap evaluation, we not only require a prediction of the presence or absence of fault seal, but also an estimate of how 'strong' the fault seal might be. That is, can the relevant parts of the fault plane hold back the excess pressures caused by a commercial hydrocarbon column? To answer this question, we require a prediction of the distribution of capillary entry pressure over the fault surface. For faults in clastic sequences without significant diagenesis, the major control on capillary entry pressure is likely to

List of Contributors

L.M. BONNELL	*Geologica A/S, P.O. Box 8034, N-4003 Stavanger, Norway* *Present address: Geocosm, Austin, TX, USA*
C. CHILDS	*Fault Analysis Group, Department of Geology, University College Dublin, Dublin, Ireland*
M.G. CONTURSI	*Statoil's Research Center, N-7005 Trondheim, Norway*
D.V. CORCORAN	*Statoil Exploration (Ireland) Ltd., Statoil House, 6, George's Dock, IFSC, Dublin 1, Ireland*
G.D. COUPLES	*Department of Petroleum Engineering, Heriot-Watt University, Edinburgh EH14 4AS, Scotland, UK*
C. DART	*Norsk Hydro ASA, P.O. Box 7190, N-5020 Bergen, Norway*
A.G. DORÉ	*Statoil (UK) Ltd., 11a, Regent Street, London SW1Y 4ST, UK*
P.E. ELIASSEN	*Statoil ASA, P.O. Box 300, N-4035 Stavanger, Norway*
Q.J. FISHER	*Rock Deformation Research Group, Earth Sciences Department, University of Leeds, Leeds, UK*
I. FRETTE	*Geologica A/S, P.O. Box 8034, N-4003 Stavanger, Norway*
M. GADING	*Statoil's Research Centre, N-7005 Trondheim, Norway*
J. GJERDE	*Norwegian Computing Center, P.O. Box 114, Blindern, N-0314 Oslo, Norway*
P.A. GILLESPIE	*Norsk Hydro Research Centre, P.O. Box 7190, N-5020 Bergen, Norway*
D. GRAULS	*Total Fina Elf e.p., Subsurface and Petrophysics Avenue Larribau, 64018 Pau, France*
T. HALVORSEN	*Department of Geology, University of Bergen, Allegt. 41, N-5007 Bergen, Norway*
A.E. HEATH	*Fault Analysis Group, Department of Earth Sciences, University of Liverpool, Liverpool L69 3BX, UK*
H.M. HELSET	*Geologica A/S, P.O. Box 8034, N-4003 Stavanger, Norway*
C. HERMANRUD	*Statoil's Research Centre, N-7005 Trondheim, Norway*
L. HOLDEN	*Norwegian Computing Center, P.O. Box 114, Blindern, N-0314 Oslo, Norway*
K. HOLLUND	*Norwegian Computing Center, P.O. Box 114, Blindern, N-0314 Oslo, Norway*
O.S. KLØVJAN	*Statoil, P.O. Box 40, N-9401 Harstad, Norway*

R.J. KNIPE	*Rock Deformation Research Group, Earth Sciences Department, University of Leeds, Leeds LS2 9JT, UK*
R.H. LANDER	*Geologica A/S, P.O. Box 8034, N-4003 Stavanger, Norway* *Present address: Geocosm, Austin, TX, USA*
G. LEWIS	*Norsk Chevron, Karenslyst Alle 2-4, P.O. Box 97 Skøyen, 0212 Oslo, Norway*
H. LEWIS	*Department of Petroleum Engineering, Heriot-Watt University, Edinburgh EH14 4AS, Scotland, UK*
A. LI	*Rock Deformation Research Group, University of Leeds, Leeds LS2 9JT, UK*
H. LØSETH	*Statoil's Research Centre, N-7005 Trondheim, Norway*
T.H. LYGREN	*Norsk Hydro, Oseberg Exploration, Bergen, Norway*
T. MANZOCCHI	*Fault Analysis Group, Department of Geology, University College Dublin, Dublin, Ireland*
J.C. MATTHEWS	*Geologica A/S, P.O. Box 8034, N-4003 Stavanger, Norway*
A.J. McCANN	*Norwegian Computing Center, P.O. Box 114, Blindern, N-0314 Oslo, Norway*
S. MORIYA	*Fault Analysis Group, Department of Geology, University College Dublin, Dublin, Ireland*
P. MOSTAD	*Norwegian Computing Center, P.O. Box 114, Blindern, N-0314 Oslo, Norway*
K. NAKAYAMA	*JGI Inc., 1-5-21 Otsuka Bunkyo-ku, Tokyo, 112-0012 Japan*
B.F. NIELSEN	*Norwegian Computing Center, P.O. Box 114, Blindern, N-0314 Oslo, Norway*
P.A.R. NELL	*Fault Analysis Group, Department of Earth Sciences, University of Liverpool, Liverpool L69 3BX, UK*
H.M. NORDGÅRD BOLÅS	*Statoil's Research Centre, N-7005 Trondheim, Norway*
P. OLDEN	*Department of Petroleum Engineering, Heriot-Watt University, Edinburgh EH14 4AS, Scotland, UK*
F. PASCAUD	*GEOGRAPH, rue Cail, 75010 Paris, France*
P. REEMST	*Geologica A/S, P.O. Box 8034, N-4003 Stavanger, Norway* *Present address: NAM, P.O. Box 28000, 9400 HH Assen, The Netherlands*
J.C. RIVENÆS	*Norsk Hydro ASA, P.O. Box 7190, N-5020 Bergen, Norway*
T. RIVES	*Total Fina Elf e.p., Structural Geology Department, Avenue Larribau, 64018 Pau, France* *Present address: Total Fina Elf Nederland, Den Haag, The Netherlands*
D. SATO	*Technical Research Center, Japan National Oil Corporation, 1-2-2 Hamada, Mihama-ku, Chiba, Japan*
S. SPERREVIK	*Department of Geology, University of Bergen, Allegt. 41, N-5007 Bergen, Norway* *Present address: Norsk Hydro Research Centre, P.O. Box 7190, N-5020 Bergen, Norway*

J.A. STRAND	*Fault Analysis Group, Department of Geology, University College Dublin, Dublin, Ireland*
E. SVERDRUP	*Roxar Software Solutions A/S, P.O. Box 165, N-0212 Skøyen, Norway*
Ø. SYLTA	*SINTEF Petroleumsforskning A/S, Trondheim, Norway*
G.M.G. TEIGE	*Statoil's Research Centre, N-7005 Trondheim, Norway*
C. TOWNSEND	*Statoil's Research Centre, N-7005 Trondheim, Norway*
J.J. WALSH	*Fault Analysis Group, Department of Geology, University College Dublin, Dublin, Ireland*
M. WANGEN	*Institute for Energy Technology, P.O. Box 40, N-2027 Kjeller, Norway*
D. WIPRUT	*Department of Geophysics, 397 Panama Mall, Stanford University, Stanford, CA 94305-2215, USA* *Present address: GeoMechanics International Inc., Parmelia House Level 1, 191 St. George's Terrace, Perth, WA 6000, Australia*
G. YIELDING	*Badleys, North Beck House, North Beck Lane, Hundleby, Spilsby, Lincolnshire PE23 5NB, UK*
M.D. ZOBACK	*Department of Geophysics, 397 Panama Mall, Stanford University, Stanford, CA 94305-2215, USA*

Contents

Shale Gouge Ratio — calibration by geohistory

Graham Yielding

At the 1996 NPF Conference on Hydrocarbon Seals we gave the first presentation of results of a fault-seal study using the Shale Gouge Ratio algorithm, describing a project undertaken in 1994 by Badleys and Norsk Hydro on the Oseberg Syd field. Over subsequent years the methodology has been applied to many tens of data sets in both exploration and production environments. This Special Publication represents an opportunity to review the performance of this fault-seal predictor. Shale Gouge Ratio, or SGR, is an estimate of the proportion of shaly material in the fault zone. This parameter is of direct importance in fault-seal prediction because the very fine-grained nature of phyllosilicates results in very small pore-throats, giving high capillary entry pressures and low permeabilities for the fault-zone material. Measurements on fault-gouge samples show that phyllosilicate content is the first-order control on their fluid-flow properties. It is used to define the fault-gouge type in mixed clastic sequences (e.g. cataclasites/framework-phyllosilicate fault rocks/clay smears). The basic assumption in the SGR algorithm is that the fault-gouge composition is governed by the bulk composition of the wall rocks that have slipped past that point on the fault. Faulting through clean sandstones generates cataclasites, whereas dragging clay beds along the fault generates clay smear. Analysis of outcrop and experimental observations suggests that the algorithm does indeed make a fair estimate of the fault-zone composition. The Oseberg Syd study suggested that an SGR value between 15 and 20% represented a threshold value between non-sealing and sealing faults, in an appraisal context. This value also represents the maximum clay content of cataclastic gouge, implying that in this field cataclasites do not form significant seals whereas more clay-rich gouges do. This threshold has proven to be surprisingly robust, not only in the Brent Province but also in other basins with mixed clastic reservoirs. Compilation of many SGR analyses with in situ pore-pressure data has allowed a better definition of the relationship between calculated SGR and maximum trapped hydrocarbon column height, i.e. the 'fault-seal failure envelope', for different geological histories (e.g. depth of burial). An advantage of the SGR method over others (e.g. 'clay smear potential') is that it predicts a physically measurable parameter (composition) and can therefore be used to predict other properties that are compositionally controlled. The most significant of these is fault-zone permeability, which may vary by many orders of magnitude between cataclasites and clay smears. If correctly calibrated, the SGR distribution on a fault plane can therefore be used as a map of fault-zone permeability, which can in turn be used to provide fault transmissibility multipliers for reservoir simulations. Case studies (e.g. as described here on the Scott Field) show that the SGR methodology can provide a very quick (and yet geologically based) route to a high-quality history match. The experience gained over the last six years shows, not surprisingly, that high-quality input data are essential to quantitative fault-seal studies, in particular good fault mapping and well-prepared Vshale (volumetric shale fraction) data. Nevertheless, Shale Gouge Ratio has proven to be a robust and quantitative predictor of fault seal in mixed clastic sequences.

Introduction

Fault seal in clastic (sand/shale) sequences is broadly predictable. Of prime importance is the juxtaposition pattern of the units at the fault. In many traps, juxtaposition seal of shale against sand is a main component of the trap geometry. However, areas of sand-against-sand juxtaposition can also contribute to the trap because of the presence of fault rocks which impede fluid flow. The generation of fault rock is intimately linked to the sliding of different lithologies past one another (Yielding et al., 1997). Mechanically derived fault rocks include clay smears, phyllosilicate-framework fault rocks, and cataclastic gouges (Fisher and Knipe, 1998). Clay-rich fault rocks tend to form the better seals because they contain finer-grained material and therefore have smaller pore-throats (Gibson, 1998). The first-order controls on fault-rock development are the lithologies (clay content) in the faulted sequence and the amount of offset on the fault. Both of these parameters are provided by routinely available data (well logs and structure maps, respectively). In exploration/appraisal settings, the capillary entry pressure of the fault-zone material is the critical parameter in determining whether a fault can successfully form a side-seal to an accumulation when sands are juxtaposed. In production, the transmissibility (permeability/thickness) of the fault zone is more important.

At the 1996 NPF Conference on Hydrocarbon Seals, Fristad et al. (1997) described how the parameter Shale Gouge Ratio (SGR) could be used to predict

Hydrocarbon Seal Quantification edited by A.G. Koestler and R. Hunsdale. NPF Special Publication 11, pp. 1–15, Published by Elsevier Science B.V., Amsterdam.

fault-seal capacity in the Oseberg Syd region of the North Sea. Yielding et al. (1997) presented further measurements of sealing faults, suggesting that it is possible to apply these quantitative predictions about the likely 'strength' of fault seals to other basins. In this context, 'strength' refers to the pore-pressure difference that can be supported at the fault between two juxtaposed reservoirs. In the few years since Fristad et al.'s presentation, Shale Gouge Ratio has rapidly become a standard methodology for fault-seal assessment: indeed it was deliberately presented as a non-proprietary algorithm.

This paper extends the work referenced above to address the following points.

- How does Shale Gouge Ratio relate to outcrop, core and experimental data?
- What threshold value of SGR is required to establish a 'static' seal, capable of maintaining trap integrity over geological time-scales?
- What is the relationship between SGR and trapped column height (different buoyancy pressures)?
- How is that relationship affected by differences in geological history such as depth at time of faulting, maximum burial depth, in situ stress?
- How should SGR be used to provide input to dynamic production models, where faults often act as low-permeability barriers?

The 'Shale Gouge Ratio' algorithm

A number of different fault-seal algorithms have been published in recent years, each attempting to predict the likely sealing capacity at reservoir–reservoir juxtapositions on a fault plane. One of these, the Shale Gouge Ratio (SGR), is an attempt to predict the proportion of shaly material in the fault zone. It was defined in publications by Fristad et al. (1997), Yielding et al. (1997) and Freeman et al. (1998). At each point on the fault, the algorithm calculates the net content of shale/clay in the volume of rock that has slipped past that point on the fault (Fig. 1). The implicit assumption in this algorithm is that material is incorporated into the fault gouge in the same proportions as it occurs in the wall rocks in the slipped interval. If this assumption is true, then SGR can provide a direct estimate of the upscaled composition of the fault zone as a result of the mechanical processes of faulting. Classification of fault rocks is fundamentally based on their composition (Fisher and Knipe, 1998), and hence *SGR can be thought of as a predictor of fault-rock types for simple fault zones*. Fault rocks with phyllosilicate content < ca. 15–20% are typically cataclasites or disaggregation zones, those with >ca. 40% phyllosilicate are clay/shale smears, and intermediate compositions are sometimes referred to as clay-matrix gouges (Gibson, 1998) or phyllosilicate-framework fault rocks (Fisher and Knipe, 1998).

Other fault-seal algorithms, for example Clay Smear Potential (CSP: Bouvier et al., 1989; Fulljames et al., 1997) and Shale Smear Factor (SSF: Lindsay et al., 1993), attempt to model the development of clay or shale smears from clay or shale beds within the faulted sequence. Clay Smear Potential was formulated after study of ductile clays, whereas Shale Smear Factor was formulated after study of lithified shales. There is some evidence from studies by Shell that Shale Gouge Ratio is a better predictor of fault-seal potential than Clay Smear Potential (Naruk and Handschy, 1997). However, the three algorithms (SGR, CSP, SSF) are not completely independent since they all relate to the amount of clay in the sequence (for a comparison see Yielding et al., 1997). Hybrid algorithms between SGR and CSP have been suggested (Knipe et al., 2000) but were not available for testing at the time of writing.

In practice, deciding which algorithm to use may depend on the format of the available input data. Clay Smear Potential and Shale Smear Factor require input of each individual clay bed. Shale Gouge Ratio can use either bed-by-bed input or zonal averages of Vshale (volumetric shale fraction), and hence incorporates the effects of clay distributed through sandstone units. It is also easier to apply to a zoned sequence (e.g. a reservoir model). A further advantage of Shale Gouge Ratio is that it is a prediction of fault-zone composition, and hence can be related to the bulk composition of fault-zone samples (core or outcrop), as discussed below. Although CSP relates to predicted clay smear thickness, the actual numbers resulting from the algorithm are not equal to the real thickness of the clay smear. SGR can therefore be compared to sample and outcrop data more easily.

Does SGR work at the outcrop scale?

An important requirement in assessing or improving the Shale Gouge Ratio algorithm is to test its prediction on faults where the deformation products can be sampled. Ideally, this should involve faults at the appropriate scale, i.e. with seismically resolvable displacements (tens or hundres of metres). However, cored fault penetrations are notoriously difficult to recover. Fault sampling may be better achieved at outcrop. It is then important to 'log' the shale content of the faulted sequence to provide input to the SGR calculation.

One location where this has been achieved is the Moab Fault zone in Utah. The Moab Fault cuts a Mesozoic aeolian-lacustrine sequence with a throw of

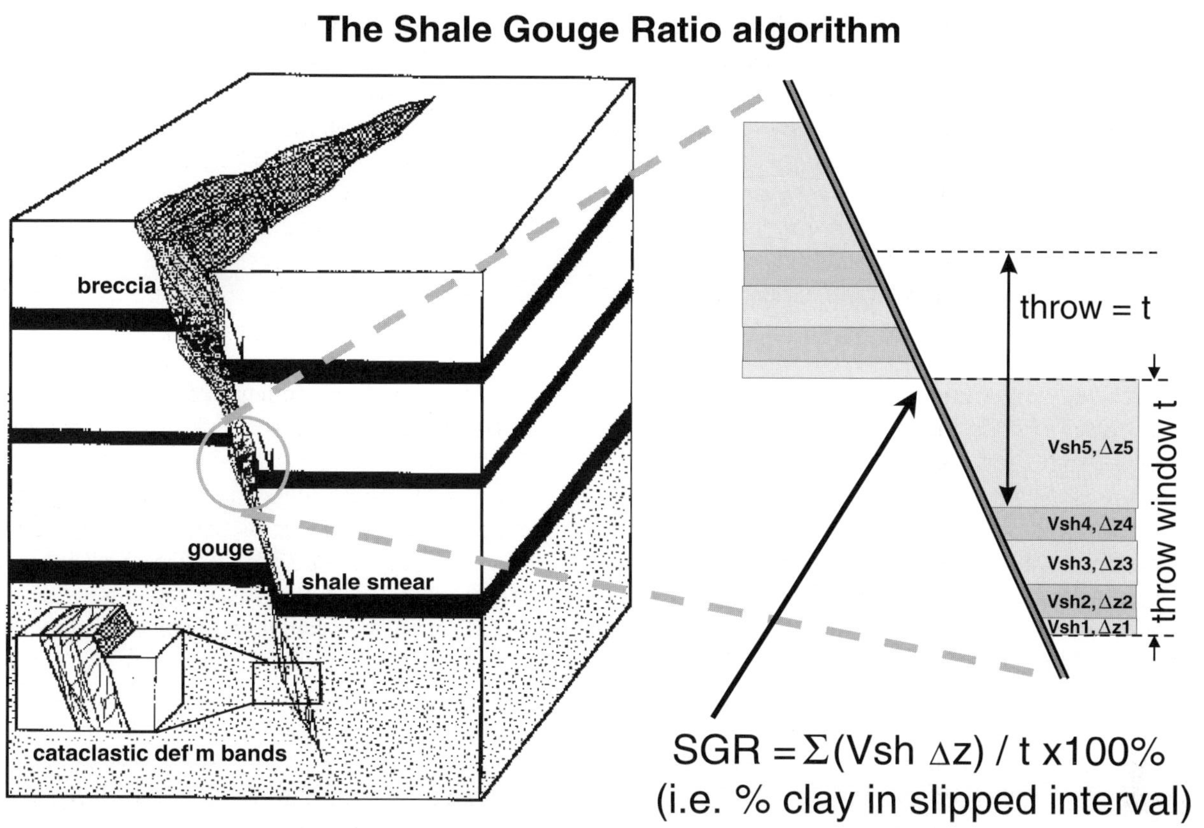

Fig. 1. Definition of the Shale Gouge Ratio, after Yielding et al. (1997) and Freeman et al. (1998). At any point on the fault surface the Shale Gouge Ratio (SGR) is equal to the net shale/clay content of the rocks that have slipped past that point. If lithotypes are incorporated into the fault zone in the same proportions as they occur in the wall rocks, then SGR is an estimate of the fault-zone composition. (Block figure after Walsh et al., 1998.)

up to 1 km. Foxford et al. (1998) provide detailed fault transects at a large number of locations, as well as calculations of Shale Gouge Ratio at the same locations (based on the faulted sequence, which is dominated by alternating mudstones and clean sands). From their transects, an estimate can be made of the proportion of 'shaly gouge' in each part of the fault zone. Fig. 2 compares the observed proportion of shaly gouge with the calculated Shale Gouge Ratio. The correlation between observed and predicted is good ($R^2 = 0.71$). At more than half the localities the calculated SGR is within 10% of the measured shale content of the fault zone. It is therefore a good predictor of average fault-zone composition.

The SGR algorithm assumes complete mixing of wall-rock components in any particular 'throw interval' (Fig. 1). An alternative end-member assumption would be that the fault-zone composition is dominated by the adjacent (juxtaposed) lithologies. This method clearly does *not* work in the Moab example, where the faulted lithologies are either clean sands (shale < 10%) or mudstones (shale > 90%): by contrast the fault-zone compositions are overwhelmingly of intermediate shale content (20–80%, Fig. 2).

Outcrop observations of faults show that in detail

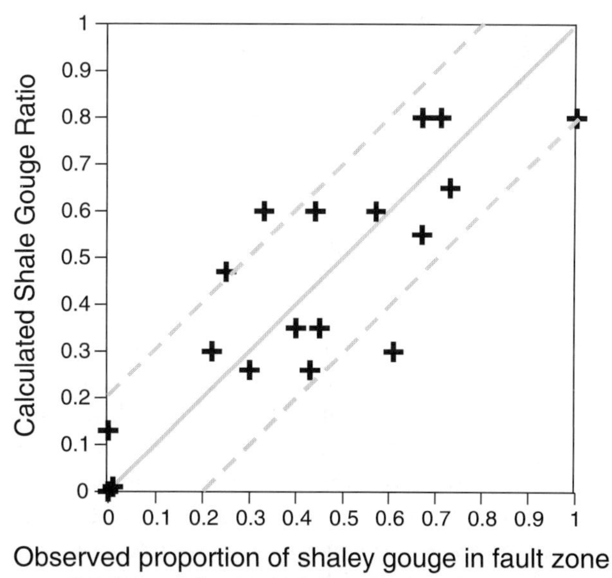

Fig. 2. Comparison of observed fault-zone compositions with calculated Shale Gouge Ratio, for locations on the Moab Fault figured by Foxford et al. (1998). Observed compositions are based on the logged transects of Foxford et al. (1998, their fig. 5) and the calculated SGRs are based on a 'triangle' juxtaposition diagram (cf. their fig. 9). Note that there is a general correlation between observed composition and calculated SGR. The dashed lines bound the field where observed and calculated values agree within 20%.

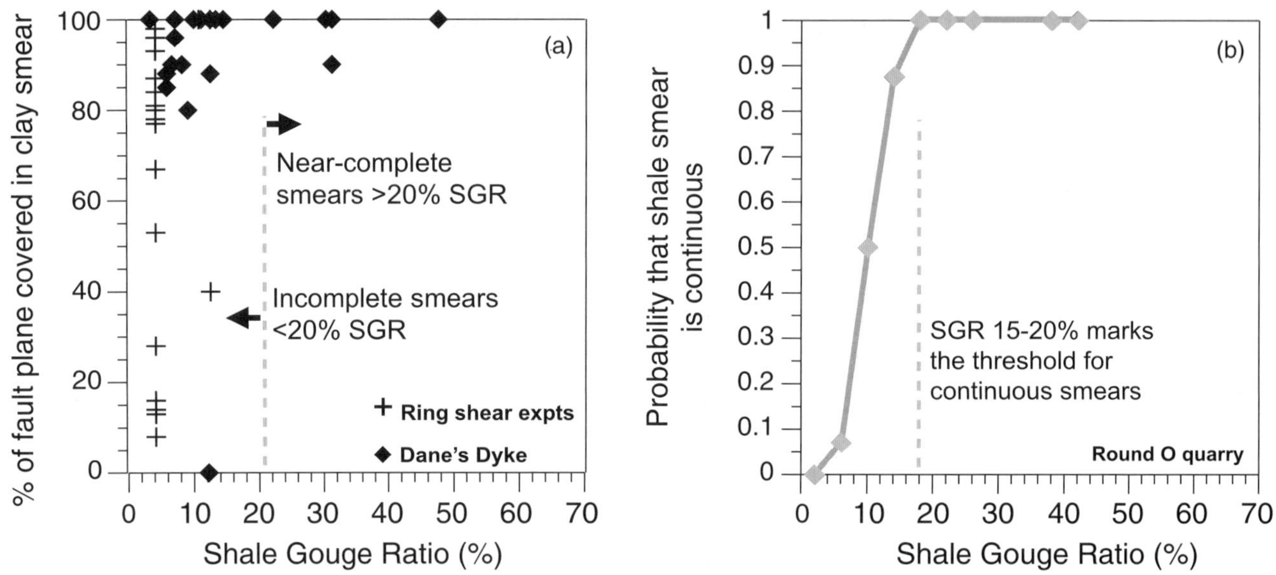

Fig. 3. Outcrop and experimental data relating to clay smear continuity. (a) Individual measurements from Dane's Dyke marls and ring-shear experiments. For each faulted clay layer, the proportion of the separation that is covered in a clay smear is recorded; 100% = smear is continuous from upthrown part of layer to downthrown part. Shale Gouge Ratio was also calculated, as a proportion of clay beds in the 'throw window' (cf. Fig. 1). Diamonds show measurements from marl smears in the Cretaceous Chalk at Dane's Dyke, Yorkshire (Childs, 2000). Maximum burial depth was ca. 1–1.5 km (Hillis, 1995). Crosses show measurements based on ring-shear experiments by Sperrevik et al. (2000), using unconsolidated sand and clay at low confining pressure (equivalent to <50 m burial depth). Both sets of data show many incomplete smears below SGR = 20%, but most smears are continuous at SGR > 20%. (b) Smear probability plot for shale smears in the Carboniferous Coal Measures at Round O Quarry (Lancashire, UK) using data collected by Childs (2000). This plot summarises observations of 80 m of fault traces on vertical quarry faces, with throws up to 5.5 m. Faulting probably occurred after lithification at 2–3 km depth (see Lindsay et al., 1993). As with the Dane's Dyke and ring-shear data in (a), SGR values of ca. 20% correspond to continuous smears.

they often have a very heterogeneous structure with a mixture of deformation products (e.g. Burhannudinnur and Morley, 1997; Foxford et al., 1998; Walsh et al., 1998; Heynekamp et al., 1999; Lewis et al., 2002). Therefore the upscaled composition predicted by the SGR algorithm may mask significant internal variation. For example, a fault in a sand-dominated sequence with only occasional clay layers may give SGR values <10% but still contain clay smears. The critical parameter here (for trap integrity) is probably the continuity of the smears. Fig. 3 summarises smear continuity data from a variety of outcrops and ring-shear experiments. The experimental data (based on work by Sperrevik et al., 2000) is for clay smears resulting from fault slip at low confining pressure (equivalent to <50 m burial depth). The outcrop data (from Childs, 2000) are for marl smears in the Cretaceous Chalk at Dane's Dyke Yorkshire (lithified before being faulted at <1.5 km depth), and for shale smears in Carboniferous Coal Measures (faulted at 2–3 km depth). All these data sets show that, although shale smears may occur when shale beds form only a small fraction of the faulted sequence, the smears tend to remain discontinuous until the proportion of shale beds reaches 15–20%. Similar observations were made at the Moab Fault by Foxford et al. (1998). On this basis a Shale Gouge Ratio of ca. 20% should be a threshold above which continuous shale smears can offer the prospect of an intact seal, and this is borne out by subsurface studies (see below). The precise value for the threshold will be affected by local conditions of the faulting. For example, in superficial near-surface slumping, sand injection processes can enhance the sand contribution into the fault zone relative to the SGR calculation, and would give a higher value for the threshold (Lewis et al., 2002).

An important point relevant to outcrop studies is that SGR is a predictor of upscaled fault-zone *composition*, not shale smear *thickness*. A number of studies have demonstrated that calculated SGR does *not* correlate with the thickness of shale smear or shaly gouge (e.g. Childs, 2000; van der Zee and Urai, 2000).

Seal strength of fault-gouge samples

For thorough trap evaluation, we not only require a prediction of the presence or absence of fault seal, but also an estimate of how 'strong' the fault seal might be. That is, can the relevant parts of the fault plane hold back the excess pressures caused by a commercial hydrocarbon column? To answer this question, we require a prediction of the distribution of capillary entry pressure over the fault surface. For faults in clastic sequences without significant diagenesis, the major control on capillary entry pressure is likely to be the composition (clay content) of the fault-zone material. Fault gouges with higher clay content have smaller pore-throat radii and higher capillary entry pressure (Fisher and Knipe, 1998; Gibson, 1998). On a broader scale, other factors also exert a control on gouge entry pressure, e.g. depth at the time of faulting and maximum depth of burial (Gibson, 1994; Fisher and Knipe, 1998; Sperrevik et al., 2002).

Samples of fault gouges provide some 'ground truth' for predictions of fault-seal capacity. They are typically very small (e.g. 1-inch plugs) and so are unlikely to be representative of millions of square metres of fault plane. Also, as pointed out by Sperrevik et al. (2002), laboratory measurements at zero confining pressure may systematically underestimate subsurface sealing properties. Nevertheless, they provide useful bounds on the behaviour of actual fault-zone material.

Fig. 4 shows a set of capillary entry and breakthrough measurements on a global data set of fault gouges, published by Gibson (1998). Sample compositions (phyllosilicate content) were determined by XRD analysis. Gibson's original results were expressed in terms of effective pore-throat radius; in the figure they have been recalculated to capillary pressure for the oil–water system at reservoir conditions (see figure caption for details). It can be seen that there is a progressive increase in minimum capillary pressure from clean cataclasites through to clay smears. Once the clay content reaches 50–60%, the capillary entry pressure does not continue to increase: this amount of clay appears sufficient to clog all the pore-throats in the material. At low phyllosilicate contents there is a much broader range of entry pressures, and this is caused by the effect of burial history on the cataclasites. It is well known that at temperatures of >90°C (typically 3 km burial depth) quartz dissolution and reprecipitation in cataclasites can destroy remaining porosity and radically reduce pore-throat diameters (e.g. Leveille et al., 1997; Fisher and Knipe, 1998; Hesthammer and Fossen, 2000). The three points in Fig. 4 labelled 'complex deformation bands' represent measurements on gouges buried to ca. 4 km, and show entry pressures 1–2 orders of magnitude higher than the other cataclasites which have seen maximum burial < 3 km.

Also shown in Fig. 4 are general ranges for North Sea data discussed by Fisher and Knipe (1998), again converted to oil–water entry pressures. These data lack detailed compositional measurements but are grouped into three categories on the basis of fault-rock type (cataclasites, phyllosilicate frameworks, and clay smears). They show good agreement with Gibson's measurements. Fisher and Knipe note that although clay content is the dominant control on these fault-rock properties, there is also an influence by maximum burial depth. For the phyllosilicate frameworks (15–40% clay content), the lower entry pressures are for samples buried to <2.5 km whereas the higher entry pressures are for samples buried to >3.5 km.

Not shown in Fig. 4 are disaggregation zones, a fault rock formed in clean sandstones at low confining pressure. Disaggregation zones are typified by grain rearrangement rather than grain breakage (Fulljames et al., 1997; Sverdrup and Bjørlykke, 1997; Fisher and Knipe, 1998), and tend to have hydraulic properties similar to the host rock, unless diagenetically altered. Crawford (1998) demonstrates how the degree of comminution increases with normal stress during the development of deformation bands in high-porosity sandstones.

The dashed line in Fig. 4 represents the *lowest* observed capillary pressures over the range 0–50% phyllosilicate. At any given clay content, there is more than one order of magnitude range in capillary pressure, *even for gouges of similar geohistory*. If this variability is representative of behaviour in actual fault zones, then the lowest entry pressures are the ones that are critical: a seal is only as strong as its weakest point. The dashed line therefore provides a prediction of 'effective seal strength' for fault zones at <3 km maximum burial depth. Fault gouge can support greater pressures (hydrocarbon columns) as clay content increases. As maximum burial depth increases beyond 3 km, effective seal strength will increase for all compositions, but more so for the clay-poor fault rocks. Thus gouge composition may become less critical for seal evaluation at great depths (4–5 km).

Subsurface calibration using in situ pressure differences

Observations of sealing faults in the subsurface provide first-hand evidence of the ability of fault zones to support pressure differences. Simple recognition of different hydrocarbon contacts across an area of reservoir juxtaposition shows that there is static pressure support, at or below the sealing capacity of the fault zone. How does this observation of a sealing or non-sealing fault compare with the 20% SGR threshold for smear continuity, noted above?

Fig. 5 shows a compilation of fault-seal observations from the Brent Province, northern North Sea. All of these faults have followed a similar 'geohistory', with faulting of Jurassic mixed clastics occurring at <500 m burial depth (e.g. Yielding et al., 1992; Roberts et al., 1995). The principal differences are the depths to which different structures have been buried during thermal subsidence, ranging from <2

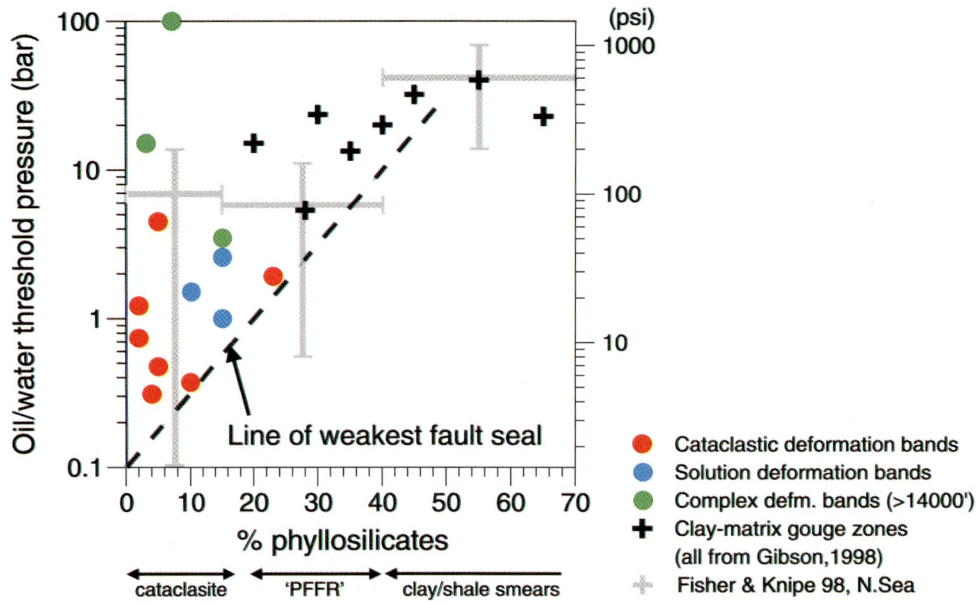

Fig. 4. Seal capacity of fault-gouge samples, using data from Gibson (1998) and Fisher and Knipe (1998). Data are capillary entry and breakthrough pressures for many different gouge samples from mixed clastic sequences. Gibson's 'effective pore-throat radius' has been converted to oil–water capillary pressure using $P_c = 2\gamma \cos\theta / R$, taking $\gamma = 40$ mN/m (Firoozabadi and Ramey, 1988) and $\cos\theta = 1$; compositions were measured by XRD analysis. Fisher and Knipe's data are for gouges from small-scale faults in North Sea cores and are grouped by fault rock: cataclasites 0–15%, phyllosilicate frameworks 15–40%, clay smears > 40%; entry pressures have been converted from quoted Hg–air values to oil–water values using typical fluid properties. The dashed line indicates the trend of weakest fault seal, which would be applicable to failure of subsurface traps (a seal is only as strong as its weakest point).

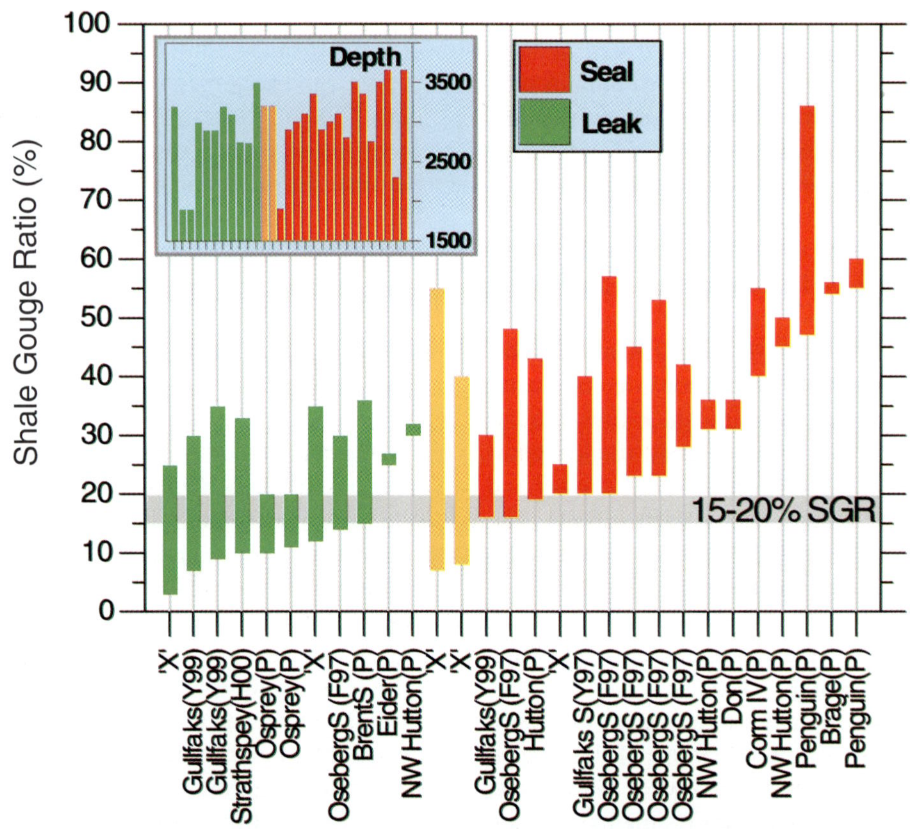

Fig. 5. Compilation of fault seal/leak observations from the Brent Province, northern North Sea. Vertical bars represent range of Shale Gouge Ratio on individual faults. Faults are characterised as 'sealing' (red) or 'leaking' (green) depending on whether there is a change of hydrocarbon contact across the fault. SGR values of 15–20% provide a threshold between sealing and leaking behaviour (if a juxtaposition window with SGR < 15% occurs, the fault leaks). Yellow bars indicate two faults which support OWC differences of <15 m, at 3200 m burial depth. The inset shows burial depths for the same sequence of faults: note the absence of any trend. References for the named faults are: F97, Fristad et al. (1997) (recalculated with updated Vshale data provided by S. Sperrevik, pers. commun.); Y97, Yielding et al. (1997); Y99, Yielding et al. (1999); P, D. Phelps, pers. commun.; H00, Harris et al. (2000).

km at Gullfaks to >3.5 km in Penguin. The vertical bars plot the range of Shale Gouge Ratio on the area of reservoir juxtapositions at the fault surface in each case. Green bars indicate a ‘leaking’ fault, as shown by common hydrocarbon contacts across the fault; red bars indicate a ‘sealing’ fault, i.e. different contacts on each side. There is an excellent correlation between minimum SGR on the fault and seal/leak behaviour. Zones on the fault surface where SGR is <15% allow leakage to occur, leading to common contacts. Where SGR is >20%, the fault is able to support differences in contact. This range of seal/leak threshold is in excellent agreement with the threshold for smear continuity described from outcrop and experiments. Where Brent Group juxtapositions have SGR < 15%, any shale smears are discontinuous, and the dominant fault-zone material (disaggregation zones and cataclasites) is generally unable to provide a recognisable seal.

The orange bars in Fig. 5 (‘field X’) indicate faults with SGR values as low as 10% and yet on these structures a difference in contact *is* observed. However, these differences in oil–water contact (OWC) are no more than 15 m, suggesting a weak seal. Furthermore, these are amongst the deeper examples included in the compilation (ca. 3200 m), and indicate that burial depth (with accompanying quartz cementation) introduces a second-order overprint onto the SGR control. Demonstration that burial depth is not the main factor influencing seal in this data set is provided by the inset, which shows the burial depths of the same sequence of fault examples: there is no trend.

If we have pore-pressure data for the two sides of a fault, a more quantitative analysis can be performed. In Fig. 6, there is a different pressure profile in each of the two wells, and reservoir overlap at the fault plane. Since isobars (like hydrocarbon contacts) are horizontal in each reservoir interval, the pressure profile can be mapped onto the fault plane from the wells on each side. Where reservoirs are juxtaposed at the fault, the difference between the two pressure profiles is the pressure difference across the fault. For common aquifers and trapped hydrocarbons, this represents a ‘static’ seal, i.e. hydrocarbon buoyancy forces are balanced by capillary seal on the fault surface. Where aquifers are at different pressures, the system may be hydrodynamic on a geological timescale, i.e. the pressure difference driving the water flow is balanced by retardation provided by the low-permeability fault zone. (See Grauls et al., 2002, for a more detailed description.)

The figure illustrates the geometry on a cross-section between two wells. On a real fault the juxtaposition geometry and depth (and hence pressure difference) will vary along the strike of the fault. A juxtaposition diagram for the fault must therefore be made first, on which to display the reservoir overlaps and pressure data (for examples see Yielding et al., 1997, 1999; Freeman et al., 1998).

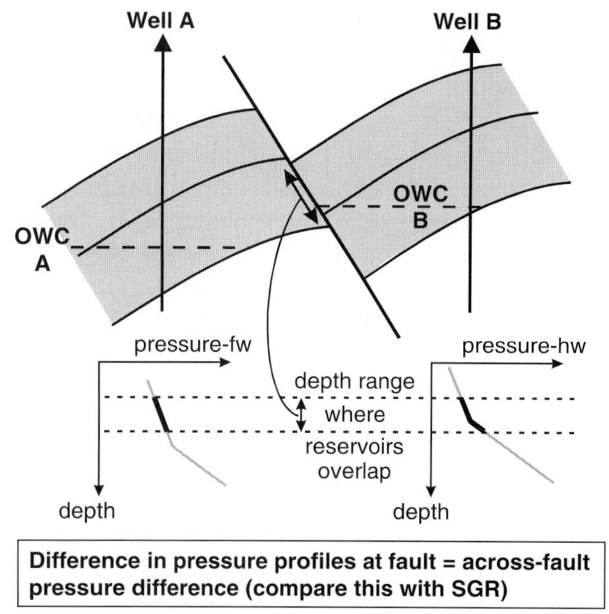

Fig. 6. Cartoon showing how well pressure data can be used to derive across-fault pressure differences. It is critical that the fault and reservoirs are accurately mapped in the subsurface in three dimensions: the reservoir juxtapositions must be constrained on a strike projection or 3D model rather than just a cross-section. The across-fault pressure differences can be compared with calculated SGR values at the same points on the juxtaposition area. If one point of the fault is at seal capacity, the rest of the juxtaposition area will show pressure differences below the seal-failure envelope (cf. Figs. 7 and 8).

At the same time, we can calculate the Shale Gouge Ratio at all points on the fault where there is reservoir overlap. Together, the SGR and ΔP for each point on the fault provide data on the probable composition of the fault zone and the pressure difference that it is currently supporting. These points are all at or below the seal-failure threshold, since the accumulations are sufficiently stable to be sampled by drilling. Without comparing SGR and ΔP over the entire reservoir juxtaposition area it is often impossible to identify the critical point that is controlling the accumulation. On a relatively uniform fault the critical leak point may be near the top where the buoyancy force is greatest; conversely, on a heterogeneous fault the weakest point may be low down in the hydrocarbon column, with the upper parts of the column more than sufficiently sealed.

Fig. 7 shows a compilation of SGR/ΔP data for Brent Province faults. Where SGR is <15% there are negligible pressure differences: these correspond to parts of fault zones that are dominated by disaggregation zones, cataclasites and discontinuous clay smears. From 15 to 40% SGR, increasing SGR allows increasing pressure differences to be supported at the fault. At any given SGR, pressure difference may range from zero up to a maximum, because most of the reservoir juxtaposition area will be below seal capacity. Many of the data points correspond to parts of faults that are not subjected to high pressure differentials. For example, these may be points that are low down in the hydrocarbon column, or come from traps that do not reach fault-seal capacity (e.g. are ultimately controlled by dip closure rather than fault seal). The maximum pressure difference seen at a particular Shale Gouge Ratio is an estimate of the seal capacity for that composition of fault rock. Points close to this line are expected to be near the capillary entry pressure for that part of the fault, i.e. may be critical in controlling the accumulation. As with Fig. 5, the data in Fig. 7 all share a common geohistory: Late Jurassic normal faulting of mixed clastic sediments at <500 m burial depth, followed by post-faulting burial to depths that depend on position within the basin (see below). The seal-failure envelope shown in Fig. 7 is in excellent agreement with the 'line of weakest fault seal' shown for gouge samples in Fig. 4 (e.g. 10 bars at SGR/phyllosilicate content of 40%). The pressure differences relate to hydrocarbon column height via the hydrocarbon–water density contrast, so the relationship shown on Fig. 7 can be used to predict potential column heights if the likely hydrocarbon density is known.

The compilation shown in Fig. 7 can be extended to other basins to explore the impact of different geohistories, see Fig. 8. Details of individual data sets are given in the figure caption. All come from basins with mixed clastic sequences with dominantly extensional faulting which occurred in the depth range 0–2 km. Maximum burial depths, however, are variable. Fig. 8b colour-codes the data sets by burial depth, and three different seal-failure envelopes have been drawn. The line labelled '<3.0 km' encloses points whose maximum burial depths are less than 3 km. Data-points from maximum burial depths between 3 and 3.5 km commonly exceed the pressure differences seen at shallower depths: the line labelled '3.0–3.5 km' encloses all of these points in addition

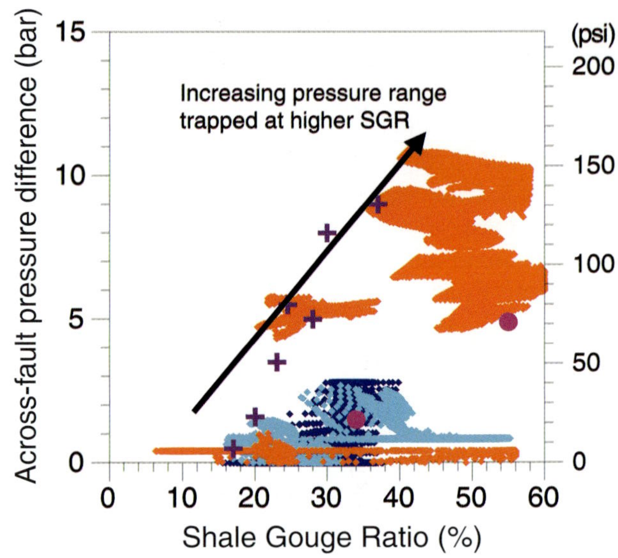

Fig. 7. Comparison of Shale Gouge Ratio and in situ across-fault pressure difference for faults in the Brent Province, northern North Sea. Data are derived as shown in Fig. 6. Clouds of small points correspond to entire reservoir juxtaposition areas. Large points correspond to 'trap-critical' locations that represent the highest pressure difference at a particular value of SGR on that fault. Includes data from Fristad et al. (1997) (recalculated with updated Vshale data provided by S. Sperrevik, pers. commun.), Yielding et al. (1997, 1999), Sverdrup et al. (2000).

to the shallower data. At still greater burial depths, data-points in the range 3.5–5.5 km depth show the highest pressure differences at a given composition, particularly at low-clay compositions. Gibson (1994) also noted the increasing efficiency of shale smears with depth in the Columbus Basin data set.

The seal-failure envelope for '<3 km' in Fig. 8b is in fact approximately equal to the 'weakest fault seal' for the fault-gouge samples (Fig. 4). Fig. 8c shows the gouge data with the seal-failure envelopes of Fig. 8b superimposed. The gouge samples lie close to or above the seal-failure envelopes over the range 0–50% clay content. This observation supports the idea that the area below the envelope represents static fault seal, whereas the area above the envelope represents seal failure of fault rocks of variable capillary entry pressure. It also supports the belief that Shale Gouge Ratio is a good predictor of average fault-zone clay content. Otherwise there would not be such consistency between the sample data plotted using

Fig. 8. (a) Comparison of Shale Gouge Ratio and in situ across-fault pressure difference for faults in a variety of extensional basins. Data are derived as shown in Fig. 6. Clouds of small points correspond to entire reservoir juxtaposition areas. Large points correspond to 'trap-critical' locations that represent the highest pressure difference at a particular value of SGR on that fault. Northern North Sea data are repeated from Fig. 7. Other data sources include Gibson (1994) (Columbus Basin), Muangsuwan (1998) (Gulf of Thailand), Davies et al. (2000) (Gulf of Mexico). 'C. North Sea' data are from the Jurassic of the Central Graben. 'Mid-Norway' data are from faults that are currently 4.2 km below the seafloor. (b) Same data as (a), but now colour-coded for depth (<3.0 km, 3.0–3.5 km, 3.5–5.5 km). The seal-failure envelopes enclose data shallower than their labelled depths, e.g. points shallower than 3 km lie below the '<3.0 km' line. (c) Seal-failure envelopes from (b) compared with the fault-gouge measurements from Fig. 4 (deeply buried gouges are denoted by green crosses). Note that the trend of seal-failure envelopes separates intact subsurface seals from the area of seal failure defined by the gouge samples.

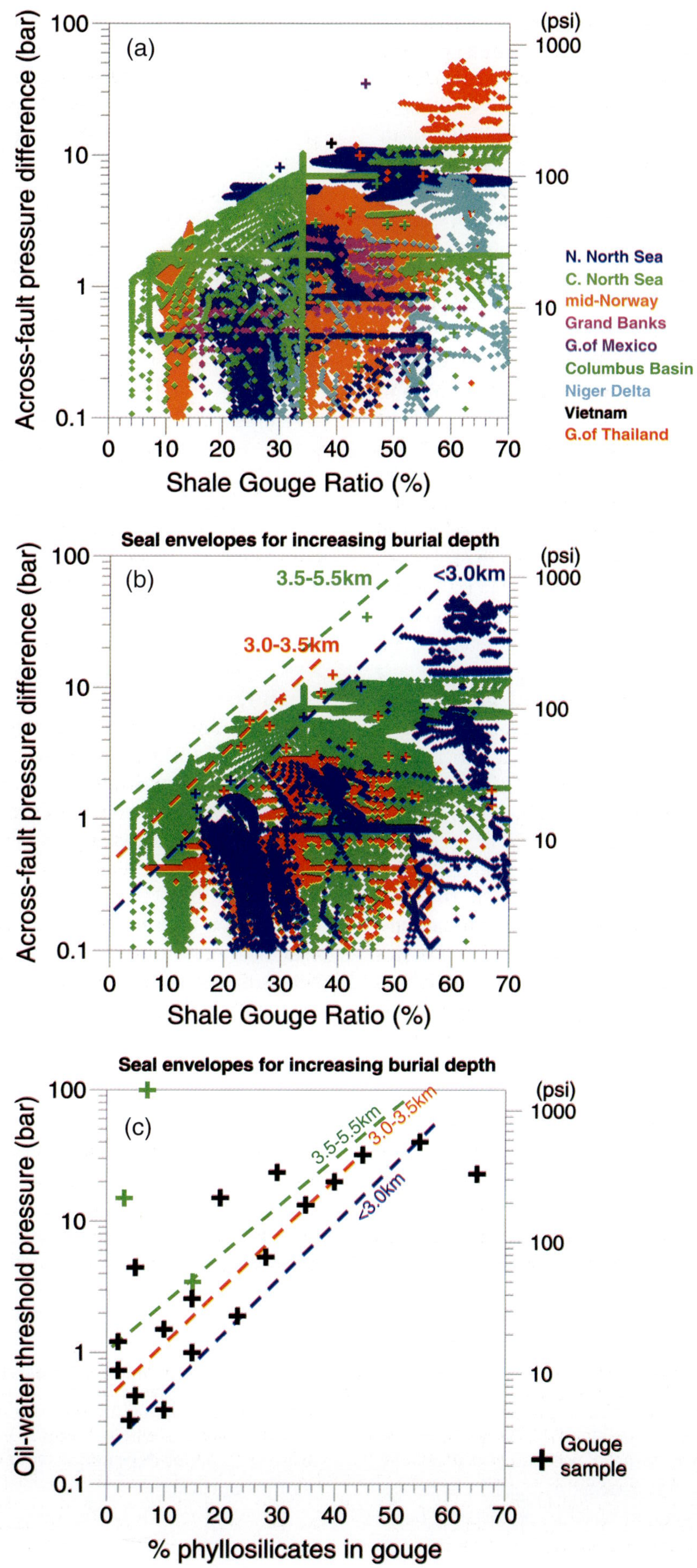
(a)
Across-fault pressure difference (bar)
Shale Gouge Ratio (%)
(psi)
N. North Sea
C. North Sea
mid-Norway
Grand Banks
G.of Mexico
Columbus Basin
Niger Delta
Vietnam
G.of Thailand
Seal envelopes for increasing burial depth
(b)
3.5-5.5km
3.0-3.5km
<3.0km
Across-fault pressure difference (bar)
Shale Gouge Ratio (%)
(psi)
Seal envelopes for increasing burial depth
(c)
3.5-5.5km
3.0-3.5km
<3.0km
Oil-water threshold pressure (bar)
% phyllosilicates in gouge
(psi)
Gouge sample

measured phyllosilicate content and the subsurface data plotted using SGR.

The position of the seal-failure envelopes in Fig. 8 shows that for burial depths < 3 km the fault-zone composition (as predicted by SGR) is the dominant control on the seal capacity, but that at depths > 3 km the burial depth has a clear second-order effect, exactly as seen for gouge samples in Fig. 4. Thus at 4–5 km burial depth the cataclastic end of the fault-rock spectrum becomes strongly affected by quartz cementation, and composition becomes a weaker discriminant in the fault-zone properties.

The data shown in Fig. 8 allow us to convert SGR values to predictions of the excess pore pressure that each part of the fault zone might be able to support, using the seal-failure envelope. To convert these excess pressures into potential hydrocarbon column heights, we need to assume a value for the hydrocarbon density, and apply it in the following equation (e.g. Schowalter, 1979; Watts, 1987):

$$h = \frac{\Delta P}{g(\rho_w - \rho_h)}$$

where ΔP is the buoyancy pressure, ρ_w is the pore-water density, ρ_h is the hydrocarbon density, and g is the acceleration due to gravity. For example, a seal capacity of 10 bar (1 MPa or 145 psi) corresponds to an oil column of up to 400 m (oil density of 0.75 g/cm^3) or a gas column of up to 130 m (gas density of 0.25 g/cm^3).

Application to production

In the production environment, hydrocarbons are removed from the reservoir, and pressure changes may be rapidly imposed on the system. Two points are important in understanding the likely impact of faults where reservoirs are juxtaposed.

- Do the induced pressure changes exceed the threshold capillary pressures for the fault-zone material? These threshold pressures can be estimated by the kind of analysis described above for exploration and appraisal.
- If threshold pressures are exceeded, the flow behaviour of the fault zones is then a function of their permeability, at some appropriate scale.

Permeability measurements on fault gouges have been published by a number of authors (e.g. Antonellini and Aydin, 1994; Crawford, 1998; Faulkner and Rutter, 1998; Fisher and Knipe, 1998; Gibson, 1998; Ottesen Ellevset et al., 1998; Sperrevik et al., 2002). A wide variety of gouges has been measured, from cataclastic deformation bands and slip planes in clean sandstones, to clay smears in mixed clastic sequences. There is a general trend of decreasing fault-gouge permeability with increasing clay content. The higher-permeability low-clay fault rocks (cataclasites and disaggregation zones) are particularly sensitive to degradation at higher temperatures and pressures. In particular, the analysis by Sperrevik et al. (2002) shows that a multi-variable relationship between clay content, depth of faulting and depth of burial can characterise much of the range of permeability behaviour of fault gouges. Their functions for fault-gouge permeability in the dynamic regime are directly analogous to the relationships described above for the static, exploration regime. For a given geohistory, SGR distribution on a subsurface fault can be used as the starting point to map fault-zone permeability.

A direct example of this is shown in Fig. 9, which uses an example from the Gullfaks field in the Brent Province, northern North Sea. Gas injection at well A-42 took a circuitous route before being recorded by the producer A-9H. The illustrated gas migration route was confirmed by 4D seismic imaging (Hesthammer and Fossen, 1997). The corresponding map of the SGR distribution on the faults clearly shows that the gas crossed the main fault between the wells at a location where the SGR is particularly low (<10%). In this case the low SGR values occur at self-juxtaposition of clean Tarbert sands near the top of the Brent Group, and correspond to a disaggregation zone (originally faulted just below seafloor). Fault-zone permeability has remained high as the maximum burial depth is <2 km. Interestingly, fault rocks formed in the same way on the neighbouring Gullfaks South field are now buried to >3 km and have much lower permeabilities as a result of quartz cementation (Hesthammer and Fossen, 2000). These examples again stress the importance of the interplay between composition/SGR and burial history.

Manzocchi et al. (1999) give a more detailed description of how this methodology can be implemented in routine reservoir simulation models. Such models typically do not incorporate fault-zone properties explicitly, but instead use 'fault transmissibility multipliers' to modify the behaviour of cell connections across faults. Generally, fault transmissibility multipliers have often been set on an ad hoc basis to achieve a match to historical production. However, a process-based approach shows that they should be calculated from the expected properties of the fault zone (thickness and permeability). Each multiplier also depends on the size and permeability of the two juxtaposed reservoir cells, since it expresses the ratio by which the slab of fault-zone material degrades the transmissibility between those cells. Multipliers are therefore model-dependent as well as dependent on fault properties.

A typical workflow for determining transmissibility multipliers is shown in Fig. 10. Since fault-zone

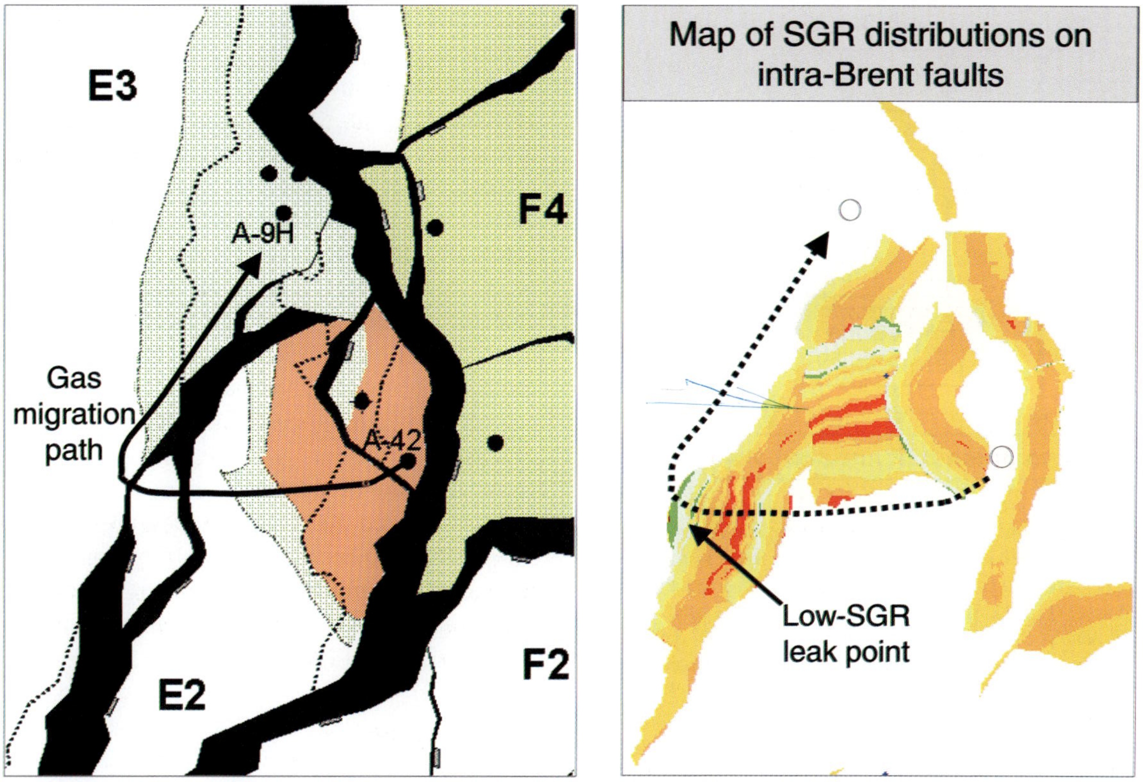

Fig. 9. Example of compositional control on dynamic behaviour during production on the Gullfaks field. In the map at left, green and red areas show Brent Group oil and gas respectively. The gas migration path from the injector A-42 to the producer A-9H crosses the fault away from the shortest route (Hesthammer and Fossen, 1997). The SGR distribution on the Brent–Brent overlaps (right map) shows that this location corresponds to the low-SGR (high-permeability) window on the fault surface (SGR colours: green = <10%, red = >30%; Yielding et al., 1999).

composition is a major factor in controlling the fault-zone permeability it is pragmatic to use Shale Gouge Ratio as an input to the fault-zone permeability calculation. However, it is clearly important to base the SGR-permeability transformation on local analogues that represent a similar geohistory. Early attempts at this application have used calibration by relevant core data (Knai and Knipe, 1998; Sverdrup et al., 2000) and in some cases have provided an excellent validation of the principle. For example, Fig. 11 shows a set of history-match curves for Block Ib of the Scott Field (North Sea). Production data for cumulative water production are shown as orange symbols and the various curves show different simulation model runs. The red and green curves show default simulator options where all faults are closed (fault transmissibility multipliers all 0, red) or where self-juxtaposed connections are open (self-juxtaposed multipliers all 1, other multipliers all 0, green). The blue curve shows the results of significant manual input to modify the individual cell–cell multipliers to achieve a better history match. The purple curve shows a first-pass model using multipliers calculated explicitly by the SGR method (workflow in Fig. 10): this is almost as good as the modified model but was achieved in a fraction of the time. Similar results are reported on the Heidrun Field (Knai and Knipe, 1998) and on the Snorre Field (Sverdrup et al., 2000). The clear message from such studies is that geologically driven transmissibility multipliers should be the first choice in reservoir simulation, allowing more time for additional studies to explore the uncertainties and sensitivities.

Despite these successes, it cannot be assumed that faulted reservoir performance can now be easily modelled. Sperrevik et al. (2002) report that simulations using fault-zone permeabilities based on core calibrations tend to give faults that are too permeable relative to the observed reservoir performance. There are a number of factors that may contribute to this bias.

(1) Measurements of permeability made on cored fault rocks are usually made at zero or low confining pressure, rather than higher-pressure reservoir conditions. The results may therefore be too high by a factor of 2 to 5 (Sperrevik et al., 2002) or perhaps higher (Morrow et al., 1984).

(2) Measurements on cored fault rocks are almost always from very small faults rather than faults with >20 m displacement which are mapped from seismic. Microfaults have simpler structure and may lack some of the low-permeability features present in a larger fault zone. For example, in clean sandstones the small structures may all be deformation bands composed of cataclasite, but with increasing displacement a polished slip surface is likely to develop and will have

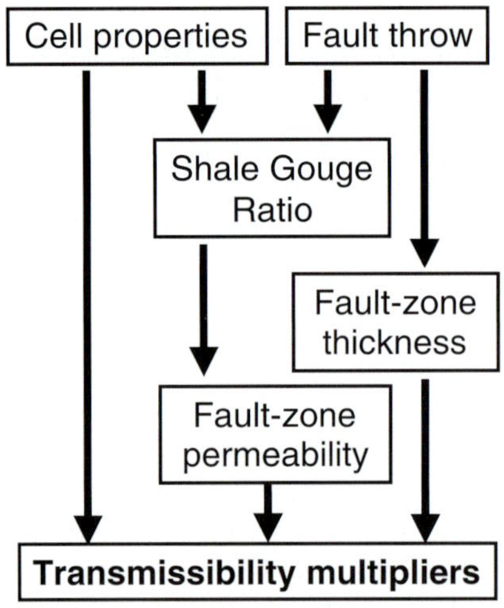

Fig. 10. Workflow for the calculation of Fault Transmissibility Multipliers for a reservoir simulation model (after Manzocchi et al., 1999). At each cell–cell connection, Shale Gouge Ratio is calculated from the local fault throw and the distribution of Vshale in the 'throw window' (Fig. 1). Shale Gouge Ratio is used to constrain upscaled fault-zone permeability, using appropriate structural data such as depth at time of faulting and maximum burial depth ('geohistory', Sperrevik et al., 2002; Knipe et al., 2000). Fault-zone thickness is estimated from the local fault displacement. Each cell–cell transmissibility multiplier is then a function of the size and permeability of the juxtaposed reservoir cells and the thickness and permeability of the fault zone.

much lower permeability (e.g. Antonellini and Aydin, 1994). Discontinuous clay smears may be present as a subordinate component at SGR < 20%, forcing tortuous flow paths even though they are unable to form a static seal worthy of an exploration trap.

(3) Recent studies (e.g. Dart and Rivenæs, 2000; Manzocchi et al., 2000) have shown that two-phase flow effects may be critical to an adequate description of fault behaviour. In many cases the fault zones may be water-wet, unflushed by the hydrocarbon charge: capillary entry effects will then be important. During production through a fault the water saturation must change, and therefore so will the relative permeabilities. As currently implemented, fault transmissibility multipliers are single-phase 'fiddle factors'; the values of such factors need to be changed as production proceeds.

As a result of the above factors, simple reliance on core-derived permeability measurements for fault rocks can be no better than a starting point. Ultimately the match to a production history is the only test that the effective permeability distribution on a subsurface fault has been estimated correctly.

Uncertainties

In both exploration and production environments, it is important to be aware of the uncertainties that feed into a 'fault-seal analysis'. Broadly these uncertainties can be considered in two groups, relating to the mapping scale and to rock/fluid properties. (See also Hesthammer and Fossen, 2000.)

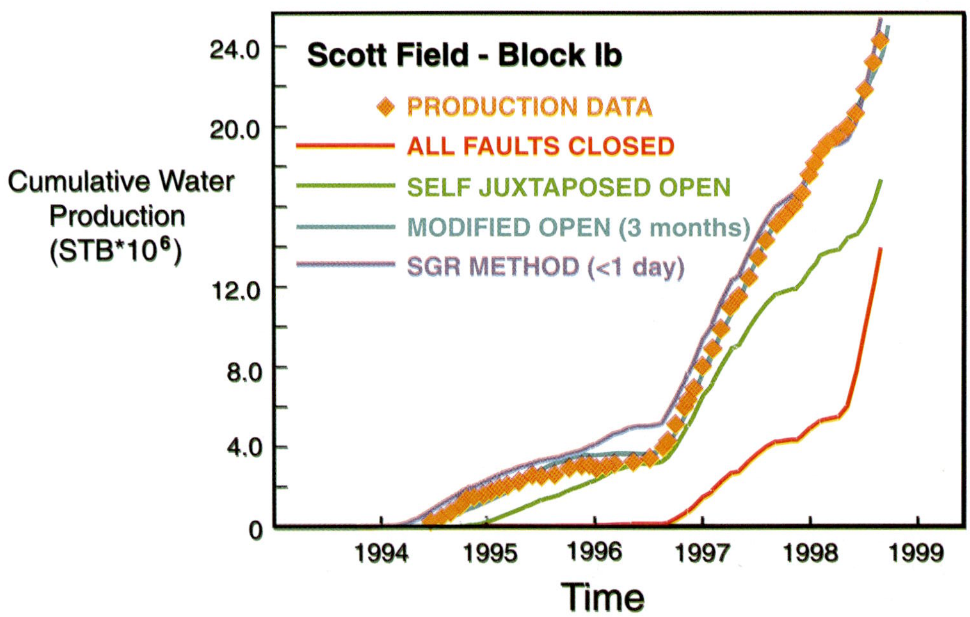

Fig. 11. Examples of reservoir simulation history-matches, using different fault properties, Scott Field, North Sea. Orange diamonds show observed cumulative water production for Block Ib of the field for 4 years from production start-up. The coloured lines show different models. The red line shows model production with all faults closed, i.e. no across-fault flow. The green line is similar but with flow allowed at connections between the same reservoir units (self-juxtapositions). The blue line ('modified open') shows the result of ca. 3 months iteration, manually adjusting transmissibilities at all the across-fault connections. The purple line ('SGR method') shows the first-pass result of calculating transmissibility multipliers using a transformation from Shale Gouge Ratio to fault-zone permeability (method of Manzocchi et al., 1999, shown in Fig. 10). Courtesy of G. Marsden, Amerada Hess.

Uncertainty at the mapping scale

(a) *Mapping uncertainty.* There is no substitute for careful, good-quality mapping of horizons AND faults. Incorrectly mapped fault geometries and fault displacements can lead to incorrect reservoir juxtapositions and incorrect calculations of Shale Gouge Ratio. If the mapping is poor or imprecise, a fault-seal analysis may give completely spurious results.

(b) *Sub-seismic relay zones* may provide unseen fluid pathways across faults that are mapped as continuous and sealed.

(c) *Sub-seismic normal drag* adjacent to the fault may mean that real displacements are smaller than mapped, affecting juxtaposition patterns and SGR calculations. (For example, see Hesthammer and Fossen, 1997, 2000.)

(d) *Sub-seismic fault strands* may partition the total displacement seen by seismic mapping. Typically two strands may separate a 'horse' of intact rock, as a result of fault-propagation processes (see Childs et al., 1996). Two lower-displacement faults will have different hydraulic properties to a single larger fault.

The above uncertainties may be investigated using 'juxtaposition' or 'triangle' plots (Bentley and Barry, 1991; Childs et al., 1997; Knipe, 1997) which show juxtapositions and fault properties at a range of fault throws. However, such plots are removed from a structural context, and ultimately the prospect evaluation or reservoir model must be based on a best estimate of the reservoir fault offsets as mapped.

Uncertainty in rock and fluid properties

(a) *Vshale determination* is a critical input to the SGR calculation (Fig. 1). This may use gamma-ray logs (poor for kaolinite) or density-neutron difference, or rely strongly on core calibration (using XRD analyses). Differences in work practice occur between and within companies, but it is important to be as consistent as possible.

(b) *Variations in behaviour between different phyllosilicate minerals* may be important (e.g. do some clays smear more easily?). There is a need for more public-domain studies (experimental, outcrop and core).

(c) *Degree of diagenetic overprint.* There is an obvious control from burial depth (hi-T) explored above, but geochemistry requires very local sampling. How representative are wells in this regard?

(d) *Hydrocarbon properties, e.g. gas vs oil.* Gas entry pressures are typically 1.5–2 times those of oil, and both are variable with depth and fluid composition (Firoozabadi and Ramey, 1988), affecting column-height estimates.

Conclusions

(1) Shale Gouge Ratio is a robust method for predicting the gross distribution of fault-rock types (cataclasites/PFFR/clay smears) on a mapped fault in mixed clastic sequences.

(2) In an Exploration/Appraisal context, higher values of SGR generally indicate the potential to hold back higher pressures (trap greater hydrocarbon columns) at sand-on-sand juxtapositions.

(3) In a Production context, higher values of SGR generally indicate lower fault-zone permeabilities, and hence more resistance to across-fault flow.

(4) In both Exploration and Production, all elements of the structural history should be considered in calibrating the calculated Shale Gouge Ratio against expected column height or fault-zone permeability. This is particularly so at lower SGR values, where different burial depths at the time of faulting can produce disaggregation zones or cataclasites, and different maximum burial depths can produce different degrees of cementation.

(5) SGR can be used, in conjunction with structural history, to produce a first-pass distribution of transmissibility multipliers for simulation, cutting months off the history-match workflow.

(6) Time needs to be invested in basin- and field-specific refinements of the relationships between SGR, entry pressure and fault-zone permeability to account for local variations (e.g. related to lithologies or burial depth), and to explore the sensitivities to unmappable features such as subseismic relay zones.

Acknowledgements

I am grateful to my colleagues at Badley Earth Sciences who have contributed to the analysis of the many data sets discussed in this study, on both the geological and software sides. Dave Phelps kindly provided many seal/leak examples from the Brent Province, and I also thank Denis Druesne, Simon Price and Arne Grønlie for data release. I am grateful to Susanne Sperrevik for providing revised Vshale data for the Oseberg Syd area, to improve on the original results reported in Fristad et al. (1997). Thanks to Gary Marsden of Amerada Hess for releasing the history match data from Scott Field. I am grateful to Conrad Childs for access to the outcrop data collected in his thesis. Discussions with Michiel Heynekamp, Laurel Goodwin, Paul Gillespie, Quentin Fisher, Dominique Grauls, Russell Davies, Tom Manzocchi and Simon Price helped to clarify many of the ideas I have tried to put in this paper. Reviews by Jim Handschy and Rob Hunsdale of a first draft of the manuscript are appreciated.

References

Antonellini, M. and Aydin, A., 1994. Effect of faulting on fluid flow in porous sandstones: petrophysical properties. Am. Assoc. Pet. Geol. Bull., 78: 355–377.

Bentley, M.R. and Barry, J.J., 1991. Representation of fault sealing in a reservoir simulation: Cormorant Block IV, UK North Sea. Society of Petroleum Engineers Reprint 22667, pp. 119–126.

Bouvier, J.D., Kaars-Sijpesteijn, C.H., Kluesner, D.F., Onyejekwe, C.C. and van der Pal, R.C., 1989. Three-dimensional seismic interpretation and fault sealing investigations, Nun River Field, Nigeria. Am. Assoc. Pet. Geol. Bull., 73: 1397–1414.

Burhannudinnur, M. and Morley, C.K., 1997. Anatomy of growth fault zones in poorly lithified sandstones and shales: implications for reservoir studies and seismic interpretation: part 1, outcrop study. Pet. Geosci., 3: 211–224.

Childs, C., Watterson, J. and Walsh, J.J., 1996. A model for the structure and development of fault zones. J. Geol. Soc., 153: 337–340.

Childs, C., Watterson, J. and Walsh, J.J., 1997. Complexity in fault zone structure and implications for fault seal prediction. In: P. Møller-Pedersen and A.G. Koestler (Editors), Hydrocarbon Seals: Importance for Exploration and Production. Norwegian Petroleum Society (NPF), Special Publication 7. Elsevier, Amsterdam, pp. 61–72.

Childs, C.J., 2000. The Structure and Hydraulic Properties of Fault Zones. Ph.D. Thesis, University of Liverpool.

Crawford, B.R., 1998. Experimental fault sealing: shear band permeability dependency on cataclastic fault gouge characteristics. In: M.P. Coward, T.S. Daltaban and H. Johnson (Editors), Structural Geology in Reservoir Characterization. Geol. Soc. Spec. Publ., 127: 27–47.

Dart, C. and Rivenæs, J.C., 2000. Evaluation of reservoir fault compartmentalisation — Do we have the tools we need? In: Hydrocarbon Seal Quantification, NPF Conference Extended Abstracts, Stavanger, October, pp. 121–124.

Davies, R., An, L., Mathis, A., Jones, P. and Cornette, C., 2000. Abstract. Am. Assoc. Pet. Geol. Meeting, New Orleans.

Faulkner, D.R. and Rutter, E.H., 1998. The gas permeability of clay-bearing fault gouge at 20°C. In: G. Jones, Q.J. Fisher and R.J. Knipe (Editors), Faulting, Fault Sealing and Fluid Flow in Hydrocarbon Reservoirs. Geol. Soc., London, Spec. Publ., 147: 147–156.

Firoozabadi, A. and Ramey, H.J., 1988. Surface tension of water–hydrocarbon systems at reservoir conditions. J. Can. Pet. Technol., 27: 41–48.

Fisher, Q.J. and Knipe, R.J., 1998. Fault sealing processes in siliciclastic sediments. In: G. Jones, Q.J. Fisher and R.J. Knipe (Editors), Faulting, Fault Sealing and Fluid Flow in Hydrocarbon Reservoirs. Geol. Soc., London, Spec. Publ., 147: 117–134.

Foxford, K.A., Walsh, J.J., Watterson, J., Garden, I.R., Guscott, S.C. and Burley, S.D., 1998. Structure and content of the Moab Fault Zone, Utah, USA, and its implications for fault seal prediction. In: G. Jones, Q.J. Fisher and R.J. Knipe (Editors), Faulting, Fault Sealing and Fluid Flow in Hydrocarbon Reservoirs. Geol. Soc., London, Spec. Publ., 147: 87–103.

Freeman, B., Yielding, G., Needham, D.T. and Badley, M.E., 1998. Fault seal prediction: the gouge ratio method. In: M.P. Coward, T.S. Daltaban and H. Johnson (Editors), Structural Geology in Reservoir Characterization. Geol. Soc., Spec. Publ., 127: 19–25.

Fristad, T., Groth, A., Yielding, G. and Freeman, B., 1997. Quantitative fault seal prediction: a case study from Oseberg Syd. In: P. Møller-Pedersen and A.G. Koestler (Editors), Hydrocarbon Seals: Importance for Exploration and Production. Norwegian Petroleum Society (NPF), Special Publication 7. Elsevier, Amsterdam, pp. 107–124.

Fulljames, J.R., Zijerveld, L.J.J. and Franssen, R.C.M.W., 1997. Fault seal processes: systematic analyses of fault seals over geological and production time scales. In: P. Møller-Pedersen and A.G. Koestler (Editors), Hydrocarbon Seals: Importance for Exploration and Production. Norwegian Petroleum Society (NPF), Special Publication 7. Elsevier, Amsterdam, pp. 51–59.

Gibson, R.G., 1994. Fault-zone seals in siliciclastic strata of the Columbus Basin, offshore Trinidad. Bull. Am. Assoc. Pet. Geol., 78: 1372–1385.

Gibson, R.G., 1998. Physical character and fluid-flow properties of sandstone-derived fault gouge. In: M.P. Coward, T.S. Daltaban and H. Johnson (Editors), Structural Geology in Reservoir Characterization. Geol. Soc., Spec. Publ., 127: 83–97.

Grauls, D., Pascaud, F. and Rives, T., 2002. Quantitative fault seal assessment in hydrocarbon-compartmentalised structures using fluid pressure data. In: A.G. Koestler and R. Hunsdale (Editors), Hydrocarbon Seal Quantification. Norwegian Petroleum Society (NPF), Special Publication 11. Elsevier, Amsterdam, pp. 141–156 (this volume).

Harris, D., Yielding, G., Levine, P., Maxwell, M. and Rose, P., 2000. Quantifying the effect of faults on flow of hydrocarbon through reservoirs: a fault seal analysis case study from the Strathspey field, North Sea. Field Rehabilitation II abstracts, Geological Society, London.

Hesthammer, J. and Fossen, H., 1997. The influence of seismic noise in structural interpretation of seismic attribute maps. First Break, 15.6, 209.

Hesthammer, J. and Fossen, H., 2000. Uncertainties associated with fault sealing analysis. Pet. Geosci., 6: 37–45.

Heynekamp, M.R., Goodwin, L.B., Mozley, P. and Haneberg, W.C., 1999. Controls on fault-zone architecture in poorly-lithified sediments, Rio Grande Rift, New Mexico: Implications for fault-zone permeability and fluid flow. In: W.C. Haneberg, P.S. Mozley, J.C. Moore and L.B. Goodwin (Editors), Faults and Subsurface Fluid Flow in the Shallow Crust. Geophysical Monograph 113, American Geophysical Union, Washington, DC.

Knai, T.A. and Knipe, R.J., 1998. The impact of faults on fluid flow in the Heidrun Field. In: G. Jones, Q.J. Fisher and R.J. Knipe (Editors), Faulting, Fault Sealing and Fluid Flow in Hydrocarbon Reservoirs. Geol. Soc., London, Spec. Publ., 147: 269–282.

Knipe, R.J., 1997. Juxtaposition and seal diagrams to help analyze fault seals in hydrocarbon reservoirs. Am. Assoc. Pet. Geol. Bull., 81: 187–195.

Knipe, R.J., Fisher, Q.J., Jones, G., McAllister, E., Needham, T., Bolton, A., Davies, R., Edwards, E., Harris, S.D., Henson, D., Li, A., Odling, N., Pecher, R., Porter, J.R., Allin, N. and White, E., 2000. Quantification and prediction of fault seal parameters: the importance of the geohistory. In: Hydrocarbon Seal Quantification, NPF Conference Extended Abstracts, Stavanger, October, pp. 39–42.

Leveille, G.P., Knipe, R., More, C., Ellis, D., Dudley, G., Jones, G., Fisher, Q.J. and Allinson, G., 1997. Compartmentalization of Rotliegendes gas reservoirs by sealing faults, Jupiter Fields area, southern North Sea. In: K. Ziegler et al. (Editors), Petroleum Geology of the Southern North Sea: Future Potential. Special Publication 123, Geological Society, London, pp. 87–104.

Lewis, G., Knipe, R. and Li, A., 2002. Fault seal analysis in unconsolidated sediments: a field study from Kentucky, USA. In: A.G. Koestler and R. Hunsdale (Editors), Hydrocarbon Seal Quantification. Norwegian Petroleum Society (NPF), Special Publication 11. Elsevier, Amsterdam, pp. 243–253 (this volume).

Lindsay, N.G., Murphy, F.C., Walsh, J.J. and Watterson, J., 1993. Outcrop studies of shale smear on fault surfaces. Spec. Publ. Int. Assoc. Sedimentol., 15: 113–123.

Manzocchi, T., Walsh, J.J., Nell, P. and Yielding, G., 1999. Fault transmissibility multipliers for flow simulation models. Pet. Geosci., 5: 53–63.

Manzocchi, T., Heath, A.E., Walsh, J. and Childs, C., 2000. Fault-rock capillary pressure: extending fault seal concepts to production simulation. In: Hydrocarbon Seal Quantification, NPF Conference Extended Abstracts, Stavanger, October, pp. 101–106.

Morrow, C., Shi, L.Q. and Byerlee, J.D., 1984. Permeability of gouge under confining pressure and shear stress. J. Geophys. Res., 89: 3193–3200.

Muangsuwan, A., 1998. Application of Geological, Geophysical and Geochemical Data to Investigate 3 Low-pay Wells in North Pailin, Pattani Basin, Gulf of Thailand. MSc thesis, Universiti Brunei Darussalam.

Naruk, S.J. and Handschy, J.W., 1997. Characterization and prediction of fault seal parameters: empirical data (abstr.). AAPG Hedberg Research Conference on 'Reservoir Scale Deformation: Characterisation and Prediction'. Bryce, Utah.

Ottesen Ellevset, S., Knipe, R.J., Olsen, T.S., Fisher, Q. and Jones, G., 1998. Fault controlled communication in the Sleipner Vest Field, Norwegian Continental Shelf: detailed, quantitative input for reservoir simulation and well planning. In: G. Jones, Q.J. Fisher and R.J. Knipe (Editors), Faulting, Fault Sealing and Fluid Flow in Hydrocarbon Reservoirs. Geol. Soc. London, Spec. Publ., 147: 283–297.

Roberts, A.M., Yielding, G., Kuznir, N.J., Walker, I.M. and Dorn-Lopez, D., 1995. Quantitative analysis of Triassic extension in the northern Viking Graben. J. Geol. Soc., 152: 15–26.

Schowalter, T.T., 1979. Mechanics of secondary hydrocarbon migration and entrapment. Am. Assoc. Pet. Geol. Bull., 63: 723–760.

Sperrevik, S., Færseth, R.B. and Gabrielsen, R.H., 2000. Experiments on clay smear formation along faults. Pet. Geosci., 6: 113–123.

Sperrevik, S., Gillespie, P.A., Fisher, Q.J., Halvorsen, T. and Knipe, R.J., 2002. Empirical estimation of fault rock properties. In: A.G. Koestler and R. Hunsdale (Editors), Hydrocarbon Seal Quantification. Norwegian Petroleum Society (NPF), Special Publication 11. Elsevier, Amsterdam, pp. 109–125 (this volume).

Sverdrup, E. and Bjørlykke, K., 1997. Fault properties and the development of cemented fault zones in sedimentary basins: field examples and predictive models. In: P. Møller-Pedersen and A.G. Koestler (Editors), Hydrocarbon Seals: Importance for Exploration and Production. Norwegian Petroleum Society (NPF), Special Publication 7. Elsevier, Amsterdam, pp. 91–106.

Sverdrup, E., Helgesen, J. and Vold, J., 2000 The influence of faults on oil recovery and water-alternating-gas (WAG) injection efficiency in the Snorre Field, northern North Sea. In: Hydrocarbon Seal Quantification, NPF Conference Extended Abstracts, Stavanger, October, pp. 3–6.

van der Zee, W. and Urai, J.L., 2000. Fault zone evolution in layered sand–mudstone sequences. In: Hydrocarbon Seal Quantification, NPF Conference Extended Abstracts, Stavanger, October, pp. 171–180.

Walsh, J.J., Watterson, J., Heath, A. and Childs, C., 1998. Representation and scaling of faults in fluid flow models. Pet. Geosci., 4: 241–251.

Watts, N., 1987. Theoretical aspects of cap-rock and fault seals for single- and two-phase hydrocarbon columns. Mar. Pet. Geol., 4: 274–307.

Yielding, G., Badley, M.E. and Roberts, A.M., 1992. The structural evolution of the Brent Province. In: A.C. Morton, R.S. Haszeldine, M.R. Giles and S. Brown (Editors), Geology of the Brent Group. Geol. Soc., Spec. Publ., 61: 27–43.

Yielding, G., Freeman, B. and Needham, T., 1997. Quantitative fault seal prediction. Am. Assoc. Pet. Geol. Bull., 81: 897–917.

Yielding, G., Overland, J.A. and Byberg, G., 1999. Characterization of fault zones for reservoir modeling: an example from the Gullfaks field, northern North Sea. Am. Assoc. Pet. Geol. Bull., 83: 925–951.

G. YIELDING *Badleys, North Beck House, North Beck Lane, Hundleby, Spilsby, Lincolnshire PE23 5NB, UK*

Rock stress in sedimentary basins — implications for trap integrity

Hege Marit Nordgård Bolås and Christian Hermanrud

Seal analysis is an important aspect of prospect evaluation, as it severely impact the prediction of hydrocarbon column heights. Seal failure is suggested to frequently result from leakage of pore fluids through faults or fractures originating from perturbations of the rock stress. In-depth knowledge of rock stress and the rocks' responses to stress changes through time should thus be included in seal evaluation. However, analyses of the complete set of interactions between stress history and leakage of hydrocarbon accumulations have apparently not been reported.

Analysis of the stress state of wells in the Norwegian North Sea, supported by public domain information, was carried out as an effort to compile such information. It was realized that the vertical (S_v) and least horizontal (S_h) stress components could typically be determined within 5–10% accuracy. Determination of the largest horizontal stress (S_H) is less well constrained, and opinions differ significantly regarding the accuracy of calculated S_H values. The processes which influence rock stress and the rocks' responses to stress perturbations seem to be well known in principle. Quantification of these processes are however scarce, a.o. due to inaccurate knowledge of the parameters which control rocks' mechanical behavior over geologic time.

As elevated temperatures can lead to diagenesis and porosity reduction, even in the presence of fluid overpressures, clastic reservoirs most often leak during subsidence. The critical factor in seal evaluation under such conditions is thus to identify the weakest spot in the pressure compartment. The location of this spot largely depends on the stress state (and lithology variations and juxtapositions across faults), and can often be identified even without accurate knowledge of the stress history. Analysis of hydrocarbon occurrence in overpressured reservoirs in the Norwegian North Sea demonstrates that fatal leakage of hydrocarbons frequently takes place from downflanks positions, often leaving hydrocarbon volumes updip. This observation suggests that shear failure and not hydrofracturing controls leakage here, and therefore that faults with certain orientations relative to the stress field are most likely to be leakage avenues.

Introduction

It has long been recognized that hydrocarbon migration, including leakage of hydrocarbon reservoirs, may take place through faults and fractures (Schnaebele, 1948; Dallmus, 1955; Snarsky, 1962; du Rouchet, 1981; Mandl and Harkness, 1987). The formation or reactivation of faults and fractures is intimately linked to rock stress, and relationships between rock stress and pore pressure have been suggested to explain migration through fractures and leakage of hydrocarbon reservoirs (Ungerer et al., 1987; Gaarenstroom et al., 1993; Grauls and Baleix, 1994). Often, leakage through fractures from overpressured reservoirs has been suggested to be due to hydrofracturing (Hubbert and Rubey, 1953; Hubbert and Willis, 1957; Secor, 1965). Such leakage has been assumed to take place when the pore pressure reaches a certain fraction of the overburden, supposedly equal to the least principal stress (Ungerer et al., 1987). Bell (1990) described the stress state of the Scotian Shelf, and suggested relationships between rock strength, hydraulic fracturing and gas migration in the area. Makurat et al. (1992) modeled the influence of Cenozoic erosion on cap rock stresses and integrity in the Barents Sea, and Linjordet and Skarpnes (1992) used caliper log data to identify the current stress state of a gas field, and thereby inferring the strike of faults which are likely to be leakage avenues for hydrocarbons. Finkbeiner et al. (1998) report a quite detailed investigation of the rock stress in the South Eugene Island oil field in the Gulf of Mexico, and use their result to illuminate the migration history of this field. Larson et al. (1993) presented a model which described tectonic fracturing from flexuring, and demonstrated that this model could be used to predict reservoir leakage if included in basin modeling software.

While all of these studies added to the general knowledge of interrelationships between stress and leakage, few of them paid much attention to the historic development of the stress state relative to the

Hydrocarbon Seal Quantification edited by A.G. Koestler and R. Hunsdale. NPF Special Publication 11, pp. 17–35, Published by Elsevier Science B.V., Amsterdam.

history of hydrocarbon supply. None of these studies were aimed at describing the full suite of processes which influence rock stress through time in sedimentary basins and the rock's responses to these processes. Accordingly, the full influence of rock stress on leakage of hydrocarbon reservoirs is yet to be revealed.

On the other hand, investigations of present day rock stresses in sedimentary basins have made substantial progress in the last decade. The compilation of the World Stress Map (Zoback and Zoback, 1989; Zoback et al., 1989; Zoback, 1992) demonstrated that the orientation of the principal stress components in sediments generally follow those in the basement, and that these orientations are generally (first order patterns) invariant with depth and consistent over large areas. Such consistencies warrant the application of rock stresses for predictive purposes, a fact that has been extensively used in prediction of wellbore stability (Morita and McLeod, 1995; Zoback et al., 1995a,b; Wiprut et al., 1997), sand production (Morita et al., 1989a,b), and productivity of reservoirs (Heffer and McLean, 1993; Teufel et al., 1991).

These observations suggest that a summary of stress generating processes in sedimentary basins and the rocks' responses to stress changes is worthwhile. Such a summary may serve as a basis for future quantification of stress-related leakage in basin modeling software. Also, quantification of the accuracy of present day stress assessments and qualitative formulations of the relationships between rock stress and leakage may result in seal evaluation guidelines.

This study describes the most common methods for stress determination in sedimentary basins. Accuracy assessments of the inferred stress is provided, a.o. based on information from exploration wells in the Norwegian North Sea. A summary of the most important stress generating mechanisms in sediments and the rocks' responses to stress changes are also covered. Finally, guidelines for seal evaluation are suggested, based on information on rock stress and variations of rock stress through time.

This paper is essentially an overview of rock stress and the fracturing of sedimentary rocks. More detailed information, which is important to the main conclusions of this study, but which breaks up the argumentation, is added in the Appendices. Appendix A provides some general, basic knowledge about rock stress and its components; Appendix B deals more specifically with maximum horizontal stress (S_H); Appendix C gives a description of North Sea pore pressures and rock stress.

Present day stress magnitudes and their accuracy

Vertical stress

Knowledge of all the three principal stress components is required to describe the stress state of the subsurface. The principal stress components are usually referred to as the vertical (S_v), the least horizontal (S_h) and the largest horizontal (S_H) stress components. Basic descriptions of the principal stress components are given in Appendix A.

The magnitude of the vertical stress is very close to the weight of the overburden. This stress is most frequently determined by integration of density logs. While this method intuitively would be expected to be accurate, significant uncertainties may exist in estimates of the overburden weight (S_v). Comparison of overburden curves for nearby wells (which have virtually identical overburden rocks) often show discrepancies of 5–10%. As an example, consider the overburden curves from three neighboring wells in the North Sea (Fig. 1) These wells have virtually identical overburden rocks at shallow depths, and should thus have identical vertical stress. The main reason for the discrepancies is probably poor log quality and the absence of log data in the shallow portions of the wells.

Due to the uncertainty related to vertical stress calculations in well locations, it is recommended to use average values based on several wells for regional extrapolations, at least in areas with large similarities in the overburden characteristics. As a crude approximation, a gradient of 2.3 g/cm^3 (1 psi/ft) is commonly used. This approximation would be accurate if the sediments had a constant density of 2.3 g/cm^3 with depth, which corresponds to an average porosity of 21% (assuming a grain density of 2.65 g/cm^3). In reality, the shallow sediments have higher porosities, while the deeper sediments have lower porosities. Accordingly, the approximation of 2.3 g/cm^3 gives too high vertical stress at shallow depths, but can be a fair approximation at greater depths (around 5 km). It is suggested that average estimates of vertical stress from several wells in areas with similar overburden is applied in seal evaluation whenever possible. An average value of 2.3 g/cm^3 can be applied at depths greater than 5 km if local data at large depths are unavailable.

The least horizontal stress

The magnitude of the least principal stress (which is horizontal in normal or strike slip stress regimes, see Appendix A) is most commonly inferred from

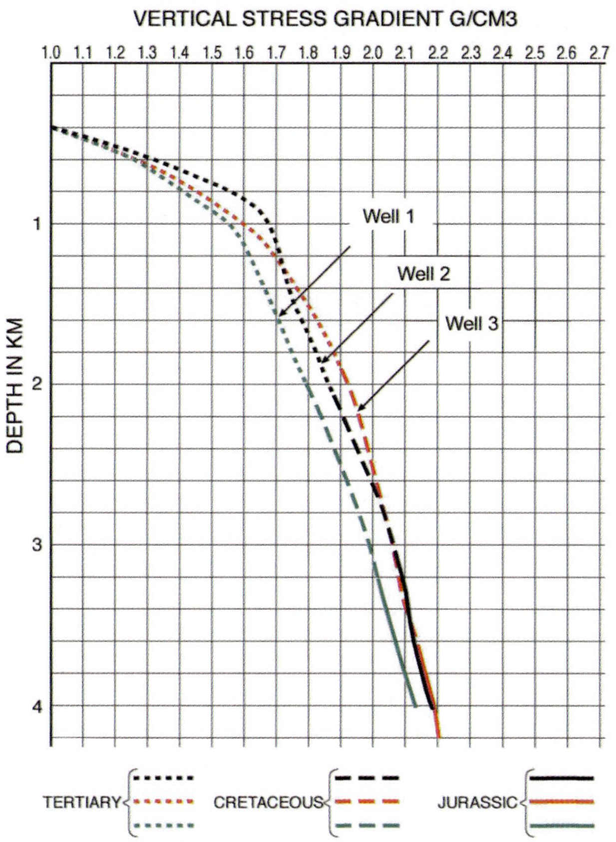

Fig. 1. Comparison of three different estimates of the vertical stress (S_v) gradient in a restricted area of the North Sea, made by three different oil field operators. (The absolute value of the vertical stress, which commonly equals the overburden weight, is found by multiplying the vertical stress gradient by the depth of interest.). The overburden rocks in this area vary insignificantly in the shallow sections, and the overburden weight (and hence the vertical stress gradient) should be identical for the uppermost 2.5 km. The differences between the curves are mainly due to inaccurate density information in the shallowest sections of the boreholes, and demonstrate the inaccuracy of vertical stress determination from individual exploration wells.

leak off tests. Here, the fluid pressure in the well is increased until fracturing is initiated, and the inference is made that this fluid pressure corresponds approximately to the pressure which is required to create a fracture normal to the least principal stress. Different operational practice through time has resulted in leak off pressure (LOP) data of mixed quality, and the least principal stress which is estimated directly from the LOP data can therefore be both over- and underestimated. Fig. 2 gives an example of possible underestimation: here, several tests were performed at virtual identical depths, giving widely differing LOP values below 3500 m. Gaarenstroom et al. (1993) suggested use of the lower envelope of the LOP's in the area (that is, a smoothed line through the lowest values which are encountered at any given depth) as an upper limit for pore pressures in an area (Fig. 3). Because of the significant errors in LOP determination, especially from older data, it is suspected that this practice often gives too low values for maximum overpressure.

Gaarenstroom et al. (1993) did not explain in depth why the lower envelope was taken as a measure of the regional S_h. Their approach appears to be valid only if errors in the LOP data always result in overestimation of the S_h from these individual measurements, a concept which seems hard to justify. If this assumption of Gaarenstroom et al. (1993) is not correct, then the lower LOP envelope suggested by these authors will underestimate the regional S_h. An averaged curve through all the individual LOP data would be appropriate if under- and overestimation of the S_h from LOP data is equally common. However, both the experiences from well 34/10-20 (Fig. 2) and the observation that the wells with the highest overpressures generally have elevated LOP values (R. Loosveld, pers. commun., 2000), suggest that an average S_h curve for the wells with the highest overpressures (and which, according to Gaarenstroom et al. (1993) are closest to leakage) should be even higher than the average curve. Hermanrud and Nordgård Bolås (2002) use the average of individual LOP data from overpressured wells only, to estimate the regional S_h in overpressured formations in the western (overpressured) part of Haltenbanken. This procedure appears to be generally appropriate for seal evaluation.

Maximum horizontal stress

The magnitude of the maximum horizontal stress cannot be measured directly in hydrocarbon exploration wells. This is unfortunate, as knowledge of all principal stress components is required to determine the stress state of the sediment. The stress state of the sediment determines which fault orientation slips first, as will be discussed below.

The maximum horizontal stress (S_H) also controls the magnitude of the stress anisotropy of the rock (which equals S_1 minus S_3 where S_1 is the largest principal stress and S_3 is the least principal stress) at conditions where the maximum horizontal stress equals the largest principal stress ($S_H = S_1$), which is crucial in determination of failure mode (shear or tensile failure, see later).

As shown in Fig. 4, the S_H may be calculated from the occurrence of borehole breakouts or tensile fractures in boreholes (Zoback et al., 1995a,b). Numerous other methods for S_H calculations also exist, and reported S_H values in the petroleum literature either stem from anelastic strain recovery methods (Teufel et al., 1991; Harper, 1995a,b), from hydraulic fracturing of previously fracture-free intervals separated by packers (Bredehoeft et al., 1976; Hickman and Zoback, 1983; Schmitt and Zoback, 1989), from

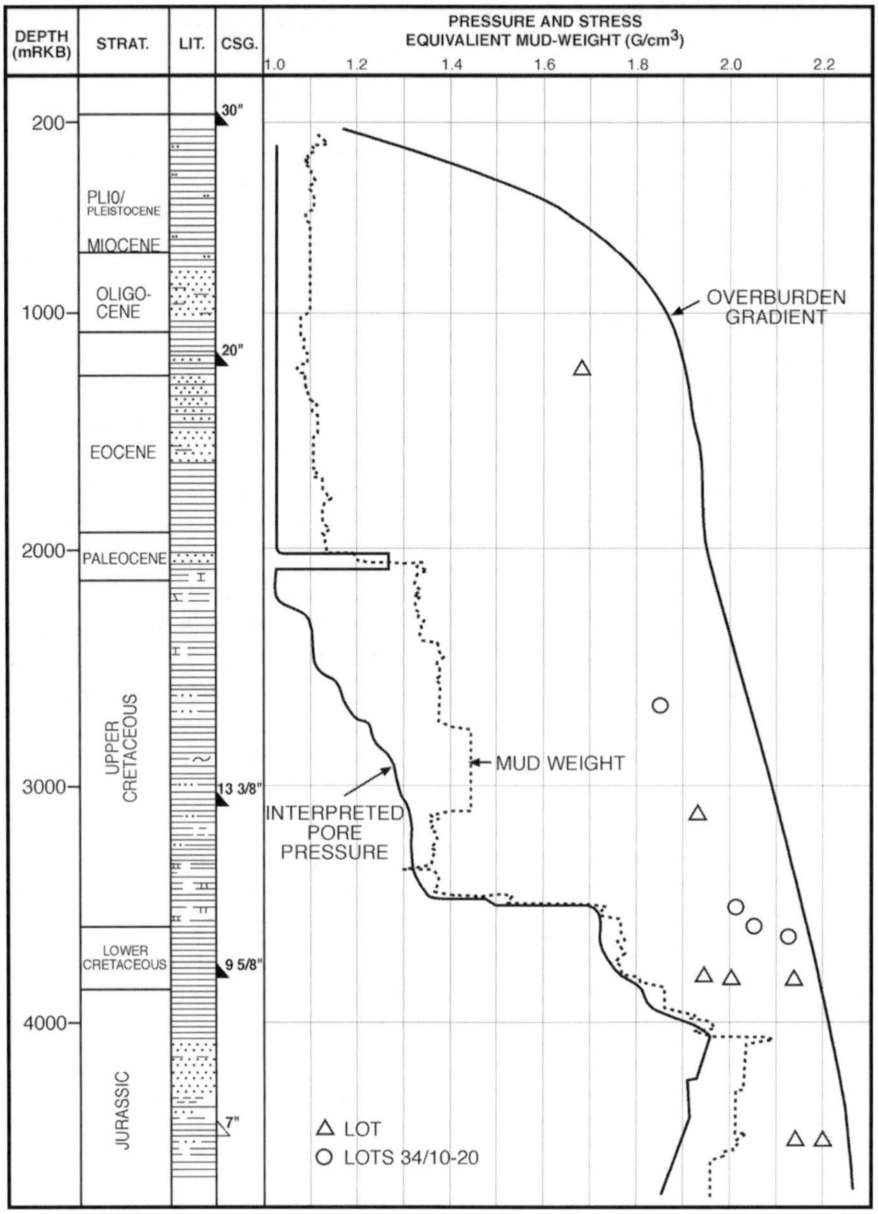

Fig. 2. Pressure, overburden and leak off test data for well 34/10-23 of the northern North Sea, and LOP data for the neighboring well 34/10-20. Note that leak off tests taken at close depths vary considerably in the 3.5 to 4.5 km depth range. These differing measurements demonstrate that errors (here: underestimation) of the least principal stress can be significant in hydrocarbon exploration wells.

simplified applications of the formula used by Bredehoeft et al. (1976) to leak off tests (Bell, 1990), from inversion of leak off pressures in several inclined wells in an area where the principal stresses do not vary laterally (Aadnøy, 1990; Aadnøy et al., 1994; Gjønnes et al., 1998), from computations based on the occurrence of borehole breakouts (Zoback et al., 1985) and tensile fractures as observed in image log data (Peska and Zoback, 1995; Zoback et al., 1995a,b), or a combination of several of these methods (Brudy et al., 1997) (see further descriptions of such methods and their uncertainties in Appendix B).

Common for most methods is that reliable assessments of their accuracy is scarce, and that the knowledge of the S_h and S_v enter into equations which are used in the S_H determinations. Errors in estimates of these two stress components will thus propagate into the S_H calculations. Underestimations of the S_H will cause the anisotropy of the rock to appear too small and hence the calculated risk of tensile failure to increase.

Relative magnitudes of principal stress components

Knowledge of the relative magnitude of the principal stress components is of significant interest, even if the magnitude of the individual stress components can not be accurately assessed. This is so because the faults which slip first strike parallel to the interme-

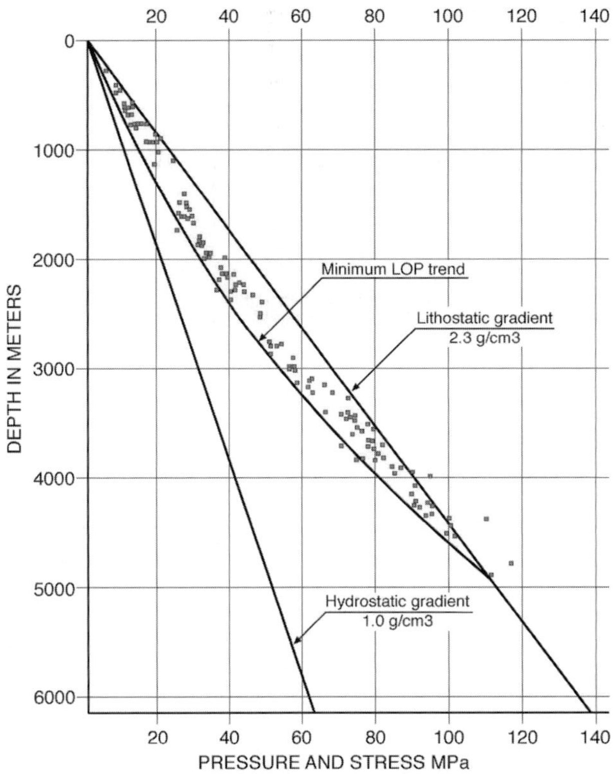

Fig. 3. Minimum LOP trend from the central North Sea. The figure is adapted from Gaarenstroom et al. (1993). It is argued that this envelope underestimates the regional least compressive stress.

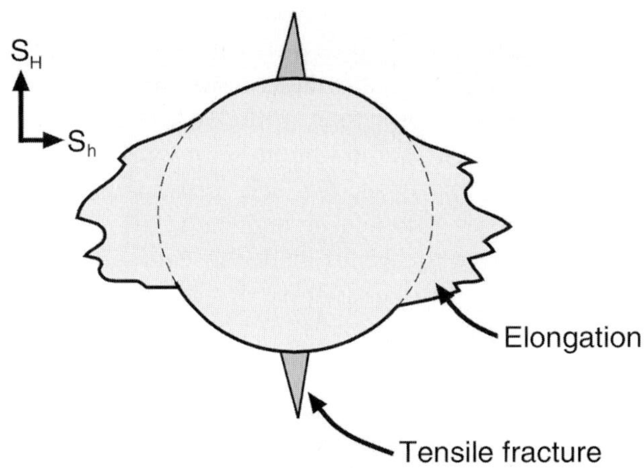

Fig. 4. A cross-section through a vertical borehole. Tensile fractures form in the direction of S_H, and wellbore elongations (borehole breakouts) form in the direction of S_h. The tensile fractures can be imaged by optical tools.

diate principal stress component S_2, forming angles of approximately 60° to S_3 and 30° to S_1 (Fig. 5). Accordingly, faults which are normal to S_h will slip first in a normal stress regime, whereas faults which strike at approximately 60° to S_h will slip first in a strike slip stress regime, and faults which are parallel to S_h will slip first in a reverse stress regime.

According to Ungerer et al. (1987), the vertical stress is most frequently the maximum principal stress in sedimentary basins. These authors also suggested that hydraulic fracturing takes place as rocks open in tension, and used these concepts to simulate hydraulic fracturing in their 2D basin model. However, worldwide compilations of stress data (the World Stress Map (WSM) database, Mueller et al., 2000), demonstrate that the three stress regimes (thrust, strike slip and normal) are all about equally common in the crust. The WSM data are dominated by earthquake focal mechanism solution data, and so overrepresent areas which are tectonically active. The WSM data also demonstrate that the orientation of the horizontal stress components as inferred from borehole breakouts (in most cases from drilling in sedimentary rocks) mimic those from earthquake focal mechanism solutions, which reflect the stress state in the basement. Accordingly, links exist between the orientation of the horizontal stress in the sediments and in the basement. This observation may indicate that stress anisotropy in the basement often is transferred to the sediments, although Bjørlykke and Høeg (1997) suggest otherwise.

Whether the relative magnitudes of the principal stress components (and thereby the stress regime) are also transferred from the basement to the sediments is less clear. Stress in sedimentary basins is redistributed through a number of processes, which operate with different intensities in various geological settings. Local analyses are thus required to determine the stress state in sedimentary rocks (which may well change with depth, as suggested by Grauls and Baleix, 1994). Quantification of stress development through time in sedimentary rocks seems to be in its infancy, and none of the different methods for stress determinations from borehole data in hydrocarbon exploration wells appear to be universally accepted. Accordingly, opinions differ concerning the relationships between the stress state in the sediments and the basement.

Wiprut and Zoback (1998) computed the largest horizontal stress by the use of borehole breakout data in the Visund area of the North Sea, and proposed S_H to be around 30% higher than the vertical stress component. This result differs significantly from those of Aadnøy et al. (1994), which suggested S_H/S_v values in the 0.8–1.0 range in the neighboring Snorre field. Aadnøy et al. (1994) also suggested decreasing anisotropy with depth, in accordance with the suggestions of Hermanrud and Nordgård Bolås (2002) for the Haltenbanken area. Further studies, including the inspection of stress-induced borehole instabilities and tensile failures in deviated wells at various azimuths seem to be a worthwhile undertaking to further constrain the relative magnitudes of the principal stress components in the area.

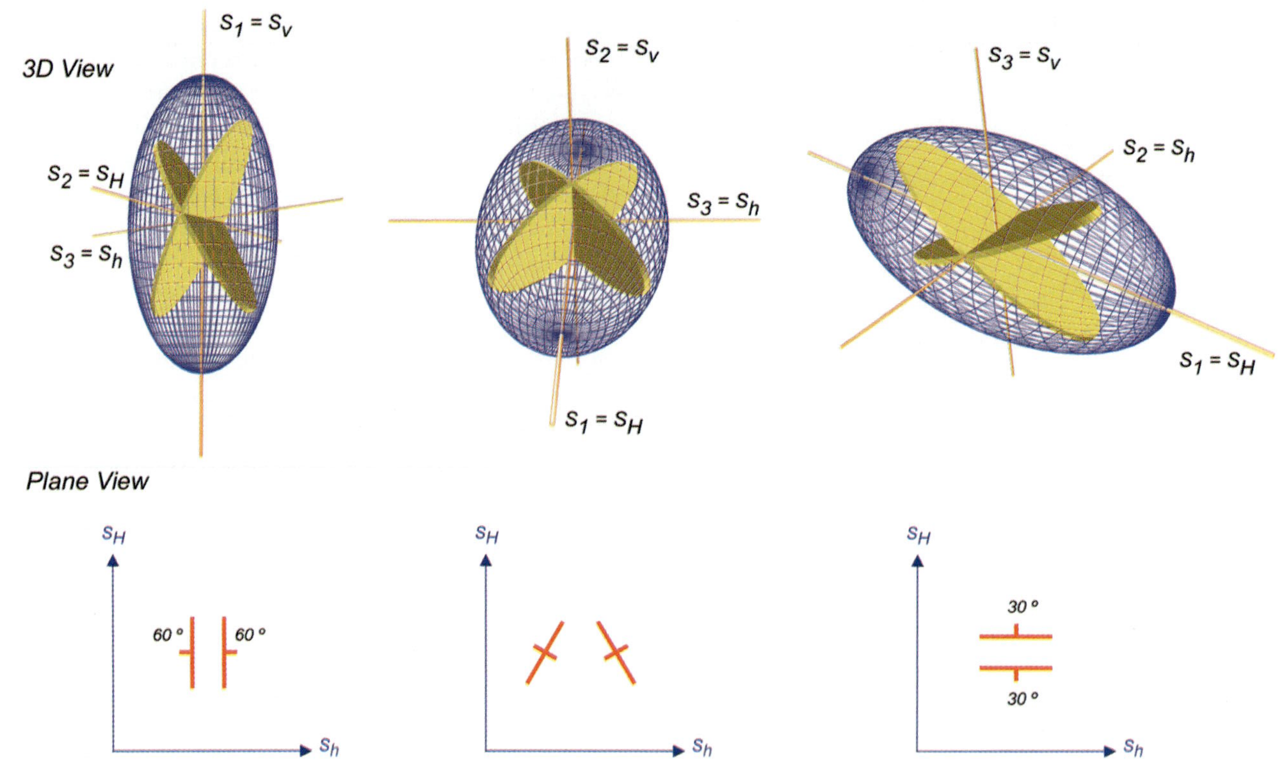

Fig. 5. Relationship between stress state and critical orientation of faults. The figures in 3D view show the stress ellipsoid, where the largest diameter corresponds to the maximum principal stress (S_1), the intermediate diameter corresponds to S_2 and the smallest diameter corresponds to S_3. The fault planes (yellow) which slip first strike parallel to S_2, at angels of approximately 30° to S_1. The figures in plane view show how the strike of critically oriented faults relates to the horizontal stress components in the three stress regimes.

Orientation of present day rock stress

Regional stress variations

Determination of horizontal stress components by the use of borehole breakout data is well documented (Zoback et al., 1995b). This approach includes observations of wellbore elongations, breakouts and tensile fracturing to determine the directions of S_H. Uncritical mapping of such data often show a wide range of scatter in the inferred stress orientations. Brudy and Kjørholt (2001) have however shown that careful inspection of borehole failures from high-resolution borehole imaging logs, combined with an extensive quality control of the input data for S_H determination, dramatically reduces the scatter in the inferred stress orientations and show remarkable consistent regional trends. Most of the North Sea appears to have a present day S_H in a ENE–WSW direction, whereas the S_H directions in the Tampen Spur areas are shifted to the NW–SE direction. The Haltenbanken area generally shows stress orientations similar to those of the Tampen Spur (Fig. 6).

As already noted, The World Stress Map effort (Zoback, 1992; Mueller et al., 2000) concluded that the direction of the largest horizontal stress (S_H) in many sedimentary basins closely mimics that in the basement (as determined from earthquake activity). This suggests that tectonic strain in basement rocks develops faster than that which can be accommodated by stress healing processes in the sediments (otherwise, the horizontal stress in the sediments would be isotropic). If this is so, not only can S_H directions from wells be extrapolated regionally, but regional interpretations may also be guided by earthquake data (provided that the stress regime of the sediments can be determined).

Local stress variations

Brudy et al. (1997) and Brudy and Kjørholt (2001) demonstrate that the regional S_H orientation is generally quite uniform with depth. However, local deviations from this consistency may occur. This was of little concern to the World Stress Map project, which goal was to describe the overall stress state of the earth. Several studies have demonstrated rotation of stresses in the vicinity of open faults (Aleksandrowski et al., 1992; Bell et al., 1992; Zoback and Healey, 1992; Yale et al., 1994; Brudy et al., 1997). These stress rotations are supposedly due to reduced shear strength along the fault planes. Such reduced shear strength will, however, not necessarily cause faults to act as open conduits, as clay smear may develop and seal the fault plane if conditions are right (Harper and Lundin,

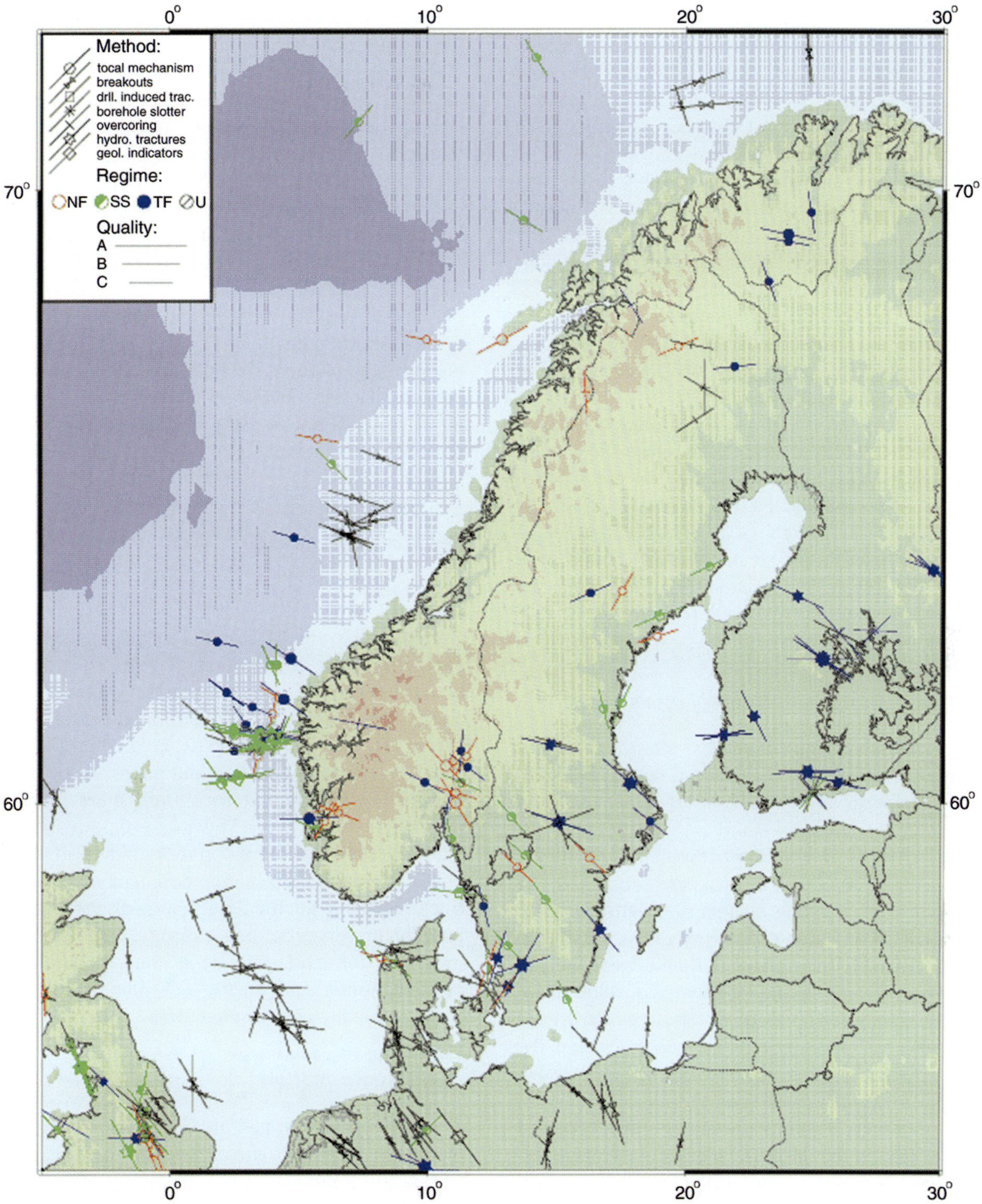

Fig. 6. Regional stress orientations in the North Sea from the World Stress Map (WSM) (see Mueller et al., 2000).

1997). Stress perturbations thus can not be used as a criterion for faults to be fluid conduits in all cases.

Local stress perturbations may also arise at layer boundaries. Sands and shales are supposed to behave differently during burial, as their elastic and viscous moduli and their thermal expansion coefficients are different. Furthermore, their pore pressures and thus their effective stress will often differ substantially. These facts lead to a number of complicated interactions at layer boundaries, and the resulting rock behavior seems hard to predict with confidence. First, the stress tensor at the lithological boundaries will be partly determined by the properties of the neighboring rocks (Bell and Lloyd, 1989; Spann et al., 1994). Secondly, situations where failure criteria are met in sandstones, but not in adjacent shales, may arise, e.g. because of increased pore pressure in the shale or in the sand (as suggested for the South Eugene Island oil field by Finkbeiner et al., 1998), or because of more pronounced viscous behavior of the shales. These observations suggest that, in general, stress concentrations and increased stress anisotropy will be expected along lithological borders. This result also applies to fault planes, where rocks with different lithological properties are juxtaposed. Such areas will thus be more prone to fracturing than areas within homogeneous rocks. As a result, renewed fracturing may occur in the immediate vicinity of preexisting fault planes, even if the fault planes themselves have been cemented and pose larger resistance to failure than the surrounding rocks.

While local variations in stress orientation and magnitudes certainly occur in the subsurface, these may often be hard to identify based on the scarcity of image logs and the natural scatter of borehole breakout data. In general, it is considered safer to follow strict criteria for breakout selection and miss some of the fine-scaled variations than to reduce the quality of the data set to the point where the validity of the individual observations can be questioned.

Origin of stress in sedimentary rocks

Integrated basin modeling most often aims at describing the fluid flow history of the rocks, with special emphasis on prediction of reservoir fluid type. Such modeling is based on quantification of the physical and chemical processes which interact in sedimentary basins. As rock stress controls the development of fractures, which has significant impact on the preservation of reservoired hydrocarbons, quantification of stress generation and dissipation in sedimentary basins should be attempted. Such quantification is outside the scope of this study. However, a summary of stress generating processes and the rocks' response to imposed stress changes is included as a basis for further research in this area.

Burial, which is a consequence of subsidence and a related overburden increase, leads to increased vertical stress. This vertical stress also leads to increased horizontal stress through various processes of thermal and mechanical origin. The increased burial leads to elevated temperatures, and thus to thermal expansion. This thermal expansion leads to increased horizontal stress, since the rocks can not expand horizontally. The vertical stress itself, however, is not increased by heating due to burial, as the rocks can accommodate the thermal stress by vertical expansion. Increased burial also leads to mechanical compression and hence increased horizontal stress due to redistribution of the imposed vertical stress. The responding rock deformation will be ductile and/or brittle, and is further described below.

Tectonic impact generally leads to anisotropy of the horizontal stress (e.g. elevated S_H values). The tectonic activity may be of either regional/global (e.g. plate tectonics) or local (e.g. diapirism) origin; see Forsyth and Uyeda (1975), Chapple and Tullis (1977), Richardsson (1992), Larson et al. (1993) and Fjeldskaar (1997) for discussions of various large-scale stress generating processes in the lithosphere. The global tectonic impact from crustal processes only influences the stress in the sediment because of crustal deformation. If the crust does not respond to the stress by faulting, then the low compressibility of the crustal rocks relative to the sediments will result in an insignificant transfer of crustal stress to these sediments, as pointed out by Bjørlykke and Høeg (1997).

Responses to stress

Principles

Rocks respond to stress changes through either ductile or brittle deformation. The deformation mode is determined by the relationship between stress anisotropy generation and stress anisotropy dissipation. Brittle deformation is favored if the stress changes are rapid and large and the rocks themselves are brittle, while softer rocks will more frequently respond to stress changes by ductile deformation.

Brittle deformation of cap rocks may lead to reservoir leakage through creation or reopening of faults or fractures, while ductile deformation of cap rocks supposedly preserves sealing capacity. The main challenge for the explorationist is thus to differentiate between brittle and ductile deformation of sealing rocks and fault zones through time.

Ductile deformation

Fig. 7 demonstrates different ductile response modes to imposed stress. Ductile deformation is caused by combinations of elastic, plastic, viscous, thermal and chemical processes (diagenesis). The illustrations portrait an initial cube of rock which receives a load. Progressing time intervals are denoted by $t = 1$, $t = 2$, etc., and arrows indicate increased stress which results as the rock is not allowed to expand laterally in the constrained cases. The dotted lines represent the shape of the rock at the previous time interval.

Elastic rock deformation results in horizontal stress increase due to vertical loading. If unconstrained, the rock expands laterally as a result of the added

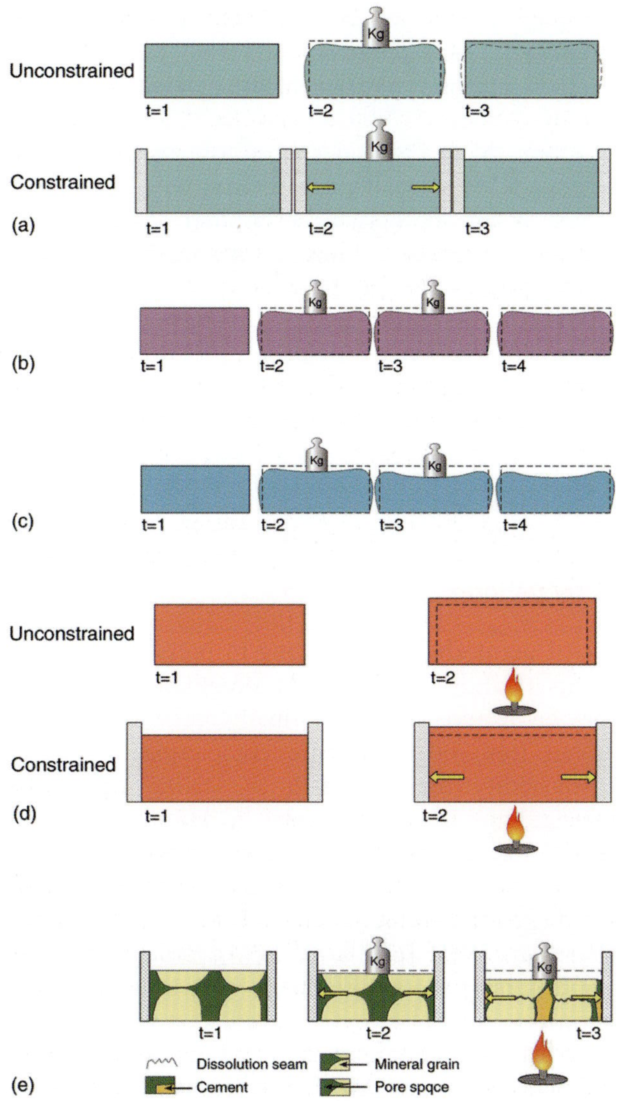

Fig. 7. Different response modes to imposed rock stress: (a) elastic rock deformation; (b) plastic rock deformation; (c) viscous rock deformation; (d) thermal rock deformation; (e) deformation by chemical processes.

overburden. The lateral expansion is determined by Poisson's ratio g (g = lateral strain/vertical strain). If lateral expansion is not possible (the constrained case), horizontal stress but no strain will result. Elastic deformation is reversible: the rock returns to its original state at $t = 3$, when the load has been removed (Fig. 7a).

Plastic rock deformation is irreversible and time-independent: this means that plastic deformation cannot be caused by prolonged exposure to stress alone if the stress remains unchanged. Shallow (mechanical) sediment compaction is mainly a plastic process (Fig. 7b).

Viscous rock deformation is irreversible and time-dependent. The rate of the viscous behavior of rocks a.o. depends on stress, temperature and material properties. Salt and clays often respond to stress changes by viscous behavior, which frequently results in drilling problems through such rocks (Fig. 7c).

Thermal rock deformation may result in stress increase due to heating. The rock seeks to expand laterally as a response to the heating. If lateral expansion is not possible (the constrained case), lateral stress will result. The magnitude of the lateral stress equals the stress that would be required to elastically compress the heated rock back to its original shape (Fig. 7d).

Chemical processes may lead to cementation of rocks at various stress states, and may impact the rock's response to further stress changes. Fig. 7e depicts a rock which receives a load, which deforms and develops increased horizontal stress as an elastic response to the vertical loading. This rock also experiences dissolution of matrix material at grain contacts, and precipitation of the same material in the pore spaces. Uplift, erosion and cooling of this rock leads to reduced vertical stress. However, the rock will not return to its original shape as a result of the uplift, contrary to what would happen if diagenesis had not taken place (Fig. 7e). The elastic stress has been arrested by the diagenesis and the strain is irreversible.

Few studies attempt to quantify the relative importance of the processes just described. The contributions from elasticity and thermal expansion in combination have been suggested to result in an approximately isotropic stress state (Voight and St. Pierre, 1974; Turcotte and Schubert, 1982). Superimposed on these processes come viscous and plastic behavior, and combinations of the above, which all act to drive the rocks towards isotropic stress states.

Quantitative modeling of the various ductile stress response modes would require knowledge of parameters which describe the various rock types' elastic, plastic and viscous behavior through geologic time. It is unclear to what extent laboratory experiments would yield reliable parameters for such modeling, and results from modeling of ductile rock deformation would have to be carefully calibrated to geological observations. Such modeling and calibration will be a major challenge, but is inevitable for successful modeling of stress-induced seal failure.

Brittle deformation

Ductile rock responses will in general lead a rock towards an isotropic stress state. Tectonic impact may, on the other hand, promote stress anisotropy. Brittle deformation is promoted by stress anisotropy, but also by rapid stress changes, high pore pressures, heterogeneous rock sequences and the rock's mechanical properties. The onset of brittle deformation will be controlled by the stress state and the rocks resistance to failure, often referred to as the failure envelope (see Fig. 8 and Appendix A).

Fracturing of sedimentary rocks takes place in the form of either tensile or shear failure. The subsiding rocks are influenced by competing processes: burial may cause the stress distribution within the rock to become more isotropic (if the ductile and thermal rock responses can accommodate the elastic and plastic responses to the increased vertical stress), and tensile failure (hydrofracturing) will be favored in highly overpressured rocks. However, tectonic impact will often result in anisotropic rock stress and hence favor shear failure. Whether or not shear failure occurs prior to hydrofracturing, depends on the rock's stress state and its resistance to failure (often described by Mohr's failure envelope; see Fig. 8 and Appendix A).

Unfortunately, most studies which suggest hydrofracturing, and thus are concerned with the failure criteria as the least effective stress approaches zero, show Mohr's failure envelope with no units on the axes. Hydrofracturing can only happen if the failure envelope intersects the x-axis at a right angle. Experimental data which reveal to what extent such failure envelopes exist are hard or impossible to obtain.

Fig. 8 shows two different and frequently applied failure envelopes. Curve 'a' intersects the x-axis at negative effective stress, implying that the rock has a tensile strength which must be overcome before fracturing occurs. The failure envelope at such low effective stress is often computed from the Griffith criteria (Jaeger and Cook, 1979). The application of these criteria requires the tensile strength of the rock as an input parameter. This strength is zero for preexisting, uncemented fractures. The magnitude of the tensile strength of shales, which are the most common cap rocks, are typically 2 MPa or less (Lockner, 1995), although individual measurements of rock samples have yielded significantly higher values.

Curve 'b' of Fig. 8 describes a linear approximation to a Mohr envelope with zero tensile strength, which is an appropriate description of the failure criteria of preexisting, uncemented fractures at high fluid pressures. This latter envelope 'b' is here shown without labels on the axes to demonstrate the qualitative difference to envelope 'a'. The location where the failure envelopes are tangent to Mohr's circle can be used to calculate the orientation of the fault where shear failure takes place (Jaeger and Cook, 1979).

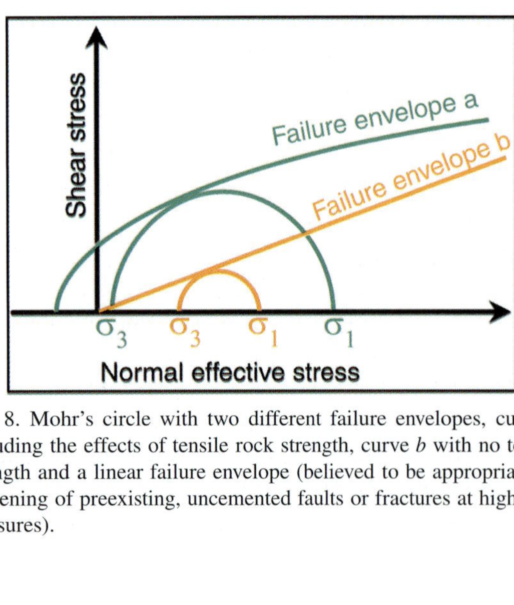

Fig. 8. Mohr's circle with two different failure envelopes, curve *a* including the effects of tensile rock strength, curve *b* with no tensile strength and a linear failure envelope (believed to be appropriate for reopening of preexisting, uncemented faults or fractures at high fluid pressures).

Application of stress analysis in seal evaluation

Fatal vs. non-fatal leakage

All permeable rocks belong to pressure compartments (Buhrig, 1989). These compartments, which may be overpressured or normally pressured, are separated from other pressure compartments by low permeability barriers (such as sealing faults and cap/sealing rocks). As these rocks subside, their porosity is reduced, and overpressures do not stop this porosity reduction once the thermal conditions for diagenetic porosity reduction have been reached (Bjørkum, 1995; Teige et al., 1999). Reduced porosity implies that excess pore fluids must leave the compartment and that all pressure compartments leak or have leaked (Hermanrud and Nordgård Bolås, 2002). Supply of hydrocarbons to the pressure compartment further increases the excess fluid volume which is expelled from the pressure compartment during burial.

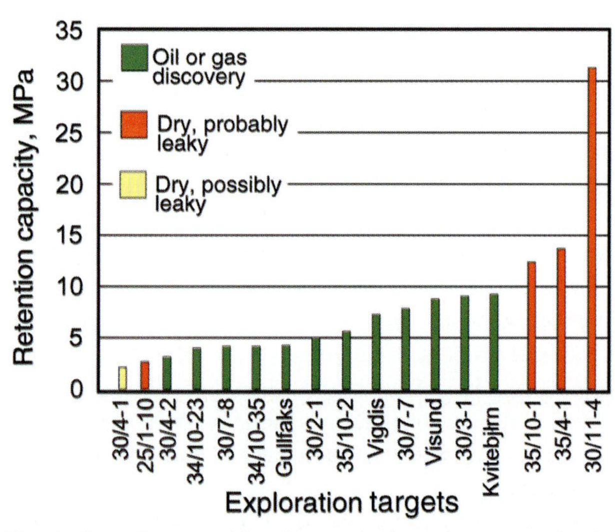

Fig. 9. Retention capacities for a selection of structures in the North Sea. Retention capacities equal LOP minus pore pressure; the LOP values were extrapolated to the depths of the pore pressure measurements in each well.

Leakage from a pressure compartment can take place by vertical and/or lateral fluid movement. When this leakage takes place below the hydrocarbon/ water contact, it has no impact on hydrocarbon occurrence. On the other hand, leakage may also be the main controlling factor of hydrocarbon column heights — even the total hydrocarbon volume may be removed from its trap by leakage mechanisms (see also Fig. 9).

We define leakage to be fatal (as opposed to non-fatal) if the leakage process has caused only residual hydrocarbons to be left in the reservoir rock pore volume affected by the leakage. Commercial volumes of hydrocarbons may or may not remain in the trap after fatal leakage, depending on the actual location of the leakage point within the compartment. In other words, fatal leakage causes hydrocarbon columns to be restricted by the leakage point (instead of any structural or stratigraphical spill). If fatal leakage occurs from the top of a hydrocarbon accumulation, only residual hydrocarbons will remain. However, if the trap experiences fatal leakage in a downflank position, updip hydrocarbons will still be preserved, and the volume of the remaining hydrocarbons will be controlled by the position of the leakage point within the pressure compartment.

Several oil accumulations in the North Sea (Gullfaks, Ekofisk, Snorre, e.g.) leak through their top seal, as is evidenced by increased concentration of hydrocarbons and seismic dim zones above the reservoirs. This leakage, which apparently is operating in significant volumes of the cap rocks, is per definition non-fatal, and the actual leakage processes are not well known. To the contrary, leakage through individual faults or fractures takes place in comparatively smaller rock volumes, but the higher flow rates lead to more efficient removal of the hydrocarbon volumes here. Excess pore fluids will leak through the pressure compartments' weakest points, and identification of these outlet locations through the whole time interval when hydrocarbons where present in the system are crucial for seal evaluation. The location of such conduits is controlled by the rocks' stress state and the way the rocks respond to the imposed stress through time. Identification of factors which favor fatal leakage is critical to seal evaluation, and is attempted in the following.

Factors which favor fatal leakage

Lithology variations in cap rocks and across faults

Failure of reservoir rocks, fault zones and sealing rocks are determined not only by the stress state, but also by the mechanical properties of these rocks. As stress is not evenly transferred in rocks with different mechanical properties, lithological boundaries will be focal points for stress concentrations and will have an increased probability for failure. As a consequence, faults with large throws and significantly different lithologies juxtaposed across the fault plane will be likely candidates for fatal leakage, and noncommercial hydrocarbon volumes may result if the faults intersect the prospect in an updip position. Fracturing of a reservoir alone will not lead to vertical leakage, as such leakage can not happen unless fluids are transported through the cap rock as well. One might guess that a cap rock with alternating sands and shales may fracture more easily than a more homogeneous shaly cap rock and therefore represent a higher exploration risk. However, no data which substantiate the validity of such an hypothesis appear to have been reported.

Upflanks position of faults which are optimally oriented for shear failure

As pointed out previously (Fig. 5), faults with certain orientations will slip first under a given stress regime. This orientation is determined by the orientation and the relative magnitudes of the principal stress components.

Faults which are preferentially oriented and intersect the pressure compartment will tend to define its weakest point and may act as the fluid outlet (leakage point) from overpressured compartments, regardless of the location of the intersection. Hence, if the intersection is located downflanks, the probability of hydrocarbon preservation updip is increased, even if faults with other strikes may intersect the crest of the trap.

Hydrofracturing

Hydrofracturing of a reservoir or cap rock takes place when the pore pressure builds up until it exceeds the least principal stress plus the tensile strength of the rock. At this stage, the rock fails in tension. Such failure will be expected to take place at the shallowest position of a pressure compartment, where the effective stress commonly is the least. This process will hence decrease the probability for preservation of a significant hydrocarbon column. Hydrofracturing is favored in rocks with isotropic stress and significant tensile strength (for fracturing criteria, see Secor, 1965). As detailed in Appendix C, it is suggested that hydrofracturing is a less common process than shear failure in sedimentary basins.

Rapid stress changes through time

Viscous rock behavior work to even out stress anisotropy. As such processes are time-dependent (Fig. 7c), it follows that rapid stress changes promote brittle rock behavior and thereby fatal leakage. Hermanrud and Nordgård Bolås (2002) describe how flexuring due to repeated glaciations and deglaciations in the Quaternary may have resulted in widespread fatal leakage of hydrocarbons in the overpressured regime in the western Haltenbanken area. In this area, both the magnitude and the orientation of the principal stress components changed during repeated cycles of glaciation and deglaciation. As a result, the most favorable orientation for fault slippage changed repeatedly, and a large number of faults were probably reactivated during this period, resulting in changing positions for fluid outlets through time. These changes had no effect on the eastern part of Haltenbanken, which is close to normally pressured, indicating less restrictions in fluid communication eastwards to the seabed.

High fluid pressures

High fluid pressures indicate that lateral fluid transport is restricted, implying that a larger fraction of the fluids may leave the pressure compartments vertically. Whether this vertical leakage restricts any in-place hydrocarbon volumes, depends on the location of vertical fluid discharge, as previously discussed.

Deeply buried reservoirs

Reservoirs become increasingly more segmented as they are buried, mainly because of cementation and reduced permeability along fault planes. These changes are results of the elevated temperatures at increased burial depth. As a consequence, the number of pressure compartments is increased, and the probability for vertical leakage in each compartment is also increased. In total, this situation results in an increased number of locations for vertical fluid outlet. This will, statistically, increase the probability for fatal leakage, but it will also result in a larger number of traps bounded by sealing faults. Once again, fatal leakage may or may not ruin the commercial value of hydrocarbon accumulations, depending on the position of the fluid outlet within each pressure compartment.

Guidelines for application of stress analysis in seal evaluation

Based on the preceding discussions, the following procedure for application of stress analysis in seal evaluation is suggested.

(1) Identify the volumetric extent of the pressure compartment which is being considered.

(2) Evaluate the probability for lateral drainage between pressure compartments. Fluid pressure interpretation and fault sealing analysis are helpful in such evaluations.

(3) Determine the orientation and relative magnitudes of the principal stress components throughout the time when hydrocarbons are believed to have been present in the reservoir. Analysis of the tectonic history of the sedimentary basin should be the cornerstone of such analyses.

(4) Identify the candidates for fluid outlet locations from the pressure compartment.

(5) Look for independent evidence of fluid outlet locations, such as vertical disturbances in seismic data (Løseth et al., 2000) and indications of hydrocarbon contact positions by analysis of direct hydrocarbon indicators (DHIs) from seismic data.

(6) Assess the relative probability of leakage from each location, and apply these results in evaluation of in-place hydrocarbon volumes and prospect risk.

Summary and conclusions

Determination of rock stress, and analyses of its influence on rock failure, can significantly aid seal evaluation. The magnitudes of the vertical and least horizontal principal stress components can often be determined to within 10% or better. Determination of the maximum horizontal stress is less straightforward, and several approaches have been suggested. The accuracy of these various approaches need further investigation.

Knowledge of the relative magnitudes of the principal stress components, and their variation through time since hydrocarbon supply to the prospect, are generally more important to seal evaluation than the magnitudes of the individual stress components. A given fault orientation will be most likely to slip under any given (anisotropic) stress state. This orientation can be determined from knowledge of the relative magnitude of the principal stress components. Quantification of the processes which result in stress anisotropy generation and dissipation in sedimentary basins would aid the interpretation of paleostress, but has apparently not been given much attention in the open literature. As a consequence, evaluation of leakage in the past is hampered with significant uncertainty.

In spite of this uncertainty, factors which promote fatal leakage from hydrocarbon reservoirs, and factors which control the actual location of leakage points within pressure compartments, can be identified. We suggest that hydrofracturing is a less common process for fluid discharge than shear failure. Accordingly, identification of the faults which are most likely to fail in shear is crucial. The existence of such faults in downflanks positions of the pressure compartments, but not at the apex of the structure, is considered positive for hydrocarbon preservation. High fluid pressures, rapid stress changes, juxtaposition of rocks with differing mechanical properties and high reservoir temperatures all increase the probability for vertical fluid outlet from a pressure compartment. However, these factors only lead to noncommercial hydrocarbon accumulation if the fatal leakage takes place at or close to the crest of the pressure compartment.

Based on these factors, a work flow for stress-related seal analysis is suggested. This work flow includes determination of the present day stress regime and the orientation of faults which are most likely to slip under that regime. This, and analysis of the tectonic history of the basin, may aid in identification of alternative locations for fluid outlet from the investigated pressure compartment through time. It is suggested that the relative probability for fluid discharge from these locations should be determined and included in the assessments of prospect risk and in-place hydrocarbon volumes.

Acknowledgements

The content of this paper was significantly improved through discussions with Arne Marius Raaen, Halvor Kjørholt and Lars Wensaas. We further appreciate the preparation of figures by Elin Storsten. Andreas G. Koestler is thanked for a constructive review of an earlier version of the manuscript.

Appendix A. Description of rock stress by Mohr's circle

The rock stress can be described by its three major components. These are most frequently referred to as S_1, S_2 and S_3 (the largest, intermediate and least of these principal stress components). It is most often assumed that one of these components is vertical, in which case the three stress components are referred to as S_v, S_h and S_H (the vertical, least horizontal and largest horizontal stress component). The three principal stress components work at right angles to each other.

Anderson (1951) described the different faulting regimes which prevail depending on the relative magnitudes of the principal stress components. When the vertical stress component is the largest, the rocks are in an extensional stress regime. When the largest stress component is S_H and the smallest stress is the S_h, the rocks will be in the strike/slip regime, whereas the rocks are in a compressive domain when the vertical stress component is the least principal stress component.

For porous rocks, the effective stresses σ_v, σ_h and σ_H (or σ_1, σ_2 and σ_3 if related to magnitudes rather than directions) describe the principal effective stress components, where the effective stress is close to $S - P$, and where P is the pore pressure. Knowledge of all the three principal effective stress components gives the complete rock stress state and enables computations of rock stability. Whether a rock is at the limit of fracturing or not is determined by relationships between the principal effective stress components and the failure criteria of the rock, as is frequently displayed by the relationship between Mohr's circle and the Mohr envelope (Fig. 8). The shear stress in this figure is $\frac{1}{2}$ $(S_2 - S_3)$. Fracturing takes place when the stress state is such that Mohr's circle intersects Mohr's envelope, and the strike of the faults which slip under a given stress regime can be identified from this diagram (e.g. Jaeger and Cook, 1979). Note that increased pore pressure will reduce the effective stress, and thus drive Mohr's circle to the left until the criteria for fracturing are reached. This fracturing may be due to shear failure or hydrofracturing. The latter involves no lateral rock movement, and requires an intersection of Mohr's envelope with Mohr's circle along the x-axis of Fig. 8.

Hydrofracturing of rocks will always happen normal to the least principal stress, and parallel to the maximum principal stress. These criteria are commonly used to identify principal stress directions from tensile fractures and borehole breakouts in wells (see Brudy and Kjørholt, 2001, for further discussions of such techniques).

Appendix B. The maximum horizontal stress (S_H)

Reported S_H values in the petroleum literature either stem from anelastic strain recovery methods (Teufel et al., 1991; Harper, 1995a,b), from hydraulic fracturing of previously fracture-free intervals separated by packers (Bredehoeft et al., 1976; Hickman and Zoback, 1983; Schmitt and Zoback, 1989), from simplified applications of the formula used by Bredehoeft et al. (1976) to leak off tests (Bell, 1990), from inversion of leak off pressures in several inclined wells in an area where the principal stresses do not vary laterally (Aadnøy, 1990; Aadnøy et al., 1994; Gjønnes et al., 1998), and from computations based on the occurrence of borehole breakouts (Zoback et al., 1985) and tensile fractures as observed in image log data (Peska and Zoback, 1995; Zoback et al., 1995a,b), or a combination of several of these methods (Brudy et al., 1997). All of these approaches have their limitations, as will be briefly discussed below.

Anelastic strain recovery (ASR) of oriented cores has been applied to determine rock stresses since the 1930s (Engelder, 1993). This method relies on measuring the expansion of a rock sample after it has been removed from its surroundings (and thus its surrounding stress field). This expansion process is far from fully understood as noted by Harper (1995a,b), who also describes experiments where rocks which are subjected to repeated cycles of stress behave rather unpredictively. Overcoring (pilot holes are drilled, strain measurement gauges are mounted inside the pilot hole, and then the core which includes the pilot hole is cut) is used largely for stress determination in tunnels and mines. While this method gives more accurate results than anelastic strain recovery as described above, the complexity and costs of retrieving such samples have prohibited extensive use in the petroleum industry. Anelastic strain recovery was used by Teufel and Farrel (1995) to investigate the stress state of the Ekofisk field of the North Sea (together with other methods). Studies which elaborate on the accuracy of this method seem to be scarce.

Hydraulic fracturing methods under controlled conditions should give quite reliable stress estimates provided that very careful test procedures are followed; see Engelder (1993) for an in-depth discussion of the application of such techniques. The methods rely on the determination of fracture reopening pressures (several pressure cycles lead to opening and closing of fractures; the pressure necessary to open the fractures the third time should be used according to Hickman and Zoback, 1983). A constant pumping flow rate is also required (Zoback and Haimson, 1982). The S_H is then computed from (Hubbert and Willis, 1957)

$$S_H = 3S_h - \text{Reopening pressure} - \text{Pore pressure} \quad (1)$$

Note that the pore pressure of the rock is needed as input to the calculations. This fact limits the accuracy of these methods in tight rocks where the pore pressure cannot be measured, but has to be inferred from indirect methods, a process which introduces quite significant uncertainties (Hermanrud et al., 1998; Teige et al., 1999). Detournay et al. (1989) demonstrated how poro-elasticity could influence the interpreted stresses from hydraulic fracturing tests, and suggest that, in certain cases, the reopening pressure may reflect the reopening some distance away from the borehole, and not, as assumed in the mathematical formulations of Hubbert and Willis (1957), at the borehole wall. Uncertainties in the determination of S_h from leak off tests further amplifies the uncertainty of S_H determination from hydraulic fracturing methods. Besides, fracture reopening pressures are most often not measured in oil wells.

Bell (1990) suggested an adaption of the hydraulic fracturing method to exploration wells. Starting up with Eq. 1, it was suggested that the instantaneous shut in pressure (ISIP) should be taken as the S_h, and that the leak off pressure (LOP) could be taken as an approximation to the reopening pressure, resulting in

$$S_H = 3(\text{ISIP}) - \text{LOP} - \text{Pore pressure} \quad (2)$$

Bell (1990) states that this equation will underestimate the S_H, especially if the rock has a significant tensile strength, but that it is likely to give reasonable results for shales. This last statement is hard to test, as the pore pressure in proper shales must be indirectly assessed and is often poorly known. Studies which describe the difference between ISIP and fracture reopening pressures appear to be missing. As a consequence the magnitude of the extra uncertainty introduces by substituting Eq. 1 by Eq. 2 cannot be readily quantified.

For the case where no pressure decline records from the leak off tests are available, the ISIP cannot be determined. In such cases, Bell (1990) suggests that a crude estimate of the S_H can be used by setting the ISIP equal to the LOP, thus giving

$$S_H = 2\text{LOP} - \text{Pore pressure} \quad (3)$$

The crudeness of this method was correctly pointed out by the author. Note, that for the cases close to hydrostatic pore pressures, this formula predicts

$$S_H \approx 1.5 S_h \quad (4)$$

since the pore pressure is approximately 0.5 times the LOP in such cases. This result seems hard to justify as a general rule.

Aadnøy (1990) and Aadnøy et al. (1994) suggested a different approach to S_H determination. Their method is based on the expectation that leak off pressures differ as a unique relationship between the stress field and the orientation of the borehole. This relationship was suggested based on the fact that after a well has been drilled, a disturbed stress field arises at each wellbore. For the case where the vertical stress S_v is larger than the horizontal stresses, these authors expected higher LOP values in vertical than in horizontal wells. With three or more wells drilled in the same stress field, least square analysis is used to optimize the (computed) principal stress orientations to minimize the misfit between observed and modeled leak off pressures.

This inversion method is based on Eq. 2 to describe the stress state of a single well, and expressions of linearized versions of the equations which describe stress around the borehole wall. The uncertainty introduced by setting LOP = fracture reopening pressure, by relying on indirect measures of pore pressures in tight rocks, and by uneven quality of S_h assessments from leak off tests also apply to this optimization method, as noted by Aadnøy (1990). Aadnøy (1990) also qualitatively discussed the sensitivity of his method to the filtration of fluid through the mud cake, the often limited number of data points, and the a priori assumed stress state (which is updated after the calculations). The sensitivity to the linearizations of the mathematical equations was investigated separately, and was shown to give significant errors for hole inclinations greater than 50°. Furthermore, the underlying assumption that the ratios between the principal stresses are in fact constant in the area of investigation (that is, independent of variations in geographical position, lithology pore pressure) also may introduce errors to the calculations. Aadnøy et al. (1994) applied their method to data from the Snorre field from the northern North Sea, and suggested that the S_H differed significantly over the field. These results are not in accordance with the later results of Brudy and Kjørholt (2001), which suggest that the S_H does not vary much in the northern North Sea. Gjønnes et al. (1998) claim that the fact that shear stress has been neglected in the inversion scheme of Aadnøy (1990) adds to the uncertainty of this model, and also present an inversion scheme which includes the shear stresses. Quantitative sensitivity analyses of the above mentioned factors needs to be performed before the robustness of the method of Aadnøy (1990) and Gjønnes et al. (1998) can be properly evaluated.

The orientation of borehole breakouts have frequently been used to determine the orientation of the horizontal stresses (see below). The existence of borehole breakouts can also be used to estimate the stress state at the borehole wall. This approach was introduced by Bell and Gough (1979, 1982), and Gough and Bell (1981, 1982) and was later refined by Zoback et al. (1985) and Barton et al. (1988), and extended to inclined boreholes by Peska and Zoback (1995). The equation suggested by Barton et al. (1988), which describes the stress state at the point of onset of the breakout on the borehole wall, relies on the following modification of the equation for stress around a hole in an elastic and isotropic medium (Kirsch, 1898):

$$\sigma_{\theta\theta}(\theta)_b = S_H + S_h - 2(S_H - S_h)\cos(2\theta_b) - \Delta P = C_{eff} \quad (5)$$

with

$$2\theta_b = \pi - 2\varphi \quad (6)$$

where $\sigma_{\theta\theta}(\theta)_b$ is the tangential stress at the borehole wall at the angle θ_b to S_H where the breakout starts to form, 2φ is the breakout opening angle, ΔP is the difference between the pressure of the drilling mud and that of the formation, and C_{eff} is the in situ compressive rock strength. As the input required for this method includes the breakout opening angle, image logs are required to use Eq. 5, a fact which limits the applicability of this approach to wells where image logs have been run. For the case where image logs are not available, the assumption that the breakout angle is a fixed value (e.g. 90°) can be applied, introducing an extra error which depends on the actual breakout angle.

Tensile fractures form in boreholes when the tangential stress at the borehole wall becomes negative. This stress can be computed by (Barton et al., 1988)

$$\sigma_{\theta\theta}(\theta)_b = S_H + S_h - 2(S_H - S_h)\cos(2\theta_b) - \Delta P + \sigma_T - 2P_0 \quad (7)$$

where $\sigma_{\theta\theta}$ is the tangential stress, θ_b is the angle with respect to the orientation of S_H, ΔP is the difference between the pressure of the drilling mud and that of the formation fluid, σ_T is the thermally induced stress, and P_0 is the pore pressure. This last term is introduced to adapt Eq. 6 (from elastic media) to porous media, where the pore pressure must be subtracted from S_H and S_h to give the effective stress of the rock.

Note that analysis of S_H from both tensile fractures and borehole breakouts requires the S_h and pore pressure as input. Generally, the pore pressures are derived from permeable rocks such as sandstones, while leak off pressures are often derived from tight (cap) rocks such as shales. As the pore pressures in the sands cannot be extrapolated to the shales with much confidence, these methods rely on the extrapolation of S_h from the shales to the nearby sands. The occurrence of tensile fractures has been compared in several wells at the Visund field in the Norwegian North Sea, and no preference of these fractures to occur in sands or interbedded shales has been found (D. Wiprut, pers. commun., 1998). This observation suggests that at least in this case, the stresses in the sands and the adjacent shales are comparable, and that the extrapolation of stresses between these lithologies is justified. The thermally induced stress is proportional to the difference between the virgin rock temperature and the temperature in the well during mud circulation. Both of these must be indirectly assessed in most cases, a fact which also introduces some uncertainty to the calculations. Chemical reactions between the formations and the drilling mud may likewise influence the stresses at the borehole wall, especially in smectite-rich formations where water/rock interactions are especially significant during drilling.

Eqs. 5–7 are based on linear elastic models, while more complex rock behavior actually takes place. The applicability of linear elastic models have been questioned by Brudy and Kjørholt (1998). These authors demonstrated that introduction of a thermo-poro-elasto-plastic model resulted in a larger range of possible stress states than what appears from analyses by linear elastic models, suggesting that the uncertainty of S_H determination by Eqs. 5–7 may have been underestimated.

As for the methods which require the breakout opening angle as input, image logs are required for analyses of tensile fractures. Such logs are normally run to achieve information of fine-scaled structures of reservoir sections (such as cross-bedding); their availability is thus limited to the reservoir sections of those wells where image logs have been run. This is unfortunate, as these logs apparently allow for more reliable stress determinations than the other methods which are available at present, and one can only hope that such logs become more widely used in the future.

The accuracy of the computed S_H thus depends both on the accuracy of input parameters (such as S_h, S_v, pore pressure, thermal cooling and rock strength), on the validity of the constitutive law (linear elastic/poroelastic/...), and on the mathematical formulation of inversion methods. Quantification of the resulting uncertainty in the calculated S_H seems a formidable task, and beyond the scope of this paper. Inspection of several boreholes in an area should reduce some of this uncertainty, provided that the local variations of the stress field are minor.

Appendix C. Retention capacities and hydrocarbon occurrence in the northern North Sea

The retention capacity (leak off pressure minus pore pressure at the same depth in a well) was suggested as a measure of sealing capacity by Gaarenstroom et al. (1993). These authors suggested that the pore pressure could reach the value of the least horizontal stress, as inferred from leak off tests, before failure was initiated. They further argued that "the formation strength of the seal has to be greater than the fluid pressure" — presumably, the reference to the formation strength points to the least principal stress plus the tensile strength of the rock. This quote, and the authors' description of the theoretical maximum pore pressure as 1 psi below the LOP, suggest that they envisage a situation where (a) the tensile strength is considered to be insignificant, and (b) fracturing occurs as the least effective stress (LOP minus pore pressure) reaches zero.

As discussed in Appendix A, the first of these conditions results in a failure envelope which intersects origo. Increasing overpressures will result in a leftwards shift of Mohr's circle, and an intersection with the failure envelope at positive least principal stress (i.e. before the circle reaches origo as it is shifted towards the left) will result. The pore pressure and the least compressive stress can be identical only when the stress state is isotropic (and the diameter of Mohr's circle becomes zero), given this failure envelope. Under such conditions, the rocks will fail by hydrofracturing and not by shear failure. Hydrofracturing will be expected to take place at the shallowest part of a pressure compartment, where the effective stress commonly is the least.

If hydrofracturing is the mode of leakage from overpressured traps, then the orientation of the sediments' principal stress components will be of little help in seal evaluation. To the contrary, shear failure will first take place along faults with certain orientations, which can be inferred from the orientation of the stress field. Identification of the fracture mode (shear or hydrofracturing) is therefore important to seal analysis.

This importance of correct fracture mode identification triggered an examination of the two fracture modes based on information from hydrocarbon exploration wells in the northern North Sea. This examination was performed by computation of the retention capacity for the most overpressured wells in the Norwegian North Sea, and also for all leaky and consequently water-bearing exploration targets in the area between 60° and 62°N that we are aware of. The results from this compilation are shown in Fig. 9.

As is seen from this figure, retention capacities below 5 MPa occur frequently. This result was obtained even though LOP measurements from individual wells (and not a minimum LOP trend, as suggested by Gaarenstroom et al., 1993) was applied. Application of a minimum LOP trend would result in negative retention capacities for several of the investigated fields and wells, including the Gullfaks field. Wells with negative retention capacities could not be drilled safely, as mud weight higher than the pore pressure, but lower than the LOP is required to prevent blow outs. The fact that these wells were safely drilled demonstrate that the pore pressures were in all cases lower than the least principal stress. These observations suggest that if fracturing took place, it happened through shear failure and not through hydrofracturing.

The lowest effective stress was observed in the wells 30/4-1 and 25/1-10. No hydrocarbon shows occurred in this former, but gas shows are generally hard to detect and existence of gas remnants in this well can not be excluded. Minor oil shows were reported from well 25/1-10. It is uncertain whether the disappointing result in well 30/4-1 was a result of leakage, whereas leakage is considered to be the most probable reason for the failure of well 25/1-10 at the 'Dyp-Frigg' trap. It is noted that these two wells have the lowest retention capacities of those which were investigated. It is presently unclear whether these wells leaked through hydrofracturing or shear failure, although their low retention capacities may be taken as supporting arguments for hydrofracturing. The other wells with low retention capacities penetrated hydrocarbon-bearing reservoirs.

Two other leaking and water-bearing reservoirs penetrated by wells in the North Sea (wells 35/4-1 and 35/10-1) have retention capacities between 10 and 15 MPa, which are similar to the retention capacities of the leaky reservoirs in the western part of Haltenbanken (Hermanrud and Nordgård Bolås, 2002). These authors suggest that these western Haltenbanken reservoirs have leaked due to stress perturbations caused by glacial flexuring. As the two North Sea wells are situated close to the hinge line of late Cenozoic uplift of mainland Norway (Doré and Jensen, 1996), flexuring may possibly also have caused failure of these reservoirs. If so, the reservoirs failed by shear, for reasons discussed in Hermanrud and Nordgård Bolås (2002).

The last possibly leaky reservoir in the study area that we are aware of is penetrated by well 30/11-4, which is normally pressured and has significant hydrocarbon shows. It is hard to understand how the stress regime in this well could result in fault failure — one possible explanation may be that the leakage was a result of rock failure in deeper strata with presumed high overpressures, and that the resulting fractures propagated through the reservoir and cap rock. This explanation should be regarded as speculative, as it is not supported by additional evidence.

In summary, there appears to be only two candidates for dry structures caused by fatal leakage of hydrocarbons through hydrofracturing in the Norwegian sector of the northern North Sea. On the other hand, the existence of numerous hydrocarbon-containing traps with low retention capacities demonstrates that these traps did not leak by hydrofracturing — such leakage would be expected to take place at the shallowest location of a pressure compartment, leaving only residual hydrocarbons in these reservoirs. However, several of the discoveries appear to be underfilled, an observation which is consistent with vertical leakage through faults which intersect the pressure compartments downflanks.

These results suggest that leakage by shear failure is a more common process than leakage by hydrofracturing in the investigated area. This observation is positive for the prospectivity of overpressured traps in the Norwegian North Sea. It also supports the arguments that seal failure can be identified in undrilled prospects if the stress regime can be identified.

References

Aadnøy, B.S., 1990. Inversion technique to determine the in situ stress field from fracturing data. J. Pet. Sci. Eng., 4: 127–141.

Aadnøy, B.S., Bratli, R.K. and Lindholm, C.D., 1994. In-situ stress modeling of the Snorre field. Eurock'94, Balkema, Rotterdam, pp. 871–878.

Aleksandrowski, P.A., Inderhaug, O.H. and Knapstad, B., 1992. Tectonic structures and wellbore breakout orientation. In: J.R. Tillerson and W.R. Wawersik (Editors), Rock Mechanics: Proceedings of the 33rd U.S. Symposium. Balkema, Rotterdam, pp. 29–37.

Anderson, E.M., 1951. The Dynamics of Faulting and Dyke Formation with Applications to Britain. Oliver and Boyd, Edinburgh.

Barton, C.A., Tesler, L.G. and Burns, K.L., 1988. In-situ stress orientation and magnitude at the Fenton Hill geothermal site, New Mexico, determined from wellbore breakouts. Geophys. Res. Lett, 15: 467–470.

Bell, J.S., 1990. Lecture: The stress regime of the Scotian Shelf offshore eastern Canada to 6 kilometers depthn and implications for rock mechanics and hydrocarbon migration. In: V. Maury and D. Fourmantray (Editors), Rock at Great Depth. Balkema, Rotterdam, pp. 1243–1265.

Bell, J.S. and Gough, D.I., 1979. Northeast–southwest compressive stress in Alberta: evidence from oil wells. Earth Planet. Sci. Lett., 45: 475–482.

Bell, J.S. and Gough, D.I., 1982. The use of borehole breakouts in the study of crustal stress. U.S. Geol. Surv. Open File Rep., 82-1075, pp. 539–557.

Bell, J.S. and Lloyd, P.F., 1989. Modeling of stress refraction in sediments around the Peace River Arch, western Canada. Current Res., Part D, Geol. Surv. Can., 89-1D, pp. 49–54.

Bell, J.S., Caillet, G. and Adams, J., 1992. Attempts to detect open fractures and non-sealing faults with dipmeter logs. In: A. Hurst, C.M. Griffiths and P.P.F. Worthington (Editors), Geological Applications of Wireline Logs II. Geol. Soc. Spec. Publ., 65: 211–220.

Bjørkum, P.A., 1995. How important is pressure solution in causing dissolution of quartz in sandstones. J. Sediment. Res., 66: 147–154.

Bjørlykke, K. and Høeg, K., 1997. Effects of burial diagenesis on stresses, compaction and fluid flow in sedimentary basins. Mar. Pet. Geol., 14(3): 267–276.

Bredehoeft, J.D., Wolff, R.G., Keys, W.S. and Shuter, E., 1976. Hydraulic fracturing to determine the regional in situ stress field, Piceance Basin, Colorado. Geol. Soc. Am. Bull., 87: 250–258.

Brudy, M. and Kjørholt, H., 1998. The initiation of drilling-induced tensile fractures and their use for estimation of the stress magnitude. Abstract volume for the Euroconference Earth Stress and the Industry — the World Stress Map and beyond, Sept. 3–5, Heidelberg, 12.

Brudy, M. and Kjørholt, H., 2001. Stress orientation on the Norwegian continental shelf derived from borehole failures observed in high-resolution borehole imaging logs. Tectonophysics, 337: 65–84.

Brudy, M., Zoback, M.D., Fuchs, K., Rummel, F. and Baumgärtner, J., 1997. Estimation of the complete stress tensor to 8 km depth in the KTB scientific drill holes: implication for crustal strength. J. Geophys. Res., 102(B8): 18453–18475.

Buhrig, C., 1989. Geopressured Jurassic reservoirs in the Viking Graben: modeling and geological significance. Mar. Pet. Geol., 6: 31–48.

Chapple, W.M. and Tullis, T., 1977. Evaluation of the forces that drive the plates. J. Geophys. Res. A, 82(14): 1967–1984.

Dallmus, K.F., 1955. Mechanics of basin evolution and its relation to the habitat of oil in the basin. In: G. Weeks (Editor), Habitat of Oil. AAPG Symp, pp. 883–893.

Detournay, E., Cheng, A.H.-D., Roegiers, J.-C. and Mclennan, J.D., 1989. Poroelasticity considerations in In situ stress determination by hydraulic fracturing. Int. J. Rock Mech. Sci. Geomech. Abstr., 26(6): 507–513.

Doré, A.G. and Jensen, L.N., 1996. The impact of late Cenozoic uplift and erosion on hydrocarbon exploration: offshore Norway and some other uplifted basins. Global Planet. Change, 12: 415–436.

du Rouchet, J., 1981. Stress fields, a key to oil migration. Am. Assoc. Pet. Geol. Bull., 65(1): 74–85.

Engelder, T., 1993. Stress Regimes in the Lithosphere. Princeton University Press, Princeton, NJ.

Finkbeiner, T., Zoback, M., Stump, B. and Flemings, P., 1998. In situ stress, pore pressure, and hydrocarbon migration in the South Eugene Island field, Gulf of Mexico. Proceedings from the workshop Overpressures in Petroleum Exploration, Pau, April 1998.

Fjeldskaar, W., 1997. Flexural rigidity of Fennoscandia inferred from the postglacial uplift. Tectonics, 16(4): 596–608.

Forsyth, D. and Uyeda, S., 1975. On the relative importance of the driving forces of plate motion. Geophys. J. R. Astron. Soc., 45: 163–200.

Gaarenstroom, L., Tromp, R.A.J., de Jong, M.C. and Brandenburg, A.M., 1993. Overpressures in the Central North Sea: implications for trap integrity and drilling safety. In: J.R. Parker (Editor), Petroleum Geology of Northwest Europe: Proceedings of the 4th Conference. The Geological Society, London, pp. 1305–1313.

Gjønnes, M., Cruz, A.M.G.L., Horsrud, P. and Holt, R.M., 1998. Leak-off tests for horizontal stress determination? J. Pet. Sci. Eng., 20: 63–71.

Gough, D.I. and Bell, J.S., 1981. Stress orientations from oil well fractures in Alberta and Texas. Can. J. Earth Sci., 18: 638–645.

Gough, D.I. and Bell, J.S., 1982. Stress orientation from borehole wall fractures with examples from Colorado, east Texas, and Northern Canada. Can. J. Earth Sci., 19: 1358–1370.

Grauls, D.J. and Baleix, J.M., 1994. Role of over pressures and in situ stresses in fault-controlled hydrocarbon migration: a case study. Mar. Pet. Geol., 11(6): 734–742.

Harper, T.R., 1995a. Dead or alive: two concepts of rock behavior. Proc. R. Inst., 66: 43–64.

Harper, T.R., 1995b. Internal forces and stress state evolution. In: M. Fejerskov and A.M. Myrvang (Editors), Proceedings from the Workshop Rock Stresses in the North Sea, Feb. 13–14, Trondheim, pp. 49–62.

Harper, T.R. and Lundin, E.R., 1997. Fault seal analysis: reducing our dependence on empiricism. In: P. Møller-Pedersen and A.G. Koestler (Editors), Hydrocarbon Seals: Importance for Exploration and Production. Norwegian Petroleum Society (NPF), Special Publication 7. Elsevier, Amsterdam, pp. 149–165.

Heffer, K.J. and McLean, J.C., 1993. Earth stress orientation — a control on, and guide to, flooding directionality in a majority of reservoirs. In: B. Linville, T.E. Burchfield and T.C. Wesson (Editors), Reservoir Characterization III. PennWell Books, Tulsa, OK, pp. 800–822.

Hermanrud, C. and Nordgård Bolås, H.M., 2002. Leakage from overpressured hydrocarbon reservoirs at Haltenbanken and in the northern North Sea. In: A.G. Koestler and R. Hunsdale (Editors), Hydrocarbon Seal Quantification. Norwegian Petroleum Society (NPF), Special Publication 11. Elsevier, Amsterdam, pp. 221–231 (this volume).

Hermanrud, C., Teige, G.M.G., Vik, E., Paasch, B., Wensaas, L. and Nordgård Bolås, H.M., 1998. Overpressures in shales — do we know what they are and why they are there? Proceedings from the Workshop Overpressures in Petroleum Exploration, Pau, April 1998.

Hickman, S.H. and Zoback, M.D., 1983. The interpretation of hydraulic fracturing pressure – time data for in-situ stress determination. Proceedings of the Workshop on Hydraulic Fracturing Stress and Measurements. U.S. National Committee on Rock Mechanics, Washington, DC, pp. 1–11.

Hubbert, M.K. and Rubey, W.W., 1953. Role of fluid pressure in mechanics of overthrust faulting. Bull. Geol. Soc. Am., 70: 115–206.

Hubbert, M.K. and Willis, D.G., 1957. Mechanics of hydraulic fracturing. J. Pet. Technol., 9: 153–168.

Jaeger, J.C. and Cook, N.G.W., 1979. Fundamentals of Rock Mechanics, 3rd ed. Champion and Hall, London.

Kirsch, G., 1898. Die Theorie der Elastizität und die Bedürfnisse der Festigkeitslehre. VDI Z, 42: 797–807.

Larson, K.W., Waples, D.W., Fu, H. and Kodama, K., 1993. Predicting tectonic fractures and fluid flow through fractures in basin modeling. In: A.G. Doré, J.H. Augustson, C. Hermanrud, D.J. Stewart and Ø. Sylta (Editors), Basin Modeling, Advances and Applications. Norwegian Petroleum Society (NPF), Special Publication 3. Elsevier, Amsterdam, pp. 373–383.

Linjordet, A. and Skarpnes, O., 1992. Application of stress directions interpreted from borehole breakouts recorded by four-arm caliper. In: T.O. Vorren, E. Bergsager, Ø.A. Dahl-Stamnes, E. Holter, B. Johansen, E. Lie and T.B. Lund (Editors), Arctic Geology and Petroleum Potential. Norwegian Petroleum Society (NPF), Special Publication 2. Elsevier, Amsterdam, pp. 681–690.

Lockner, D.A., 1995. Rock failure. In: Rock Physics and Phase Relations, a Handbook of Physical Constants. AGU Reference Shelf 3, pp. 127–147.

Løseth, H., Gading, M. and Wensaas, L., 2000. Location of leakage points and timing of leakage from seismic data. In: Hydrocarbon Seal Quantification, Norwegian Petroleum Society (NPF) Conference, 16–18 October, Extended Abstracts Volume, pp. 187–190.

Makurat, A., Tørudbakken, B., Monsen, K. and Rawlings, C., 1992. Cenozoic uplift and caprock seal in the Barents sea: fracture modeling and seal risk evaluation. Soc. Pet. Eng., Pap. 24740, pp. 821–830.

Mandl, G. and Harkness, R.M., 1987. Hydrocarbon migration by hydraulic fracturing. In: M.E. Jones and R.M.F. Preston (Editors), Deformation of Sediments and Sedimentary Rocks. Geol. Soc. London Spec. Publ., 29: 39–53.

Morita, N. and McLeod, H., 1995. Oriented perforation to prevent casing collapse for highly inclined wells. SPE Drilling and Completion (Sept.), pp. 139–145.

Morita, N., Whitfil, D.L., Fedde, O.P. and Levik, T.H., 1989a. Parametric study of sand-production prediction: analytical approach. SPE Production Engineering, pp. 25–33.

Morita, N., Whitfil, D.L., Massie, I. and Knudsen, T.W., 1989b. Realistic sand-production: numerical approach. SPE Production Engineering, pp. 15–24.

Mueller, B., Reinecher, J. and Fuchs, K., 2000. The 2000 release of the World Stress Map (available online at http://www-wsm.physik.uni-karlsruhe.de/pub/stress_data/stress_data_frame.html).

Peska, P. and Zoback, M.D., 1995. Compressive and tensile failure of inclined well bores and determination of in situ stress and rock strength. J. Geophys. Res., 100(B7): 12791–12811.

Richardsson, R.M., 1992. Ridge forces, absolute plate motions, and the intraplate stress field. J. Geophys. Res., 17: 478–490.

Schmitt, D.R. and Zoback, M.D., 1989. Poroelastic effects in the determination of the maximum horizontal principal stress in hydraulic fracturing tests — a proposed breakdown equation employing a modified effective stress relation for tensile failure. Int. J. Rock Mech. Min. Sci. Geomech. Abstr., 26(6): 499–506.

Schnaebele, R., 1948. Monographie Géologique du Champ Pétrolifère de Pechelbronn, Strasbourg. Mém. Carte Géol. Alsace Lorraine, 7.

Secor, D.T., 1965. Role of fluid pressure in jointing. Am. J. Sci., 263: 633–646.

Snarsky, A.N., 1962. Die primäre Migration des Erdöls. Freiburger Forschungsh. C, 123.

Spann, H., Mueller, B. and Fuchs, K., 1994. Interpretation of anomalies in observed stress data at the Central Graben (North Sea) — numerical and analytical approach. Soil Dyn. Earthquake Eng., 13: 1–11.

Teige, G.M.G., Hermanrud, C., Wensaas, L. and Nordgård Bolås, H.M., 1999. Overpressure versus shale porosity: no relationship in North Sea and Haltenbanken shales. Mar. Pet. Geol., 16(4): 321–335.

Teufel, L.W. and Farrel, H.E., 1995. Interrelationship between in situ stress, natural fractures, and reservoir permeability anisotropy — A case study of the Ekofisk Field, North Sea. In: L.R. Myer, N.G.W. Cook, R.E. Goodman and C.F. Tsang (Editors), Fractured and Jointed Rock Masses. Balkema, Rotterdam, pp. 573–578.

Teufel, L.W., Rhett, D.W. and Farrel, H.E., 1991. Effect of reservoir depletion and pore pressure drawdown on in-situ stress and deformation in the Ekofisk field, North Sea. In: J.C. Roegiers (Editor), Rock Mechanics as a Multidisciplinary Science. Proceedings of the 32nd U.S. Rock Mechanics Symposium. Balkema, Rotterdam, pp. 63–72.

Turcotte, D.L. and Schubert, G., 1982. Geodynamics. Application of Continuum Physics to Geological Problems. Wiley, New York.

Ungerer, P., Doligez, B., Chénet, P.Y., Bessis, F., Lafargue, E., Giroir, G., Heum, O.R. and Eggen, S., 1987. A 2-D model of basin scale petroleum migration by two-phase fluid flow. Application to some case studies. In: B. Doligez (Editor), Migration of Hydrocarbons in Sedimentary Basins. Technip, Paris, pp. 415–456.

Voight, B. and St. Pierre, B.H.P., 1974. Stress history and rock stress. Int. Soc. Rock Mech., Proc 3rd Congr., 2, pp. 580–582

Wiprut, D.J. and Zoback, M.D., 1998. High horizontal stress in the Visund field, Norwegian North Sea: consequences for borehole stability and sand production. SPE/ISRM Pap. 47244, Soc. Pet. Eng., pp. 199–208.

Wiprut, D., Zoback, M., Hanse, Tor-Harald and Peska, P., 1997. Constraining the full stress tensor from observations of drilling-induced tensile fractures and leak-off tests: application to borehole stability and sand production on the Norwegian margin. Int. J. Rock Mech. Min. Sci, 34(3–4), paper 00365.

Yale, D.P., Rodrigues, J.M. and Mercer, T.B., 1994. In-situ stress orientation and the effects of local structure — Scott Field, North Sea. Eurock'94, Balkema, Rotterdam.

Zoback, M.L., 1992. First and second order patterns of tectonic stress: the World Stress Map project. J. Geophys. Res., 97(11): 703–711, 728.

Zoback, M.D. and Haimson, B.C., 1982. Issues in rock mechanics. Proc. 23rd U.S. Symp. Rock Mechanics, pp. 143–156.

Zoback, M.D. and Healey, J.H., 1992. In situ stress measurements to 3.5 km depth in the Cajon Pass scientific research borehole: implications for the mechanics of crustal faulting. J. Geophys. Res., 97: 5039–5057.

Zoback, M.L. and Zoback, M.D., 1989. Tectonic stress field of the continental U.S. In: L.M. Pakiser and W.D. Mooney (Editors), Geophysical Framework of the Continental United States. Geol. Soc. Am. Mem., 172: 523–539.

Zoback, M.D., Moos, D., Mastin, D. and Anderson, R.N., 1985. Wellbore breakouts and in situ stress. J. Geophys. Res., 90(B7): 5523–5530.

Zoback, M.L., Zoback, M.D., Adams, J., Assumpaco, M., Bell, S., Bergman, E.A., Blüming, P., Brereton, N.R., Denham, D., Ding, D., Fuchs, K., Gay, N., Gregersen, S., Gupta, H.K., Gvishiani, A., Jacob, K., Klein, R., Knoll, P., Magee, M., Mercier, J.L., Mueller, B., Paquin, C., Rajendran, K., Stephansson, O., Suarez, G., Suter, M., Udias, A., Xu, Z.H. and Zhizhin, M., 1989. Global patterns of tectonic stress. Nature, 341: 291–298.

Zoback, M.D., Barton, C., Brudy, M., Chang, C., Moos, D., Peska, P. and Vernik, L., 1995a. A review of some new methods for determining the in situ stress state from observations of borehole failure with application to borehole stability and enhanced production in the North Sea. In: M. Fejerskov and A.M. Myrvang (Editors), Proceedings from the Workshop Rock Stresses in the North Sea, Feb. 13–14, Trondheim, pp. 6–21.

Zoback, M.D., Barton, C., Moos, D., Peska, P. and Vernik, L., 1995b. Utilization and analysis of multiple modes of borehole failure for estimation of in situ stress magnitudes. Proc. 8th Int. Symp. Rock Mechanics, Sept. 25–30, Tokyo.

H.M. NORDGÅRD BOLÅS *Statoil's Research Centre, N-7005 Trondheim, Norway*
C. HERMANRUD *Statoil's Research Centre, N-7005 Trondheim, Norway*

The role of diagenesis in the formation of fluid overpressures in clastic rocks

Hans Martin Helset, Robert H. Lander, James C. Matthews, Paul Reemst, Linda M. Bonnell and Inge Frette

We have developed a model of fluid flow and pressure development in sedimentary basins that incorporates pore volume loss due to mechanical compaction and to chemical diagenesis (quartz cementation, grain contact quartz dissolution and illitization). Mechanical compaction is modeled to be a function of effective stress. In this model, pore volume loss due to mechanical compaction will be retarded when overpressure develops. The diagenetic processes are modeled as being kinetically controlled and the reaction progress depends only on the temperature history. Hence pore volume loss due to chemical compaction is not retarded by overpressure. By including diagenetic effects on overpressure development, the pressure model should be more generally applicable than models that consider mechanical compaction to be the sole process that reduces porosity. To demonstrate the potential importance of chemical compaction in the formation of fluid overpressures in different settings, we calibrated our model with data obtained from the Halten Terrace offshore mid-Norway and from the Gulf of Mexico. In both cases, the diagenetic processes have the potential to control on the timing and magnitude of overpressuring. From 25% and up to 80% of the present-day overpressure may be caused by pore volume loss resulting from diagenetic reactions. Pressure build-up from diagenetic processes also potentially controls the timing of hydraulic fracturing. If diagenetic processes are actively contributing to overpressure generation, then unrealistically low shale permeabilities are not needed to retain overpressures for geologic time periods (>10 My).

Introduction

The build-up of abnormally high reservoir fluid pressures (overpressures) has important implications for fluid flow and seal integrity, as well as for drilling safety. Most existing models use compaction disequilibrium as the primary cause of overpressure development. Effects of diagenetic reactions on pore volume reduction are often neglected. Compaction disequilibrium models can be made to provide predictions of overpressure that match pressure measurements in pre-Tertiary sediments, but only by assuming shale permeabilities that are significantly lower than measured values (Mann and Mackenzie, 1990; Deming, 1994). Models based primarily on compaction disequilibrium often suggest that present-day overpressures are 'fossil', and are related to early burial that occurred tens to hundreds of million years ago. In order to conserve overpressures over such time periods, the shale units need to have extremely low permeabilities. Shale porosity data from Haltenbanken (Hermanrud et al., 1998) show no indication of excess fluid pressure, suggesting late overpressure development. The assumption that overpressures have been conserved over long time periods is also not compatible with our own observations of intergranular volume (IGV) values in overpressured sandstones from the North Sea.

Quartz cementation has been suggested as a process that may generate overpressure (Bjørkum and Nadeau, 1998). Our forward modeling of porosity loss in sandstones has allowed us to study the relative importance of mechanical compaction and quartz cementation through time. Porosity loss due to mechanical compaction is most important during the early phases of burial. On the other hand, quartz cementation typically starts after most mechanical compaction is complete, and will reach a peak rate at even greater depth (Lander et al., 1999). In many basin settings, quartz cementation may be an ongoing process. At the depths at which relatively rapid quartz cementation occurs in the sandstones, the overlying shale units will have low permeabilities that help conserve the overpressures generated by quartz cementation.

An important implication of models incorporating diagenetic effects is that hydraulic fracturing could potentially occur at much greater depths than would be expected for models that rely largely on com-

Hydrocarbon Seal Quantification edited by A.G. Koestler and R. Hunsdale. NPF Special Publication 11, pp. 37–50, Published by Elsevier Science B.V., Amsterdam.

paction disequilibrium as the generation mechanism for overpressure development. Based on published data on fracture pressures, fluid retention must begin at depths no greater than around 1200 m in order for compaction disequilibrium to cause hydraulic fracturing (Gaarenstroom et al., 1993; Osborne and Swarbrick, 1997). The deeper occurrence of hydraulic fracturing is possible because, unlike mechanical compaction, diagenetic reactions such as quartz cementation are comparatively insensitive to fluid pressures within the range normally encountered in sedimentary basins (Bjørkum, 1996).

We have developed a model of fluid pressure that incorporates loss of pore volume due to both mechanical compaction and to diagenesis. Published kinetic models for quartz cementation and smectite illitization have been implemented, as described later. The model is a forward model of sedimentation, temperature and pressure evolution. It has been used in several studies to predict fluid overpressure in undrilled prospects.

The main purpose of this paper is to evaluate the relative importance of the different pressure generating mechanisms. In order to achieve this, we first present an overview of processes and mechanisms that may contribute to the development of overpressure in sedimentary basins. Next, we describe the processes that we have chosen to include in our model together with the mathematical formulations used.

Finally, we present results from two case studies, together with an evaluation of the model performance, and of the relative importance of the different pressure generating mechanisms. In the case study from Halten Terrace we constrain the pressure modeling by honoring available well data such as temperature, quartz cement volumes, smectite content and porosities. We then predict the relative contribution from each of the different pressure generating mechanisms. No model calibration was performed in the case study from the Gulf of Mexico.

Processes generating overpressure

Overpressure, or geopressure, is defined as the pore pressure exceeding the hydrostatic pressure. Fig. 1 shows fluid pressure vs. depth for a hypothetical overpressured well. The hydrostatic and lithostatic gradients are included as straight lines. The amount of overpressure is denoted by Φ. Effective vertical stress σ_e is defined as the difference between fluid pressure and lithostatic pressure.

Overpressures in sedimentary basins form when the rate of pore volume reduction is fast relative to the rate of pore fluid release, or when the rate of

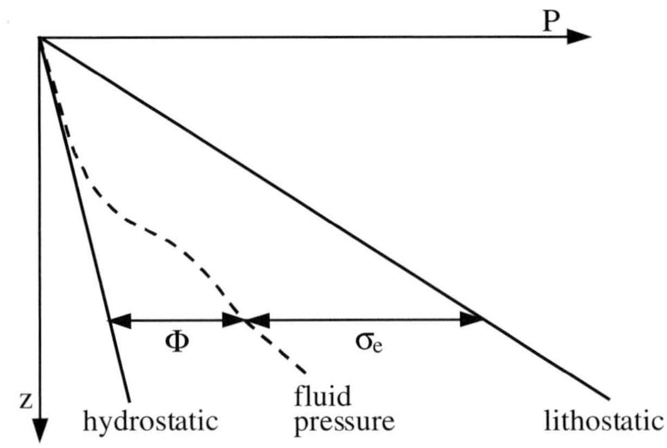

Fig. 1. Fluid pressure vs. depth for a hypothetical overpressured well, P = pressure, z = depth. The fluid pressure profile is shown as a dashed line. The hydrostatic and lithostatic pressure gradients are shown for reference. Overpressure, Φ, is defined as the difference between fluid pressure and hydrostatic pressure. Effective stress, σ_e, is defined as the difference between lithostatic pressure and fluid pressure.

pore fluid expansion is fast relative to the rate of fluid release. Large-scale flow systems can also influence the overpressure development, but this is generally only important in mountainous regions (Bachu and Underschultz, 1993) and will be neglected here. As discussed in the next section, both pore volume reduction and pore fluid expansion act as sources for overpressure.

Mechanisms that cause pore volume decrease include:

- mechanical compaction
- diagenesis (or chemical compaction).

Fluid expansion can be caused by:

- aquathermal expansion
- smectite dehydration and illitization
- kerogen maturation
- oil-to-gas cracking.

A sediment unit will compact mechanically due to the load of the overlaying sediment column. Microscopically, mechanical compaction occurs by grain rearrangement, fracturing or ductile deformation. Empirical relations between porosity and depth have been widely used to model sediment compaction (Athy, 1930; Bethke, 1985). More general relations between porosity and effective stress have been derived to take into account the effect of overpressure on compaction (Bethke, 1985). Following the Terzaghi effective stress relationship, the vertical effective stress experienced by sediment grains is the difference between lithostatic pressure and fluid pressure. When fluid pressure increases above hydrostatic pressure, effective stress may be reduced and part of the overburden is supported by the pore fluids (see Fig. 1). Generation of overpressure will therefore retard the effective stress driven mechanical compaction of the sediments. This effect is included through a porosity–effective stress model.

Quartz cementation is an important process causing pore volume loss in sandstones. Traditionally, quartz dissolution has been described as a pressure solution process (McBride, 1989). The reaction rate is often assumed to be proportional to effective stress (Rutter, 1983). Consequently, it would be expected that quartz dissolution would be retarded when overpressures develop and effective stress is reduced.

Recent research suggests that the coupled processes of grain contact quartz dissolution and quartz cementation are controlled by temperature, and are insensitive to effective stress under conditions that occur within hydrocarbon reservoirs (Walderhaug, 1994; Bjørkum, 1996). Quartz precipitation is considered to be the rate-limiting factor. Quartz dissolves at stylolites (mica/quartz or illite/quartz interfaces), and is transported to the precipitation sites by diffusion (Bjørkum and Nadeau, 1998). The dissolution of quartz grains causes compaction and pore volume reduction of the sedimentary column corresponding to the thickness of the dissolved region. In addition, pore volume is lost in the adjacent sediment due to quartz precipitation. When pore volume is lost, water must escape from the system, giving a contribution to overpressure. The implication of this approach is that quartz cementation will not be retarded as overpressure develops. The reaction may, therefore, continue to generate overpressure even during periods of erosion.

Aquathermal expansion of the pore water has been suggested to be an important pressure generating mechanism. As temperature increases, the volume of pore fluid increases, leading to overpressuring. The rate of fluid volume increase is, however, fairly low compared with the rate of fluid release. Except when seals have extremely low permeability, the effect of aquathermal expansion on overpressure seems to be minimal (Bethke, 1985; Luo and Vasseur, 1992; Osborne and Swarbrick, 1997). For example, Bethke (1985) suggested that a pressure increase of 0.7% is produced by aquathermal expansion in slowly subsiding basins.

Smectite illitization reactions in clays/shales have long been suggested to contribute to overpressure development (Powers, 1967; Burst, 1969; Perry and Hower, 1972; Bruce, 1984). Most of the early papers that discuss smectite dehydration are concerned with water released during illitization of smectite. In this paper we distinguish between smectite dehydration and illitization. Smectite dehydration will be inhibited by overpressure and is considered to be less important (Colten-Bradley, 1987). Pressure generation due to smectite illitization results from pore volume reduction and fluid volume increase. In addition, the smectite illitization reaction has been suggested to create textural changes within shales, which reduce permeability (Bjørkum and Nadeau, 1998).

Hydrocarbon generation and cracking are potential mechanisms for generation of overpressure. When solid kerogen is converted into liquid or gaseous hydrocarbons, the volume of pore fluids increases. Depending on the rate of fluid volume increase, this may lead to increase in fluid pressure. Hydrocarbons remain largely insoluble in the pore water. The presence of hydrocarbons will also decrease the relative permeability of rocks to water, thus promoting overpressure. Luo and Vasseur (1996) performed numerical modeling to assess the significance of organic matter cracking as a mechanism to generate overpressure. They found that oil generation is important only when the organic matter content is fairly large (>5%). The generation of gas has a larger effect on overpressure because of the large volume expansion involved. Their numerical modeling also showed that oil-to-gas cracking is important at relatively great depth where the permeabilities are sufficiently low to retain the pore fluids. Using a temperature gradient of 35°C/km they find that the maximum effect of gas generation on overpressure occurs at a depth of 5.5 km. Burrus (1998) presented results of pressure modeling from the Gulf of Mexico. In his simulations, hydrocarbon generation was responsible for 10% or less of the overpressure.

We have developed a numerical model of pressure generation and dissipation in a sedimentary column. Of the processes described above, we have included pore volume loss due to mechanical compaction, quartz cementation in sandstones and smectite illitization in shales. We have chosen to neglect the effect of aquathermal expansion since this seems to be of little importance compared to other pressure generating mechanisms. The effects of hydrocarbon generation and oil–gas cracking have also been neglected. Nevertheless, these processes may in some cases be important, at least on a local scale.

Mathematical formulation

Our numerical model simulates the development of overpressure in a basin through time. The differential equation describing compaction driven fluid flow is outlined below (Eq. 1). Compaction of the sediment column is caused by mechanical compaction and by diagenesis (chemical compaction). Mechanical compaction is modeled using porosity–effective stress relations (Bethke, 1986; Lander and Walderhaug, 1999). Chemical compaction includes quartz cementation in sandstones and smectite illitization in shales. The chemical compaction processes are simulated using kinetic models that take temperature history as input. A standard formulation of conductive heat flow through the sediments is used (Bethke, 1985), neglecting convective heat flow.

Dissipation of fluid pressure through the rock matrix is modeled using a Darcy flow formalism where sediment permeability controls the rate of fluid flow. When high overpressures develop, hydraulic fracturing may occur causing additional pressure dissipation. A fracture pressure curve can be manually defined in the model to resemble observed fracture pressures (see e.g. Gaarenstroom et al., 1993).

We first describe the differential equation governing the fluid pressure, and then continue with describing the processes that contribute in the pressure equation. We introduce the hydraulic potential (overpressure) as the dependent variable in our modeling. The overpressure is $\Phi = P - \rho_f g z$ where P is fluid pressure, ρ_f is fluid density, g is the constant of gravity, and z is the depth (positive downwards) (see Fig. 1). For hydrostatic conditions we have $P = \rho_f g z$ and $\Phi = 0$. The one-dimensional equation of compaction-driven groundwater flow can be written (Bethke, 1985, 1986).

$$A\frac{\partial \Phi}{\partial t} = \frac{\partial}{\partial z}\left(K_z \frac{\partial \Phi}{\partial z}\right) + B$$

$$A = \phi \beta_f$$

$$K_z = \frac{k}{\mu}$$

$$B = -\phi \beta_f \rho_f g \frac{\partial z}{\partial t} + \left(K_z \frac{\partial \Phi}{\partial z}\right)\frac{1}{\rho_f}\frac{\partial \rho_f}{\partial z} - \frac{1}{1-\phi}\frac{\partial \phi}{\partial t} + Q_w \quad (1)$$

Mechanical/chemical compaction

Clay mineral reactions

where ϕ is porosity, k is permeability, β_f is fluid compressibility, μ is fluid viscosity, and ρ_f is fluid density. The symbols are further explained in the nomenclature list. We assume incompressible rock grains and neglect aquathermal expansion of the pore fluids. We have chosen to write the differential equation governing the fluid flow in terms of overpressure Φ. The equations could also be written in terms of fluid pressure (Bethke, 1986).

Hydrostatic boundary conditions are imposed at the top of the sedimentary column, $\Phi = 0$. At the bottom of the column (basement) there is a no-flow boundary, $K_z(\partial\Phi/\partial z) = 0$. Since the equation is written in terms of the overpressure Φ, the water depth does not enter the equation. However, to calculate the absolute fluid pressure, the water depth must be accounted for.

Eq. 1 has the form of a diffusion or heat equation. The coefficient A controls the rate of pressure build-up and is the product of the fluid compressibility and the volume of pore fluid (porosity). The coefficient K_z is equal to permeability divided by viscosity; it is analogous to the diffusion coefficient in the standard diffusion equation, and controls the rate at which the overpressure dissipates. The fluid flow rate increases with increasing permeability and decreasing viscosity. We assume constant viscosity in the present implementation of the model. The B-term has the form of a source term and is particularly interesting in the case of a compacting sedimentary column in that it contains terms related both to pore-volume loss and to increase in fluid volume. Both mechanical compaction and diagenetic reactions will act as sources of overpressure through the porosity loss terms in B. Mathematically, the loss in pore volume acts as a source for overpressure. Mechanical and chemical compaction, both in shales and sandstones, contribute to the coefficients A and B through the porosity (ϕ), the porosity loss rate ($\partial\phi/\partial t$) and the change in unit thicknesses ($\partial z/\partial t$). The smectite illitization reaction will in addition contribute directly as a fluid source term (Q_w) since bound interlayer water is released in the reaction.

The permeability of both shales and sandstones decreases with decreasing porosity, and hence with increasing burial depth. Shale permeability, being orders of magnitude lower than sandstone permeability, has the greatest influence on the overpressure development. Permeabilities are modeled using a Carman-Kozeny model (Lake, 1989). Permeability is a function of both porosity and specific surface area. We model a variation in shale specific surface area resulting from the conversion of smectite to illite.

The frame of reference for the pressure model as presented here is a one-dimensional vertical column. For situations where faults or sediment heterogeneities limit the lateral communication and define distinct pressure compartments, a one-dimensional treatment will give a good description of the fluid flow and the pressure development. The model is being extended to include lateral fluid flow.

Quartz cementation in sandstones

In quartz-rich sandstones, quartz dissolution and precipitation is the main cause of chemical compaction. Quartz cementation is modeled as a kinetic process (Walderhaug, 1996), where precipitation is considered to be the rate-limiting factor.

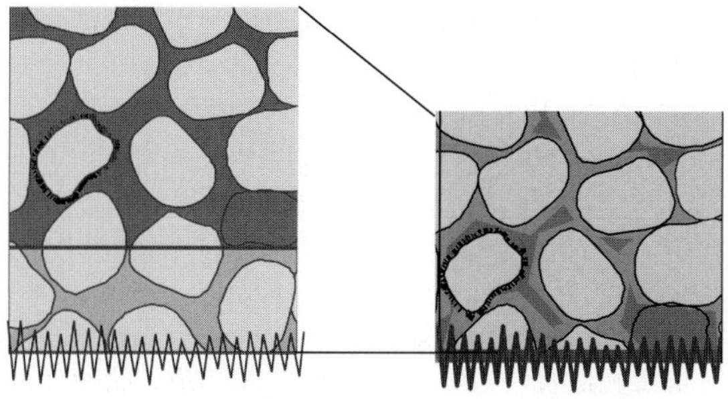

Fig. 2. Schematic illustration of the model for chemical compaction in sandstones. Quartz grains dissolve at stylolites. The dissolved quartz precipitates in the pore space of the undissolved region. Pore volume is lost due to quartz cementation. In addition, the pore space in the dissolved region (light gray) is lost.

Whereas mechanical compaction is most rapid at shallow burial depths, the rate of quartz cementation increases markedly at temperatures above 60–70°C (Bjørkum and Nadeau, 1998). The two processes usually overlap in time, with the relative importance of each process depending on the burial history at any given location (Lander et al., 1999). Sandstone compaction is schematically illustrated in Fig. 2. During intermediate and deep burial, quartz grains are dissolved and pore space is lost at stylolites. The dissolved quartz precipitates in the surrounding sandstone. The chemical compaction is the sum of the pore volume lost due to quartz cementation and the pore volume lost because of dissolution and shortening of the column.

Chemical compaction in shales

We model the reaction progress of the smectite to illite transformation using a kinetic formulation published by Huang et al. (1993). An important input parameter is the illitizable smectite content in the shales at deposition. We assume, in the work presented here, that the potassium concentration in shale pore fluids is constant and is buffered by the presence of K-feldspar. The stoichiometry of the smectite–illite reaction is based on a montmorillonitic smectite as a starting composition (Guven, 1988). From the stoichiometry, a volume balance for the reaction is derived. Smectite illitization has the potential to contribute to overpressure development in two ways: loss of pore volume and reduction in permeability (Bjørkum and Nadeau, 1998). The transformation from smectite to illite implies a reduction both in bulk volume and in pore volume. Pore volume is also lost because of precipitation of quartz released by the reaction. We assume that all silica released by the reaction precipitates locally at the same rate as it is released by the clay mineral reactions. In addition, bound interlayer water is released, increasing the fluid volume in the pores. Permeability may be significantly reduced due to rearrangement of the pore microstructure and increase in specific surface area during the reaction. Based on Nadeau's fundamental particle model (Nadeau et al., 1984) and the limited amount of data from the literature on the physical size of clay crystallites in the illitization reaction series, we use a maximum in specific surface area at a reaction progress of 50%.

Results

To assess the importance of the different pressure generating mechanisms in different basin settings, we chose case studies from the Halten Terrace area offshore mid-Norway and the northern Gulf of Mexico. The pressure model is a forward model of sedimentation, temperature and pressure and can be used to predict fluid pressures away from well control. The main purpose in this paper was, however, to evaluate the relative importance of the different pressure generating mechanisms.

In the case study from Halten Terrace we constrained the pressure modeling by honoring available well data, i.e. temperature, quartz cement abundance, illite fraction in the smectite/illite phase, log and core-derived porosities, and fluid pressures. We then used the calibrated model to evaluate the relative contribution from the different pressure generating mechanisms. In the case from Gulf of Mexico, the model was used in a predictive mode, and no calibration was performed.

Halten Terrace

Well 6506/11-1 is located in a highly overpressured region, close to the major faults that separate overpressured from hydrostatically pressured reservoirs (Fig. 3). The stratigraphic column is displayed in Fig. 4. The sediment column from the Halten Terrace mainly consists of shales, siltstones and sandstones. The sandstones in the Jurassic interval are the Garn, Ile, Tofte and Tilje formations. The area experienced rapid burial in the Jurassic and early Cretaceous, uplift in the Neogene, followed by rapid Pliocene burial.

Point-count data were available for sandstones from the Garn, Ile and Tilje formations. Data from several thin-sections were averaged for each formation for use in the pressure modeling. Sandstone in the Garn and Ile has medium grain size, whereas

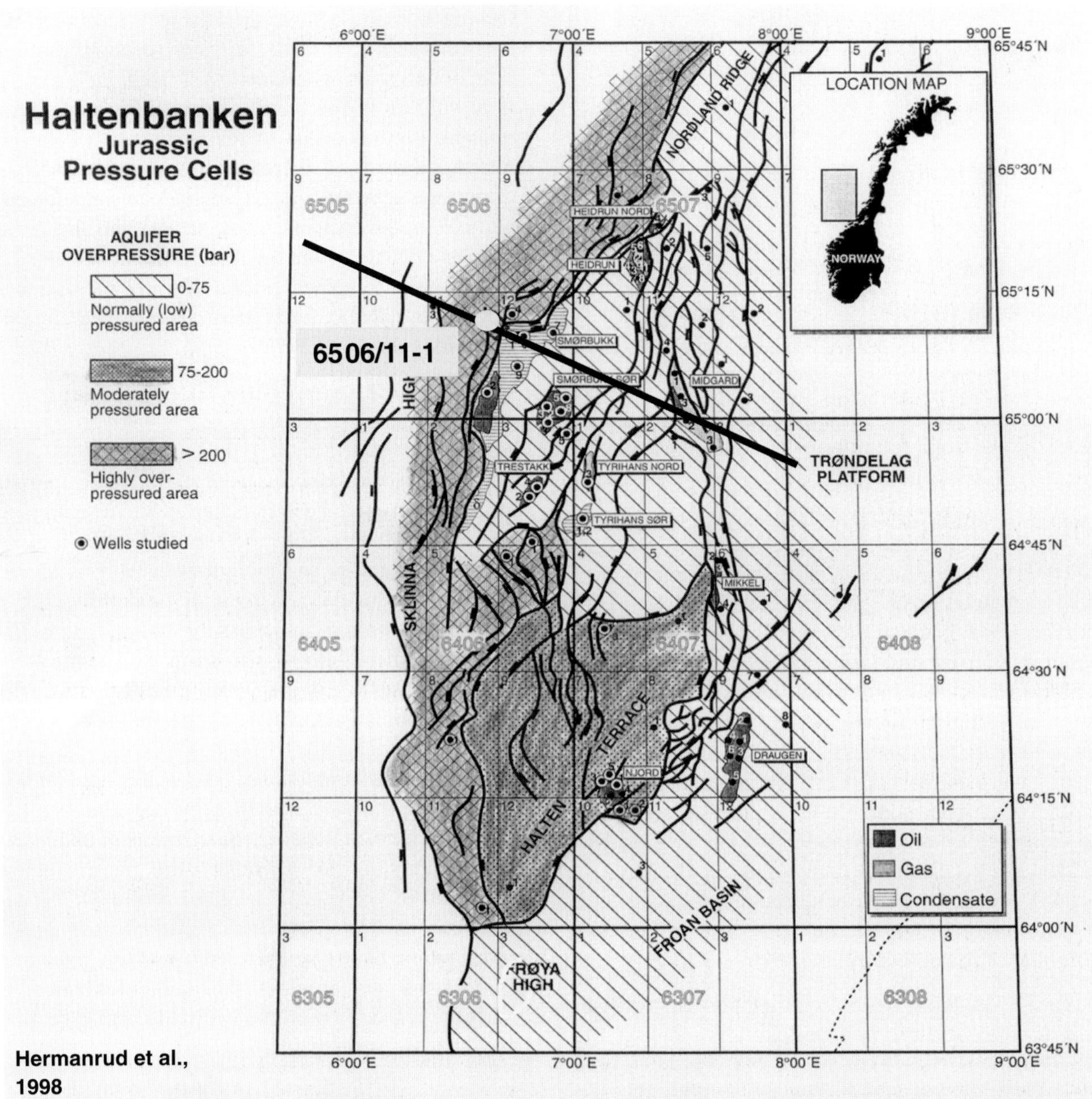

Fig. 3. Map showing the regional setting of the Haltenbanken area. The high-pressure region is located on the western flank. The well 6506/11-1 is identified as a dot on the map. The well is located in the high-pressure region close to the border of the hydrostatically pressured region.

Tilje sandstone is coarse grained. All three formations are moderately sorted. The percentage of grain coating coverage for the Garn and the Ile formations is fairly low (15–16%); however, the Tilje formation has a high percentage of grain coating coverage (75%). Consequently, the Garn and Ile formations have around 10% quartz cement, whereas the Tilje formation has only 3.4% quartz cement. For the Lysing formation, Garn sandstone data were used, while Ile sandstone data were used for the Tofte formation.

We estimated thermal conductivities for the sediments from an in-house database. The temperature calculations included the modeling of basement processes. The temperature model was calibrated to measured well temperatures by adjusting the lithospheric thickness.

The thermal model serves as a basis for modeling quartz cementation in sandstones. The activation energy for quartz cementation was optimized to produce a good fit to the observed quartz cement abundances. The modeled amounts of quartz cement are within 2% of the measured data (Table 1).

A total of 10 shale samples from different depths were submitted for quantitative XRD analysis. Quantitative phase analysis was done by a reference intensity ratio method (Hillier, 2000). The mineralogy of the clay-sized fractions is broadly similar in all sam-

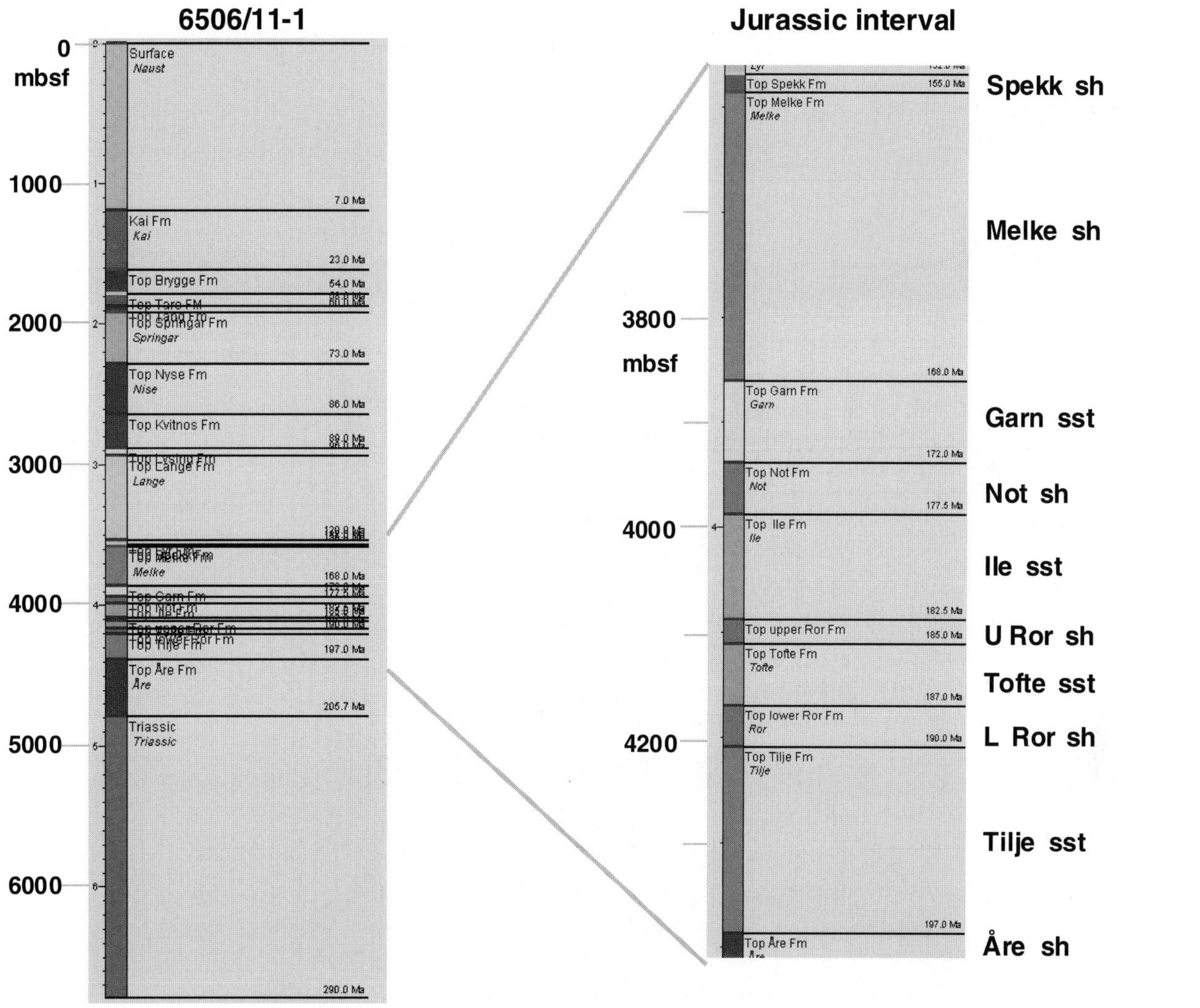

Fig. 4. Stratigraphic column for the well 6506/11-1. The Jurassic interval is magnified. The Jurassic sandstone units are the Garn, Ile, Tofte and Tilje formations.

ples. Generally, they are all dominated by mixed-layer illite–smectite with lesser amounts of kaolin (presumably kaolinite), illite and minor to trace amounts of chlorite. The mixed-layer illite–smectite varies from a composition of about 20% illite–80% smectite with random mixed-layering in the shallowest samples, to 70% illite–30% smectite with ordered mixed-layering in the deepest. Rather than showing a gradual transition, the samples appear to form two groups clustered around the two compositions with no indication of intermediate compositions. The data set indicates that the smectite–illite transformation occurs at around 2750 m. The predicted illite fraction vs. depth is plotted in Fig. 5. The observed fractions are plotted as squares. Observations are only qualitative; they are reported as randomly interstratified illite–smectite and ordered illite–smectite. These two classes are assigned illite fractions of 0.2 and 0.7 respectively. No further attempt was made to tune the kinetic model of the illitization reaction. The predicted 50% illite in the illite–smectite is at 2350 m, i.e. 400 m shallower than the transition indicated by the observations.

TABLE 1

Modeled amount of quartz cement in the sandstone lithologies together with the measured quartz cement abundance from thin-section analysis

	Measured (%)	Modeled (%)
Lysing		2.2
Garn	9.6	10.2
Ile	10.7	11.8
Tofte		13.0
Tilje	3.4	5.2

The values are averages for the units.

Porosities are dependent both on chemical and mechanical compaction. A satisfactory calibration was obtained for the chemical reactions, i.e. quartz cementation in sandstones and smectite illitization in

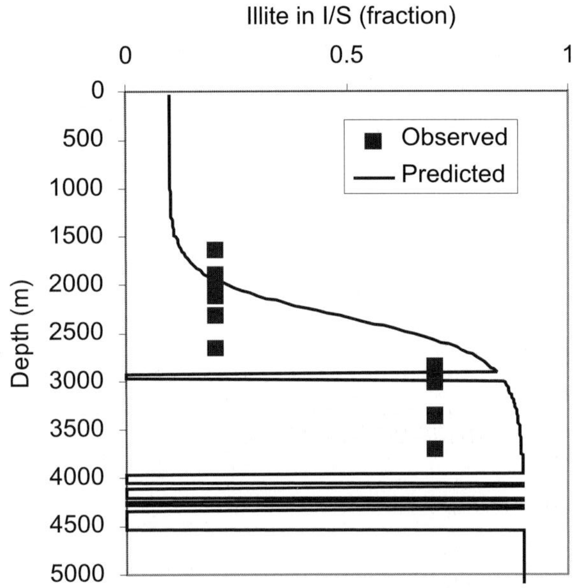

Fig. 5. Predicted fraction of illite in the illite–smectite vs. depth together with observed qualitative fractions. The qualitative fractions (low and high) are indicated as 0.2 and 0.7 respectively.

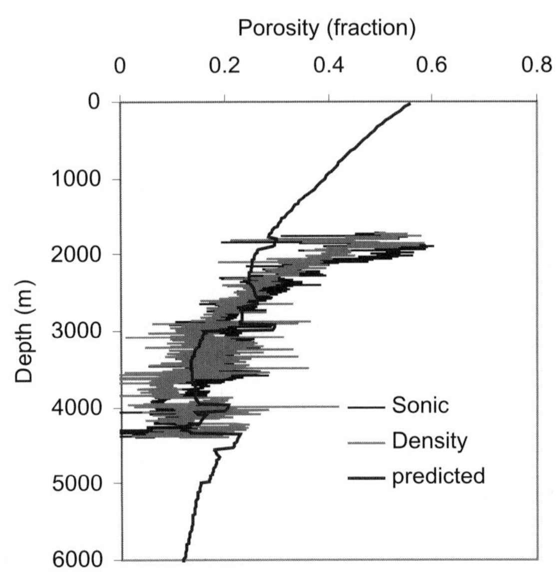

Fig. 6. Predicted present-day porosity together with log-derived porosities from sonic and density logs.

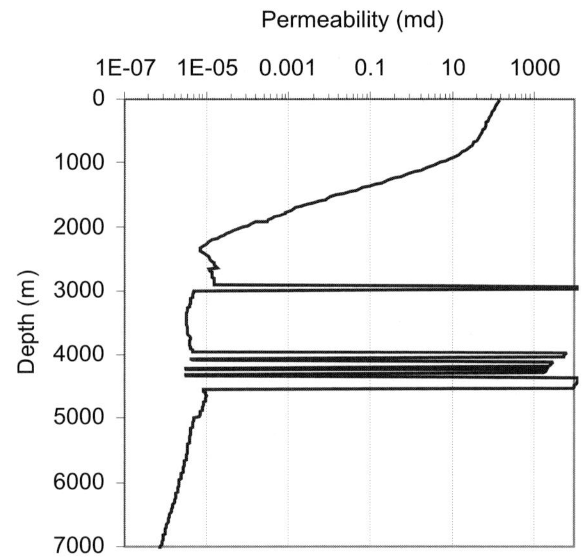

Fig. 7. Present-day predicted permeability (mD) plotted vs. depth (m).

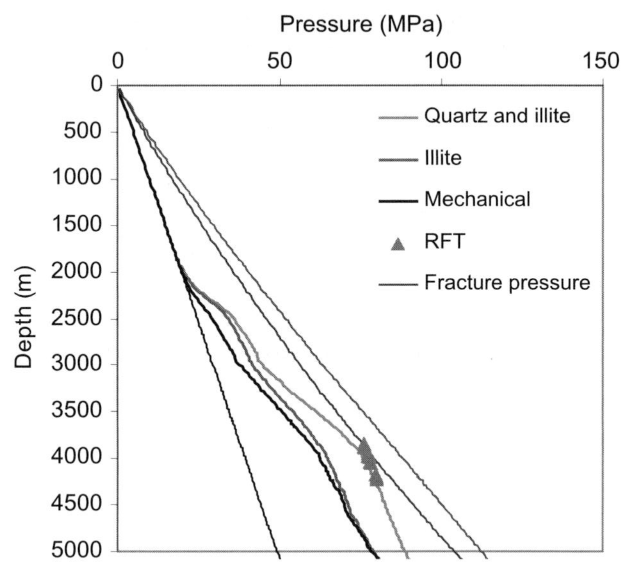

Fig. 8. Modeled fluid pressure (MPa) vs. depth (m) for the 6506/11-1 well. Results from two simulation runs are shown, one with quartz cementation in sandstones, and one where the sandstone diagenesis is turned off. Pressure data from RFT are included around 4000 m. The fracture pressure curve, together with hydrostatic and lithostatic pressures are included for reference.

shales, as described above. After calibrating the diagenetic reactions, mechanical compaction was tuned to match the measured present-day porosities. Porosities for the 6506/11-1 well obtained from sonic and density logs along with the predicted present-day porosities are shown in Fig. 6. By adjusting mechanical compaction parameters, a good match is obtained for depths below 2000 m.

Calculated permeabilities for the present-day are plotted vs. depth in Fig. 7. Some adjustment to the shale permeabilities was made in order to match the present-day measured overpressures. The resulting permeabilities are greater than 10^{-9} mD (10^{-21} m^2) in all parts of the column. This is within the range of measured shale permeabilities of 10^{-7} to 10^{-10} mD (10^{-19} to 10^{-22} m^2) (Katsube and Connell, 1998).

Measured fluid pressure data from RFT (Repeat Formation Test) were available in the 3845–4231 m depth interval. Calculated fluid pressures vs. depth at present-day are shown in Fig. 8 together with fluid pressure data from RFT measurements. Three simulation runs were performed: (1) with mechanical compaction, quartz cementation and illitization, (2) with mechanical compaction and illitization, and (3) with only mechanical compaction. Case 1 (with mechanical compaction, quartz cementation and il-

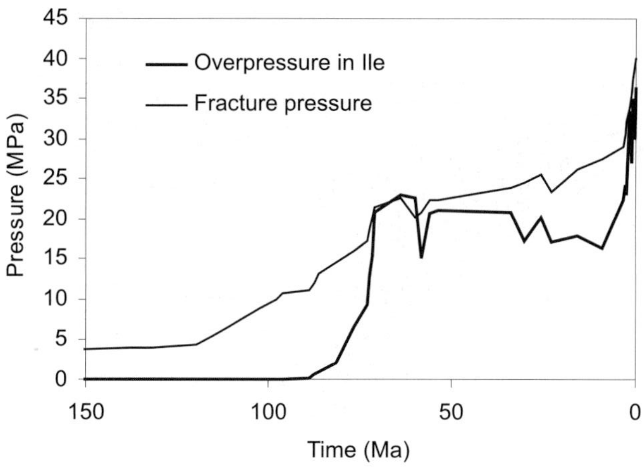

Fig. 9. Overpressure (MPa) vs. time (Ma) for top of the Ile formation.

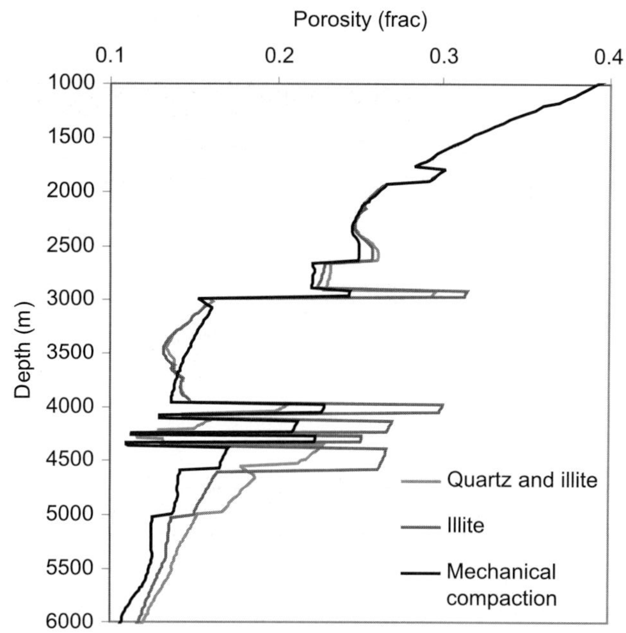

Fig. 10. Porosity–depth curves for the cases with and without quartz cementation in sandstones.

litization) shows the highest overpressures. Distinct pressure build-up is seen in the Jurassic interval, and modeled pressures are close to those measured using RFT. At around 4000 m depth, approximately 25% of the overpressure is attributed to quartz cementation. Illitization has less impact on present-day overpressure. The most important contribution is seen in the depth interval between 2000 and 3000 m where the smectite illitization is actively ongoing (cf. Fig. 5). In this interval, smectite illitization is responsible for about 40% of the present-day overpressure.

The development of fluid overpressure through time in the Ile formation is shown in Fig. 9. Mechanical compaction, quartz cementation and illitization are included in the simulation. Also shown in the figure is the fluid overpressure needed to reach fracture pressure. The model predicts a rapid pressure increase occurring at 90 Ma. Hydraulic fracturing is predicted to occur in the Ile formation between 71 and 60 Ma. Fracturing is also predicted in shallower shale layers in the time interval 71 to 60 Ma. In addition, we predict fracturing in the shallower shale layers at 54 Ma, while at this time there is no fracturing in the Ile formation. The increase in pressure during the last few million years is associated with the rapid burial starting at 3.5 Ma. This rapid burial causes the fluid pressure to approach fracture pressure. The predicted overpressure at present day is 36 MPa, which corresponds well with the measured overpressure of 38 MPa.

The present-day porosities vs. depth for the three cases — with quartz cementation and illitization, with illitization, and only mechanical compaction — are shown in Fig. 10. Highest overpressure is experienced when the effect of quartz cementation is included (Fig. 8). The additional overpressure leads to reduced mechanical compaction and higher porosities in the deeper part of the column below the Jurassic sand units, i.e. between 4600 and 6000 m. A maximum porosity difference of 4% is observed when including both quartz cementation and illitization compared to mechanical compaction. Only minor differences in porosities are seen in the shallower shale units.

Gulf of Mexico

A limited data set was available for the Gulf of Mexico case study. We have chosen to model a well from the Lake Sand field on the Southern Louisiana Gulf Coast, and use petrographic and shale XRD data presented by Freeman (1990) to evaluate the model performance. We used the model parameters obtained in the Halten Terrace case study. No further model calibration was performed for the Lake Sand example.

The basin setting is shown schematically in Fig. 11. Several faults separate the lithologic units into compartments. The Lake Sand well used in this case study is indicated to the right in the cross-section. The stratigraphic column for the Lake Sand well is shown in Fig. 12. The sediment column consists of middle-Miocene and younger interbedded muds and sands. The burial has been almost linear over 23 Ma to the present-day depth of 6 km. No pressure data could be obtained for this well. Nevertheless, regional trends indicate that hard overpressure starts at around 3900 m.

A satisfactory prediction was obtained to measured well temperatures, sandstone and shale porosities and quartz cement abundance (Table 2). The illitization reaction is correctly predicted to be complete at

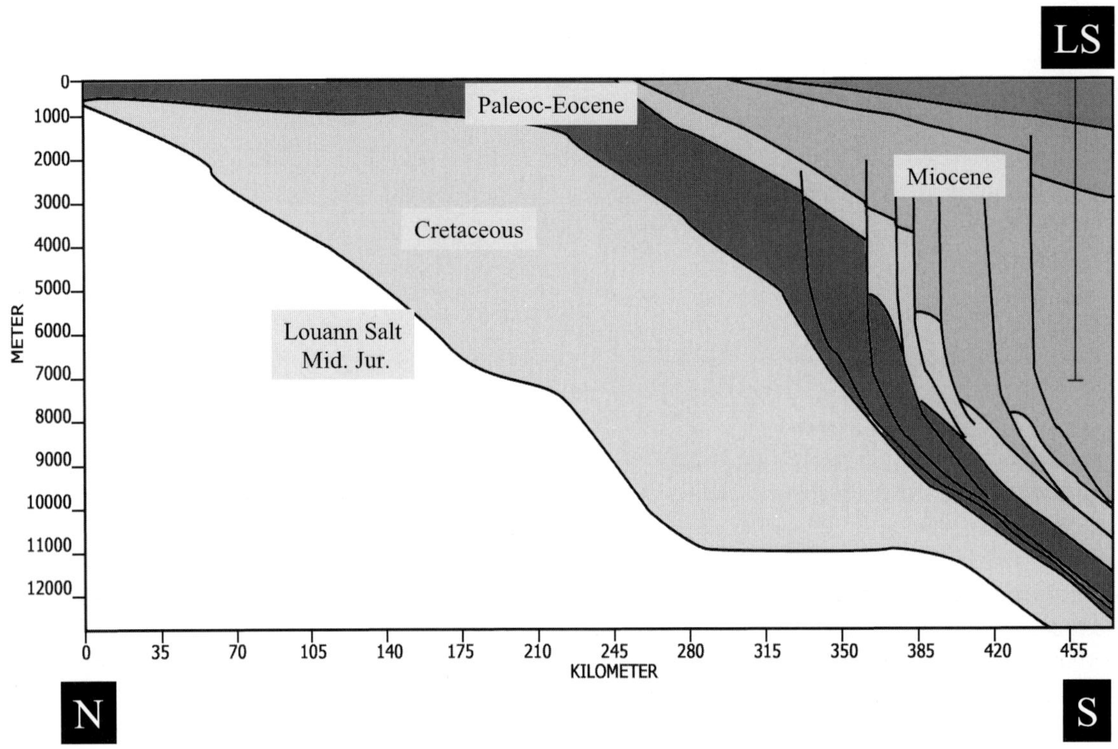

Fig. 11. Cross-section showing the basin setting. Several faults separate the units into compartments. The Lake Sand well (*LS*) is seen to the right in the cross-section.

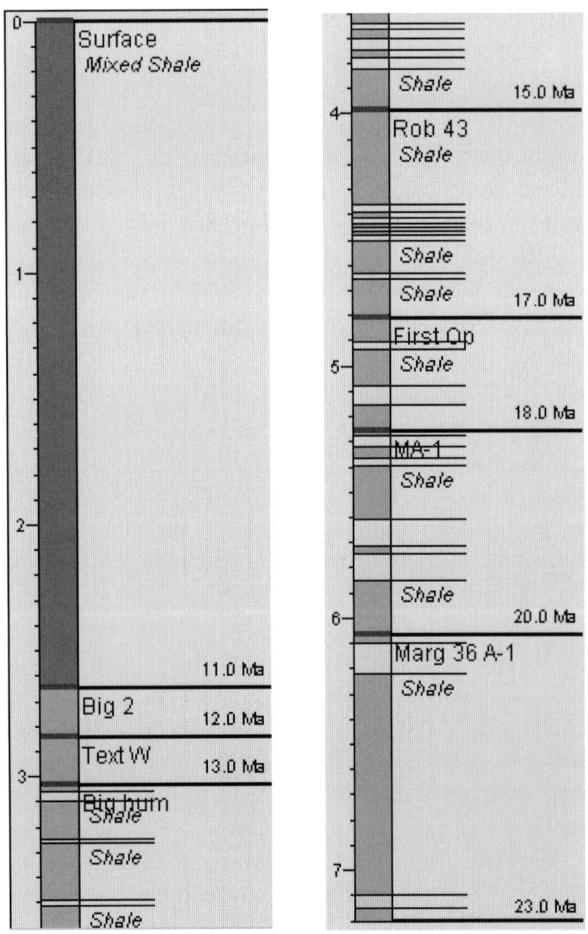

Fig. 12. Stratigraphic column for the Lake Sand well.

the depth of 7160 m. The transition zone of the reaction at present day is predicted to be at around 4000 m.

The present-day predicted porosity is plotted vs. depth in Fig. 13. The shale porosity declines rapidly, and below 3800 m is constant at 5%, which is used as the minimum porosity limit. Calculated permeabilities are plotted in Fig. 14. The decline in permeability coincides with the decline in porosity, and also with the illitization reaction progress. The increase in permeability between 4000 and 6000 m is due to hydraulic fracturing, which is modeled by increasing the permeability by a factor of 100. The minimum permeability calculated in the column is 3.5×10^{-6} mD (3.5×10^{-21} m^2), which is inside the range of measured shale permeabilities (Neuzil, 1986; Katsube and Connell, 1998).

To evaluate the importance of quartz cementation in the build-up of overpressure, three models were run: (1) with mechanical compaction, quartz cementation and illitization, (2) with mechanical compaction and illitization, and (3) with only mechanical compaction. Present-day calculated fluid pressures are plotted vs. depth for all cases in Fig. 15. The fracture pressure, together with hydrostatic and lithostatic pressures are included for reference. Top of modeled overpressure is at around 3400 m. The model suggests that fluid pressures for the case with quartz cementation are at fracture pressure in the depth interval between 3800 m

TABLE 2

Summary of well data and model results for the Lake Sand well

	Well data (at 7160 m)	Model results (at 7160 m)
Average sandstone porosity (%)	3	2.7
Average sandstone quartz cement (%)	16	19
Illitization reaction progress (%)	100	100
Log-derived shale porosities (from literature) (%)	8–10	5
Measured shale porosities (from literature) (%)	5	5
Depth of overpressured zone (m)	3900	3400

Well data for porosities, quartz cement abundance and illitization reaction progress are taken at 7160 m depth.

and 6000 m. The fluid pressures for the other cases are lower, and are everywhere below fracture pressure. The model suggests that quartz cementation and illitization contribute roughly 40% each to the maximum predicted present-day overpressure.

Fig. 16 shows the fluid overpressures plotted vs. time for the present-day depth of 6100 m. Pressure histories for all three cases are shown. Overpressure starts to build up around 11 Ma. The onset of overpressuring is slightly later when only mechanical compaction is included. Fluid pressure reaches fracture pressure at 6.6 Ma in the case with quartz cementation and illitization. Several fracturing episodes have occurred throughout the burial history. No fracturing occurs for the other runs. At 6100 m depth the difference in pressure between the highest (quartz cementation and illitization) and lowest (mechanical) prediction is 56 MPa.

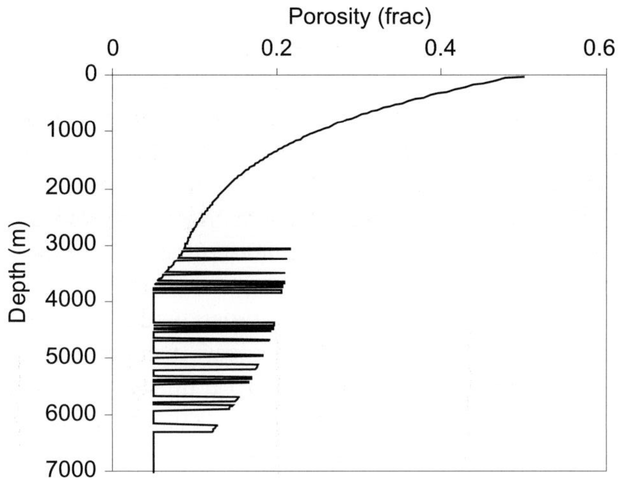

Fig. 13. Predicted present-day porosity (frac.) vs. depth (m).

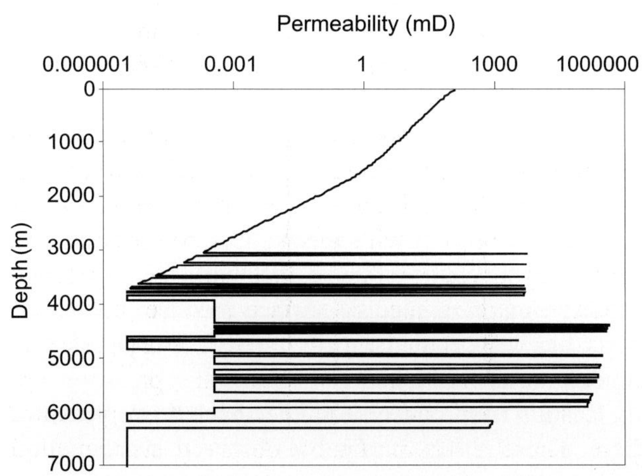

Fig. 14. Predicted present-day permeability (mD) plotted vs. depth (m).

Discussion

The model of fluid flow and overpressuring presented here includes pressure generation due to both mechanical compaction and diagenesis. We have chosen to incorporate the most important diagenetic porosity reducing mechanisms, i.e. quartz cementation in sandstones and smectite illitization in shales. These diagenetic reactions are assumed to be ki-

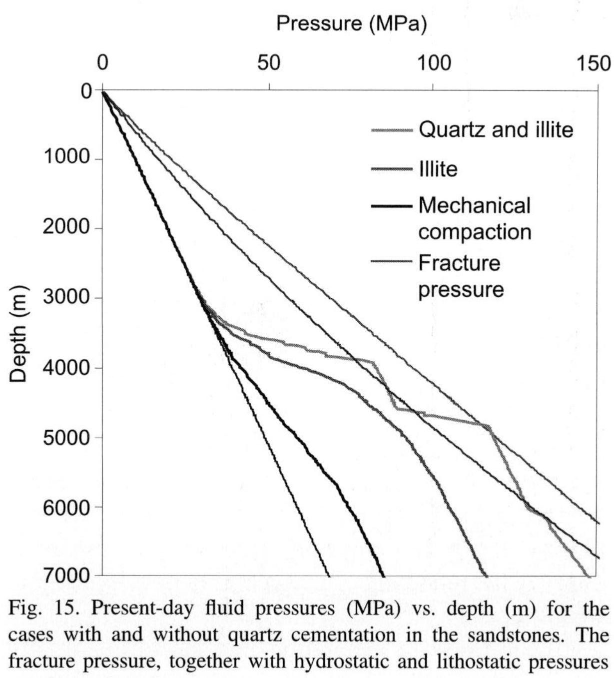

Fig. 15. Present-day fluid pressures (MPa) vs. depth (m) for the cases with and without quartz cementation in the sandstones. The fracture pressure, together with hydrostatic and lithostatic pressures are shown for reference.

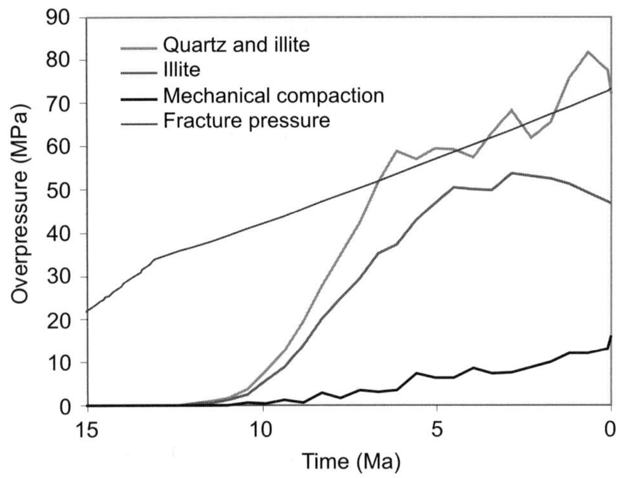

Fig. 16. Fluid overpressures (MPa) vs. time (Ma) for the present-day 6100 m, showing the cases with and without quartz cementation. The fracture pressure (overpressure needed to cause hydraulic fracturing) is included for reference.

netically controlled; they will not be retarded by development of overpressures.

Results from two case studies were presented. In the Halten Terrace case we chose to calibrate the model to honor available well data. This gives a more accurate description of the diagenetic processes and of the pore volume loss rates. However, the calibration procedure does not constrain our evaluation of the relative contribution from the different pressure generating effects. In the Gulf of Mexico case no calibration was performed.

Case study results indicate that quartz cementation and smectite illitization could contribute significantly to the build-up of overpressure. Diagenetic processes are more important in generating overpressure that mechanical compaction in deeper prospects where mechanical compaction is nearly complete under hydrostatic conditions. Permeabilities in overlaying shales decrease with depth, efficiently retaining any overpressure generated in the deeper parts of the basin.

The model suggests that smectite illitization could contribute to overpressuring through pore volume loss and fluid release. More importantly, it may control the permeability of shale as suggested by other workers (Bjørkum and Nadeau, 1998). The least permeable units control the release of overpressure. The evolution of shale permeability is therefore very important when predicting overpressure development. Our modeling results indicate that the depth at which overpressuring starts is determined by shale permeabilities in the overlaying units, which we relate to the illitization reaction progress. More work is needed to improve the modeling of permeability development in shales.

Of the pressure generating mechanisms currently not included in the fluid flow model, organic matter cracking may be the most important, as demonstrated by Luo and Vasseur (1996). The volume expansion by kerogen cracking, and in particular gas generation will lead to increase in fluid pressures that can be significant, at least on a local scale.

The model frame of reference is presently a 1D vertical column. Lateral fluid flow will in many settings contribute to pressure build-up and dissipation. Lateral fluid flow occurs mainly within sand units. For situations where faults limit the lateral communication and define distinct pressure compartments, a one-dimensional treatment will give a good description of the fluid flow and the pressure development. The sand unit is then representative for a pressure compartment. A natural extension of the model is to include lateral fluid flow.

Conclusions

Overpressures form in compacting sedimentary basins when the rate of pore volume reduction is faster than the rate of pore fluid release, or when the rate of pore fluid expansion is faster than the rate of pore fluid release. We have developed a model of fluid flow and pressure development that includes pore volume loss due to both mechanical compaction and chemical diagenesis. The chemical processes are modeled as being kinetically controlled, and are not retarded by reduction in effective stress due to pressure build-up. In other words, diagenetically controlled pore volume loss can therefore continue, even during periods of uplift.

Dissolution of quartz grains at stylolites implies a reduction both in bulk volume and pore volume of the sandstone. Precipitation of the dissolved quartz in the surrounding regions of the sandstone causes additional pore volume loss.

Our modeling assumes that smectite illitization leads to a reduction in bulk volume and pore volume of the shale. This is based on a reasonable reaction stoichiometry and a consideration of the result of mass transfer in shales under confining stress. Water is released in the reaction, increasing the total volume of pore fluids and causing additional overpressuring. Even more important for build-up and retention of overpressures is the relation between smectite illitization and shale permeability. Our model assumes that textural changes associated with the illitization reaction increase specific surface area and tortuosity. This has important controls on the development of shale permeabilities.

Results from two case studies have been presented here. These results imply that observed overpressures were generated from a combination of processes and mechanisms. The relative importance of each pressure generating mechanism is controlled by the basin setting. In the Halten Terrace case, 25% of the present-day overpressures may be related to chemical compaction. Surprisingly, in the Miocene delta section of the Gulf of Mexico, the chemical compaction processes could be even more important. In our calculations, up to 80% of the present-day overpressure is related to chemical processes. In both cases the diagenetic processes may cause pressures to increase up to fracture pressure inducing hydraulic fracturing. The modeling indicates that the timing of overpressuring due to chemical compaction is therefore an important control on the timing of hydraulic fracturing.

Shale permeabilities within the range of measured values are sufficiently low to explain the development of high overpressures. Since pressure generation due to diagenetic reactions may be an ongoing process even in the deeper parts of the basin, no unrealistically low permeabilities are needed to retain overpressures for long periods of time.

Nomenclature

Units: L = length, t = time, m = mass, T = temperature, E = energy

g = constant of gravity = 9.81 m/s^2
k = permeability [L^2]
P = fluid pressure [m/Lt^2]
Q_w = fluid source [L^3/t]
t = time [t]
z = spatial coordinate [L]
β = coefficient of compressibility [Lt^2/m]
ϕ = porosity [dimensionless]
Φ = overpressure (hydraulic potential) = $P - \rho_f g z$ [m/Lt^2]
μ = fluid viscosity [m/Lt]
ρ = density [m/L^3]
σ_e = effective stress = $\rho_{sm} g z - P$

Subscripts
f = fluid
sm = saturated medium

Acknowledgements

Quentin Fisher reviewed the original manuscript and suggested a number of ways to improve the paper. We thank BP–Amoco, Norsk Hydro (Saga), and Statoil for supporting this applied research project.

References

Athy, L.F., 1930. Compaction and oil migration. Am. Assoc. Pet. Geol. Bull., 14: 25–35.

Bachu, S. and Underschultz, J.R., 1993. Hydrogeology of formation waters, northeastern Alberta basin. Am. Assoc. Pet. Geol. Bull., 77: 1745–1768.

Bethke, C.M., 1985. A numerical model of compaction-driven groundwater flow and heat transfer and its application to the paleohydrology of intracratonic sedimentary basins. J. Geophys. Res., 90 (B8): 6817–6828.

Bethke, C.M., 1986. Inverse hydrologic analysis of the distribution and origin of Gulf Coast-type geopressured zones. J. Geophys. Res., 91 (B6): 6535–6545.

Bjørkum, P.A., 1996. How important is pressure in causing dissolution of quartz in sandstones? J. Sediment. Res., A66: 147–154.

Bjørkum, P.A. and Nadeau, P.H., 1998. Temperature controlled porosity/permeability reduction, fluid migration, and petroleum exploration in sedimentary basins. Aust. Pet. Produc. Explor. Assoc. J., 38: 453–464.

Bruce, C.H., 1984. Smectite dehydration — its relation to structural development and hydrocarbon accumulation in northern Gulf of Mexico Basin. Am. Assoc. Pet. Geol. Bull., 68: 673–683.

Burrus, J., 1998. Overpressure models for clastic rocks, their relation to hydrocarbon expulsion: a critical reevaluation. In: B.E. Law, G.F. Ulmishek and V.I. Slavin (Editors), Abnormal Pressures in Hydrocarbon Environments. Am. Assoc. Pet. Geol. Mem., 70: 35–63.

Burst, J.F., 1969. Diagenesis of Gulf Coast clayey sediments and its possible relation to petroleum migration. Am. Assoc. Pet. Geol. Bull., 53 (1): 73–93.

Colten-Bradley, V.A., 1987. Role of pressure in smectite dehydration; effects on geopressure and smectite-to-illite transformation. Am. Assoc. Pet. Geol. Bull., 71 (11): 1414–1427.

Deming, D., 1994. Factors necessary to define a pressure seal. Am. Assoc. Pet. Geol. Bull., 78 (6): 1005–1009.

Freeman, C.W., 1990. Diagenesis of Miocene Sandstones and Shales, Southern Louisiana Gulf Coast. University of Missouri, Columbia, 94 pp.

Gaarenstroom, L., Tromp, R.A.J., de Jong, M.C. and Brandenburg, A.M., 1993. Overpressures in the Central North Sea: implications for trap integrity and drilling safety. Petroleum Geology of Northwest Europe: Proceedings of the 4th Conference, Barbican Centre, London, 29 March–1 April 1992, The Geological Society, London.

Guven, N., 1988. Smectites. In: S.W. Bailey (Editor), Hydrous Phyllosilicates (Exclusive of Micas), Vol. 19. Mineralogical Society of America, Washington, DC, pp. 497–552.

Hermanrud, C., Wensaas, L., Teige, G.M.G., Vik, E., Nordgård Bolås, H.M. and Hansen, S., 1998. Shale porosities from the well logs on Haltenbanken (offshore mid-Norway) show no influence of overpressuring. In: B.E. Law, G.F. Ulmishek and V.I. Slavin (Editors), Abnormal Pressures in Hydrocarbon Environments. Am. Assoc. Pet. Geol. Mem., 70: 65–85.

Hillier, S., 2000. Accurate quantitative analysis of clay and other minerals in sandstones by XRD: comparison of a Reitveld and a reference intensity ratio (RIR) method and the importance of sample preparation. Clay Miner., 35: 291–302.

Huang, W.L., Longo, J.M. and Pevear, D.R., 1993. An experimentally derived kinetic model for smectite-to-illite conversion and its use as a geothermometer. Clays Clay Miner., 41 (2): 162–177.

Katsube, T.J. and Connell, S., 1998. Shale permeability characteristics. Geol. Surv. Can., Curr. Res., 1998-E: 183–192.

Lake, L.W., 1989. Enhanced Oil Recovery. Prentice Hall, Englewood Cliffs, NJ.

Lander, R.H. and Walderhaug, O., 1999. Porosity prediction through simulation of sandstone compaction and quartz cementation. Am. Assoc. Pet. Geol. Bull., 83 (3): 433–449.

Lander, R.H., Helset, H.M., Matthews, J.C. and Bonnell, L.M., 1999. Can sandstone diagenesis induce fluid overpressure? AAPG Hedberg Research Conference on Multi-Dimensional Basin Modeling, Colorado Springs, CO.

Luo, X. and Vasseur, G., 1992. Contributions of compaction and aquathermal pressuring to geopressure and the influence of environmental conditions. Am. Assoc. Pet. Geol. Bull., 76 (10): 1550–1559.

Luo, X. and Vasseur, G., 1996. Geopressuring mechanism of organic matter cracking: numerical modeling. Am. Assoc. Pet. Geol. Bull., 80 (6): 856–874.

Mann, D.M. and Mackenzie, A.S., 1990. Prediction of pore fluid pressures in sedimentary basins. Mar. Pet. Geol., 7: 55–65.

McBride, E.F., 1989. Quartz cement in sandstones. Earth Sci. Rev., 26: 69–112.

Nadeau, P.H., Tait, J.M., McHardy, W.J. and Wilson, M.J., 1984. Interstratified XRD characteristics of physical mixtures of elementary clay particles. Clay Miner., 19: 67–76.

Neuzil, C.E., 1986. Groundwater flow in low-permeability environments. Water Resour. Res., 22: 1163–1195.

Osborne, M.J. and Swarbrick, R.E., 1997. Mechanisms for generating overpressure in sedimentary basins: a reevaluation. Am. Assoc. Pet. Geol. Bull., 81 (6): 1023–1041.

Perry, E.A.J. and Hower, J., 1972. Late-stage dehydration in deeply buried pelitic sediments. Am. Assoc. Pet. Geol. Bull., 56: 2013–2021.

Powers, M.C., 1967. Fluid release mechanisms in compacting marine mudrocks and their importance in oil exploration. Am. Assoc. Pet. Geol. Bull., 51: 1240–1254.

Rutter, E.H., 1983. Pressure solution in nature, theory and experiment. J. Geol. Soc. London, 140: 725–740.

Walderhaug, O., 1994. Precipitation rates for quartz cement in sandstones determined by fluid-inclusion microthermometry and temperature history modelling. J. Sediment. Res., A64: 324–333.

Walderhaug, O., 1996. Kinetic modelling of quartz cementation and porosity loss in deeply buried sandstone reservoirs. Am. Assoc. Pet. Geol. Bull., 80: 731–745.

H.M. HELSET *Geologica A/S, P.O. Box 8034, N-4003 Stavanger, Norway*
E-mail: hmh@geologica.no

R.H. LANDER *Geologica A/S, P.O. Box 8034, N-4003 Stavanger, Norway*
Present address: Geocosm, Austin, TX, USA

J.C. MATTHEWS *Geologica A/S, P.O. Box 8034, N-4003 Stavanger, Norway*

P. REEMST *Geologica A/S, P.O. Box 8034, N-4003 Stavanger, Norway*
Present address: NAM, P.O. Box 28000, 9400 HH Assen, The Netherlands

L.M. BONNELL *Geologica A/S, P.O. Box 8034, N-4003 Stavanger, Norway*
Present address: Geocosm, Austin, TX, USA

I. FRETTE *Geologica A/S, P.O. Box 8034, N-4003 Stavanger, Norway*

Prediction of sealing capacity by the equivalent grain size method

Kazuo Nakayama and Daichi Sato

Cap rocks do not hold infinite amounts of oil/gas so there must be a physical limitation of the column height of oil/gas that a trap can hold. The sealing capacity of a trap is one of the most important conditions that needs to be evaluated quantitatively. In this paper, we review the basic phenomenon of hydrocarbon trapping by capillary pressure. By evaluating the observed hydrocarbon column height from a field, we can quantify the sealing capacity in terms of pore throat radius. The pore throat radius, however, is not a convenient parameter for geologists, so we convert it into grain size (assuming that the sediments consist of equal-sized spheres). In this conversion, we use a theoretical relation of COEF (ratio of pore throat to grain size) as a function of porosity. We improve this relationship by conducting a physical experiment where we measure the capillary pressure for an artificial seal to derive an empirical relationship between COEF and porosity. Thus we can evaluate top seal capacity in terms of imaginary grain size (this is named 'equivalent grain size method'). Then we propose a 'practical' flow chart for evaluating sealing capacity in a quantitative way. One of the biggest problems for seal evaluation is that it was usually based on the structural and stratigraphical interpretations. For the effective seal evaluation, we propose an integrated approach such as doing all the interpretations at the same time. This 'multi-disciplinary' approach would solve the debatable phenomena on the sealing/non-sealing problems to give us a deeper understanding of the petroleum systems.

Introduction

Estimation of sealing capacity for cap rock usually requires laboratory measurement of pore-throat size for shale samples, such as by mercury-injection capillary pressure determinations. After the capillary pressure curve is obtained, sealing capacity of the trap can be discussed. The conditions for realization of this method are two-fold: (1) the seal sample must be obtained prior to estimation, and (2) the measurement of pore distribution on a sample is regarded to be representative of the whole trap, even if the sample area is one to million of the real sealing area (Downey, 1994). To overcome these conditions, we develop a method that would predict sealing capacity of a cap rock over a certain trap in a macro-view from the original oil/gas column height, assuming that an enormous amount of hydrocarbons has been migrated. In this way, we can evaluate a top seal.

To expand this method to a practical seal evaluation, we need to consider characteristics of the migration pathway to conclude a generic classification of traps. Then together with a possible sealing fault as a lateral seal, we propose simplified criteria for practical seal evaluation as described by the flow chart. The problems for seal evaluation are also discussed.

Development of the equivalent grain size method for top seal evaluation

For the geological time scale, where the hydrocarbon migration process is in mind, the 'static' fluid flow model, which is controlled only by capillary pressure, is necessary for understanding the phenomenon. On the contrary, for the production time scale, where pressure drops occur by production of oil as an artificial effect by human beings, the dynamic model treatment is required for simulating the phenomenon (Yielding et al., 2000). The permeability is a key control for the fluid movement in this case. In the following section, we limit our discussions only to the geologic time scale, so that we can make the things simple and clear for 'practical' purposes.

Another thing to be noted here is that the amount of hydrocarbons generated in the kitchen area of the basin is usually quite huge compared with that in traps of a geologic nature. This is true for most basins according to our experience from basin modeling, which we have applied to many basins over the world. Therefore, there will be an equilibrium situation attained at the trap, into which some hydrocarbons migrated and some out of it. With these conditions in mind, we think and try to quantify the things that are happening in the traps.

Hydrocarbon Seal Quantification edited by A.G. Koestler and R. Hunsdale. NPF Special Publication 11, pp. 51–60, Published by Elsevier Science B.V., Amsterdam.

Hydrostatic trap equilibrium equation

Recent progress in geochemistry suggests that hydrocarbons have migrated in a separate oil/gas phase in most of the economical accumulations (Barker, 1978; Hunt, 1979). Physical constraint for hydrocarbons to be trapped at the top of a reservoir is a restraining force, exerted downward by capillary pressure, so that the hydrocarbons can be trapped until the upward force by buoyancy (associated with a growing column) equals the capillary force. At the maximum, the hydrostatic equilibrium condition is attained (Fig. 1).

$$\frac{2\gamma(\cos\theta)}{R} = gH_c\,(\rho_w - \rho_h) \qquad (1)$$

where γ = interfacial tension, θ = wettability (interfaced angle), R = pore-throat radius, g = acceleration of gravity, ρ_w, ρ_h = densities of water and hydrocarbons, H_c = maximum hydrocarbon column height.

As postulated in the previous section, most of the traps with high relief can be filled up to a maximum with migrated hydrocarbons. If we collect such cases of observed original hydrocarbon columns, we can evaluate or back-calculate a pore throat radius (R) assuming a proper consideration of interfacial tension (γ), wettability (θ) and densities of fluids (ρ_o and ρ_h). This tells us that hydrocarbons cannot leak until a certain amount is trapped at the top of reservoirs assuming a water-wet condition in cap rocks (Berg, 1975; Schowalter, 1979). On the contrary, water as a wetting phase can go through its permeable media. Sealing capacity for top seal can be quantified by the hydrostatic equilibrium equation (Fig. 1), therefore it should be related to the pore-throat size in the cap rock.

Equivalent grain size estimation

The radius of the pore is now a candidate as a term to evaluate sealing capacity of the trap; however, it is not a common parameter among geologists. Instead, we introduce grain size, which seems to be more acceptable for geology-related researchers or explorationists. Now let us assume that a cap rock consists of equal-sized grains. Then, the pore-throat can be regarded as a function of grain size and porosity if the geometrical ratio of pore-throat and grain size is considered. Since porosity is to follow a regional compaction curve, sealing capacity should also be related to the grain size. We call this grain size 'equivalent grain size', which is not a real one in nature, but effective in the evaluation of seal capacity. The pore-throat size can be changeable according to the type of packing. Although porosity does not change as grain size changes, it does change according to the packing (Fig. 2). In fact, there is a relationship between the ratio of pore-throat size to grain size (hereafter termed COEF) and porosity (ϕ) according to theoretical types of packing (Fig. 3).

If we follow this theoretical relationship, we can get an approximation as follows (Nakayama and Van Siclen, 1981):

$$\mathrm{COEF} = 1.92(\phi)^2 - 0.0882(\phi) \qquad (2)$$

and

$$R = (1/2)(\mathrm{COEF})d_e \qquad (3)$$

where COEF = ratio of pore throat to grain size, ϕ = porosity, d_e = grain diameter.

Concept of the experiment

To evaluate the theoretical relationship, we conduct physical experiments to measure the maximum oil

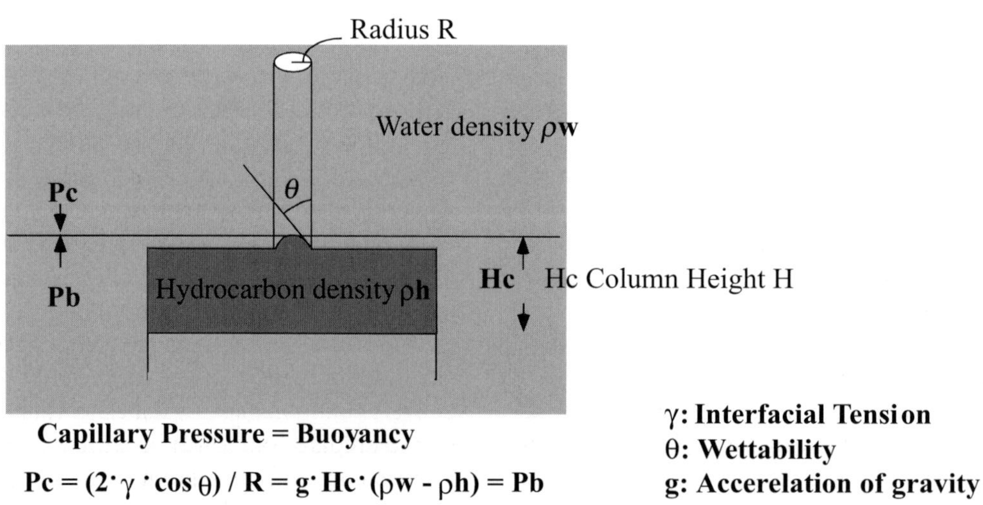

Fig. 1. Assuming that the migration of hydrocarbons is taking place in a separate hydrocarbon phase, the necessary condition for hydrocarbons to be trapped is that the upward buoyancy must be smaller than the downward capillary pressure. At the maximum, the buoyancy should be equal to the capillary pressure, which is called the 'hydrostatic equilibrium trap equation'.

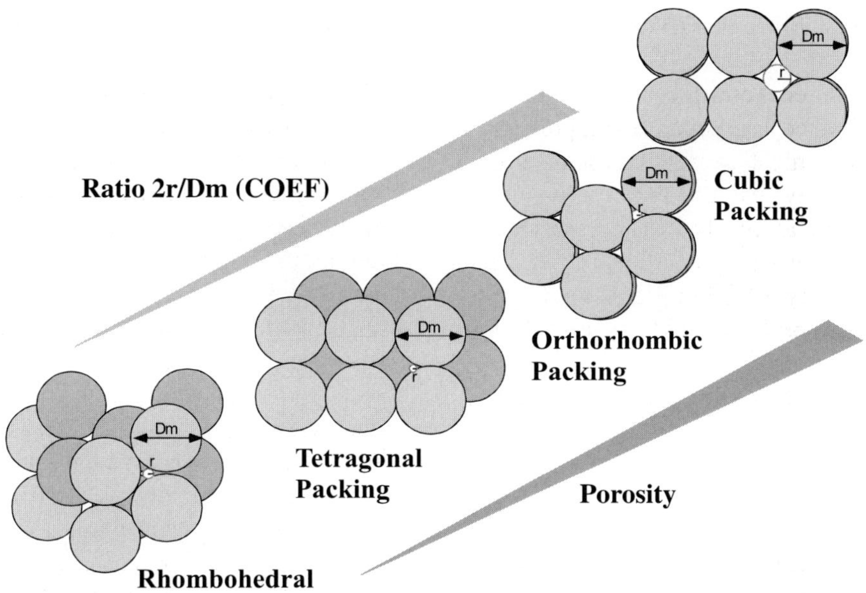

Fig. 2. Theoretical relation of pore throat to porosity according to the packing types. The ratio of pore throat diameter (2× radius) to grain size (called 'COEF') is a function of the porosity according to the geometrical calculation if the grains consist of equal-sized spheres.

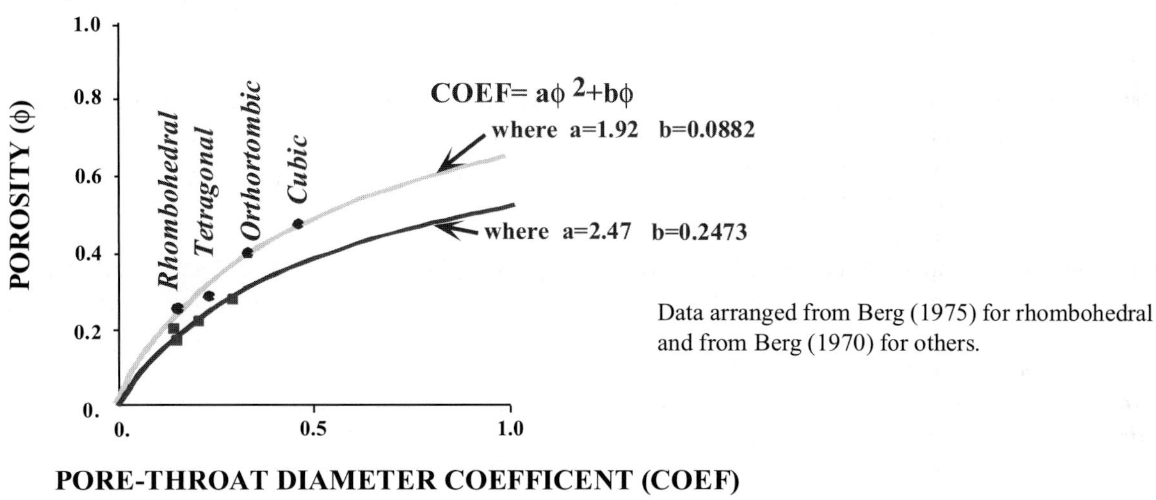

Fig. 3. Approximations of the relationship between porosity and COEF obtained for the theoretical case using the different packing types ($a = 1.92$, $b = 0.0882$) and for the case of the physical experiment ($a = 2.47$, $b = 0.247$).

column in a sample of artificial silica beads with a known grain size distribution (Fig. 4). The apparatus consists of pipelines and a glass cylinder with the artificial seal disk in the center, which is made of artificial glass beads. Two inlet tubes (one for oil, the other for water) are attached to the bottom of the cylinder, by which we can control the amount to be filled into the system (Fig. 5).

At first, the whole system is occupied by water, then the oil line is opened to create some accumulation under the seal disk. For the oil, we use dodecane as a measurable oil sample. Since the theoretical column height is still too high for this apparatus, we push the water line with some column height of water (manometer) so that it can control the pressure at the bottom of this artificial trap (Fig. 5). The water column height of the upper part of the seal disk, that is a function of the pressure of the manometer, is recorded as the leakage of oil occurs at the top of the seal. The column height measurements produce a graph as shown in Fig. 6. The first part of increasing pressure corresponds to the stage in which oil is getting into the pore space in the artificial seal, the second part (plateau) corresponds to the time of sealing, and the last overpressure corresponds to leakage. Thus we can measure the pressure of leakage that must be converted into the equivalent oil column height.

We prepare four kinds of artificial seal disks, each representing different grain sizes such as 100, 200, 300, and 400 μm. The grains are not strictly equal-sized, but have some distribution around the representative size. We use a percentile score of 90 for

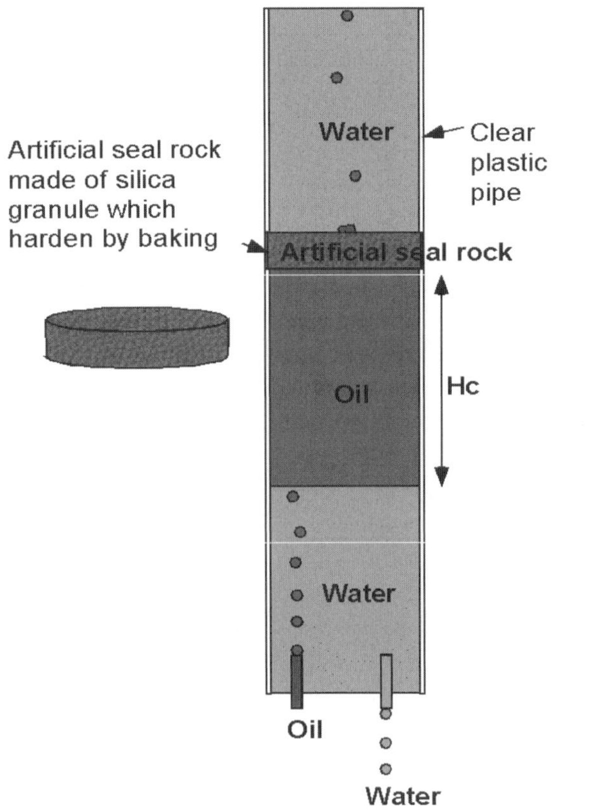

Fig. 4. Schematic diagram showing the inside of the main cylinder of the physical experimental apparatus for the artificially created sealing condition. The seal disk is made of equal-sized glass beads. Oil is gradually pumped into the cylinder where the water is present (as simulating a water-wetting condition). It is designed so that the hydrocarbon column height (*Hc*) can be measured when the oil starts to leak.

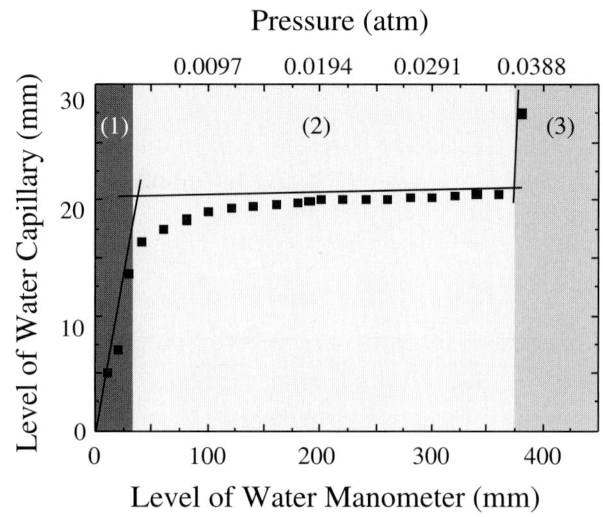

Fig. 6. Graph showing a typical result of the experiment. As h_{WM} increases gradually, h_C increases correspondingly: area (1) = oil migrating into the pore space of the seal; area (2) = oil accumulating; area (3) = oil leaking.

the grain size as representative (Brooks and Purcell, 1952; Berg, 1975). From Eq. 1, the observed capillary pressure values, measured by the physical experiments, can be converted to the pore throat (R), hence we calculate the COEF values from the experiments (Table 1). Thus by controlling the grain size distribution, we measure the column height and successfully derive the corrected relationship between pore-throat radius and equivalent grain diameter (Fig. 3). We derive the new function as follows:

$$\mathrm{COEF} = 2.47(\phi)^2 - 0.247(\phi) \tag{4}$$

Once this relationship is determined for the larger grain sizes that can be investigated by this experiment apparatus, we extrapolate it into the domain of small grain sizes of natural shale.

The practical advantage of this method is that sealing capacity is related to the grain size (not the pore size), which we can imagine as a rock facies from

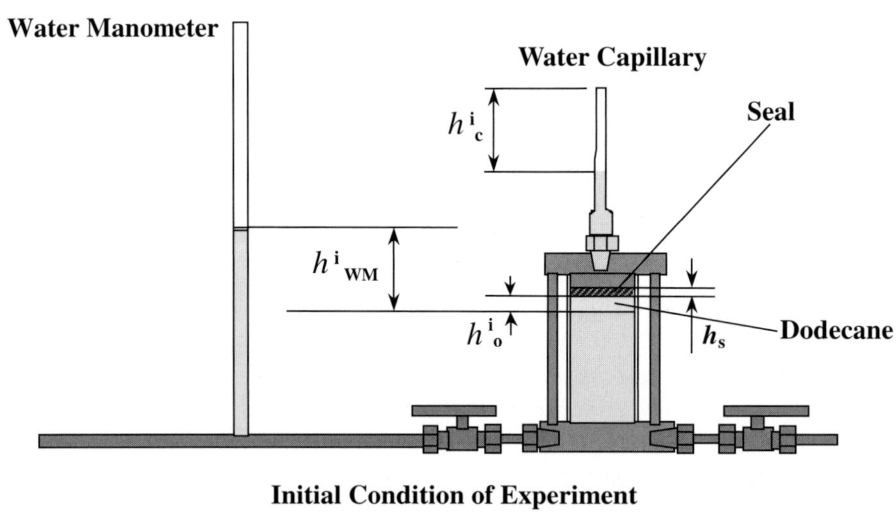

Fig. 5. Schematic diagram showing the concept of the experiment for artificial seal measurement. For the actual measurement, the hydrodynamic force pushes the separate oil column as far as necessary. Setting up the seal/oil (dodecane)/water sequence (from the upper position) in the cylinder, the water column (h_{WM}) increases gradually in the water nanometer. To measure the leaking, the water column in the water capillary (h_C) above the seal is measured.

TABLE 1

Result of the pore throat estimation with four kinds of artificial seal material consisting of different-sized glass beads

Sample	Porosity (%)	Average grain size (μm)	90% grain size (μm)	Pore throat (μm)	Theoretical capillary pressure (atm)	Experimental capillary pressure (atm)	Calculated pore throat estimated from sealing pressure (μm)
A (100)	0.28	79.33 ± 10.78	91.4	8.008	0.1025	0.0674	12.204
B (200)	0.224	185.90 ± 16.28	203.6	11.818	0.0695	0.0431	18.985
C (300)	0.172	280.88 ± 28.71	322	11.587	0.0708	0.0379	21.601
D (400)	0.203	357.12 ± 22.94	392.8	19.056	0.0431	0.0327	24.991

The pore throat is experimentally estimated from the capillary failure by the pressure created by the water nanometer. Since there are some ranges in the grain size distribution, the percentile of 90 was used as an effective grain size.

a sedimentological view. Developing this method, therefore, quantifies the estimation of sealing capacity even before drilling (prediction), if the equivalent grain size could be characterized by the depositional environments. Although the equivalent grain size is different from the real grain size of cap rock, the estimated oil/gas column height by this method is a more realistic prediction for the whole trap scale.

Calculation of top seal evaluation

Assuming the relation between porosity and COEF that is derived from our experiment, we can easily calculate the estimated oil/gas column height. A spread sheet has been developed for this purpose (Fig. 7). Physical conditions such as temperature, density, interfacial tension, and wetting angle are automatically determined according to a rather simple function of the given depth that is programmed. Equivalent grain size is the only unknown for the column height calculation. Therefore, we can use this sheet by trial and error to find out the suitable grain size for trapping the observed oil/gas column height.

If we have set up the suitable grain size for a field, we can estimate the maximum column height for the other traps within or near this field.

Statistics from 90 fields

We apply this concept to about 90 fields over the world where the original hydrocarbon columns of the top-sealed traps are reported (AAPG Treatise of Petroleum Geology series, see Beaumont and Foster, 1990a, 1990b, 1990c, 1990d, 1992; Foster and Beaumont, 1990a, 1991b, 1992a, 1992b, 1993). We try to figure out the equivalent grain size for each layer of the field. Our case studies suggest that an average equivalent grain size for usual clastic shale is about

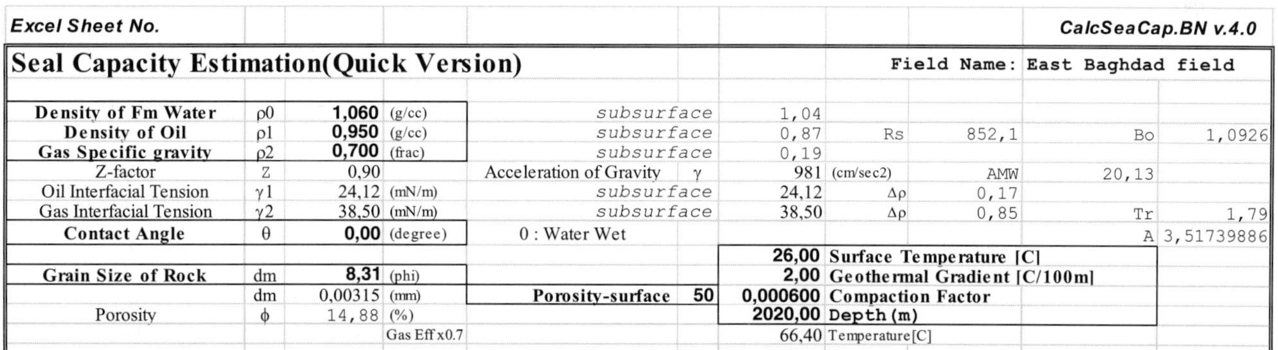

Fig. 7. Spread sheet (Excel) for calculating the oil/gas column height from the equivalent grain size. The given fluid densities at the surface are converted into the subsurface condition, which is calculated from the depth, geothermal gradient, surface temperature, etc. The porosity is calculated from the exponential function of depth provided the compaction factor is given. The other parameters (interfacial tensions, contact angle and Z-factor) are also estimated for the subsurface condition. The sheet calculates the oil/gas column height corresponding to the given equivalent grain size. If the observed maximum column height is known, we can find a suitable equivalent grain size using this sheet by trial and error. (The sheet is available upon request from the authors.)

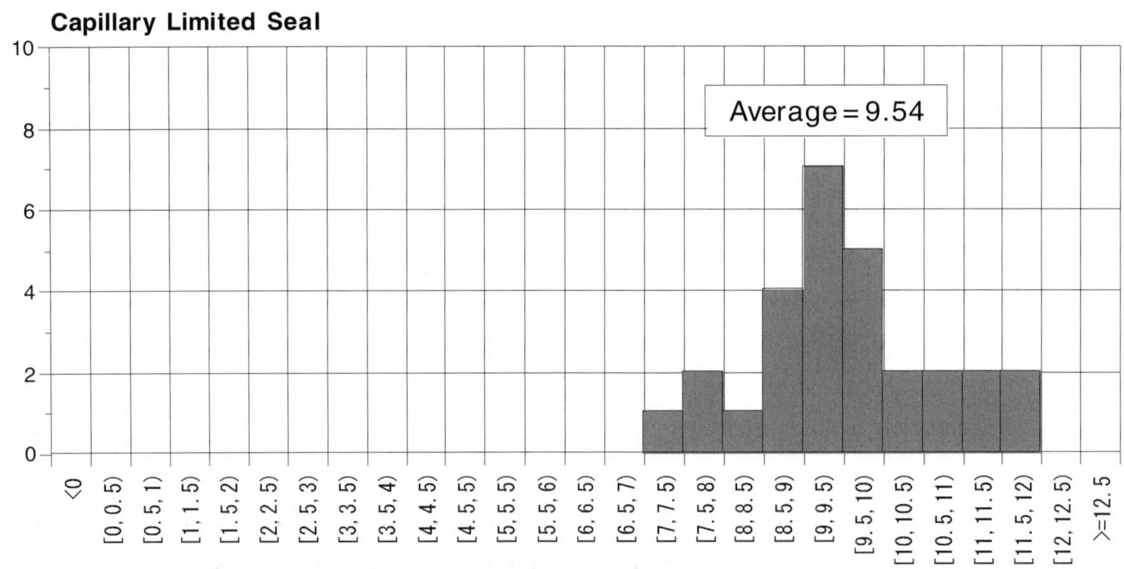

Fig. 8. Statistics of estimated equivalent grain size for 90 oil/gas fields over the world. The cases are carefully selected only for the capillary limited traps from the AAPG Treatise of Petroleum Geology series (see Beaumont and Foster, 1990).

9–10 on the phi-scale (= 2^{-9} to 2^{-10} mm), which is equivalent to 'silt' size (Fig. 8). This indicates a rather coarser sediment than the actual mudstones or shales as a cap rock, because the largest pore throat seems to act as a real effective seal.

If the shale is calcareous, it may have less sealing capacity equivalent to a larger equivalent grain size than clastic shale. Carbonate cementation probably increases brittleness that is equivalent to making the equivalent grain size larger. This is our preliminary result or expectation from a small number of field observations of carbonate reservoirs. Thus we can possibly evaluate sealing capacity of the trap prior to drilling or without samples.

Practical seal evaluation

Capillary limited traps vs spill-point limited traps

How does the concept of 'equivalent grain size' work for the real exploration problems? The above-mentioned estimation of maximum oil/gas column height by 'equivalent grain size' method is applicable only for the traps where some leakage occurs through the top seal (capillary limited traps). If the leakage occurs through the spill-point, the trap is called 'spill-point limited' and the evaluation should be made in a different way. Therefore, we review the generic classification of traps, and how fault seal can be interpreted before we discuss the procedure of practical seal evaluation.

Capillary limited traps

Traps should be generically classified according to their sealing mechanism. One typical case is named here as 'capillary limited trap' (Fig. 9). In this trap, the structural relief is high enough to attain the maximum hydrocarbon column height that can be held by the top seal regardless of the lateral seal (fault or stratigraphic boundaries). The excess hydrocarbons over the maximum column height should leak through the top of the trap.

It is suggested in general that oil had been accumulated first, then gas has migrated into from the deeper part of basin center. Based on this suggestion, a special tendency is observed in the vertical distribution of oil/gas pools according to this type of traps. The noticeable effect of the 'capillary limited trap' is that gas, if it has migrated into the trap after the trap has been already filled up with oil, tends to leak selectively into upper reservoirs (Sales, 1997). This is because the gas is more buoyant than oil, and the leakage of this type of trap would occur at the top of the reservoir. Gas is rather distributed in the shallower part while oil stays in the rather deeper part (but not as deep as the gas cracking zone).

Spill-point limited traps

The second type is named as 'spill-point limited trap', where the hydrocarbons is leaking laterally from the spill-point; it is either a synclinal spill-point or cross-fault spill-point (Fig. 10). In a spill-point limited trap, the level that can be held by the top seal is deeper than that by the spill-point. Therefore, the original hydrocarbon column height is not the maximum column height that should be held by the top seal capacity. If we calculate the equivalent grain size from the observed hydrocarbon column, the grain size is larger than in reality.

In this type, the gas from a later migration tends

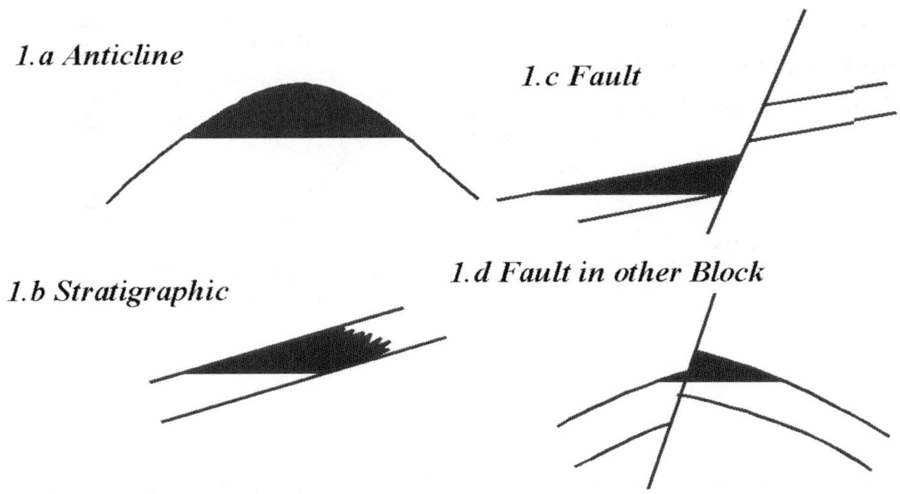

Fig. 9. Capillary limited traps. Trap capacity is limited by the top seal. If gas migrates into the trap where oil is already accumulated, the gas has a tendency to go up through the accumulated oil and to leak easily from the top of the structure. As a result, oil still remains in the trap.

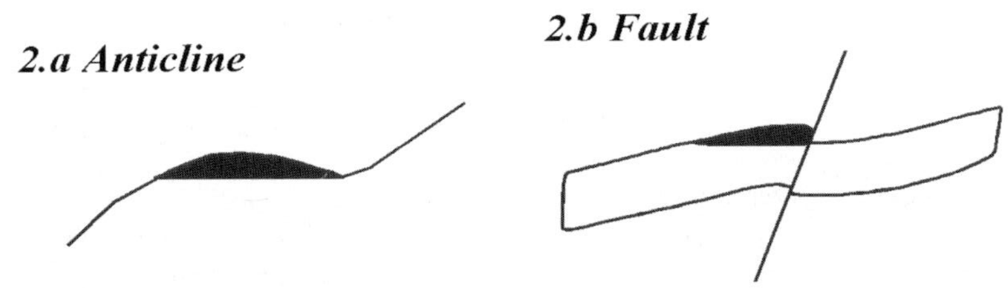

Fig. 10. Spill-point limited traps. Trap capacity is limited by the spill-point. If gas migrates into the trap where oil is already accumulated, the gas has a tendency to push down the accumulated oil and the oil is leaking easily from the spill-point of the structure. As a result, gas will be replaced by oil.

to squeeze out the oil, because the gas in the upper part pushes down the oil reserves so that the oil migrates into the adjacent reservoirs in the up-dip direction. The famous migration theory that Gussow proposed (Gussow, 1954) is of this type. If this type of migration occurred through a cross-fault spill-point within a field, oil is finally spilled out so that only gas occupies all the traps. Typical examples of this type of migration can be seen in most fields of the Gulf of Thailand according to our special experience in the area.

It is important to note that we could determine the type of trapping mechanism of the field from the vertical distribution of the oil/gas pools within a field or near fields.

Three possible states of fault functions

It may be necessary to evaluate sealing capacity in the lateral direction, especially when the trap is bounded by faults. There are three possible states in which a fault can deal with hydrocarbon migration through it: (1) non-sealing, in the case of sand-to-sand juxtaposition; (2) sealing in the case of sand-to-shale juxtaposition, or of shale smear within the fault; (3) acting as a conduit, in the case of a fault in an abnormally pressured zone, and of a fault that is re-activated.

These states of sealing, non-sealing or acting as a conduit really present a debatable issue, but we simply define that the three possible states can occur only in the cases described above. Many discussions have been made on the fault as a conduit for leakage, but we postulate here that the fault is just a geometrical boundary as Allan (1989) defined, that cannot be a conduit except when abnormal pressure re-activates the fault plane.

Non-sealing/sealing: fault-smear development and the fault as a conduit

If the trap is bounded by a fault, Allan's juxtaposition model is applied and the cross-fault spill-point should be detected for evaluation. But in the case that shale smear may have been developed (usually the fault throw is greater than 100 m, and often can be seen in the case of strike-slip faulting), the fault seal acts as a top seal. Even in this case, the equivalent grain size for a fault seal must be smaller than usual because injection or abrasion of shale may have decreased the grain size (Fig. 11). Then the top seal is still effective for the whole trap where the bounded

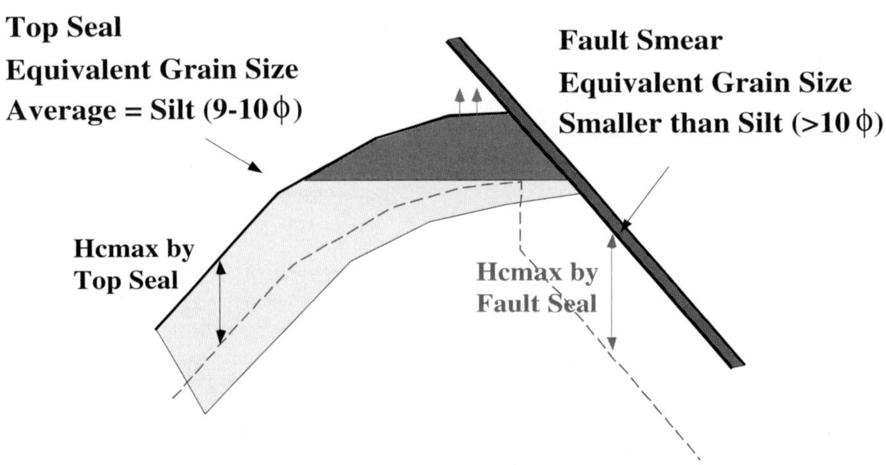

Fig. 11. Diagram showing how the potential sealing surface occurs within the trap across a fault. The original column height of a trap is determined by the shallowest level of the potential sealing surface. At the fault plane, the potential sealing surface is lowered as the sealing capacity increases because of the partial development of a smear. As a result, the excess hydrocarbons will be leaking from the top of the structure.

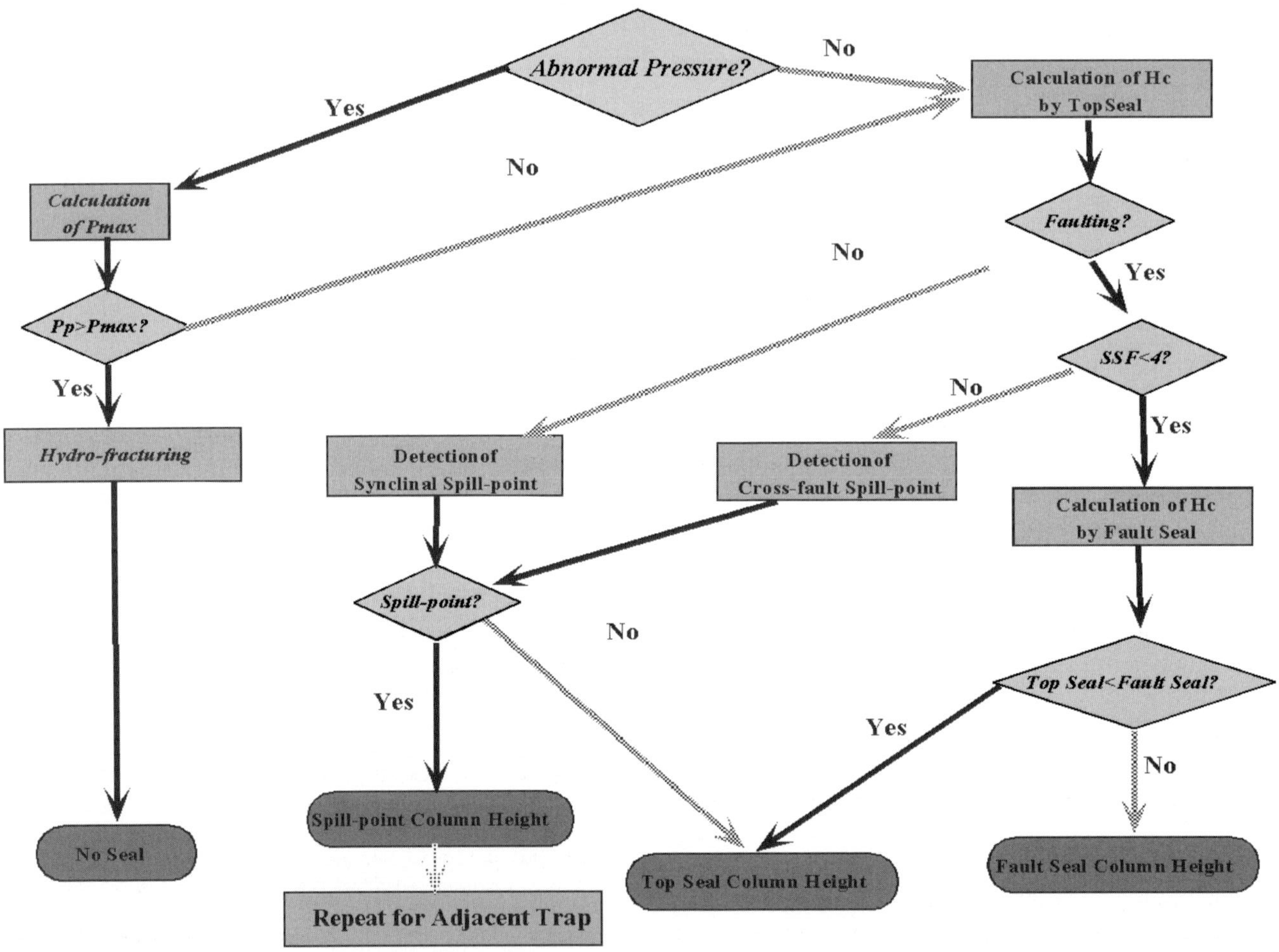

Fig. 12. Flow chart of the practical seal evaluation. The seals are evaluated first if enough overpressure exists to keep a fault open. If not, make a calculation of the column height held by the top seal. Also either the level of the structural spill-point or the cross-fault spill-point should be examined to compare the level with the level held by the top seal. The shallowest potential sealing surface is the effective seal capacity of the trap.

fault is smeared. The observed hydrocarbon column height for the trap is controlled by the highest level of the sealing potential surface. Sand injection might occur to make an open conduit within the fault plane, but such a case may occur at much shallower depths. In the case of abnormally high pressure within fault-bounded traps, such a high pressure may reactivate the fault so that the fault plane is open and may act as a conduit.

Flow chart of seal evaluation and problem on seal evaluation

With the above-mentioned considerations in mind, we can evaluate a sealing capacity for the field. For this purpose a flow chart for such a evaluation is constructed (Fig. 12). One can follow this chart for a practical seal evaluation. Note here that the chart is not complete, but it is a working hypothesis. In most of the cases, one would have tried to interpret sealing mechanisms including migration pathways using an already existing geological interpretation, where some misinterpretations might have been included. The most serious problem on evaluating seals is using data for analysis that had already been interpreted, and not the raw data. We may need to modify the previous interpretation so that the sealing mechanism keeps right.

How can we do it? 'Parallel thinking' or a 'multi-disciplinary approach' is the solution. Re-interpretation of previous contour maps may be necessary, and structural and/or stratigraphical interpretations and seal evaluation should be done at the same time, so that we can find the best and most reasonable interpretation.

Thus we can quantify the effective seal over the whole trap. The first advantage of this method is that we can estimate the maximum column height expected for the target traps prior to drilling (or before accessing to the sample). Second, the result will be concordant with the observation of the original column height because the method is based on a macro-view of the seal phenomenon rather than on a micro-scale with which only small samples can be measured.

Conclusions

Top seal capacity can be predicted by the equivalent grain size method, for which we use an imaginary grain size instead of pore-throat radius to represent the maximum column height that the cap rock can reach. Our applications of this method for about 90 fields over the world suggest that an average equivalent grain size for clastic shales is estimated at 9–10 on the phi-scale, which is rather coarser than we expected.

The trapping mechanism is very important for evaluating a trap. In this paper we classify the traps into two categories: capillary limited traps and spill-point limited traps. Gas tends to be distributed in shallower reservoirs in the case of capillary limited traps, whereas oil tends to migrate into deeper reservoirs in the case of spill-point limited traps.

Simplified criteria were introduced for evaluating a fault as sealing or non-sealing. They are: (1) apply Allan's juxtaposition model assuming the fault plane is nothing but a geometrical boundary; (2) calculate the shale smear factor for possible sealing; (3) consider possible leakage by abnormally high pressure or fault re-activation. A flow chart for 'practical seal evaluation' is developed following these considerations (Fig. 12).

For a reasonable understanding of petroleum systems (oil/gas migration and accumulation), structural, stratigraphical, and sealing evaluations have to be combined and a multi-disciplinary interpretation is necessary.

Acknowledgements

The authors thank TRC-JNOC (Technology Research Center of Japan National Oil Corporation) for their support of this research and for the permission to publish this article. We also thank Dr. Gary D. Couples for his critical review of this article.

References

Allan, U.S., 1989. Model for hydrocarbon migration and entrapment within faulted structures. Am. Assoc. Pet. Geol. Bull., 73: 803–811.

Barker, C., 1978. Primary migration — the importance of water–mineral–organic matter interactions in the source rock. In: Physical and Chemical Constraints on Petroleum Migration, AAPG Continuing Education Course Note Series, 8, pp. D1–D19.

Beaumont, E.A. and Foster, M.H. (Compilers), 1990a. AAPG Treatise of Petroleum Geology: Atlas of Oil and Gas Fields, 1990, Structural Traps I, 232 pp.

Beaumont, E.A. and Foster, M.H. (Compilers), 1990b. AAPG Treatise of Petroleum Geology: Atlas of Oil and Gas Fields, 1990, Structural Traps II, 267 pp.

Beaumont, E.A. and Foster, M.H. (Compilers), 1990c. AAPG Treatise of Petroleum Geology: Atlas of Oil and Gas Fields, 1990, Structural Traps III, 355 pp.

Beaumont, E.A. and Foster, M.H. (Compilers), 1990d. AAPG Treatise of Petroleum Geology: Atlas of Oil and Gas Fields, 1990, Structural Traps IV, 382 pp.

Beaumont, E.A. and Foster, M.H. (Compilers), 1992. AAPG Treatise of Petroleum Geology: Atlas of Oil and Gas Fields, 1992, Structural Traps VII, 347 pp.

Berg, R.R., 1975. Capillary pressure in stratigraphic traps. Am. Assoc. Pet. Geol. Bull., 59: 939–956.

Brooks, C.S. and Purcell, W.R., 1952. Surface area measurements on sedimentary rocks. Pet. Trans., AIME, 195, pp. 289–296 (T.P. 3458).

Downey, M.W., 1994. Hydrocarbon seal rocks. In: L.B. Magoon and W.G. Dow (Editors), The Petroleum System — from Source to Trap. Am. Assoc. Pet. Geol. Mem., 60: 159–164.

Foster, M.H. and Beaumont, E.A. (Compilers), 1991a. AAPG Treatise of Petroleum Geology: Atlas of Oil and Gas Fields, 1991, Stratigraphic Traps II, 360 pp.

Foster, M.H. and Beaumont, E.A. (Compilers), 1991b. AAPG Treatise of Petroleum Geology: Atlas of Oil and Gas Fields, 1991, Structural Traps V, 305 pp.
Foster, M.H. and Beaumont, E.A. (Compilers), 1992a. AAPG Treatise of Petroleum Geology: Atlas of Oil and Gas Fields, 1992, Stratigraphic Traps III, 445 pp.
Foster, M.H. and Beaumont, E.A. (Compilers), 1992b. AAPG Treatise of Petroleum Geology: Atlas of Oil and Gas Fields, 1992, Structural Traps VI, 304 pp.
Foster, M.H. and Beaumont, E.A. (Compilers), 1993. AAPG Treatise of Petroleum Geology: Atlas of Oil and Gas Fields, 1993, Structural Traps VIII, 328 pp.
Hunt, J.M., 1979. Petroleum Geochemistry and Geology. Freeman, San Francisco, CA, 617 pp.
Gussow, W.G., 1954. Differential entrapment of oil and gas — a fundamental principle. Am. Assoc. Pet. Geol. Bull., 38: 816–853.
Nakayama, K. and Van Siclen, D.C., 1981. Simulation model for petroleum exploration. Am. Assoc. Pet. Geol. Bull., 65: 1230–1255.
Sales, J.K., 1997. Seal strength vs.trap closure — a fundamental control on the distribution of oil and gas. In: R.C. Surdam (Editor), Seals, Traps, and the Petroleum System. Am. Assoc. Pet. Geol. Mem., 67, pp. 57–83.
Schowalter, T.T., 1979. Mechanics of secondary hydrocarbon migration and entrapment. Am. Assoc. Pet. Geol. Bull., 63: 723–760.
Yielding, G., Breton, P. and Marsden, G., 2000. Fault-seal evaluation in exploration and production environments. Abstracts AAPG Annual Convention, New Orleans, p. A163.

K. NAKAYAMA *JGI, Inc., 1-5-21 Otsuka, Bunkyo-ku, Tokyo 112-0012, Japan*
E-mail: naka-kaz@jgi.co.jp

D. SATO *Technical Research Center, Japan National Oil Corporation, 1-2-2 Hamada, Mihama-ku, Chiba, Japan*

Effective permeability of hydrofractured sedimentary rocks

Magnus Wangen

Natural hydrofracturing is studied numerically with a 1-D model where overpressure build-up is due to expulsion of fluids from porosity reduction. The porosity is assumed given as a function of depth. This assumption covers compaction processes controlled by temperature. It is therefore nothing in the model, except for hydrofracturing, that prevents the fluid pressure from exceeding the lithostatic pressure because there is no feedback from the pressure on the compaction. Three aspects of the hydrofracture process are accounted for in the model: (1) a fracture criterion, (2) permeability enhancement, and (3) fracture sealing. The hydrofracturing is triggered when the fluid pressure exceeds a given fraction of the lithostatic pressure. Hydrofracturing is modeled by increasing the bulk permeability of the rock by a given factor (the fracture factor). Sealing of the fractures is modeled by letting the fracture factor decay towards 1 by a given half life. Hydrofracturing is studied with three cases which are different with respect to the maximum fracture factor and the half life for fracture healing. The cases show that a large maximum fracture factor gives larger amplitude in the oscillatory fluid pressure than a smaller maximum fracture factor. A long half life is shown to produce smaller amplitudes in the pressure oscillations than a short half life, because the permeability enhancement lasts longer. An important result in this study is that the fluid pressure is intermittent during hydrofracturing, but not the fluid flow. This is a direct consequence of the compaction model used, where the porosity is controlled by depth. The numerical case studies show that hydrofractured depth intervals are characterized by a gravity number close to 1. This observation is used to derive an estimate for the permeability of hydrofractured sediments.

Introduction

Brittle sediments hydrofracture when the pore fluid pressure becomes sufficiently large. The pore fluid will carry the weight of the entire basin above when the pore fluid pressure equals the lithostatic pressure. This may be the reason why pore fluid pressures beyond the lithostatic pressure are never observed. However, certain models for pore pressure build-up during burial can easily predict pressures above the lithostatic. Such models for overpressure build-up require a mechanism like hydrofracturing to enhance the permeability, thereby keeping the pressure below the lithostatic pressure. A common rule is to say that hydrofracturing takes place when the pore fluid pressure exceeds 75% of the lithostatic pressure. High overpressure at this level is observed in several reservoirs, especially those at depths greater than ~2.5 km.

Fluid pressure was shown to be an important controlling factor for fracturing by Hubbert and Rubey (1959). A large number of observations of fractures and faults have after Hubbert and Rubey (1959) been interpreted in light of fluid pressures (Davis and Reynolds, 1996). The work of Hubbert and Rubey (1959) and a number of later authors (see Davis and Reynolds, 1996) have focused on the criterions for fracturing and the associated styles of fracturing. The fracture criterions are often conveniently expressed as failure envelopes, where the envelope relates the shear and the normal stresses acting on the fracture surface. Fracturing triggered by pressure build-up creates pathways for the fluid, which limit further pressure build-up. The permeability enhancement of natural hydrofracturing and its implications for the fluid flow is the aim of this paper. Although these are important issues, they are little studied and poorly understood, even though some work has been done.

Doligez et al. (1988) apply hydrofracturing in two 2-D case studies, one from the North Sea and another from the Gulf of Lions. Hydrofracturing leads to an enhanced permeability by use of a permeability factor which is a function of the positive difference of the fluid pressure and the least compressive stress. A similar approach is also taken by Lerche (1990), where the permeability factor is a function of the positive difference between the fluid pressure and a fracture threshold. Roberts and Nunn (1994) models episodic fluid flow by use of a fracture permeability obtained from a parallel plate model, where the aperture is dependent on the pressure. This dependence is a linear function of the positive difference of the

Hydrocarbon Seal Quantification edited by A.G. Koestler and R. Hunsdale. NPF Special Publication 11, pp. 61–74, Published by Elsevier Science B.V., Amsterdam.

fluid pressure and fracture threshold. A somewhat different approach is chosen by Wang and Xie (1998) and L'Heureux and Fowler (2000). These authors allow the fracture permeability to last after a fracture event independent of the fluid pressure. L'Heureux and Fowler (2000) have introduced a time constant for fracture sealing. The approach chosen here to model hydraulic fracturing is similar to the approach of L'Heureux and Fowler (2000) by the use of a maximum fracture factor and the use of a half life for the decay of the fracture factor. However, the model for pressure build-up is different. L'Heureux and Fowler (2000) compute porosity reduction as a function of effective vertical stress, while the porosity in this study is a function of depth.

It should be mentioned that hydrofracturing is an often used method to increase the permeability around wells (Engelder, 1993). It is also studied and modeled at a pore scale (Tzschicholz and Wangen, 1999). However, these processes do not generate hydrofractures from a fluid pressure generated from within the system, but from a fluid pressure applied to the system. They are therefore not natural hydrofracturing.

Models for pressure build-up

The fluid pressure is (normally) bounded below by the hydrostatic pressure. The hydrostatic pressure is simply the pressure from the weight of the water column. Fluid pressures are often much higher than the hydrostatic pressure, especially below 2 to 3 km depth in many sedimentary basins. However, the fluid pressure seems always bounded above by the lithostatic pressure. The lithostatic pressure is the pressure from the weight of the sedimentary column above a point, including the weight from both the sediment matrix and the pore fluid. The fluid pressure in excess of the hydrostatic pressure is called the overpressure, see Fig. 1. Overpressure develops when fluids are expelled. Fluid expulsion is driven by the gradient of the overpressure, and a large overpressure gradient is needed to drive fluids through low permeable rocks, like for instance seals.

There are several processes that lead to expulsion of fluids in sedimentary basins (Osborne and Swarbrick, 1997). Three examples are: (1) reduction of the porosity, (2) decreasing fluid density, and (3) generation of fluids. Causes for porosity reduction can be both mechanical and chemical. Unlithified young sediments compact under increasing effective stress. However, lithified sediments are more likely to compact chemically by local dissolution and reprecipitation of minerals. The minerals are precipitated as cement in the pore space and the pore fluid is thereby expelled. The second type of expulsion process is due to an expanding fluid volume, and thereby a decreasing density. An example of a process leading to decreasing fluid density is cracking of oil to gas. Expansion of the pore fluid due to increasing temperature has also been suggested as an overpressure generating mechanism (Shi and Wang, 1986; Lou and Vasseur, 1992). The third type of expulsion process, generation of fluids, can be dehydration of clays (Colton-Bradley, 1987; Audet, 1995) or oil generation (Lou and Vasseur, 1996).

Mechanical compaction

The usual way to model compaction and overpressure has been to assume that the porosity is a function of the effective vertical stress, σ'. The effective vertical stress is the overburden p_b minus the fluid pressure p_f

$$\sigma' = p_b - p_f \tag{1}$$

Effective vertical stress dates back to Terzaghi (Terzaghi, 1943), who successfully modeled compaction of soils based on this quantity. The porosity of clays is measured in the laboratory as a function of effective stress and it is observed to fit the following relationship (Skempton, 1970),

$$e = e_0 - C_v \log(\sigma') \tag{2}$$

where e is the void ratio and C_v is the compression index. This relationship has also been applied to compaction of sediments at a basin scale (Audet and Fowler, 1992; Audet, 1995, 1996). Another relationship for porosity as a function of the effective vertical stress is the Athy-type of porosity function

$$\phi = \phi_0 \exp(-\alpha\sigma') \tag{3}$$

Athy (1930) fitted porosity observations to a function decreasing exponentially with depth. An empirical porosity–effective stress relationship can be made by rewriting Athy's porosity–depth relationship in terms of the effective vertical stress. This approach, which unifies compaction as a function of the effective vertical stress with the porosity observations of Athy, was first suggested by Smith (1971).

Porosity functions like Eqs. 2 and 3 imply a strong coupling between compaction and fluid pressure. Maximum compaction is obtained at a minimum fluid pressure which is normally the hydrostatic pressure. The maximum vertical effective stress is therefore the overburden minus the hydrostatic pressure. The other extreme is when the fluid pressure approaches the overburden, in which case the effective vertical stress approaches zero. In this limit, where the effective vertical stress approaches zero, there are

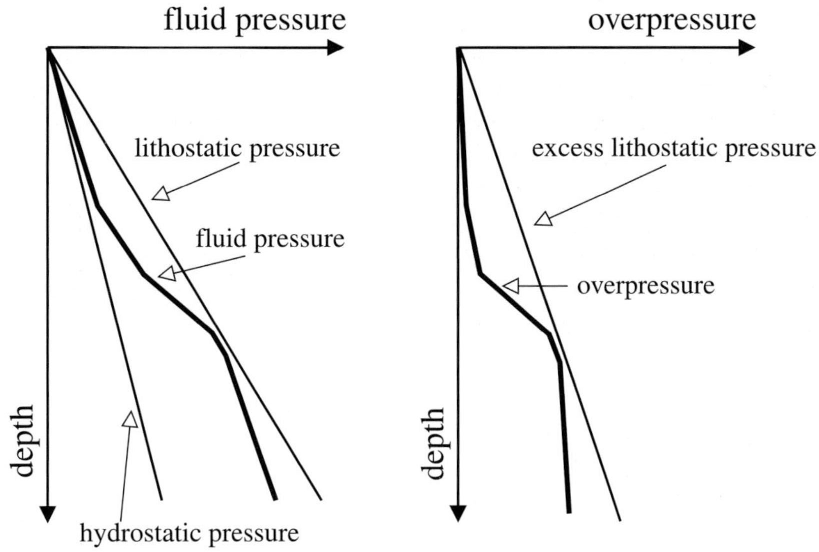

Fig. 1. The fluid pressure is limited below by the hydrostatic pressure and above by the lithostatic pressure. The hydrostatic pressure is the weight of the water column, and the lithostatic pressure is due to the weight of the entire sedimentary column (with both sediments and fluid). The fluid pressure minus the hydrostatic pressure is the overpressure.

effectively no compaction, and therefore no expulsion of fluids. This implies that models for overpressure build-up based on the compaction as function of effective vertical stress cannot model fluid pressures beyond the lithostatic stress.

It should be mentioned that there are compaction models based on a viscoelastic description of the rocks (Schneider et al., 1996; Connolly and Podladchikov, 1998). These models also imply porosity reduction given as a function of the fluid pressure.

Chemical compaction

Compaction as a function of effective vertical stress is not the only mechanism for porosity reduction in sediments. Actually, there are authors that claim that chemical processes like cementation of the pore space do control the compaction of lithified sedimentary rocks (Bjørlykke and Høeg, 1997; Bjørlykke, 1999a,b). It is difficult to model chemical compaction in a general manner, covering a wide range of lithologies and mineral reactions. Many aspects of the dissolution and reprecipitation processes are poorly understood, except for simple systems such as quartz cementation in quartzose sandstones.

This latter system is characterized by dissolution of quartz along stylolites and other mica–quartz contacts. The stylolites and the mica–quartz contacts create a supersaturation which leads to local dissolution, transport of the silica by diffusion into to the pore space where it precipitates as cement. The precipitation step is the slowest step in the process and the entire local pore space becomes supersaturated (Walderhaug, 1994a,b; Aase et al., 1996; Bjørkum, 1996; Oelkers et al., 1996). An important petrographical observation with respect to quartz cementation is that the process is controlled by temperature and not by fluid pressure (Bjørkum, 1994, 1996). This observation might apply to cementation of other types of rock as well.

We are interested in the pore space reduction on a basin scale. The details at a pore space level have therefore to be averaged to a length scale of several meters or tens of meters. The porosity must be interpreted as an average porosity which spans an entire lithological unit or may be several units.

A simple way to model chemical compaction controlled by temperature is to assign porosity–depth curves to each lithology. This approach works for burial histories with deposition at a constant rate, because all sediment grains experience the same temperature history down to the same depth. It must, of course, be assumed that the thermal gradient stays constant through the burial history. The use of burial at a constant rate together with a constant thermal gradient applies to the test cases shown later to study hydrofracturing. The uppermost, youngest and unlithified sediments do not compact chemically, but mechanically with a fluid pressure which is often hydrostatic. Hydrostatic compaction of the uppermost and unlithified sediments implies porosity as function of depth. The porosity can therefore, given this assumption, be specified as a porosity–depth curve from the basin surface to the basement.

An important point with respect to pressure modeling based on porosity–depth curves is that there is no feedback from pressure on the porosity reduction. Therefore, nothing can prevent the pressure build-up to be limited by lithostatic pressure.

Hydrofracturing

There are three elements of the hydrofractured process which are needed in order to model hydrofracturing coupled to fluid pressure build-up. These are: (1) a fracture criterion, (2) a way to enhance the permeability of the cells which are hydrofractured, and (3) a way to heal the fractures, and thereby reduce the enhanced permeability of the fractured rock.

A fracture criterion

A usual fracture criterion is simply to say that hydrofracturing takes place when the fluid pressure reaches a certain fraction of the lithostatic pressure (Davis and Reynolds, 1996). Typical pressure thresholds suggested in the literature lay in between 70% and 85% of the lithostatic pressure.

This fracture criterion originates from fracture mechanics where stress is replaced by effective stress. Effective stress, denoted σ', is the confining stress minus the fluid pressure. The shear and the normal stress acting on a plane making an angle θ with the largest principal stress, σ_1, is conveniently plotted as a Mohr circle (Davis and Reynolds, 1996). The largest principal stress σ_1 is now assumed to be the lithostatic stress. The stress state in the horizontal plane is assumed isotropic, which implies that the two least principal stresses, σ_2 and σ_3, are equal. The subtraction of the fluid pressure from a stress state has no effect on the shear stress, but it becomes subtracted from the normal stress (see Figs. 2 and 3). The Mohr circle for the effective stress state is therefore the Mohr circle for the confining stress state shifted to the left by the value of the fluid pressure.

The shear stress and the corresponding normal stress required for fracturing along a plane can be approximated by a line. This linear relationship between the shear and the normal stresses is the Mohr-Coloumb failure criterion,

$$\tau = \sigma_0 + \mu\sigma \quad (4)$$

where σ_0 is the cohesive strength of the rock and μ is the coefficient of internal friction. Fig. 2 illustrates hydrofracturing when the horizontal stress σ_3 is much less than the lithostatic stress σ_1. The diameter in the Mohr circle is $\sigma_1 - \sigma_3$. A large diameter require less fluid pressure before the circle is shifted enough to the left so that it touches the fracture envelope. Hydrostatic fluid pressure alone could be enough to cause fracturing if σ_3 is sufficiently low. This happens for example during extension and faulting. The angle of the fractures with the vertical (the largest principal stress σ_1) is $\theta_v = \pi/2 - \theta$.

A situation with stresses close to an isotropic state is shown in Fig. 3. The diameter of the Mohr circle is then small. The Mohr circle cannot be made touching the fracture envelope in this case because the radius is less than the cohesive strength σ_0. A fluid pressure almost as large as the lithostatic is needed to shift the Mohr circle so far to the left that the least principal effective stress, $\sigma_3' = \sigma_3 - p_f$, becomes negative. Tension fracturing or jointing happens when σ_3' becomes negative. The criterion for hydrofracturing is therefore that the fluid pressure must become as large as the least compressive stress, σ_3.

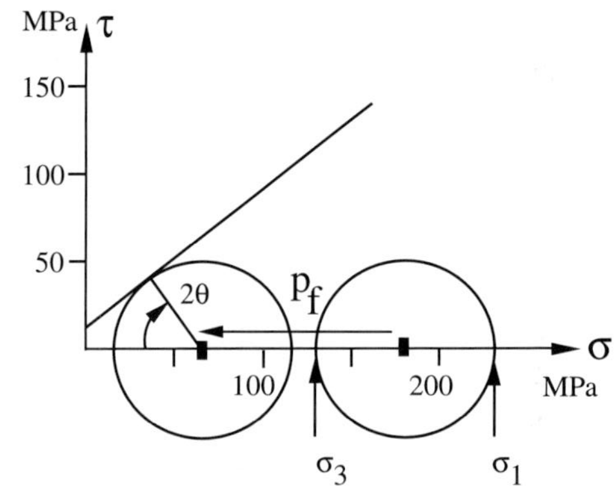

Fig. 2. The largest principal stress σ_1 is the lithostatic stress, and the least principal stress σ_3 is horizontal. Fluid pressure shifts the Mohr circle until its touches the fracture envelop. The fluid pressure required for fracturing is much less than the lithostatic (σ_1) when there the least principal stress is much less than the lithostatic. The fracture is a shear fracture making an angle θ with the normal.

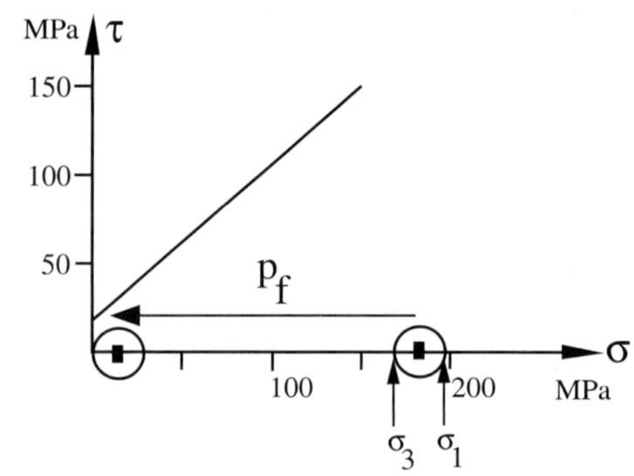

Fig. 3. The figure shows a stress situation which is close to isotropic. The least principal stress (σ_3) and the largest principal stress (the lithostatic, σ_1) are close. This situation requires a fluid pressure close to the lithostatic pressure for fracturing to happen. The fractures become tensile fractures or joints.

The linear Mohr-Coloumb fracture envelope is the simplest way to give a fracture envelope. A commonly used alternative to the Mohr-Coloumb failure envelop is the parabolic Griffith failure envelope. The Griffith envelope incorporates the tensile strength of the rock, and thereby provides an envelop for the negative normal stresses σ.

Fracture enhancement of the permeability

The fluid pressure build-up is reduced by hydrofracturing. Hydrofracturing therefore has the effect of creating an enhanced average permeability of the fractured rock. It is impossible on a basin scale to take into account the details of fractures and fracture networks. The impact of the fractures is therefore sought through a factor, termed the fracture factor, which gives how much the unfractured bulk permeability has been increased.

It is difficult to know in advance how much the permeability of a computational cell should be increased with as a result of fracturing. The approach chosen here is to try a moderate fracture factor by doing a small time step. If the fluid pressure is still above the threshold for fracturing, then the moderate factor is multiplied once more, before yet another small time step is done. The simulator proceeds like this until there are no more cells left where the fracture criterion evaluates to true. The factor, which is multiplied to the permeability of a fractured cell at each small time step, is a numerical parameter. It is used simply as a means to approximate the permeability enhancement. However, the permeability enhancement is not allowed to increase unlimited. There is an upper bound for the permeability enhancement which is the maximum fracture factor, which will be denoted f_{max}.

An important point here is that the fractures create an enhanced permeability of the rocks which after fracturing are not sensitive to the fluid pressure. The permeability enhancement is permanent, unless the fractures are sealed by for instance cementation. The fracture process is assumed to create permanent damages on the fracture walls, caused for instance by loose fragments. The fracture walls will therefore not fit exactly back together, and there will be a permanent fracture porosity along the fracture. The fractures can therefore not become sealing by compressive stresses, unless the compressive stress acts for such a long time that the fractures become sealed by creep processes.

This approach is different from the use of idealized parallel plate fractures where the fracture apertures are pressure dependent. This latter approach does not only assume that pressure build-up can cause fracturing, but also that the fractures are kept open for fluid flow by the fluid pressure. The parallel plate fracture where the fracture aperture is pressure dependent operates like a valve. The pressure has to be higher than the threshold to open the value. The parallel plates seal completely when the pressure drops below the threshold, and there is therefore no fracture permeability when the pressure is below the threshold.

Healing of fractures

Hydrofracturing causes an increased bulk permeability. However, this increased permeability will not last forever. Especially precipitation of cement in the voids along the fracture will in the end seal the fracture. This is a likely fate of fractures at depths where the temperature is sufficiently high for diagenesis to operate effectively. It is also at such depths where cemented and lithified rocks are brittle. Creep processes can also be envisaged to seal fractures in brittle rocks. The seal processes are here modeled by assuming that the fracture factors decrease exponentially with time towards unity. The time constant for the decay, the half life for the fracture factor, is an input parameter to the fracture model. This time constant is denoted t_0.

A comment to the fracture model

The hydrofracture model suggested here is simple. It is deliberately simple, because the interaction of the fluid pressure with the flow properties of sedimentary rocks is still poorly understood. A simple as possible empirical model is therefore sought, where there is a minimum number of parameters. There are effectively only two parameters in the proposed model: the maximum fracture factor and the half life for the fracture factor once acquired. A similar approach has been suggested by L'Heureux and Fowler (2000). L'Heureux and Fowler multiply the permeability with a maximum fracture factor in all cells where the fluid pressure exceeds the threshold, and the fracture factor will then decay exponentially with a given half life. There is no feedback from the pressure on the fracture permeability, unless the pressure build-up causes refracturing.

Numerical examples

Hydrofracturing is demonstrated in a simple 1-D basin model. The model is simple in the sense that it has only one lithology, and that it models pressure build-up during burial at a constant rate of sediment deposition. The use of 1-D models with only one lithology is a common setting to study basic behavior of pressure build-up during burial (Gibson, 1958; Bredehoeft and Hanshaw, 1968; Smith, 1971; Sharp and Domenico, 1976; Bethke and Corbet, 1988; Audet and Fowler, 1992; Wangen, 1992; Birchwood and Turcotte, 1994; Schneider et al., 1996; Fowler and Yang, 1999). The lithological properties must therefore be conceived as average or effective properties.

The pressure build-up is caused by porosity reduction, and the porosity is assumed known as a function of depth. An Athy porosity–depth relationship is chosen,

$$\phi = \phi_0 \exp\left(-\frac{z_{net}}{z_{net,0}}\right) \tag{5}$$

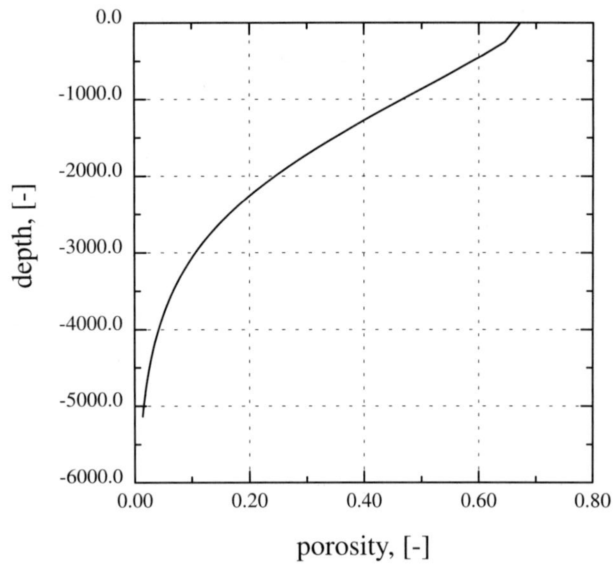

Fig. 4. The porosity–depth curve used in the numerical case studies.

where depth z_{net} is given as fully compacted (porosity free) rock. The surface porosity is $\phi_0 = 0.7$ and the depth $z_{net} = 1000$ m. The porosity is plotted in Fig. 4 as a function of real depth. The permeability k is given as a power law of the porosity,

$$k(\phi) = k_0 \exp(b\phi) \tag{6}$$

where the parameters are $k_0 = 1 \times 10^{-23}$ m^2 and $b = 27$. It is seen from the permeability relationship (Eq. 6) that k_0 is the permeability at zero porosity and that the permeability at surface porosity is $k(\phi_0) = 1.6 \times 10^{-15}$ m^2 when $\phi_0 = 0.7$. The permeability is plotted as a function of porosity in Fig. 5a and as a function of depth in Fig. 5b. The plot in Fig. 5b is obtained by combining the permeability function Eq. 6 with the porosity–depth curve shown in Fig. 4.

TABLE 1

Fracture parameters

Case	f_{max}	t_0 (Ma)
1	25	10
2	25	100
3	250	10

It should be noted that expulsion is directly related to a porosity as a function of depth. This implies that there is no feedback from the fluid pressure on the compaction and the rates of fluid expulsion. Therefore, no mechanisms prevent the fluid pressure to be limited by the lithostatic pressure, except for hydrofracturing. The pressure equation is given in Wangen (1993, 1997).

The burial history is deposition of 4000 m net (porosity) free sediments at a constant rate during 120 Ma. Fracturing is triggered when the fluid pressure exceeds 90% of the lithostatic pressure. Hydrofracturing is studied with three different cases with respect to the fracture parameters f_{max} and t_0 (Table 1). These cases demonstrate how these two parameters control the pressure build-up during hydrofracturing.

The pressure is plotted as overpressure in this section. The overpressure is obtained by subtracting the hydrostatic pressure from the fluid pressure. The overpressure is therefore limited below by the zero and above by the lithostatic pressure minus the hydrostatic pressure. The latter pressure in the following is termed the excess lithostatic pressure.

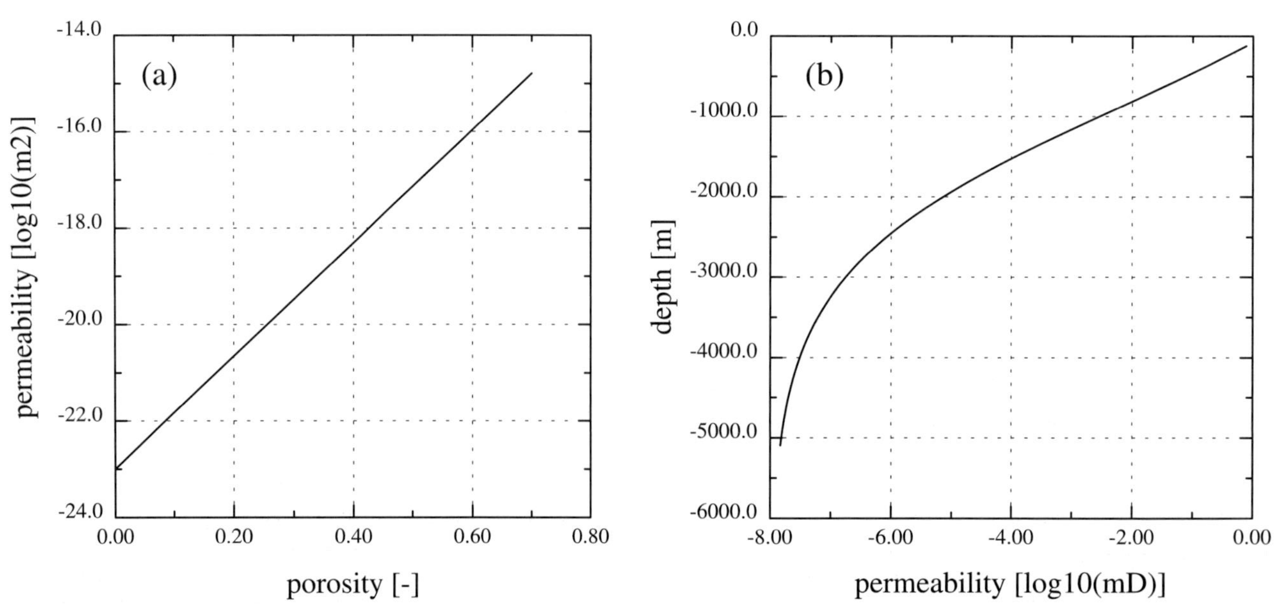

Fig. 5. (a) Permeability as a function of porosity. Permeability function (Eq. 6) is plotted with $b = 27$ and $k_0 = 1 \times 10^{-23}$ m^2. (b) Permeability as a function of depth. The curve is obtained by combining the porosity–depth curve in Fig. 4 with the permeability–porosity curve in (a).

Case 1

Case 1 combines a max fracture factor $f_{max} = 25$ with a half life of fracture healing $t_0 = 10$ Ma. The pressure at the present time is shown in Fig. 6a. The excess lithostatic pressure is shown as a dotted line. It is seen that the upper 2 km of the basin is almost hydrostatic. Then, below 2 km depth, a steep pressure raise brings the fluid pressure sufficiently close to the lithostatic pressure to exceed the fracture threshold. This is seen from Fig. 2b, where the effective permeability is plotted. (The effective permeability is the unfractured bulk permeability multiplied with the fracture factor.) The point of hydrofracturing is shown as a step-like increase in the permeability by a factor roughly equal to the max fracture factor f_{max}. The fracture factor at present time is shown in Fig. 6c. The point where the fracture factor raises steeply from 1 to its maximum value $f_{max} = 25$ shows the depth where hydrofracturing starts. Fig. 6c

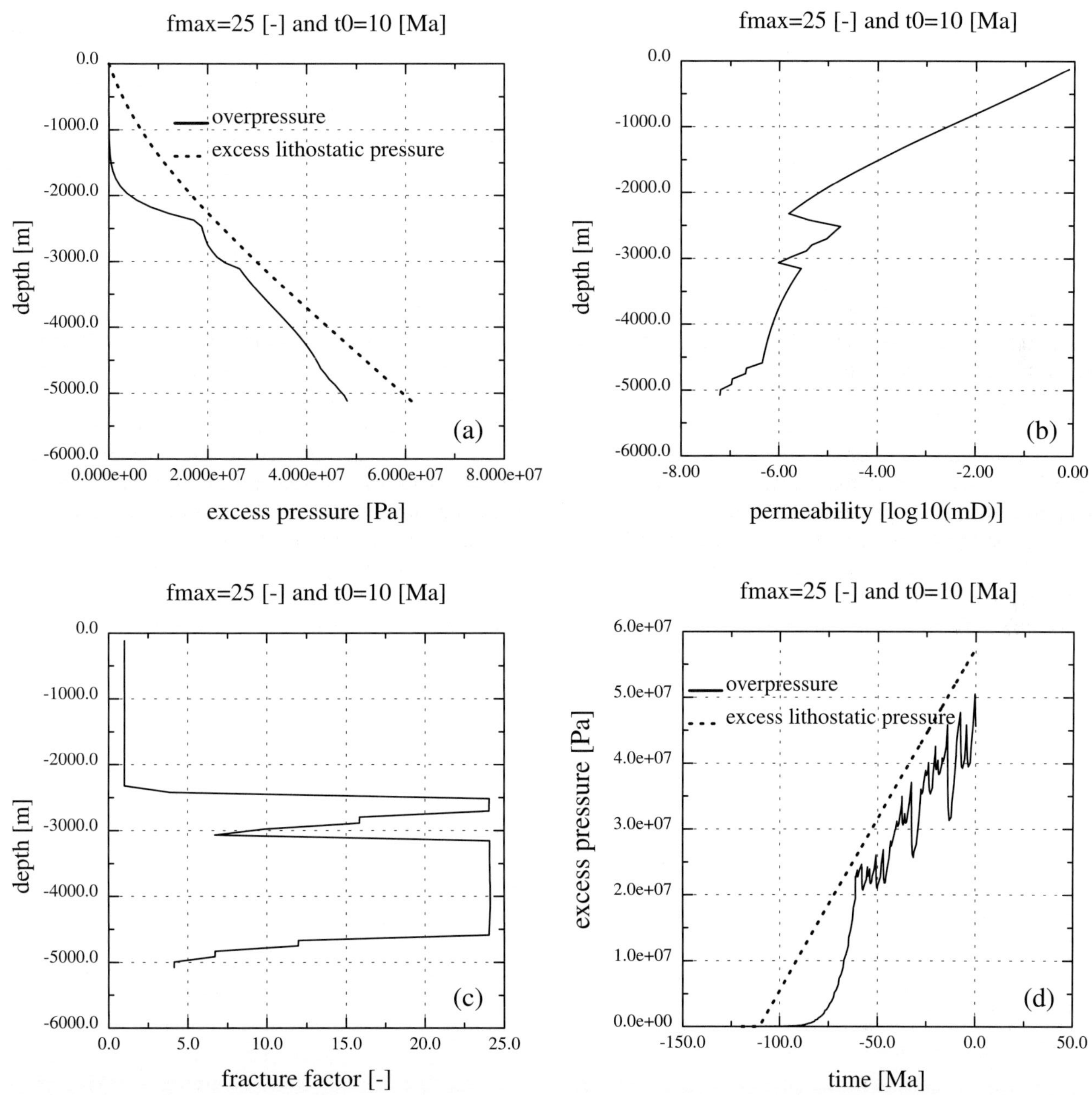

Fig. 6. (a) The overpressure as a function of depth at the present time for case 1. The uppermost 2 km is hydrostatic, and then comes a pressure seal. Hydrofracturing limits the pressure build-up at the base of seal at ~2.5 km. (b) The permeability as a function of depth at present time for case 1. The permeability is seen to increase below ~2.5 km due to hydrofracturing. (c) The fracture factor as a function of depth at present time for case 1. Most of the depth interval below ~2.5 km has enhanced permeability. (d) The overpressure through time for a cell close to the base of the sedimentary column in case 1. There are hydrostatic conditions until ~75 Ma, when the overpressure starts to rise. Hydrofracturing is triggered when the overpressure is close to the excess lithostatic pressure. The overpressure is then oscillating from then on.

shows that the permeability is enhanced by the maximum fracture factor in most of the depth interval of hydrofracturing.

The pressure plot shown in Fig. 6a is typical of that found in many sedimentary basins. The young and unlithified sediments are almost hydrostatic. Then, below the hydrostatic upper part, there is a pressure seal where the overpressure raises steeply. The overpressure often stays high below the seal (Gaarenstroom et al., 1993; Leonard, 1993; Darby et al., 1996).

The pressure through time is shown in Fig. 6d for a cell which is close to the base of the sedimentary column. This cell is deposited early in the burial history, -120 Ma, and it stays almost hydrostatic for the first 55 Ma after deposition. Then, there is a raise in the fluid pressure, which brings the fluid pressure to the threshold for fracturing. From then on there are several fracture events. The fluid pressure drops right after a fracture event, before it starts to raise as the fracture permeability decreases due to fracture healing. An important point here is that the fluid flux is not oscillatory. The fluid flux is caused by expulsion of fluid because of porosity reduction, when the porosity is given as a function of depth. There is therefore no feedback from the fluid pressure on the rate of fluid expulsion in this model. However, the fluid pressure is oscillating because the permeability is rapidly changing due to the fracture events. This model therefore predicts oscillatory pressures, but not episodic fluid flow or fluid pulses.

It should be mentioned that there is a small pulse of fluid associated with a pressure drop because of the compressibility of the pore fluid. This episodic fluid flux was not observed in these cases, probably because it was less than the Darcy velocity. The flux is dependent on the minimum time step. It could be made a significant spike in the Darcy velocity if the minimum time is made sufficiently small and the permeability enhancement is made sufficiently large. However, the volume of fluid expelled in one spike is much less than the Darcy flux integrated over a time span of 1 Ma because of the low compressibility β of brine ($\beta \approx 1 \times 10^{-10}$ Pa^{-1}). The volume of fluid ΔV_f compressed in a fluid volume V_f under the pressure difference Δp is

$$\frac{\Delta V_f}{V_f} = \beta \Delta p \qquad (7)$$

This follows directly from the definition of compressibility. A pressure change $\Delta p = 1 \times 10^7$ Pa (equal to 1 km column of water), implies a change in the fluid volume $\Delta V_f / V_f \approx 1 \times 10^{-3}$. This change in fluid volume corresponds to a porosity change equal to $\Delta\phi = 1 \times 10^{-4}$ for a rock with the porosity $\phi = 0.1$.

The non-episodic fracture flow shown here is different from the models of Roberts and Nunn (1994) and Wang and Xie (1998). These latter models show episodic fluid flow because porosity is a function of effective vertical stress in the entire basin. Overpressured sediments then have a larger porosity than hydrostatic sediments, and overpressured sediments are therefore a reservoir of fluids. Hydrofracturing and the associated pressure drop therefore lead to rapid (episodic) expulsion of fluids from the overpressured sediments.

Case 2

Case 2 is similar to case 1 except that the half life for sealing of fractures is increased from 10 Ma to 100 Ma. The maximum fracture factor is still 25. The overpressure at present time is shown in Fig. 7a, and it has the same characteristics as in case 1. The uppermost 2 km of the basin is hydrostatic, and then comes a sealing interval where the fluid pressure is raising from a hydrostatic level to a level close to the lithostatic. The pressure raise is limited by hydrofracturing, but the overpressure is seen to remain high below the point of hydrofracturing. The depth of hydrofracturing and the enhanced bulk permeability is shown in Fig. 7b. There is clearly a jump in the permeability corresponding to the maximum fracture factor, $f_{max} = 25$. The fracture factor at present time is shown in Fig. 7c, which shows that a cell experiences only one fracture event during the burial history. The cells fracture only once, and that is when they are brought down to the depth ~2.5 km. There is no need for refracturing in this case because the time for fracture healing is long ($t_0 = 100$ Ma). The fracture factor is seen to drop slowly from its maximum value $f_{max} = 25$ with depth and age. The pressure in the cell close to the base of the basin is shown in Fig. 7d, and it is seen to fluctuate less than in case 1. This can be explained with the lesser number of fracture events, and that fracturing occurs only once in the life of a cell.

Case 3

Case 3 is different from case 1 by the value of the maximum fracture factor, which is now $f_{max} = 250$. The overpressure at present time is shown in Fig. 8a. The overpressure plot is now characterized by two pressure seals. There is one at ~2.5 km and another at ~4.5 km. Fracturing is taking place at the base of both seals, which is seen from the permeability plot, Fig. 8b, where the permeability is making a jump of two orders of magnitude. The plot of the fracture factor, Fig. 8, shows that it is below the two seals where the permeability is

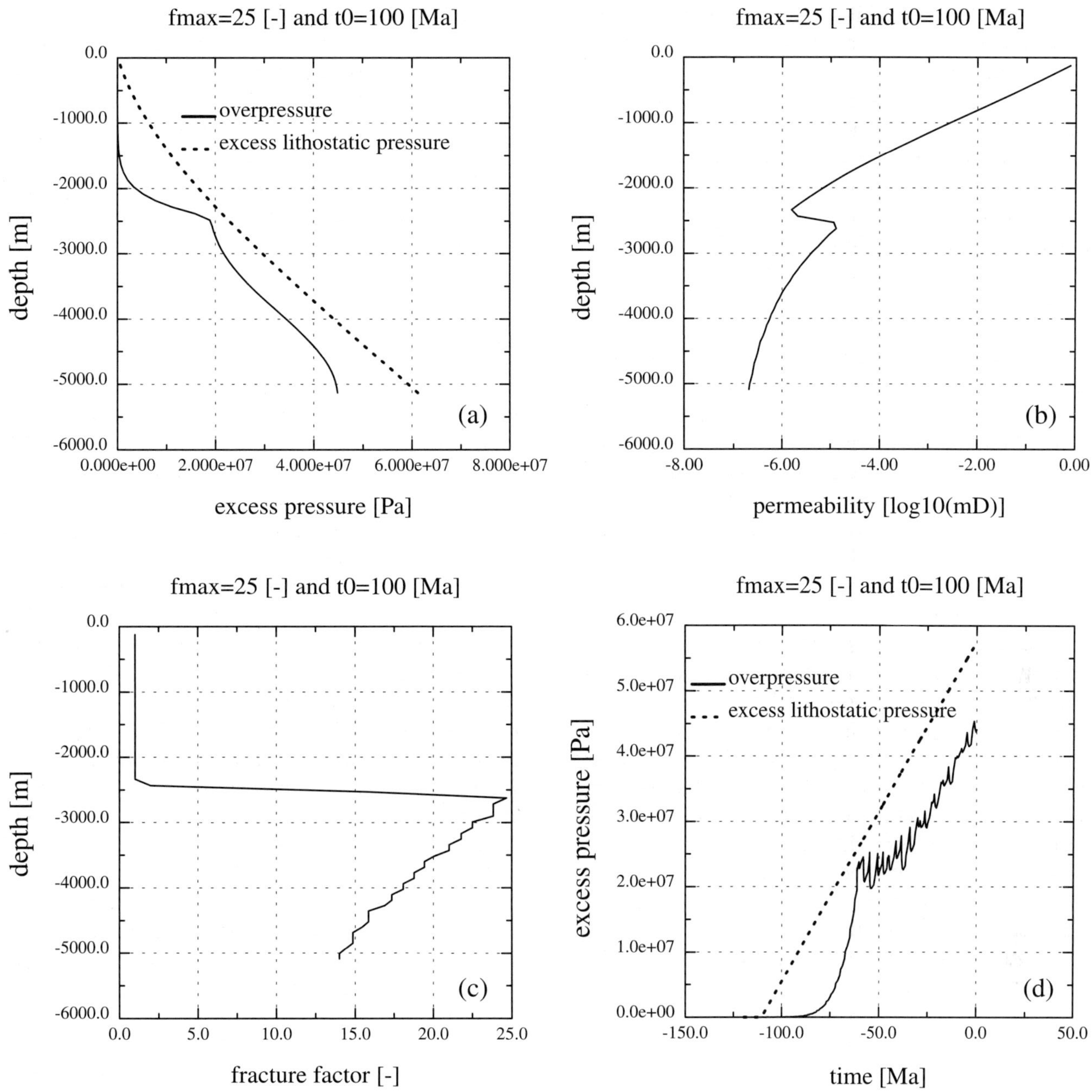

Fig. 7. (a) The overpressure as a function of depth at the present time for case 2. (b) The permeability as a function of depth at present time for case 2. (c) The fracture factor as a function of depth at present time for case 2. The fracture factor is decreasing below its maximum value at the base of the seal, which indicates that the cells hydrofractures only once. (d) The overpressure through time for a cell close to the base of the sedimentary column in case 2.

enhanced most. A cell will therefore experience two fracture events in this case, if it is buried deeply enough, and that is at the base of the two seals. A large maximum fracture factor implies that there is a noticeable bulk permeability enhancement although the half life for fracture sealing is the same as in case 1. The overpressure in a cell close to the base of the basin is plotted in Fig. 8d. The overpressure is now fluctuating more than in case 1 (as seen from Fig. 6d), because a fracture event changes the a cells bulk permeability with a factor $f_{\max} = 250$, which is one order of magnitude larger than in case 1.

Fracture permeability of seals

The fluid flux (Darcy velocity) generated by porosity reduction during burial and deposition is shown in Fig. 9. The porosity is a function of depth, where depth is given as net (porosity free) sediments. The Darcy velocity, v_D, in this case is given by the following simple and exact expression

$$v_D = \omega \, (e - e_{bot}) \tag{8}$$

where e is the void ratio and where e_{bot} is the void ratio at the base of the sedimentary column (Wangen,

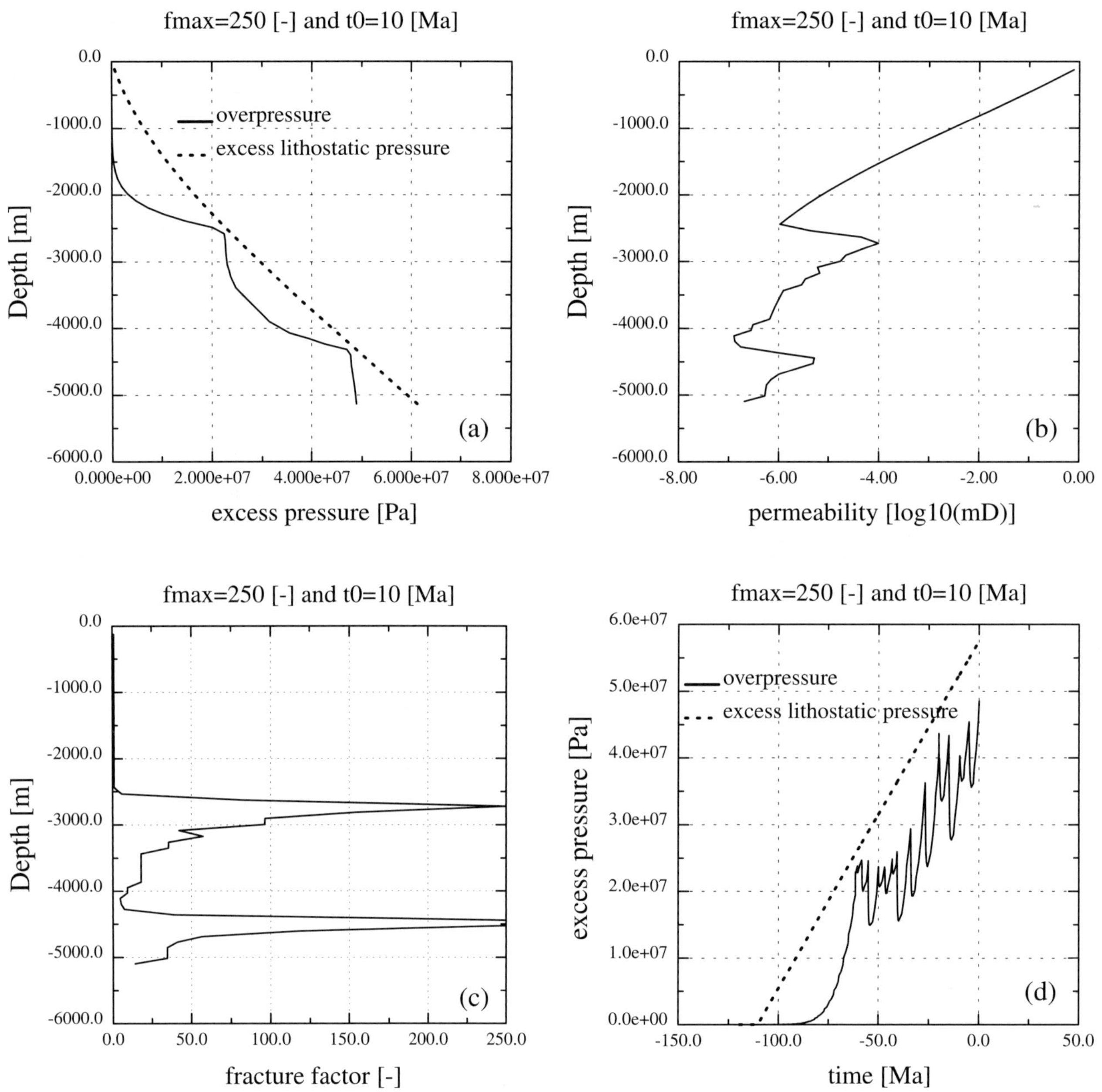

Fig. 8. (a) The overpressure as a function of depth at the present time for case 3. The overpressure is approaching the excess lithostatic pressure at two depth positions. (b) The permeability as a function of depth at present time for case 3. (c) The fracture factor as a function of depth at present time for case 3. The permeability is enhanced most below the two depth positions where the overpressure is close to the excess lithostatic pressure. (d) The overpressure through time for a cell close to the base of the sedimentary column in case 3.

1997). The Darcy velocity is $v_D \approx \omega e$ when the basin is sufficiently deep for $e_{bot} \approx 0$. The Darcy velocity is $v_D \approx \omega$ towards the surface if the surface void ratio is $e \approx 1$ ($\phi_0 \approx 0.5$). Expression 8 is usually a good estimate for the Darcy velocity, even if the porosity is not an explicit function of depth.

The Darcy velocity is proportional to the gradient of the overpressure in 1-D when the basin surface is at hydrostatic pressure. This is written

$$v_D = -\frac{k}{\mu}\frac{\partial p_e}{\partial z} \tag{9}$$

where p_e is the overpressure, k is the permeability and μ is the viscosity. An expression for the gradient $\partial p_e / \partial z$ is obtained from expression 8

$$\frac{\partial p_e}{\partial z} = \frac{\mu}{k}\omega(e - e_{bot}) \tag{10}$$

The overpressure is bounded above by the excess lithostatic pressure, $p_{ex,lith}$, and the gradient of the excess lithostatic pressure can be written

$$\frac{\partial p_{ex,lith}}{\partial z} = (\rho_s - \rho_w)(1 - \phi)g \tag{11}$$

where ρ_s and ρ_w are the densities of solid and fluid, respectively. The gradent of the overpressure must be greater than $p_{ex,lith}$ to form a pressure seal. (See

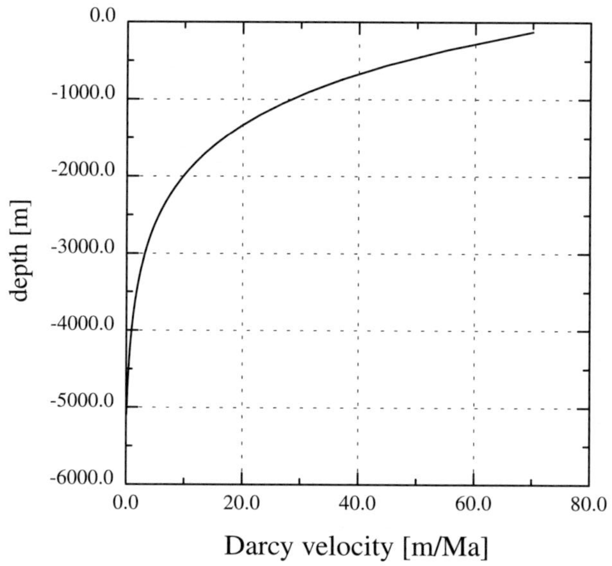

Fig. 9. The Darcy velocity as a function of depth at present time. The Darcy velocity is caused by expulsion of fluid given a porosity–depth trend as shown in Fig. 4, when burial is at a constant rate 33.3 m/Ma measured as net (porosity free) rock. Notice that this Darcy flux is the same for all cases regardless of hydrofracturing (see more explanation in the text).

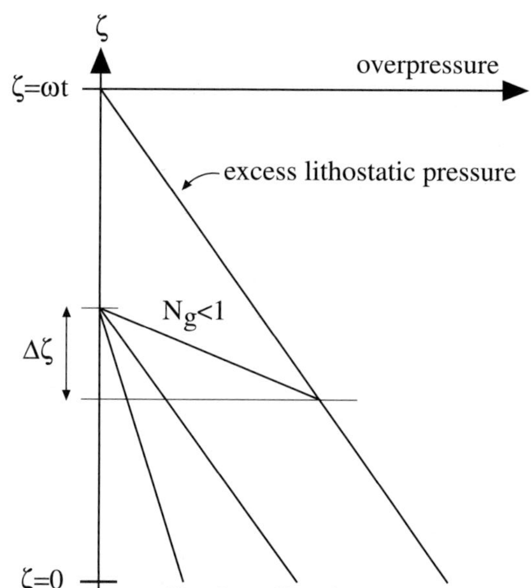

Fig. 10. The overpressure across a seal is shown for three cases of the gravity number N_g. The vertical coordinate denoted ζ is related to z by $d\zeta = (1-\phi)\,dz$, and measures the net amount of sediments. The overpressure gradient $\partial p_e/\partial\zeta$ is larger than $\partial p_{ex,lith}/\partial\zeta$ when $N_g < 1$. These gradients are equal when $N_g = 1$, and $\partial p_e/\partial\zeta < \partial p_{ex,lith}/\partial\zeta$ when $N_g > 1$.

Fig. 10, where the $\partial p_e/\partial z$ is plotted when it is larger, equal to and less than $\partial p_{ex,lith}/\partial z$.) An overpressure gradient being equal to $\partial p_{ex,lith}/\partial z$ is therefore a natural choice for a reference overpressure gradient. If the overpressure gradient, Eq. 10, is scaled with the reference gradient (Eq. 11) the condition for high overpressure (or an overpressured seal) is

$$\frac{\partial p_e/\partial z}{(\rho_s-\rho_w)(1-\phi)g} > \frac{1}{(1-\phi)N_g} \tag{12}$$

where the gravity number N_g is

$$N_g = \frac{k(\rho_s-\rho_w)g}{\mu\omega(e-e_{bot})} \tag{13}$$

The condition for a pressure seal can then be written in terms of the gravity number as $N_g < 1$. This is shown in Fig. 10 where the depth is plotted as net (porosity free) sediments, denoted ζ.

It is interesting to see what the gravity number really is in the cases already presented. These gravity numbers are shown in Fig. 11. Common for all three cases is a gravity number that starts out as a large value close to 10^4. The gravity number stays above 1 in the upper 2 km of the basin which is hydrostatic. It is seen that N_g decreases from little above 1 to a little below 1 in the depth interval of the seal. N_g then stays a little above 1 for the hydrofractured depth interval below the seal. This observation that $N_g \approx 1$ in the depth interval of hydrofractured sediments can be used to make an estimate for the average permeability of hydrofractured sediments during burial

$$k_{frac} \approx \frac{\mu\omega(e-e_{bot})}{(\rho_s-\rho_w)g} \tag{14}$$

This estimate relates the fracture permeability to the burial rate. The burial rate is a parameter which can vary quite a bit from a value close to zero up to 1000 m/Ma. The value of the difference $e - e_{bot}$ can be a small value if we are close to the base of the basin. It should be noted that the estimate for the fracture permeability applies only if the permeability is so low that the gravity number N_g is less than one.

A similar gravity number

$$N = \frac{k(\rho_s-\rho_w)g}{\mu\omega} \tag{15}$$

is shown to appear as the relevant parameter to characterize pressure build-up during burial in several basic models (Audet and Fowler, 1992; Wangen, 1992, 1997, 2000). There are also two analytical solutions for pressure build-up during constant burial. One is the Gibson solution where the porosity is decreasing linearly with increasing effective stress (Gibson, 1958; Wangen, 1992, 1993), and the other solution is based on porosity as an explicit given function of depth (Wangen, 1997). These analytical solutions, which both have N as the only parameter, behave according to the interpretation of N_g given above. The solutions yield fluid pressure comparable to the lithostatic pressure for $N \ll 1$ and fluid pressures close to hydrostatic for $N \gg 1$.

The gravity number shows the relative importance of both the burial rate and the permeability. There

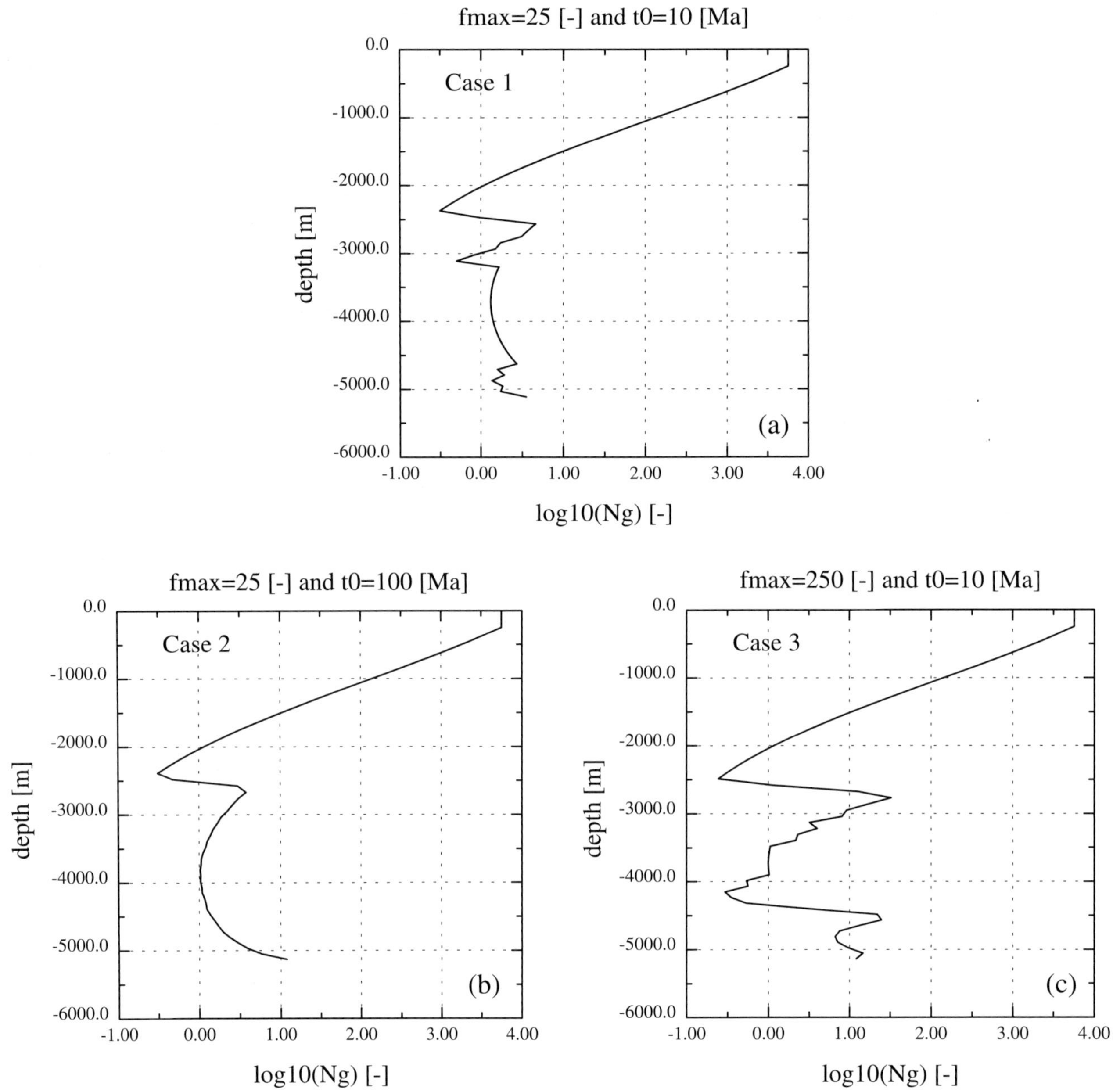

Fig. 11. The gravity number as a function of depth at present time for (a) case 1, (b) case 2, and (c) case 3.

must be burial to have overpressure build-up. Furthermore, the permeability must be sufficiently low if pressure build-up can be comparable with the lithostatic pressure.

Summary

Natural hydrofracturing is studied in a model where overpressure build-up is due to expulsion of fluid from porosity reduction. The porosity is assumed given as a function of depth. This is the case for compaction processes controlled by temperature during constant burial with a constant thermal gradient through the geohistory. There is no feedback from the pressure build-up to the expulsion process. There is therefore nothing that prevents the fluid pressure from exceeding the lithostatic pressure, except for hydrofracturing. This is different from models based on compaction controlled by effective vertical stress. In the latter models the compaction and fluid expulsion is reduced, and therefore also the overpressure build-up, by increasing fluid pressure.

Three aspects to the hydrofracture process are needed in a model for overpressure build-up: (1) fracture criterion, (2) permeability enhancement, (3) fracture sealing. Hydrofracturing is triggered by fluid pressures larger than a threshold value, which is given as a fraction of the lithostatic pressure, and it leads to an instant larger bulk permeability. The permeability enhancement is given as a factor by which the unfractured permeability is multiplied. The fracture factor is allowed to increase in steps up to a maximum value called the maximum fracture factor. The fracture permeability (due to a fracture factor larger than 1) will not last forever after a fracture event. It decays towards 1 exponentially with time by a given half life.

Hydrofracturing is studied in three cases which are different with respect to the maximum fracture factor and the half life for fracture healing. The cases show that a large maximum fracture factor gives larger amplitude in the oscillatory fluid pressure than a smaller maximum fracture factor. A large fracture factor implies also that the permeability enhancement lasts longer than for a small maximum fracture factor given the same half life for fracture sealing. Therefore, refracturing occurs less frequently with a large max fracture factor than a small max fracture factor. The half life for fracture sealing also controls pressure oscillations. A long half life produces smaller amplitudes in the pressure oscillations than a short half life, because the permeability enhancement lasts longer, which also reduces the need for refracturing.

The build-up of overpressure is characterized by a gravity number which is proportional to the burial rate and inverse proportional to the permeability. Hydrostatic conditions correspond to a gravity number larger than one, and overpressure comparable to the lithostatic pressure corresponds to a gravity number much less than one. Seals are places where the pressure goes from low (hydrostatic) to high (lithostatic) values. These depth intervals are characterized by gravity numbers decreasing from a value larger than one to a value less than one at the base of the seal. Furthermore, the numerical case studies show that hydrofractured depth intervals are characterized by a gravity number close to 1. This observation is used to derive an estimate for the permeability of hydrofractured sediments.

An important result in this study is that the fluid pressure is intermittent during hydrofracturing, but not the fluid flow. This is a direct consequence of the compaction model used, where the porosity is controlled by depth. Hydrostatic compaction of the upper unlithified sediments and the chemical compaction controlled by temperature of the deeper lithified sediments can be represented by a porosity–depth curve. There is therefore no episodic fluid flow caused by hydrofracturing in this model, because the porosity is not a function of the pressure. However, the fluid pressure shows oscillations due permeability changes caused by hydrofracture events.

References

Aase, N.E., Bjørkum, P.A. and Nadeau, P., 1996. The effect of grain coating micro-quartz on preservation of reservoir porosity. Am. Assoc. Pet. Geol. Bull., 80 (10): 1654–1673.

Athy, L.F., 1930. Density, porosity, and compaction of sedimentary rocks. Bull. Am. Assoc. Pet. Geol., 14: 1–24.

Audet, D.M., 1995. Mathematical modelling of gravitational compaction and clay dehydration in thick sediment layers. Geophys. J. Int., 122: 283–298.

Audet, D.M., 1996. Compaction and overpressure in Pleistocene sediments on the Louisiana Shelf, Gulf of Mexico. Mar. Pet. Geol., 13: 467–474.

Audet, D.M. and Fowler, A.C., 1992. A mathematical model for compaction in sedimentary basins. Geophys. J. Int., 110: 577–590.

Bethke, C.M. and Corbet, T.F., 1988. Linear and nonlinear solutions for one-dimensional compaction flow in sedimentary basins. Water Resour. Res., 24 (3): 461–467.

Birchwood, R.A. and Turcotte, D.L., 1994. A unified approach to geopressuring, low-permeability zone formation, and secondary porosity generation in sedimentary basins. J. Geophys. Res., 99 (B10): 20051–20058.

Bjørkum, P.A., 1994. How important is pressure in causing dissolution of quartz in sandstones? (abstr.) Am. Assoc. Pet. Geol., Annu. Meet. Progr. Abstr., 3: 105.

Bjørkum, P.A., 1996. How important is pressure in causing dissolution of quartz in sandstones? J. Sediment. Res., 66 (1): 147–154.

Bjørlykke, K., 1999a. Principal aspects of compaction and fluid flow in muds and mudstones. In: A.C. Aplin, A.J. Fleet and J.H.S. Macquaker (Editors), Physical and Fluid Flow Properties. Geol. Soc. London, Spec. Publ., 158: 73–78.

Bjørlykke, K., 1999b. An overview of factors controlling rates of compaction, fluid generation and flow in sedimentary basins. In: B. Jamtveit and P. Meakin (Editors), Growth, Dissolution and Pattern Formation in Geosystems. Kluwer, Dordrecht, pp. 381–404.

Bjørlykke, K. and Høeg, K., 1997. Effects on burial diagenesis on stress, compaction and fluid flow in sedimentary basins. Mar. Pet. Geol., 13: 267–276.

Bredehoeft, J.D. and Hanshaw, B.B., 1968. On the maintenance of anomalous fluid pressures, I. Thick sedimentary sequences. Geol. Soc. Am. Bull., 79: 1097–1106.

Colton-Bradley, V.A.C., 1987. Role of pressure in smectite dehydration — effects on geopressure and smectite-to-illite transition. Am. Assoc. Pet. Geol. Bull., 71: 1414–1427.

Connolly, J.A.D. and Podladchikov, Y.Y., 1998. Compaction-driven fluid flow in viscoelastic rock. Geodin. Acta, 11: 55–84.

Darby, D., Haszeldine, R.S. and Couples, G.D., 1996. Pressure cells and pressure seals in the UK central graben. Mar. Pet. Geol., 13 (8): 865–878.

Davis, G.H. and Reynolds, S.J., 1996. Structural Geology of Rocks and Regions. 2nd edition, Wiley, Chichester, pp. 245–251.

Doligez, B., Burrus, J. and Ungerer, P., 1988. Hydraulic fracturing during basin scale fluid migration: An integrated approach. In: B. Hitchon and S. Bachu (Editors), Proceedings 4th Canadian/American Conference on Hydrogeology. National Water Well Association, Dublin, OH, pp. 251–259.

Engelder, T., 1993. Stress Regimes in the Lithosphere. Princeton University Press, Princeton, NJ, pp. 131–170.

Fowler, A.C. and Yang, X.-S., 1999. Pressure solution and viscous compaction in sedimentary basins. J. Geophys. Res., 104 (B6): 12989–12997.

Gaarenstroom, L., Tromp, R.A.J., de Jong, M.C. and Brandenburg, A.M., 1993. Overpressure in the Central North Sea: implications for trap integrity and drilling safety. In: J.R. Parker (Editor), Petroleum Geology of Northwest Europe: Proceedings of the 4th Conference. Geological Society of London, London, pp. 1305–1313.

Gibson, R.G., 1958. The progress of consolidation in a clay layer increasing in thickness with time. Geotechnique, 8: 171–182.

Hubbert, M.K. and Rubey, W.W., 1959. Role of fluid pressure in mechanics of overthrust faulting, Part 1. Geol. Soc. Am. Bull., 70: 115–166.

Leonard, R.C., 1993. Distribution of subsurface pressure in the Norwegian Central Graben and applications for exploration. In: J.R. Parker (Editor), Petroleum Geology of Northwest Europe: Proceedings of the 4th Conference. Geological Society of London, London, pp. 1295–1303.

Lerche, I., 1990. Basin Analysis. Quantitative Methods, Vol. 1. Academic Press, London, pp. 208–245.

L'Heureux, I. and Fowler, A.D., 2000. A simple model of flow patterns in overpressured sedimentary basins with heat transport and fracturing. J. Geophys. Res., 105 (B10): 23741–23752.

Lou, X. and Vasseur, G., 1992. Contribution of compaction and aquathermal pressuring to geopressure and the influence of environmental conditions. Am. Assoc. Pet. Geol. Bull., 76: 1550–1559.

Lou, X. and Vasseur, G., 1996. Geopressuring mechanism of organic matter cracking: Numerical modeling. Am. Assoc. Pet. Geol. Bull., 80: 856–874.

Oelkers, E.H., Bjørkum, P.A. and Murphy, W.M., 1996. A petrographic and computational investigation of quartz cementation and porosity reduction in North Sea sandstones. Am. J. Sci., 296: 420–452.

Osborne, M.J. and Swarbrick, R.E., 1997. Mechanisms for generating overpressure in sedimentary basins: A reevaluation. Am. Assoc. Pet. Geol. Bull., 81 (6): 1023–1041.

Roberts, S.J. and Nunn, J.A., 1994. Episodic fluid expulsion from geopressured sediments. Mar. Pet. Geol., 12: 195–204.

Schneider, F., Potdevin, J.L., Wolf, S. and Faille, I., 1996. Mechanical and chemical compaction model for sedimentary basin simulators. Tectonophysics, 263: 307–317.

Sharp, J.M. and Domenico, P.A., 1976. Energy transport in thick sequences of compacting sediment. Geol. Soc. Am. Bull., 87: 390–400.

Shi, Y. and Wang, C.-Y., 1986. Pore pressure generation in sedimentary basins, overloading versus aquathermal. J. Geophys. Res., 91 (B2): 2153–2162.

Skempton, A.W., 1970. The consolidation of clays under gravitational compaction. Q. J. Geol. Soc. London, 125: 373–409.

Smith, J.E., 1971. Dynamics of shale compaction and evolution of pore-fluid pressure. Math. Geol., 3 (3): 239–263.

Terzaghi, K., 1943. Theoretical Soil Mechanics. Wiley, New York, NY.

Tzschicholz, F. and Wangen, M., 1999. Modelling of hydraulic fracturing of porous materials. In: M.H. Aliabadi (Editor), Fracture of Rock. Computational Mechanics Publications, WITpress, Southampton, pp. 227–260.

Walderhaug, O., 1994a. Temperatures of quartz cementation in Jurassic sandstones from the Norwegian continental shelf — evidence from fluid inclusions. J. Sediment. Res., A64 (2): 311–323.

Walderhaug, O., 1994b. Precipitation rates for quartz cement in sandstones determined by fluid-inclusion microthermometry and temperature history modeling. J. Sediment. Res., A64 (2): 324–333.

Wang, C.-Y. and Xie, X., 1998. Hydrofracturing and episodic fluid flow in shale-rich basins — A numerical study. Am. Assoc. Pet. Geol. Bull., 82 (10): 1857–1869.

Wangen, M., 1992. Pressure and temperature evolution in sedimentary basins. Geophys. J. Int., 110: 601–613.

Wangen, M., 1993. A finite element formulation in Lagrangian coordinates for heat and fluid flow in compacting sedimentary basins International. J. Numer. Anal. Methods Geomech., 15: 705–733.

Wangen, M., 1997. A simple model for overpressure build-up. Geophys. J. Int., 130: 757–764.

Wangen, M., 2000. Generation of overpressure by cementation of pore space in sedimentary rocks. Geophys. J. Int., 143: 608–620.

M. WANGEN *Institute for Energy Technology, P.O. Box 40, N-2027 Kjeller, Norway*
E-mail: magnus@ife.no

Geomechanical simulations of top seal integrity

Helen Lewis, Peter Olden and Gary D. Couples

The integrity of the top seal of a reservoir may be compromised when that sealing layer is deformed, either as a consequence of compaction of the underlying materials, or by a tectonic event. Here we assess, using a geomechanical simulator, the stress state that can develop in a simple fold–fault system where the seal and its surrounding materials are folded, or flexed, as a consequence of incremental displacement on an underlying fault, into a gentle monoclinal shape. We have conducted an extensive set of numerical simulations to consider the effects on the resulting deformation of variations in fault displacement, of material-property contrasts between layers, and the specification of boundary conditions. The stress states within the seal and its surrounding layers are dominated by the bending process, even at very small fault displacements. All our simulations create along-seal variations in minimum principal stress (σ_3) that are a consequence of bending. These σ_3 variations can exceed 10 MPa where the seal and its enclosing shales have a strong strength contrast, but still reach 1–2 MPa when the seal and its surrounding rocks have identical properties. However, the details of the variations are strongly dependent on the specifics of the model, so quantification requires that a model be related to a particular set of properties and constraints. If our results are applicable to natural events, leak-off test (LOT) values (as a proxy for the minimum stress) that are measured in seals above reservoirs can be expected to indicate significant, deformation-induced variations over short lateral distances. By considering the timing of deformation with respect to initial hydrocarbon retention and/or overpressure, together with alteration of the seal's petrophysical properties during deformation, our results can be used to make general predictions about the sealing capacity.

Introduction

Most approaches to seal-capacity prediction focus on identifying the conditions for seal failure in situations with assumed, simple stress states. In analyses of this type, mechanical seal failure is implicitly associated with an increased permeability, with a decrease in entry pressure, or with both. Simple cases include prediction of a Mohr–Coulomb type failure, or of hydraulic fracture if the fluid pressure in the underlying rocks increases (assuming classical effective stress). Grauls (1998) advances this idea by considering the effects of assuming different far-field stress states. On the other hand, Couples (1999) highlights the possibility that failures in seal-like rocks can occur in a compactional mode, potentially increasing sealing capacity. This point of view emphasises the changes in seal capacity that can occur as fluid pressure and stress states evolve during basin development (beyond initial compaction). Clearly, it is necessary to consider both the failure mode and the post-failure behaviour of seals to make a robust prediction of seal capacity during the basin history. More detailed studies of the mechanics of seal deformation are required to fully understand the conditions that lead to seal failure, and the (positive or negative) changes in seal capacity that occur as the seal deforms.

In this paper we consider a specific case in which a top seal is deformed above a reactivated fault that bounds a reservoir block (Fig. 1). The results indicate that important lateral and vertical variations in stress magnitude commonly occur in such systems, and should be incorporated in strategies for predicting sealing. We use geomechanical simulations to assess the evolving stress distributions in and around a seal layer, and consider the implications for top-seal integrity.

A 2D geomechanical modelling approach is used to simulate the deformation of sealing shales above a lower, already-faulted reservoir. Our 2D model represents a vertical cross-section 10 km long and 2.5 km

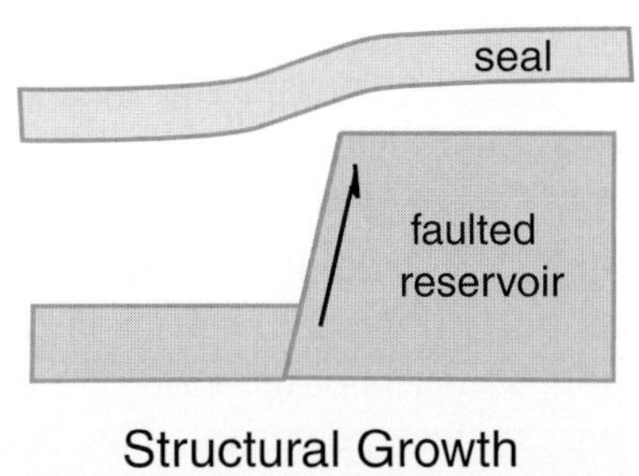

Fig. 1. Cartoon cross-section illustrating a simple flexural deformation affecting a seal layer above a faulted reservoir undergoing renewed uplift.

Hydrocarbon Seal Quantification edited by A.G. Koestler and R. Hunsdale. NPF Special Publication 11, pp. 75–87, Published by Elsevier Science B.V., Amsterdam.

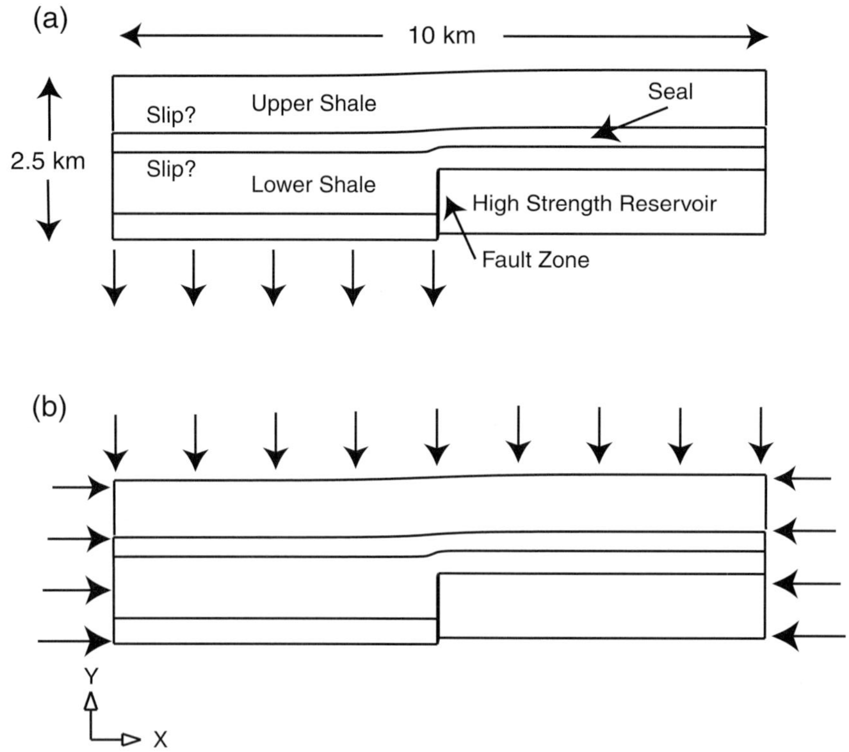

Fig. 2. (a) Geometry of geomechanical model, showing dimensions, distribution of rock types and imposed displacement of downthrown reservoir block. For details of lateral displacement constraints, see text. (b) Stress boundary conditions imposed on models. 10 MPa on top representing overburden, and loads due to side-burdens equal to 10 MPa plus $\rho g z$.

high (Fig. 2). The cover package above the reservoir consists of a lower shale unit, a 200 m thick seal unit, and an upper shale unit. The lower shale is thicker on the downthrown side of the vertical fault, indicating that the seal and upper shale intervals (of uniform thickness) were deposited well after the major structural events that created the reservoir structure. This trap geometry and geohistory are similar to that of many North Sea fields, and fields in other basins.

We use a finite-element, continuum approach (VISAGE^{TM}) to calculate the stress and strain histories of the model when another, later, faulting event (in the form of uniform downward movement of the downthrown block) is imposed. This loading has an obvious correlation to natural events in which a trap may undergo later rejuvenation. The loading also approximates (for the seal layer, at least) the deformation that might occur if the thick lower shale experiences compaction, with more subsidence of the thicker rocks on the downthrown side of the fault.

Our series of simulations investigate the degree to which variations in finite strain, material properties and layer contrasts influence stress and deformation state evolution. Two sets of models are used to reveal the active mechanical processes. The first set is a general scoping study to identify those parameters that have significant effects on the simulated stress and deformation state evolution within the models. The second set focuses on the effects of material property variations.

In order to study the effects of finite permanent deformation, we use an elastic–plastic material description in which the yield conditions are determined by a Mohr–Coulomb failure curve. Post-failure (plastic) behaviour of the materials can be either neutral-hardening or strain-hardening. Strain-softening is another possible post-failure mode, but this behaviour requires a quite different numerical approach to include in simulations, and we did not pursue that avenue here. Suites of material properties for the seal and the two shale units represent a range of scenarios that are discussed later. For most of the comparisons shown in this paper, we imposed a tension cut-off on the failure curve to prevent tensile stresses developing. This choice has only a minor impact, since the region of tension allowed under the Mohr–Coulomb curve is quite small anyway.

Note that we do not include poro-elastic or poro-plastic effects in the simple continuum models described here. As a result of this choice we must post facto associate the bulk behaviours of the materials, and the consequent stress states, with likely fluid-pressure and fluid-flow behaviours. Changes in seal capacity during deformation are not explicitly predicted by the model and must be inferred (see below).

TABLE 1

Mechanical properties used in modelling

	Cohesion (S_o) for M–C failure (MPa)	Friction angle (ϕ) for M–C failure (MPa)
Weak	2	15
Intermediate	5	20
Strong	10	25

Refer to Fig. 3 for definitions. Information derived from Horsrud et al. (1998).

The models

Setting up appropriate models requires both knowledge of the important elements of the problem to be simulated, and making choices of material properties, and the initial and boundary conditions that approximate the real problem as best as possible. In practice, simplifications must be made. Here we are concerned with deformation, particularly the minimum principal stress evolution during finite, permanent straining, of a seal layer contained in a shale sequence overlying a reactivated vertical fault. The initial and boundary conditions (including the applied displacements to produce fault incremental motions) are relatively simple. The main focus is the behaviour of both the seal layer and its surrounding shales. The details of the reservoir's deformation are not considered here.

Material parameters of the model

In most of the models that we have considered, we differentiate the seal layer from its enclosing shales. However, all our seal and shale material descriptions fit within the 'shales' category sensu lato and are discussed as a group here. We have selected a simplified isotropic, elastic–plastic material description that is just complex enough to capture the first-order behaviours of these materials. We model the reservoir that underlies the shales and seal as a purely elastic material. Fig. 3 summarises the essential elements of this description, derived from laboratory results reported by Horsrud et al. (1998), who investigated muddy rocks derived from North Sea settings (see also Table 1). Our shales are assumed to behave elastically upon initial loading, and to deform permanently once the 'yield stress' is reached.

We have considered two post-yield failure rules that specify the degree to which strength properties change during permanent deformation. These are strain-neutral, which assumes no change of load capacity with increased strain, and strain-hardening, where increased load-carrying capacity develops with increased strain. Strain-hardening material descriptions are appropriate to describe materials undergoing compaction and its accompanying porosity loss. The greater solidity that is associated with compaction can lead to a stronger material that requires incrementally more energy to cause more deformation. Strain-neutral materials can be thought of as deforming in a condition much like that envisaged in critical-state mechanics (Jones and Addiss, 1986; Couples, 1999). By varying both the yield stress and the 'hardening' laws we successfully capture a wide range of mechanical responses that we believe represent key behaviours of natural top seals.

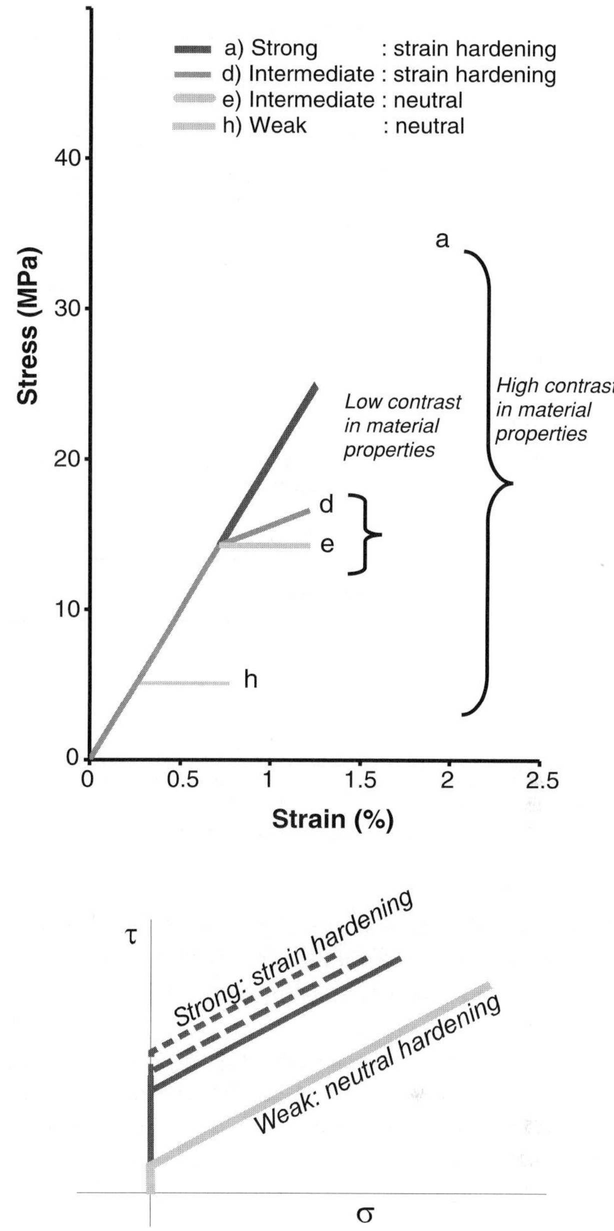

Fig. 3. Elastic–plastic behaviours of materials used in simulations. Top shows stress–strain responses of four materials, including post-yield hardening characteristics. Lower shows tension cut-off applied to the Mohr–Coulomb yield criterion. See Table 1 for the parameters of muddy rocks derived from Horsrud et al. (1998).

Initial and boundary conditions

For the models we report in this paper, we assume an initial isotropic stress state with the vertical stress increasing with depth due to body forces (i.e. gravity). We impose a vertical normal stress of 10 MPa on the top of the model to represent overburden loads. The density assigned to the materials is approximately half of the true density to represent effective stress values if the pore fluids are hydrostatically pressured. We hold the right-side reservoir unit fixed while moving the left-side reservoir unit down. Total imposed (downward) displacements range up to 250 m and are applied incrementally. We allow the ends of the model to move vertically in all cases. Horizontal (outward or inward) movement of the ends was allowed only where layer-parallel slip was introduced. We assume plane-strain in all models discussed here. The top of the model is free to move in both x- and z-directions.

Layer-parallel slip has been observed in a large number of natural equivalents of the model (e.g. Couples et al., 1998) and is incorporated into some of the model sets. Studies of mechanically effective layer-parallel slip (Couples et al., 1998; Couples and Lewis, 2000) suggest that its presence produces changes in fold shape, in the stress state developed, and in the type and distribution of deformation — as compared to cases without layer-parallel slip. An approximation to frictional layer-parallel slip is introduced into some of our models by adding a very thin (less than 1 m), weak layer on the top and bottom of the seal layer, between it and the adjacent shale units (Fig. 2). Simple-shear deformation of these thin layers enables a localisation of deformation that is somewhat like layer-parallel slip (while retaining the continuum approach adopted here). However, only small amounts of layer-parallel motion are permissible when using this approximation.

The simulations

The simulation results for all our models have a number of similarities, including the universal development of a bending stress state. We first present the results, including the bending stress state, for a simple 'base case' to illustrate these common responses. We then consider in more detail the responses of the models to changes in material properties. The results shown in the accompanying figures are, as appropriate, the magnitudes of maximum principal stress σ_1, minimum principal stress σ_3, vertical stress σ_v, and VISAGE^{TM} failure modes.

The base case

Recognition of the bending stress state is fundamental to interpreting the results of all of our models. Bending stress state evolution and characteristics are addressed first. This understanding then provides the framework to assess the effects, on the resultant deformation, of several factors important in model design: (1) changes in the magnitude of faulting and 'rate' of faulting; (2) presence or absence of layer-parallel slip either side of the seal; and (3) changes in model constraints. Material properties, which also have a significant effect on the simulation results, are covered in the 'Detailed comparisons' section below.

Bending stress state

The stress state that evolves at the smallest fault displacement in our simulations is predominantly that of bending, and not one that is typically associated with continued faulting. The bending stress state pattern is characteristic of all of our models at all fault displacements, with the possible exception of the highest (250 m) fault displacements where faulting-associated deformation becomes more obvious.

A flexural deformation creates a spatially variable stress state, or bending stress state (Figs. 7 and 8). Characteristically, a bending stress state exhibits a pattern in which the maximum principal stresses are at high angles to the layering in the outer arcs of the antiform and synform, while the inner arcs have maximum principal stresses that are sub-parallel to the layering (Fig. 4). The stress trajectories curve both within and between these domains. Note that this state is quite different from that which is imagined for the concept of folding in which it is assumed that a neutral surface exists within the layer (see discussion in Couples et al., 1998). This bending stress state can be seen, with variations, in all of our simulation results.

Magnitude and 'rate' of faulting

A series of simulations were run with progressively larger imposed fault displacements. Fig. 5 shows the minimum principal stress (σ_3) at 50 m and 200 m fault displacements when the displacement increment is 5 m. The primary effects of increased fault movement are (1) to change the shape of the flexure, particularly the amplitude and curvature of the seal layer, and (2) to increase the magnitudes of differential stress throughout the model, while retaining the

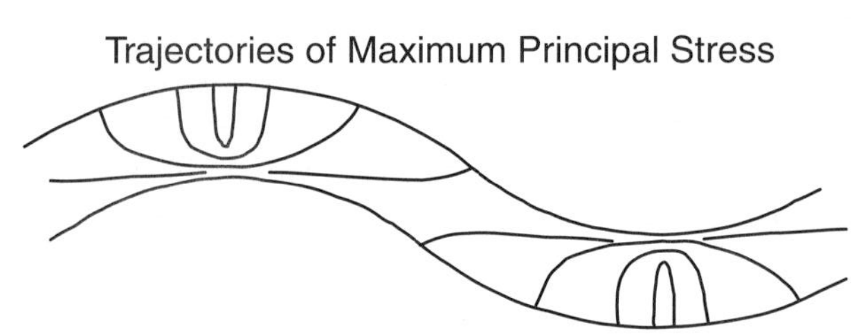

Fig. 4. Trajectories of maximum principal stress, σ_1, in a simple flexure (after Couples et al., 1998).

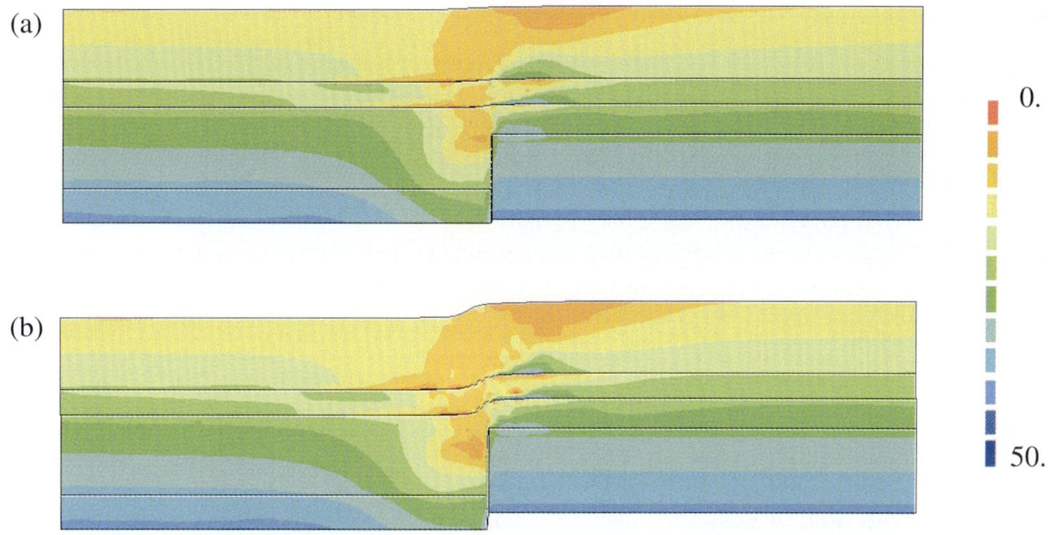

Fig. 5. Changes in stress state (here represented by plot of σ_3 magnitude) associated with increases in imposed displacement and development of flexural deformation. (a) 50 m imposed displacement. (b) 200 m imposed displacement. Values in MPa from 0 (red) to 50 (dark blue). Note that the differences due to increased deformation are only slight, with the pattern being established early in the flexural process.

pattern of the stress distributions. The distribution of failed zones (see later figures) also changes during the simulation sequence. During early stages of fault movement, the extent of the failed material increases with fault movement, although it is mostly confined to a zone above the fault. However, after the initial creation and propagation of the failed zone, it scarcely expands with increased faulting. Portions of this zone later develop stresses that drop below their yield stress, and so have returned to an elastic response.

When the fault displacement is imposed in smaller steps (e.g. 1 m instead of 5 m increments), the zone of deformation is smaller for the same accumulated amount of faulting, and the yielding 'spike' is more pronounced in the seal layer. These differences suggest that these artefacts are related to the numerical implementation, since we have not used a time-dependent rheological description. To allow easy comparison the results described in the subsequent section are all based on a displacement increment of 5 m, even though we do not know which displacement increment gives the 'right' results.

Layer-parallel slip

An approximation to frictional layer-parallel slip is introduced into some models by adding a very thin, weak layer on the top and bottom of the seal layer, between it and the adjacent shale units (Fig. 2). Fig. 6 shows minimum principal stress and strain in the x-direction with (a and c), and without (b and d), layer-parallel slip. These results are typical of virtually all our simulations and show the following.

(1) With slip the upper shale, seal and lower shale each develop a separate bending stress state and associated strain state. There are sharp stress gradients approaching the layer boundaries, and changes from layer-parallel to layer-normal orientations of the maximum and minimum principal stresses across these boundaries.

(2) With slip the overall stress differences and stress magnitudes are, generally, smaller than in comparable simulations without the ability to slip along the layer interfaces.

(3) The monoclinal shape is broader when there is layer-parallel slip. The broader shapes are associated with less-concentrated bending, and this relationship is thought to be responsible for the less-severe stress anomalies.

(4) These effects become more pronounced as the structural relief increases (i.e. as the fault displacement becomes larger).

The general behaviours appear robust and agree with those produced in physical models (Couples and Lewis, 2000), and as inferred from observations in the field (Couples et al., 1998).

Model constraints

Free (x-direction movement allowed) versus fixed (no x-direction movement allowed) end constraints have a discernable impact in those simulations which included our approximation to layer-parallel slip. But in simulations without the layer-parallel slip, there is virtually no difference between fixed- and free-end models. If the ends of the model had been closer to the flexure location, the choice of end constraints might have been considerably more significant. Couples and Lewis (1998) show that in laboratory-deformed rock models with a very similar

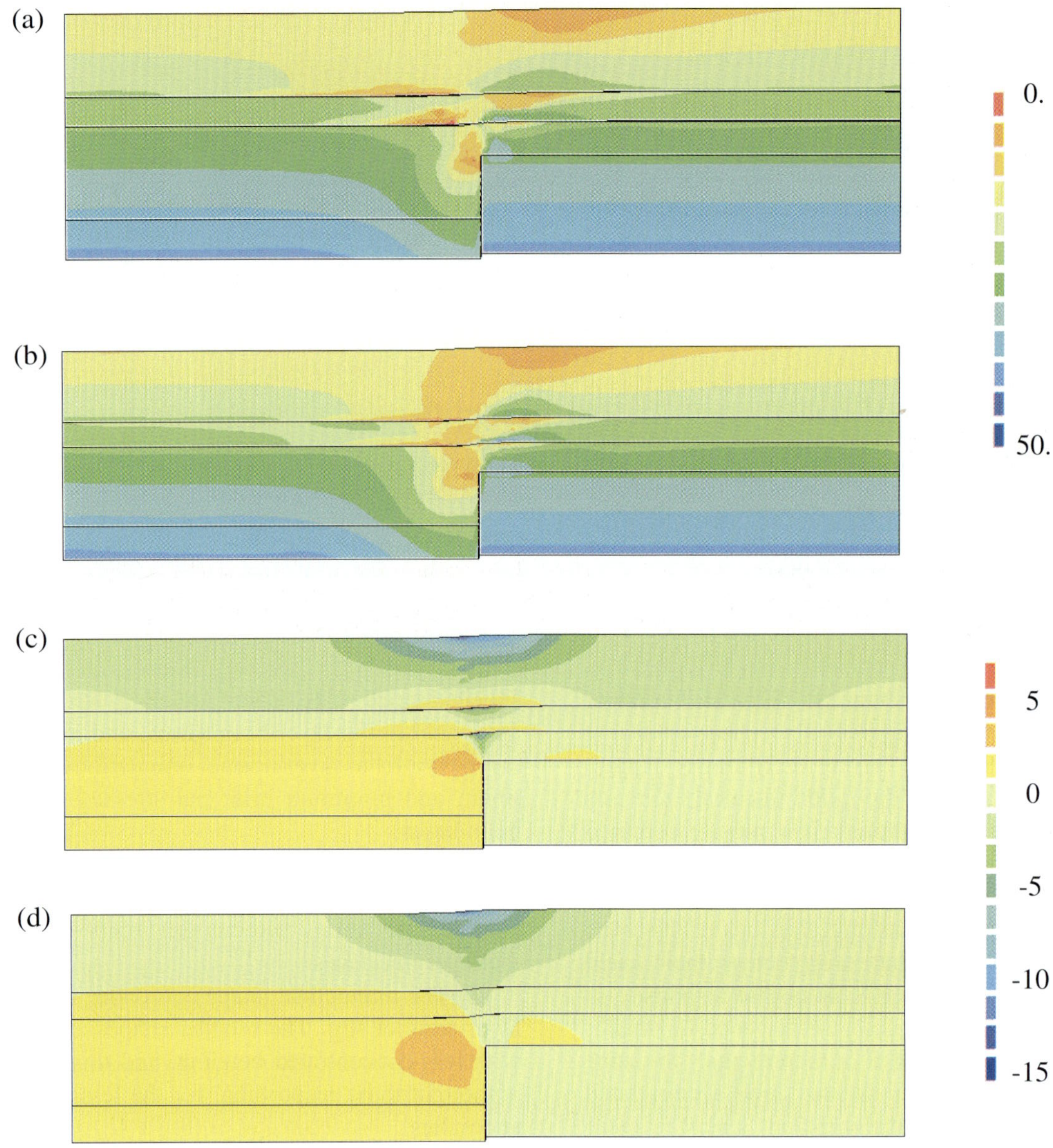

Fig. 6. Comparison of models with and without approximation to frictional layer-parallel slip. (a,b) Changes in stress state (σ_3 magnitude), (a) with slip, (b) without slip. Stress magnitudes shown in MPa, ranging from 0 (red) to 50 (dark blue). (c,d) Contours of finite strain in the x-direction, (c) with slip and (d) without slip. Key shows units of strain (expressed as %) in the x-direction, with extensional strains in blue, and shortening strains in red.

configuration to that used here, layer-parallel displacement is greatest near the flexure and decreases away from it. In natural deformations where flexures are closely spaced, the constraints that this spatial arrangement places on lateral movements could cause conflicts between the material balance requirements for each fold.

Detailed comparison of the effects of material properties

These models are used to simulate changes in the deformation when the relative strengths of the seal and shales are changed, and when the material properties assigned to the seal and shales are changed. With three layers and six material types (three yield-stress values, plus hardening or neutral behaviour for each, as illustrated in Fig. 3), the number of potential models is quite large. Here we present results from two model cases using eight arrangements of upper shale, seal and lower shale material properties. Case 1 uses two materials, one with a high yield strength and with strain-hardening, and the other with a low yield strength and strain-neutral behaviour. Case 1 captures the maximum material variability (Fig. 7). Case 2 also uses two materials, both with intermediate yield strength values but with different hardening laws. Case 2 represents the minimum variability in rock

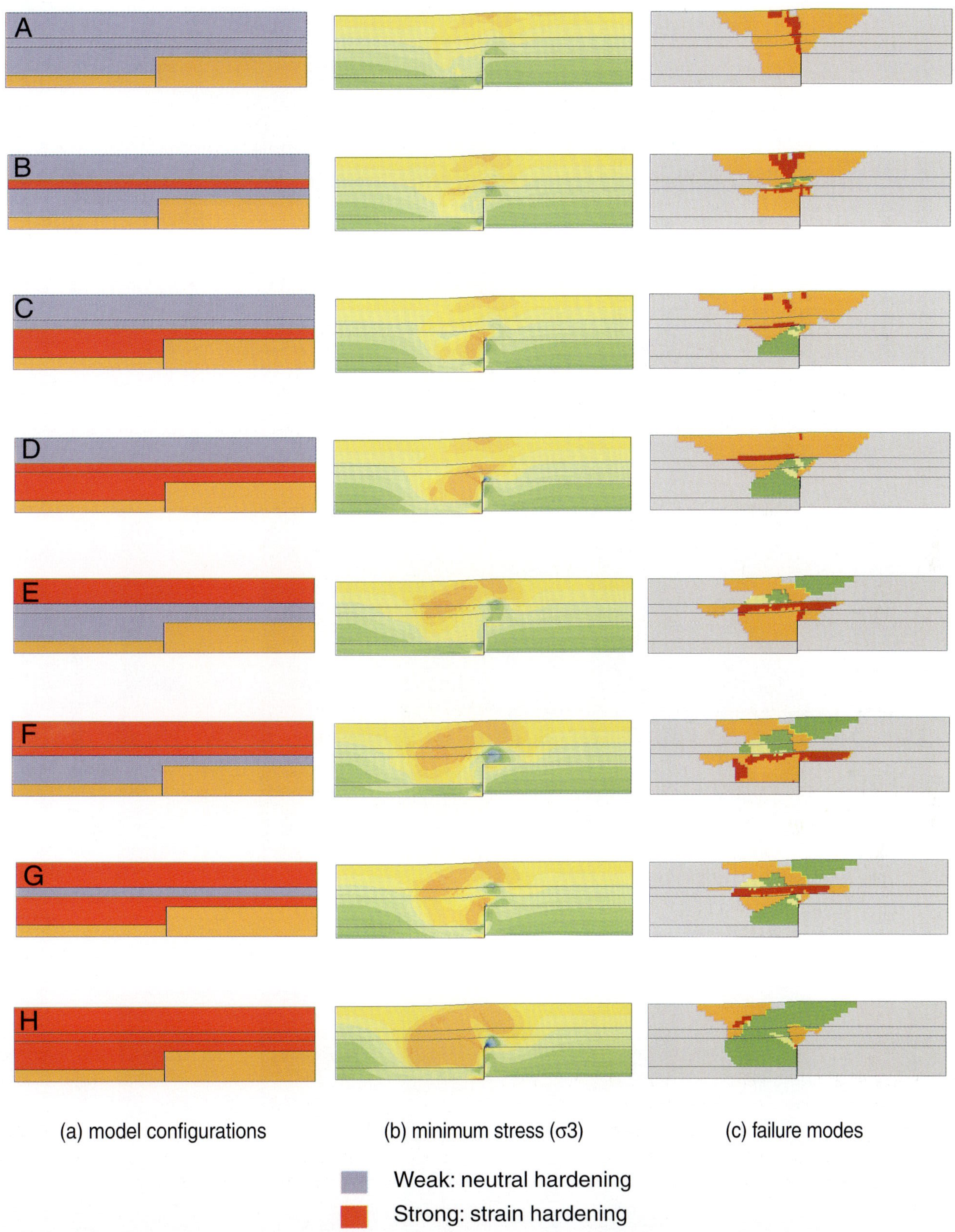

Fig. 7. Comparison of models with maximum material-property contrasts. (a) Model configurations A–H with (b) their minimum principal stress (σ_3) values, and (c) failure modes. Stress magnitude values in MPa from −10 (red, tensile) to 100 (dark blue, compressive). See Fig. 8 for explanation of failure-mode colours.

properties. Each model from these groups is loaded identically to a 50 m total displacement, achieved in ten increments of 5 m. None of these models have layer-parallel slip, and the ends cannot move in the x-direction.

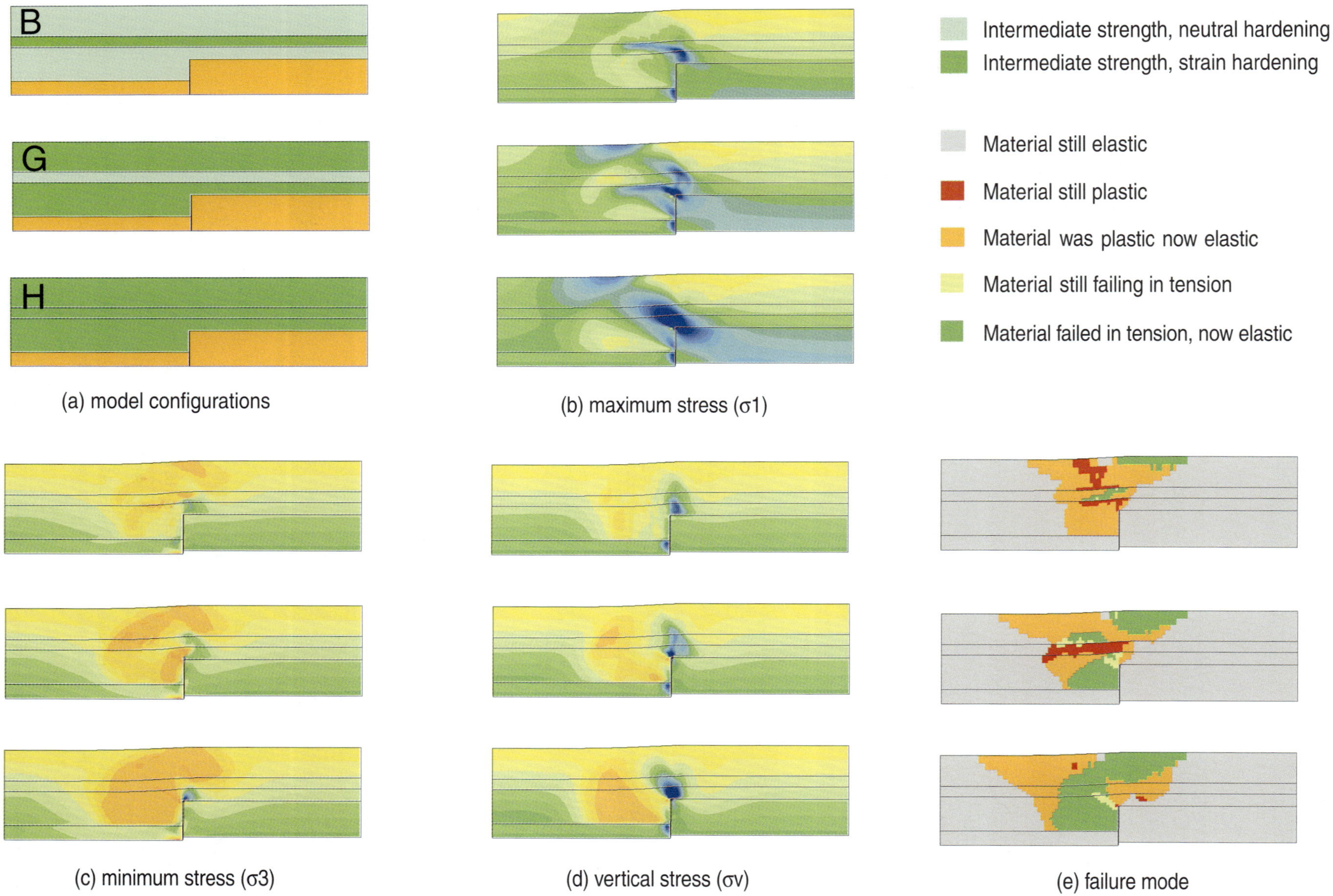

Fig. 8. Comparison of models with minimum material-property contrasts. (a) Model configurations B, G and H (which are similar to similar configurations depicted in Fig. 7); (b) maximum principal stress (σ_1) values; (c) minimum principal stress (σ_3) values; (d) vertical stress (σ_v) values; and (e) failure modes. Stress magnitude values in MPa from -10 (red, tensile) to 100 (dark blue, compressive). Failure mode colours indicating state at this stage of deformation.

The models

Each case contains eight separate models (Fig. 7):

A. Seal, and both shales of weaker material;
B. Lower shale weaker, seal stronger and upper shale weaker material;
C. Lower shale stronger, seal and upper shale weaker;
D. Lower shale and seal stronger, upper shale weaker;
E. Lower shale and seal weaker, upper shale stronger;
F. Lower shale weaker, seal and upper shale stronger;
G. Lower shale stronger, seal weaker, upper shale stronger;
H. Seal and both shales stronger.

N.B. In the text our use of the terms maximum principal stress (σ_1) and minimum principal stress (σ_3) follows the geological convention of compressive stress as positive, so σ_1 is the most compressive stress. VISAGE^{TM} uses the engineering convention (compressive stress negative). This manifests itself in the figures where the contoured compressive stresses have negative values: it has no other consequences.

Observations

The most important observation is that all models respond in a flexural sense and develop a bending stress state. The following comments refer to variations of this theme that are related to changes in material properties (yield strengths and post-failure behaviour) of the seal and shales.

(1) Within each case, the greatest differences in both stress magnitudes and in failure modes is between the end-member configurations H and A (Fig. 7). The different combinations of layer strength properties introduce perturbations into these basic stress and failure patterns. The most perturbed case is for a relatively weak seal embedded in relatively strong shales (model G, Fig. 7).

(2) The stronger the materials, the greater are the variations in magnitude of the stress differences throughout the models (compare models B, G and H in Figs. 7 and 8). This response is reasonable in light of the fact that stress is the reaction — the dependent variable. However, variation in material yield strength is only part of the story. Work hardening has a more significant effect on these differences than does initial yield strength alone. This is because finite deformation of such materials produces a stronger material.

(3) Sharper gradients in both σ_3 (models B, G and H in Figs. 7 and 8) and σ_1 (models B and G vs. H in Fig. 8) are created as the layers are given more contrasting material properties. These abrupt changes of state are often localised at the material interfaces.

(4) Yielded regions are more localized to the layer interfaces when there are strength differences across that interface (Fig. 7c and Fig. 8e).

(5) The vertical stress is considerably different from the usually assumed $\rho g z$ (where ρ = density, g = gravitational constant and z = depth of burial). Fig. 8d shows three examples. No simulations showed a vertical stress that follows the $\rho g z$ pattern around the flexures, even at 1 m fault displacement. This situation occurs because of the load redistribution that takes place in the models. This process is something akin to the way that an arch alters how loads are transmitted. Lower stresses in one location are typically associated with greater stresses somewhere else.

(6) Typically, the lowest magnitudes of σ_3 in the seal occur within and to the left of the synclinal area (Fig. 7b and Fig. 8c). However when the seal is 'strong', and particularly, when the underlying shale is also 'strong', the lowest σ_3 values occur at the top of the anticlinal portion of the seal. Models B and H of Fig. 7 show this effect. This anticlinal area always has somewhat lowered σ_3, but this effect is enhanced in the cases cited.

(7) The differences between corresponding models (e.g. two G configurations) in the high-contrast (Case 1) and low-contrast (Case 2) examples are small. The change in material-property contrast does affect the stress magnitudes somewhat, but the overall stress distributions are the same. This also applies to the failure-mode patterns.

Geomechanical generalisations concerning flexural systems

We have simulated the deformation of a layered fold–fault system, albeit one with an irregular starting shape for one of the layers. The fault displacement at the reservoir level induces flexural deformation in the seal and surrounding shales. The stress state that develops in the seal and shales is that of bending associated with a flexure, or with a series of stacked flexures where layer-parallel slip is allowed. The details of the bending stress state are dependent on all of the factors investigated here, but the main aspects are consistent from model to model. The bending stress patterns appear very early in the flexural process, and these patterns are retained as the deformation increases.

The following are the main generalisations.

(1) Although the general stress pattern is established early in the deformation, stress magnitudes do increase somewhat with both strain-hardening and strain-neutral post-failure behaviours. In some models, very large stresses develop at material interfaces. These are very probably not realistic, but instead are

a consequence of a combination of the overly simple, elastic–plastic rheology and the welded (non-slipping) interfaces.

(2) Layer-parallel slip, as approximated in our models, alters both the stress field and the strains/displacements. In general, layer-parallel slip results in smaller stress magnitudes and stress differences, smaller strains, and changes from layer-parallel to layer-normal maximum principal stress orientations across slipping boundaries. More of the bulk strain is achieved by discrete (or, in our case, localised) slip.

(3) The vertical stress, which is initially equal to ρgz, is quickly altered as the flexures develop. This behaviour is common to all models, including those with much-lower and much-higher yield strengths than those that have been presented here. The explanation for this result is that the simulations are merely an illustration of how loads become re-distributed throughout the whole system as a consequence of the bending distortion and gravity combined. The horizontal stresses (σ_x) are also altered by the bending. If this situation is common in the subsurface, then the normal assumption of vertical stress being sensitive only to depth warrants serious re-examination.

(4) The minimum principal stress develops spatially varying magnitudes whose differences can be on the order of 10 MPa along or through the seal layer. To a lesser extent, minimum principal stress magnitudes change temporally as the structure evolves.

(5) The stress pattern observed in the models is dominantly associated with the flexural process, but at fault displacements of 200 m or more, a zone of distributed shear can develop immediately above the fault and extending towards the antiform crest. This could represent the 'beginning of the end' of the folding process, and may be indicative of how faults can propagate through an overlying sequence of layered rocks.

These results are representative of an important behaviour mode of a fold–fault system, but they should still be treated with caution as they do not address all material responses. In particular they do not include a strain-softening response, which should result in localisation of strain. We anticipate that strain softening would tend to focus deformation into zones, in many cases onto the material boundaries, tending to reduce the stress magnitudes and differences, while increasing strains in the strain-softening zones and decreasing it elsewhere. Strain softening is also typically associated with dilatancy, and this response would be important if we were coupling the models with fluid flow simulations.

Implications for the prediction of top-seal capacity

In studies that consider the capacity of top seals (e.g.: Eaton, 1969; Grauls and Baleix, 1993; Osborne and Swarbrick, 1997; Grauls, 1999; Schneider et al., 1999), the primary focus is typically on the stress state in the seal, specifically the least stress (σ_3), which is often assumed to be horizontal (in which case it would be designated as σ_h). The rationale for this approach is based on a simple mechanical analysis that predicts a hydrofracture type of failure if the pore-fluid pressure of a rock increases to equal, or slightly exceed, the value of the least stress. Such a failure mechanism is implicitly assumed to have a high hydraulic conductivity, and so it is usually inferred that any such event will lead to a release of the high pore pressure, or at least a valving action to limit the pressure increase.

If we interpret our simulation results in a similar fashion, then we simply need to consider the distributions of the least stress that are produced by the flexural deformations of the various material configurations. Each of the simulations of flexural deformation produces a reduction in the magnitude of σ_3 in and along the seal layer on the downthrown side of the fault block (i.e. in the general synclinal area of the monocline). On the contour plots (Figs. 7 and 8), the seal layer has σ_3 magnitudes represented by light-green colours at the ends of each model, and the stress values at these locations approximate the pre-flexure conditions. The reductions of σ_3 in the synclinal area are at least one colour band ($\approx$4 MPa) in all models, and as much as three colour bands ($\approx$12 MPa) in some. The smallest values created by the deformation are on the order of 1–2 MPa. Clearly, the synclinal regions of such flexures represent locations where leaks might occur if overpressures increase to reach the magnitude of the least stress.

In the anticlinal area of each model there is also a reduction in the magnitude of σ_3. This reduction occurs across a distance of a few hundred metres extending to the right (as we see it) of the crestal positions. In those models where the seal is a stronger material, the reduction in stress magnitude is greater. In models where the seal is weaker, or where it is overlain by a stronger material, a small region of increased σ_3 occurs in the immediate crestal location. The regions of lowered σ_3 might represent possible leak points on the upthrown portions of structures (like those noted above on the downthrown side), but there might also be a local crestal area where leaks are less likely. It seems that flexural deformation of a seal layer alters the stress state such that there is a high potential for leaks, but that identifying the exact position of leakage (which might be important if the layers were tilted) requires specific knowledge about the mechanical details of the rock units.

Are there indications of such real-world spatial variability of minimum stress? The principal evidence to partially answer this question consists of LOT pressure data. Compilations of LOT data from most regions, when plotted as a function of depth, indicate a progressive increase of least stress magnitude with depth. However, recent studies (e.g. White and Swarbrick, 2001) show considerable variations (of 10 MPa or more) in the inferred magnitude of σ_3 at depths of 2 to 4+ km, and, within particular stratigraphic units, as a function of lateral position (A.J. White, pers. commun., 2001). Such variations could well be the result of spatially variable stress states caused by flexural processes like those simulated here.

Can the simulations be interpreted further? What needs to be done to enhance the value of the geomechanical simulation approach? Here we seek to expand the simple concept of sealing capacity adopted above, applying it to two, potentially unrelated, topics. In one, the focus is on the ability of a layer (a potential top seal) to retain a hydrocarbon column. Although the permeability of the layer might be important in a very dynamic situation, when hydrocarbon charge is keeping up with losses through the seal, it is the seal's entry pressure that determines whether any hydrocarbons can be trapped. The maximum column of hydrocarbon that can be retained is the one that produces a top-of-column buoyancy pressure that is equal to the least stress (which is the same analysis as described above). The second topic focuses on the ability of a top seal to retain overpressure, and here it is the permeability that matters, along with consideration of the least stress (representing the leakage analysis described above). We consider these two, different aspects of top-seal capacity (hydrocarbons and overpressure) in two, contrasting geological scenarios.

In the first scenario, the flexural deformation of the 'seal' (which, at the time of deformation, may not have been operating as a seal) occurs at some time in the geological past. The question to answer is: What aspects of the simulation results can be applied to help us understand what happens when new fluids encounter the seal at a later time? Will the seal layer trap newly arriving hydrocarbons? This prediction requires a model that relates changes in entry-pressure to the deformation process (i.e. $Pc = f(\varepsilon)$). Alternatively, will the seal be able to retain overpressure (perhaps as imposed through a lateral transfer process; see Yardley and Swarbrick, 2000)? This prediction depends on associating permeability changes with deformation (i.e. $k = f(\varepsilon)$). But in both cases, the stress state is also important, particularly the magnitude of the least stress. This interest arises because σ_3 may limit the maximum fluid pressure that can be attained (assuming that a hydrofracture-like failure mechanism occurs when the fluid pressure approaches the least stress magnitude). Each of these questions requires us to consider if time-dependent deformation occurs after the flexure, as a consequence of the presence of differential stresses that were created by that flexure, and how that deformation, if it occurs, might further change the petrophysical properties.

In the second scenario, deformation occurs when the seal is already acting to retain fluids. We can imagine a case where there are hydrocarbons that are nearly normally pressured, or one without hydrocarbons, but with high water pressures, or another possibility is where hydrocarbons and overpressure occur together. The case of hydrocarbons without overpressure is essentially identical to that of hydrocarbon retention in the preceding scenario. However, when there is overpressure, there will be an associated reduction of effective stress. Simulations of this complex situation need to explicitly couple geomechanics and fluid flow. Because the geomechanical simulations reported here are not coupled in that way, they can only be used to infer the likely general behaviour of a scenario involving concurrent deformation and overpressure. It is worth noting, however, that such inferences depend on models that link deformation with the evolution of permeability, as was noted in the previous scenario.

The more-advanced material models that are required to address these issues need to be able to consider major volumetric changes during deformation, and the way that the deformation mode varies as a function of the mean effective stress. The poro-visco-plastic (PVP) material description outlined by Couples (1999) meets these requirements. For many rocks that are important to the questions raised here, it is appropriate to associate strain-hardening behaviour (like we assumed for some of our elastic–plastic materials) with compactional volume strains (porosity losses), strain-softening with dilational volume strains, and strain-neutral deformations with the 'critical state' (i.e. the crest of the yield surface). It is also reasonable to assume that when permeability decreases as a consequence of deformation, there is an associated increase in entry-pressure, and that increased permeability is associated with a decreased entry pressure. These petrophysical changes can be attached to the PVP geomechanical model, so the additional issues raised here can be addressed by a further simulation effort.

It is worth noting that the stress patterns developed during flexural deformations are not likely to be significantly different in that next generation of simulations (based on comparisons of simulations using several approaches; not shown here), but that the stress and strain paths will differ considerably in some regions of the models. Those differences will possibly alter some of the conclusions that have been reached here.

Summary

The conceptual model of flexural deformation of layers has been shown to be very robust relative to variations in material properties, model constraints, and phenomena such as layer-parallel slip. The geomechanical simulations indicate significant spatial variations of stresses (including the magnitudes of σ_3 and σ_v) that are a direct consequence of the flexural processes. Lateral σ_3 variations in a layer representing a seal can exceed 10 MPa if there are mechanical strength differences within the layered rock succession, but even where the mechanical properties of the layers are identical, there are still variations of 1–2 MPa in the magnitude of σ_3. The flexural processes examined in this paper represent a possible explanation of the observed spatial variability of σ_3 as determined from LOT data.

Areas at high risk of seal failure — that is, locations where σ_3 magnitudes are low — typically occur in and to the left of the synclinal area, but also within the upthrown region, away from the crest. In simulations where the seal layer is weak, and/or where it is overlain by a stronger layer, there is a local region of increased σ_3 in the crestal region. When both the seal layer and the lower shale are 'strong', the lowest σ_3 values in the seal occur at the top of the anticline.

Although this study has developed a process-based explanation for the creation of stress state anomalies in seal layers, we caution against applying these results widely without consideration of the details of each case. In these models, flexural processes proved sensitive to factors such as the thickness of the lower shale, and its strength. There is also very likely to be a strong dependence on the seal's location relative to other major mechanical boundaries and a dependence on the seal layer's thickness, especially when its internal mechanical properties are heterogeneous. The models demonstrate that operative layer-parallel slip will very likely also have a strong impact on how the mechanical state evolves.

In brief, this study should be taken as demonstrating that numerical simulation can be employed to address the capacity of seals. However, prediction is sensitive to local circumstances, and generalisations should be treated with caution. Nevertheless, the work provides a rational basis for understanding one possible cause of the observed variability in LOT information.

Acknowledgements

We thank the GeoPOP II sponsor companies — Amerada Hess, BG, BP Amoco, Conoco, Enterprise Oil, ExxonMobil, JNOC, Norsk HYDRO, Phillips Petroleum, Statoil, Texaco and TotalFinaElf — for enabling this work, and for the feedback provided. We also thank Dr. N. Koutsabeloulis of VIPS Ltd. for guidance in the use of VISAGE^{TM}. We are also grateful to an anonymous reviewer whose suggestions improved the manuscript considerably.

References

Couples, G.D., 1999. A hydro-geomechanical view of seal formation and failure in overpressured basins. Oil Gas Sci. Technol. Rev. IFP, 54: 785–795.

Couples, G.D. and Lewis, H., 1998. Lateral variations of strain in experimental forced folds. Tectonophysics, 295: 79–91.

Couples, G.D. and Lewis, H., 2000. Effects of interlayer slip in model forced folds. In: J.H. Cosgrove and M.S. Ameen (Editors), Forced (Drape) Folds and Associated Fractures. Geol. Soc. London Spec. Publ., 169: 129–144.

Couples, G.D., Lewis, H. and Tanner, P.W.G., 1998. Strain partitioning during flexural slip folding. In: M.P. Coward, H. Johnson and T.S. Daltaban (Editors), Structural Geology in Reservoir Characterization and Field Development. Geol. Soc. London Spec. Publ., 127: 149–165.

Eaton, B.A., 1969. Fracture gradient prediction and its application in oil field operations. J. Pet. Technol., 21: 1353–1360.

Grauls, D., 1998. Overpressure assessment using a minimum principal stress approach. In: Overpressures in Petroleum Exploration, Proc. Workshop, Pau, April 1998. Bull. Centre Rech. Elf Explor. Prod. Mem., 22: 137–147.

Grauls, D., 1999. Overpressures: causal mechanisms, conventional and hydromechanical approaches. Oil Gas Sci. Technol. Rev., 54 (6): 667–678.

Grauls, D. and Baleix, J.M., 1993. Role of overpressure and in situ stresses in fault controlled hydrocarbon migration. Mar. Pet. Geol., 11: 6.

Horsrud, P., Sønstebø, E.F. and Bøe, R., 1998. Mechanical and petrophysical properties of North Sea shales. Int. J. Rock Mech. Min. Sci., 35 (8): 1009–1020.

Jones, M.E. and Addiss, M.A., 1986. The application of stress path and critical state analysis to sediment deformation. J. Struct. Geol., 8: 575–580.

H. LEWIS *Department of Petroleum Engineering, Heriot-Watt University, Edinburgh EH14 4AS, Scotland, UK*
E-mail: helen.lewis@pet.hw.uk

P. OLDEN *Department of Petroleum Engineering, Heriot-Watt University, Edinburgh EH14 4AS, Scotland, UK*

G.D. COUPLES *Department of Petroleum Engineering, Heriot-Watt University, Edinburgh EH14 4AS, Scotland, UK*

Osborne, M.J. and Swarbrick, R.E., 1997. Mechanisms for generating overpressure in sedimentary basins: a re-evaluation. Am. Assoc. Pet. Geol. Bull., 81: 1023–1041.

Schneider, F., Bouteca, M. and Sarda, J.-P., 1999. Hydraulic fracturing at sedimentary basin scale. Oil Gas Sci. Technol. Rev., 54 (6): 797–805.

White, A.J. and Swarbrick, R., 2001. LOPs: 'Fracture gradients' and pore-pressure in-situ stress coupling. Extended Abstract, 63rd EAGE Meeting, Amsterdam, June 2001.

Yardley, G.S. and Swarbrick, R.E., 2000. Lateral transfer: a source of additional overpressure? Mar. Pet. Geol., 17: 523–537.

Top seal assessment in exhumed basin settings — some insights from Atlantic Margin and borderland basins

D.V. Corcoran and A.G. Doré

Empirical observation indicates that hydrocarbon accumulations in exhumed basins of the Atlantic Margin are commonly characterised by underfilled traps and hydrostatically pressured or modestly overpressured reservoirs. These observations are reviewed in the context of the generic mechanisms by which top-seals leak, the properties of cap-rocks and the physical processes which occur during exhumation. The fluid retention capacity of any cap-rock lithology during exhumation is dependent upon the physical and mechanical characteristics of the cap-rock at the time of exhumation and the timing and conditions of the associated deformation relative to the timing of hydrocarbon emplacement. The permeability and deformational characteristics of halite renders it an excellent cap-rock with a high retention capacity, even under conditions of exhumation. However, mudrocks may also form effective cap-rocks in exhumed basins when the deformation associated with exhumation occurs prior to embrittlement and the shale cap-rock exhibits ductile behaviour. Shale and evaporite cap-rocks form the main regional seals to hydrocarbon accumulations in exhumed basins of the Atlantic Margin and borderlands. *Syn-exhumation* top-seal efficiency (fluid retention capacity) is a major exploration risk in these basins, though *post-exhumation* top-seal integrity in these basins may be relatively high under certain conditions. Consequently, a major exploration risk factor in exhumed basin settings pertains to the limited hydrocarbon budget available post-regional uplift and the efficiency of the re-migration process.

Introduction

On a geologic time-scale the evolution of oil and gas accumulations is a dynamic process which is a function of the rates of ingress and egress of petroleum from the hydrocarbon trap. Top-seals are rocks which prevent the vertical migration of hydrocarbons out of traps. Therefore, an effective regional cap-rock constitutes an essential element of all petroleum systems.

During the evolution of a petroliferous basin any lithology can serve as a top-seal for a hydrocarbon accumulation provided that its capillary entry pressure exceeds the upwards buoyancy pressure exerted by the hydrocarbon column in the underlying accumulation. In practice, however, the vast majority of effective seal rocks are evaporites, fine-grained clastics and organic-rich mudrocks (Downey, 1994). Worldwide empirical observation of the 25 largest oil and gas accumulations indicates that they all depend on shale or evaporite seals (Fig. 1) (Grunau, 1987).

The basic physical principles governing the effectiveness of petroleum cap-rocks are well established (Hubbert, 1953; Berg, 1975; Schowalter, 1979; Watts, 1987). Lithology, uniformity of stratigraphy and thickness are factors which influence seal capacity (Downey, 1984). However, the fundamental rock properties which control seal performance are the capillary entry pressure of the seal (dominantly controlled by pore throat diameter) and the ductility of the seal rock which is a function of pressure, temperature and lithology. Although top-seal integrity is recognised as a major exploration risk factor in exhumed basins few systematic studies of top-seal performance in these settings have been published to date (Gabrielsen and Kløvjan, 1997; Seedhouse and Racey, 1997; Spain and Conrad, 1997; Cowan et al., 1999).

In all petroliferous basins the adequacy of the hydrocarbon charge together with the timing and rate of fill, spill and vertical or lateral leakage are key determinants of the present-day in-place oil and gas volumes preserved in hydrocarbon traps. However, in exhumed basins, the interplay between top-seal performance and hydrocarbon fill, spill and leakage is more critical as the 'switching off' of hydrocarbon generation during regional uplift may result in a lower probability of trap replenishment post-exhumation (Doré and Jensen, 1996). Physical processes which may impact cap-rocks during exhumation include erosion, tectonic deformation, shear failure, hydrofracturing due to disequilibrium pore pressure

Hydrocarbon Seal Quantification edited by A.G. Koestler and R. Hunsdale. NPF Special Publication 11, pp. 89–107, Published by Elsevier Science B.V., Amsterdam.

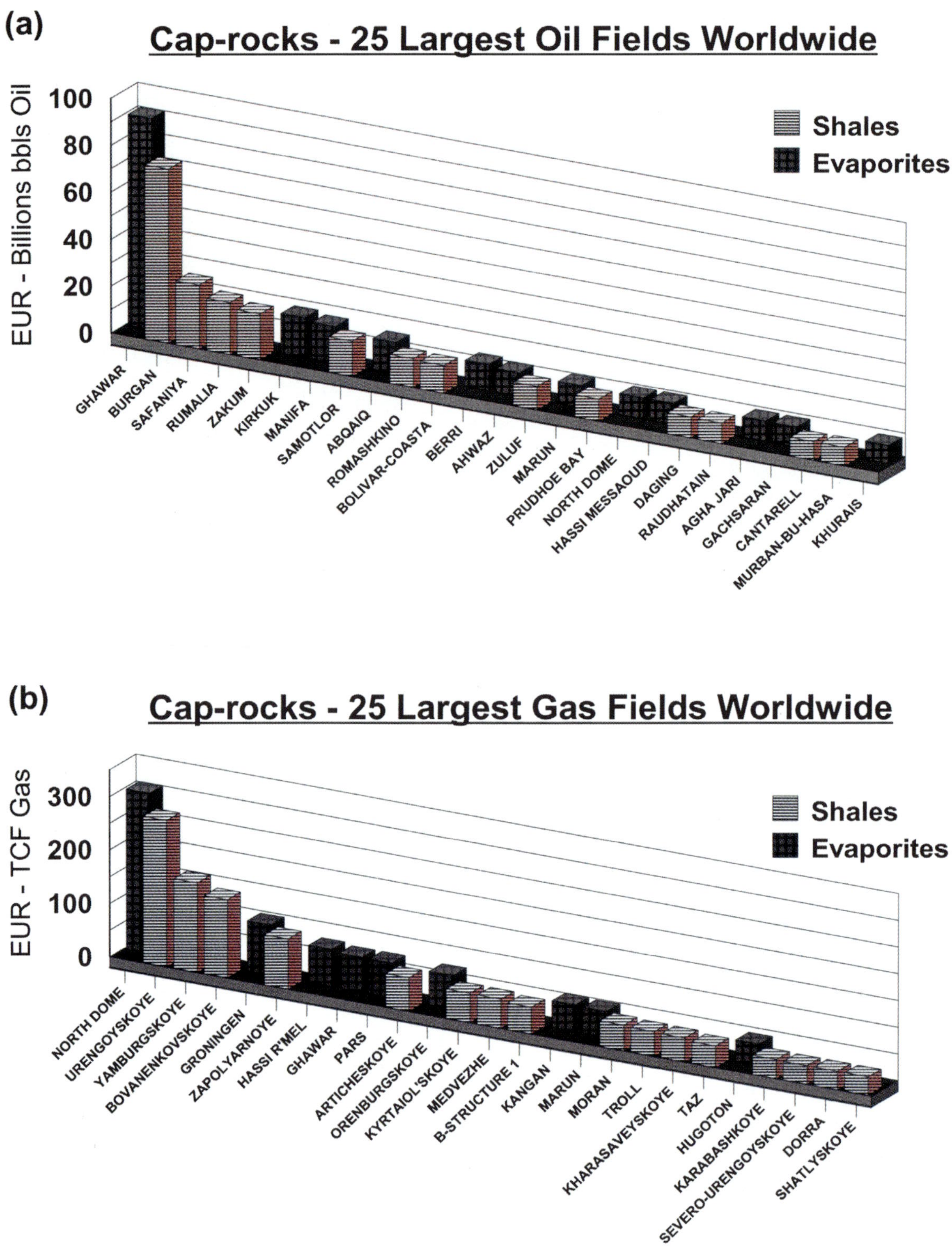

Fig. 1. Mudrock and evaporite cap-rocks are the most common seals to hydrocarbon accumulations worldwide. Estimated ultimate recovery (EUR) from (a) the 25 largest oilfields and (b) the 25 largest gasfields, all of which depend upon shale or evaporite top-seals (TCF = trillion cubic feet). Adapted from Grunau (1987).

conditions and a changing hydrodynamic regime. Furthermore, there is an increased risk of net hydrocarbon losses due to diffusion where gas accumulations are dependent upon porous and permeable shale seals.

The aim of this paper is to offer some insights with respect to the physical properties and processes which impact top-seal performance in exhumed basin settings. Empirical evidence from exhumed basins of the Atlantic Margin and borderland basins is then discussed in the context of these observations.

Top-seal leakage mechanisms — a summary

There are four generic mechanisms by which top-seals leak: tectonic breaching, capillary leakage, hydraulic leakage, molecular transport (diffusion). Fig. 2 is a summary of the physical principles governing these mechanisms, which have been articulated by numerous authors (Schowalter, 1979; Gretener, 1981; Krooss et al., 1992; Davis and Reynolds, 1996; and others). These four generic leakage mechanisms are here reviewed in the context of exhumed basin settings of the Atlantic Margin.

Tectonic breaching

Where deformation of a cap-rock occurs post-emplacement of hydrocarbons there is an increased risk of tectonic breaching and cap-rock leakage. Top-seal failure via tectonic breaching is the most readily recognised form of seal failure at the scale of seismically defined hydrocarbon traps. In addition, tectonic breaching may be facilitated by sub-seismic resolution faulting where the top-seal consists of mudrock layers interbedded with permeable siltstones and sandstones. The style and magnitude of tectonic deformation in any sedimentary basin is influenced by a number of factors, including plate tectonic setting, pre-existing structural grain and the presence or absence of detachment layers. During basin inversion, compressional, transpressional or reactivated extensional deformation may result in leakage through cross-fault juxtaposition of reservoirs from different stratigraphic levels (Fig. 2a (i)) or through the development of a connected network of juxtaposed leaky beds within the cap-rock interval (Fig. 2a (ii)). In addition, radial extension fractures will develop above the neutral surface of inversion folds and may result in leakage into the overlying sediments (Fig. 2a (iii)). Hall et al. (1997) have shown from case studies in the deep Central Graben of the North Sea that reservoir objectives lying above or close to the neutral surface of an inversion fold have a higher probability of being breached than reservoirs lying below.

Tectonic breaching is an important leakage mechanism in the exhumed basin settings of the Atlantic Margin where syn-exhumation extensional fault re-activations are probable in addition to the overprint of compressional deformation resulting from the far-field signature of Alpine orogenesis and ridge push phenomena (Murdoch et al., 1995; Doré et al., 1999).

Capillary leakage

The driving force for petroleum movement in the sub-surface is buoyancy influenced by overpressure and hydrodynamics. The force opposing the movement of petroleum is the capillary resistance of porous rocks (Fig. 2b). Standard equations have been developed to describe these opposing forces P_{buoy} and P_{cap} at the interface of a hydrocarbon reservoir and cap-rock (Hubbert, 1953; Berg, 1975; Schowalter, 1979; Watts, 1987; Clayton and Hay, 1994). Theoretically, for capillary leakage to occur the upwards buoyancy pressure of a hydrocarbon column plus any excess overpressure or hydraulic head must exceed the P_{cap} of the top-seal. Clayton and Hay (1994) have modelled the capillary seal capacity of a mudstone seal in a continually subsiding basin, based on appropriate figures for interfacial tension, contact angle, largest interconnected pore throat radii and subsurface density difference of gas and water (Fig. 3). The computed seal capacity curve for methane indicates that the modelled mudstone would retain a gas column of 500–1000 m depending upon depth.

In an exhumed basin the predicted capillary retention capacity of an average mudstone is likely to be higher at any present-day depth as the higher compaction state of the exhumed mudstone will result in smaller interconnected pore throats. With respect to the exhumed basins of the Atlantic Margin the magnitude of all known gas columns, discovered to date, is less than 500 m, considerably less than the capillary seal retention capacity modelled for an average mudstone (Fig. 3).

Hydraulic leakage

Where the capillary entry pressures to a cap-rock (evaporite or super-tight shale) are so high that capillary failure is implausible, hydraulic leakage may occur through brittle top-seals due to the generation of new tension fractures (hydrofractures), shear fractures or the dilation of pre-existing fault planes (Fig. 2c). Hydraulic fracturing can occur independent of tectonic breaching and results from changes in effective stress conditions in the cap-rock. These changes may be induced by the development of disequilibrium pore pressure conditions or by changes in the tectonic load (Fig. 4). For example, a reduction in the minimum compressive stress (σ_3), induced by *extension* during regional uplift, may per se result in the formation of dilatant shear fractures of certain orientation within the cap-rock (Fig. 4a). Shear fractures will also be formed when the prevailing stress field results in conditions of high differential stress in the cap-rock (large $\sigma_1 - \sigma_3$). Under this scenario the gradual elevation of pore fluid pressures prior to exhumation, or the removal of overburden without the re-equilibration of elevated pore fluid pressures, will result in Coulomb failure along planes in the rock

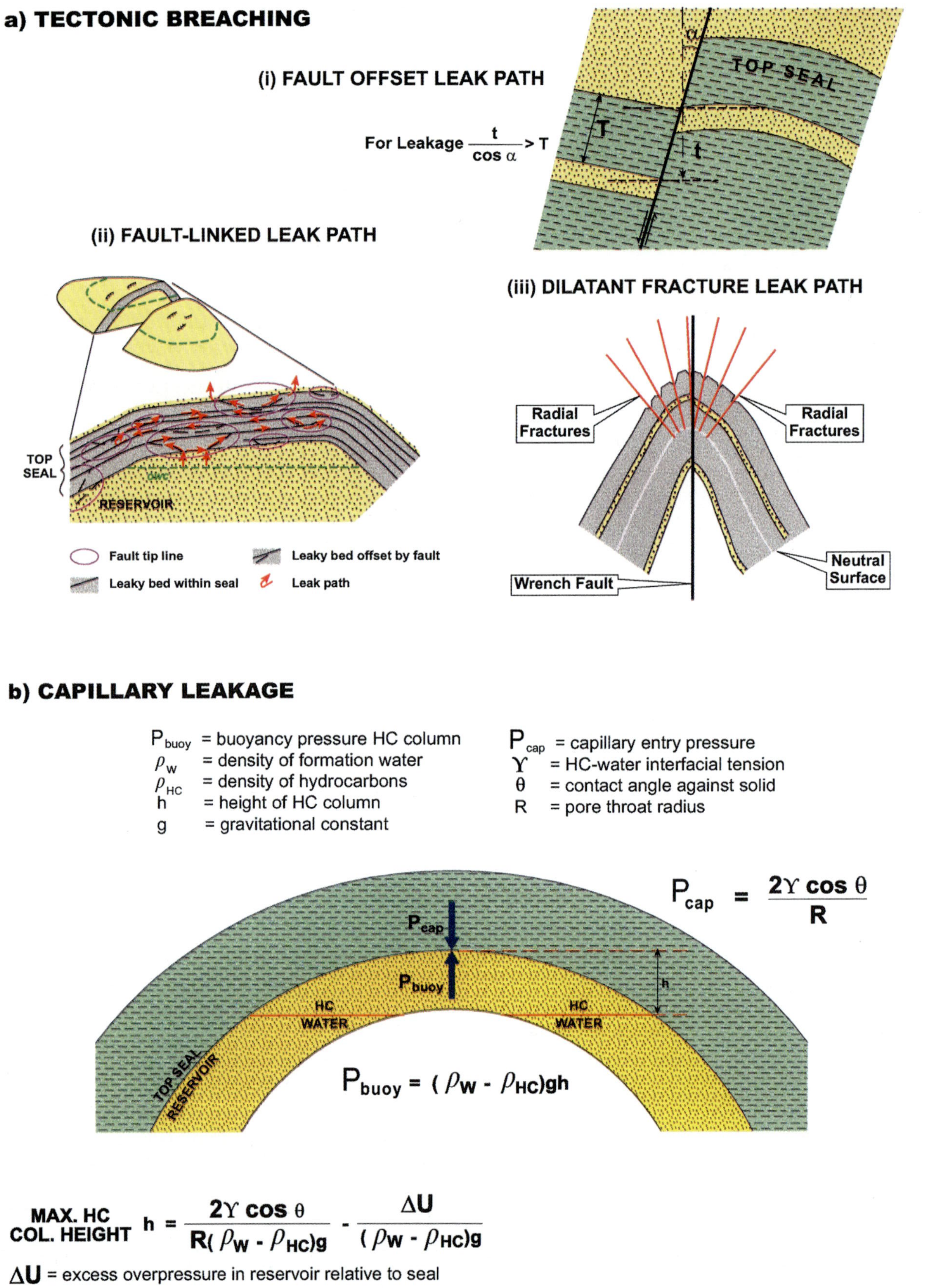

Fig. 2. The four generic mechanisms by which top-seals leak. (a) Tectonic breaching (3 common modes): (i) fault offset leak path; (ii) fault-linked leak path; (iii) dilatant fracture leak path. (b) Capillary leakage: for capillary leakage to occur the buoyancy force generated by the hydrocarbon column (P_{buoy}), plus any excess overpressure (ΔU) in the reservoir relative to the seal, must exceed the capillary resistance of the porous, water-wet, cap-rocks (P_{cap}).

c) HYDRAULIC LEAKAGE

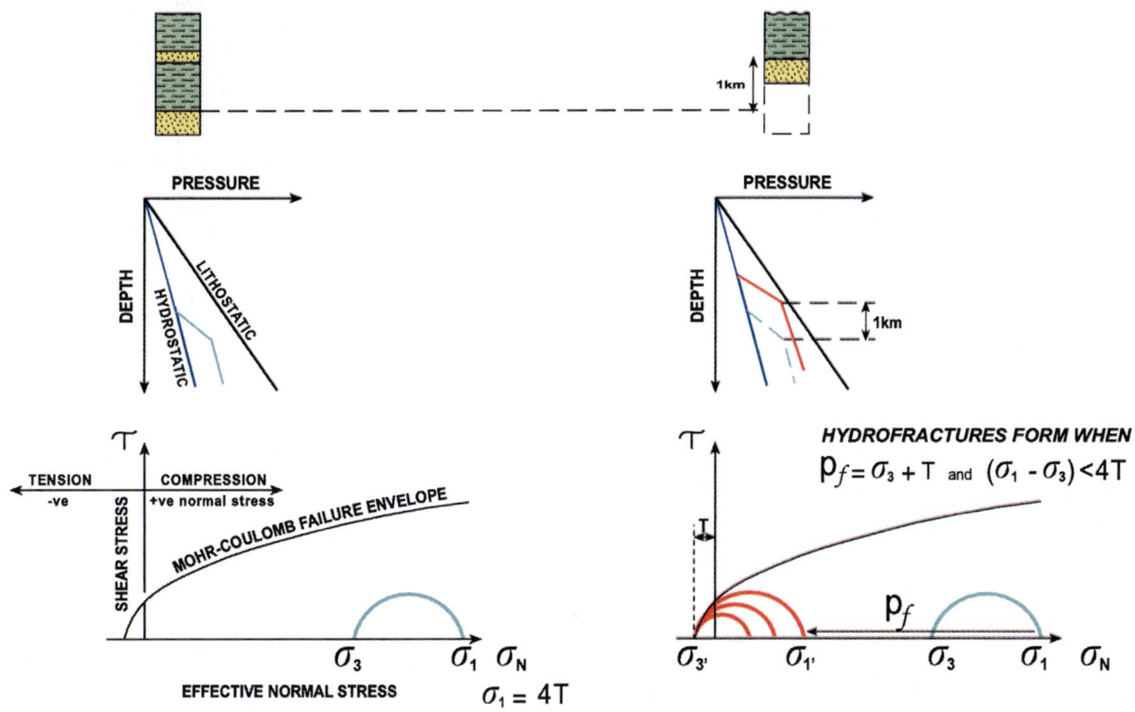

d) MOLECULAR TRANSPORT

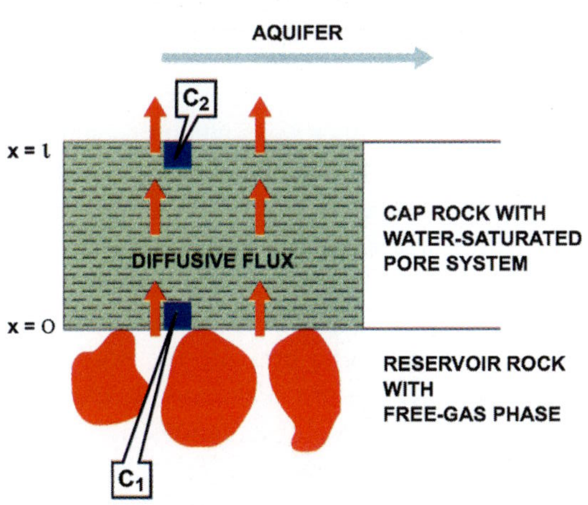

$$t_{1/2(IN)} \approx \left(\frac{Q_{1/2}}{l C_1} - \frac{1}{3}\right) l^2/D$$

$t_{1/2(IN)}$ = TIME REQUIRED FOR DIFFUSIVE LOSS OF 1/2 RESERVOIR GAS ($Q_{1/2}$) INTO CAPROCK

l = AVERAGE CAPROCK THICKNESS

C_1 = BOUNDARY CONCENTRATION CH_4 AT CAPROCK/RESERVOIR INTERFACE

D = DIFFUSION COEFFICIENT METHANE

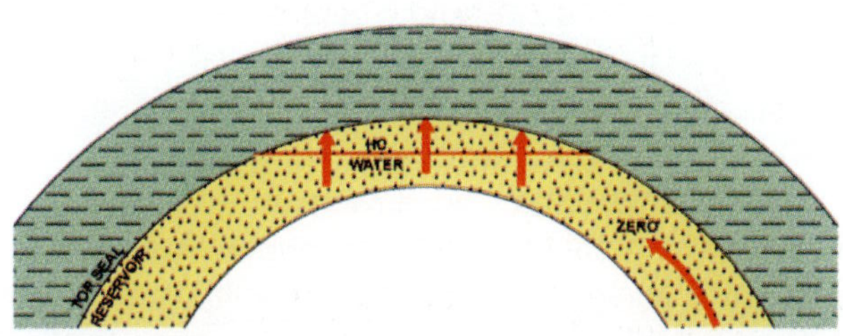

Fig. 2 (continued). (c) Hydraulic leakage: can result from the development of tension fractures (hydrofractures) which arise from changing effective stress conditions. Tension fractures occur under conditions of low differential stress (small $\sigma_1 - \sigma_3$), when pore fluid pressure in the cap-rock reduces the minimum effective horizontal stress below zero to the tensile strength of the rock. (d) Molecular transport: primarily the diffusion of methane through water-saturated shaley cap-rocks. Adapted from Krooss et al. (1992), Davis and Reynolds (1996), Hall et al. (1997) and Ingram and Urai (1999). Capillary leakage equations from Schowalter (1979) and Clayton and Hay (1994).

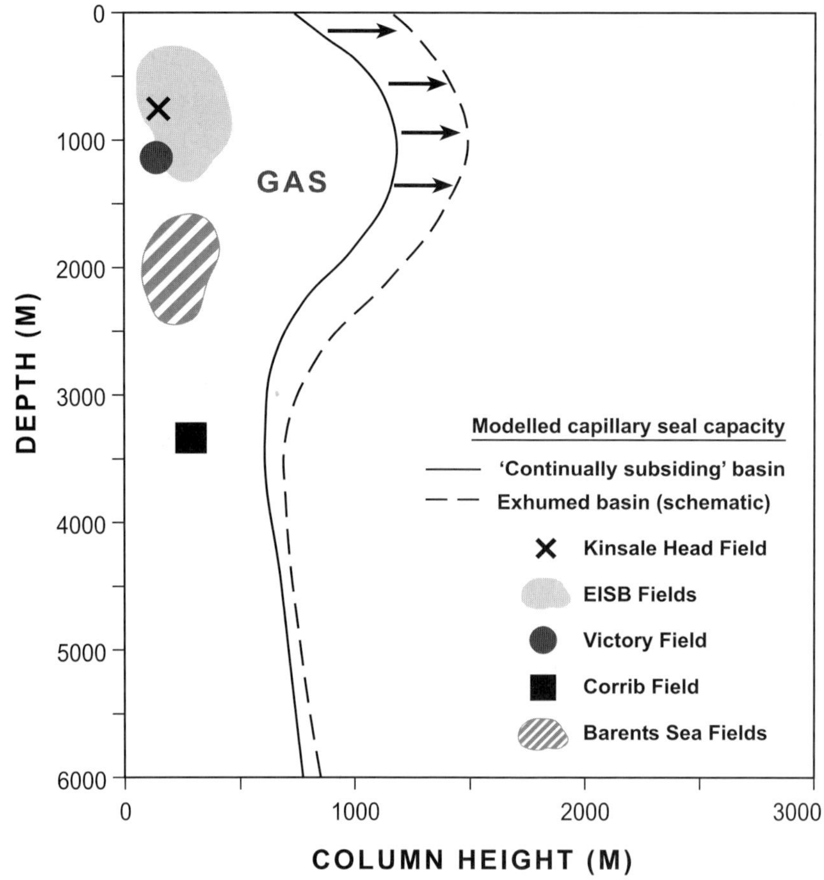

Fig. 3. Modelled capillary seal capacity with depth for a mudstone in a continually subsiding basin (from Clayton and Hay, 1994). The inferred equivalent curve for an exhumed basin setting is schematically shown (dashed line) together with the magnitude of the maximum hydrocarbon columns in gas accumulations from some exhumed Atlantic Margin and borderland basins.

which make appropriate angles with σ_1 (Fig. 4b). Pre-existing fractures, joints or faults (in fact any planes of reduced cohesion) have important implications for the mechanical behaviour of cap-rocks during exhumation. When fractures are present their physical characteristics and their orientation must be known in order to evaluate their structural significance (Gretener, 1981).

Hydrofractures occur under conditions of low differential stress when pore fluid pressure at the cap-rock–reservoir interface reduces the minimum effective horizontal stress below zero to the tensile strength of the rock (Fig. 2c). In extensional basins, where the minimum compressive stress (σ_3) is significantly less than the maximum compressive stress (σ_1), these hydrofractures are invariably vertical to semi-vertical in orientation and form perpendicular to the minimum horizontal stress (σ_3). For hydrofractures to develop in preference to shear fractures the conditions $P_f = \sigma_3 + T$ and $\sigma_1 - \sigma_3 < 4T$ must be satisfied (P_f is the pore fluid pressure, σ_1 is the maximum compressive stress, σ_3 is the minimum compressive stress and T is the tensile strength of the cap-rock) (Hubbert and Rubey, 1959; Secor, 1965; Sibson, 1995). These conditions can occur in highly overpressured systems undergoing continual subsidence or during exhumation when rapid denudation, without re-equilibration of overpressures, results in tensile failure. In either case, when this state prevails, pervasive tension fractures may develop in the cap-rock which will result in a catastrophic loss of any pre-existing hydrocarbon fill.

Hubbert and Rubey (1959) demonstrated that when the pore fluid pressure in a sedimentary basin approaches the lithostatic pressure the fluid pressure is released by rock failure. Palciauskas and Domenico (1980) supported these observations by showing theoretically that microfractures can develop in overpressured sedimentary beds while undergoing burial. Capuano (1993) provided direct petrographic evidence of the occurrence of microfractures in situ at depths of 3–5 km, in geopressured Oligocene shales of the Gulf Coast Basin. Furthermore, the computed fracture permeabilities (order of 10^{-13} m^2) in these shales combined with paragenetic relationships indicate that fluid flow occurred preferentially through these microfractures rather than through the matrix of these shales. The development of these hydrofractures during burial is facilitated by a mechanism of episodic tensile failure in a low differential stress

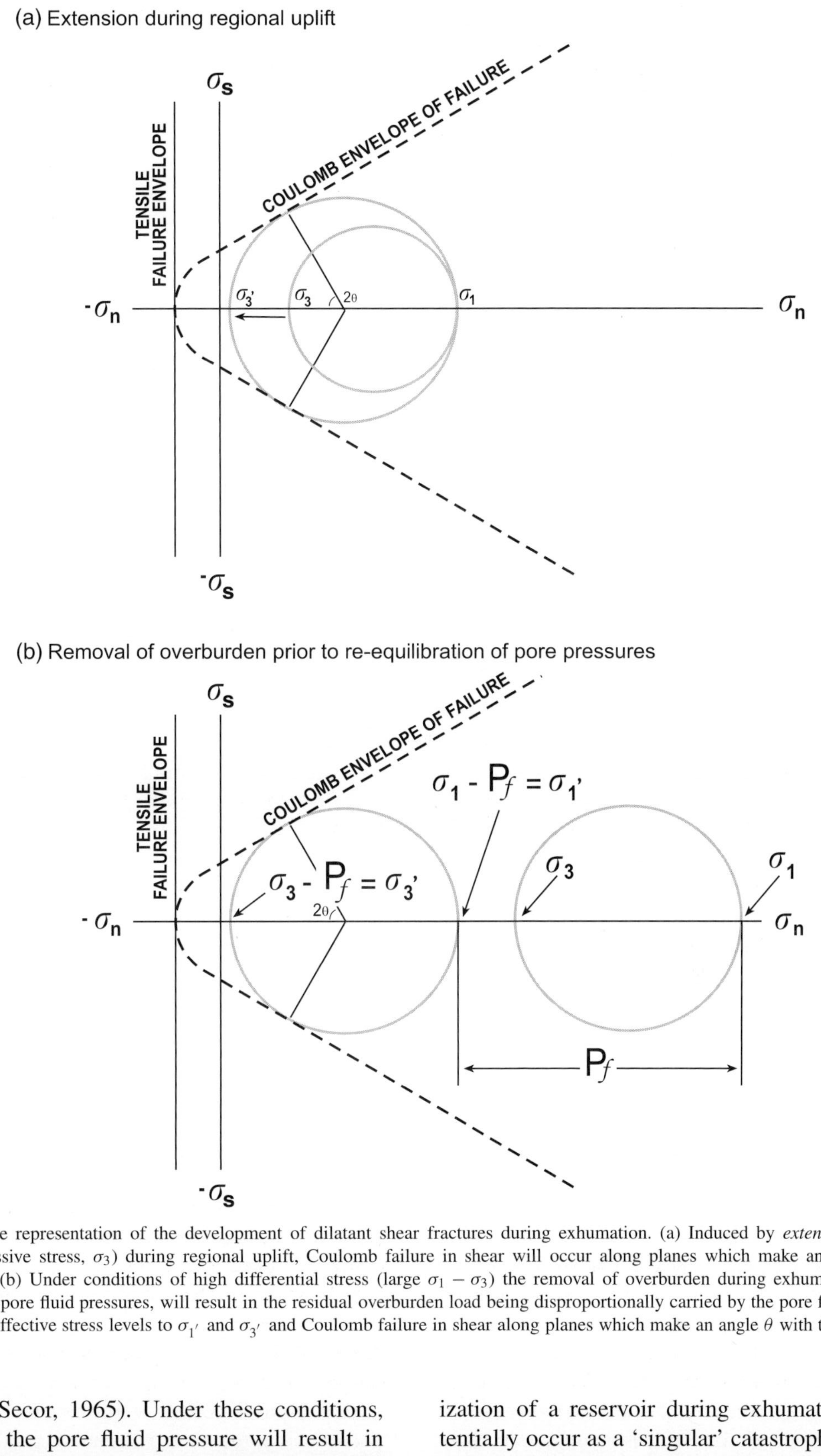

Fig. 4. Mohr circle representation of the development of dilatant shear fractures during exhumation. (a) Induced by *extension* (reduction of minimum compressive stress, σ_3) during regional uplift, Coulomb failure in shear will occur along planes which make an angle θ with the orientation of σ_1. (b) Under conditions of high differential stress (large $\sigma_1 - \sigma_3$) the removal of overburden during exhumation, without the re-equilibration of pore fluid pressures, will result in the residual overburden load being disproportionally carried by the pore fluid pressure (P_f), a lowering of the effective stress levels to $\sigma_{1'}$ and $\sigma_{3'}$ and Coulomb failure in shear along planes which make an angle θ with the σ_1 axis.

environment (Secor, 1965). Under these conditions, the raising of the pore fluid pressure will result in the cracking of the rock in tension and the release of the fluid pressure, followed again by a raising of fluid pressure and a repetition of this cycle (Fig. 2c).

In contrast with this pressure cyclicity which is manifest in a continually subsiding basin, depressurization of a reservoir during exhumation could potentially occur as a 'singular' catastrophic event. This may arise because some of the processes which produce overpressure (Osborne and Swarbrick, 1997), such as disequilibrium compaction, dehydration reactions and kerogen transformation will have been arrested once exhumation begins. Once catastrophic failure via fracturing has occurred during exhumation, the cap-rock unit can only regain 'seal status' when the high-permeability open fractures are healed or annealed. Fracture closure, during or post-exhumation, can occur through a range of mechanisms, including cementation and increased horizontal compressive stress. Cementation in the fracture may be caused by the cooling of upwards flowing fluids with the resulting redistribution of silica and other mineral phases or, in the absence of fluid flow, by the chemical diffusion of solids into the fracture driven by local thermodynamic potentials (Pedersen and Bjørlykke, 1994).

Molecular transport (diffusion)

Diffusion is a continual and ubiquitous process in sedimentary basins and its role in hydrocarbon migration has been analysed by several workers (Fig. 2d) (Krooss et al., 1992; Montel et al., 1993; Schloemer and Krooss, 1997). The diffusive transport mechanism primarily pertains to the dismigration of natural gas accumulations in certain circumstances and has little relevance for oil dismigration due to the increased size of oil molecules relative to shale pore throat dimensions. Although the modelling studies of Kettel (1997) indicate that methane diffusion constants for rock salt are non-zero over geological time-scales, diffusion losses from gas accumulations capped by thick evaporitic seals are considered minimal. Empirical observation of long-lived gas accumulations in Upper Proterozoic reservoirs sealed beneath Lower Cambrian salt in the extensively exhumed Lena-Tunguska province of the former Soviet Union supports this conclusion (Kontorovitch et al., 1990).

In contrast, Leythaeuser et al. (1982) have demonstrated that gas may diffuse through water-saturated cap-rocks over geological time-scales. This diffusion model suggests that the evolution and preservation of natural gas accumulations is dependent upon the ratio of gas supply to the trap and gas losses through the cap-rock. For example, a case study of the Harlingen gas field, offshore Netherlands, indicated that half the 68 bcf contained in the Lower Cretaceous reservoir would be lost by diffusion through the shales and marls of the Hauterivian cap-rock (390 m thick) in 4.5 million years (Leythaeuser et al., 1982). Subsequent re-evaluation of these estimates by Krooss et al. (1992), suggests that the rate of diffusive hydrocarbon losses through the cap-rock at Harlingen are an order of magnitude lower (approximately 70 million years to dissipate half the in place gas via diffusion through the cap-rock).

In a petroliferous basin that is characterised by continual subsidence, hydrocarbon escape by diffusion and other processes can be wholly or partly offset by an active generation and migration system. However, in an exhumed basin setting, where the hydrocarbon generation and migration system is 'switched off' during regional uplift, diffusive losses through water saturated shaley cap-rocks will increase with time since uplift and may be significant.

Cap-rocks — some physical properties

The examination of cap-rocks to hydrocarbon accumulations is primarily concerned with the properties of the weakest point of the reservoir–top-seal interface. As highlighted by Downey (1994) measured properties of a random core sample may not be relevant to the physical properties of the cap-rock at the leak point. Furthermore, geohistory is an important control on the sealing properties of top-seals (Knipe et al., 2000) and the location of the potential leak point throughout the evolution of the hydrocarbon trap. Some of the petrophysical and mechanical properties which most influence top-seal performance are summarised below.

Lithology, porosity and permeability

Evaporites and mudrocks are commonly found as effective top-seals to hydrocarbon accumulations because they typically possess very low porosity and permeability, high capillary entry pressures, are relatively ductile and are often laterally continuous at the basin scale. However, other lithologies, such as siltstones and sandstones, have been identified as having capillary retention capacity and can form the top-seal to a hydrocarbon column (Spain and Conrad, 1997).

In most sedimentary basins, mudstone porosities range from 5 to 80%, depending upon compaction state (Sclater and Christie, 1980). Mudstone permeabilities vary by ten orders of magnitude (10^{-4} to 10^{-15} mD) and by three orders of magnitude at a single porosity, primarily due to grain size variations (Dewhurst et al., 1999). However, the largest interconnected pore throat diameter is the critical factor with respect to the capillary entry pressure of the mudstone. In tight mudrocks (permeability 10^{-9} D range) the risk of capillary failure and Darcy flow through the matrix is low, as the capillary entry pressure commonly exceeds the buoyancy force of any potential hydrocarbon column (Fig. 3). In this case, the top-seal retention capacity of the mudrock is a function of the ductility of the mudstone and the potential for the formation of dilatant fractures under tectonic deformation or changing pore fluid pressure conditions.

Halite forms an excellent top-seal as a result of two characteristics: it has a practically infinite capillary entry pressure and it flows plastically under deformation. When it forms a continuous layer over the potential hydrocarbon trap and is immediately juxtaposed above the hydrocarbon-bearing reservoir, the seal risk for that trap is considerably reduced, even where the trap has experienced post-emplacement tectonic deformation and exhumation.

At both the hydrocarbon trap and basin scale, cap-rock units can manifest vertical and lateral heterogeneities. Internal lithostratigraphy of the unit can vary, with mudrocks often interbedded with large amounts of leaky strata such as siltstones or sandstones. These lithologies will be more prone to leakage than the mudrocks and can result in multimodal pore throat diameter distributions within the cap-rock interval and the development of 'waste zones', if the siltstones are located immediately above the reservoir unit. In a study of hydrocarbon seals in the exhumed East Irish Sea Basin, Seedhouse and Racey (1997) utilised mercury injection porosimetry to identify pore throat distributions in the Mercia Mudstone Group (MMG) seal and to describe some of the heterogeneities observed in this cap-rock interval. These authors identified the presence of halite immediately above the reservoir as a key factor in the retention of hydrocarbon columns in the exhumed Triassic Sherwood Sandstone Group reservoir. In addition, their study found that Sherwood Sandstone accumulations with significant hydrocarbon columns, but which are directly capped by mudrocks, invariably manifested hydrocarbon shows within the cap-rock interval up to the level of the first halite bed encountered above the reservoir. This suggested that the buoyancy forces exerted by the individual hydrocarbon columns were sufficient to overcome the capillary entry pressure of the heterogeneous MMG at these locations, but not the capillary resistance of the halite units.

Strength, ductility and brittleness

The mechanical response of rocks to an applied stress varies under different conditions so a valid comparison of the strength and ductility of rocks can only be made if the conditions of deformation are also known. Ductility is a rock property which pertains to the amount of strain that a material can withstand prior to brittle failure if it undergoes brittle failure at all. Ductile rocks respond to an applied stress by an initial, though limited, elastic deformation, followed by sustained plastic deformation before failure. Brittle rocks respond to an applied stress by first shortening elastically and then failing by the formation of discrete fractures and faults. A rock is considered ductile

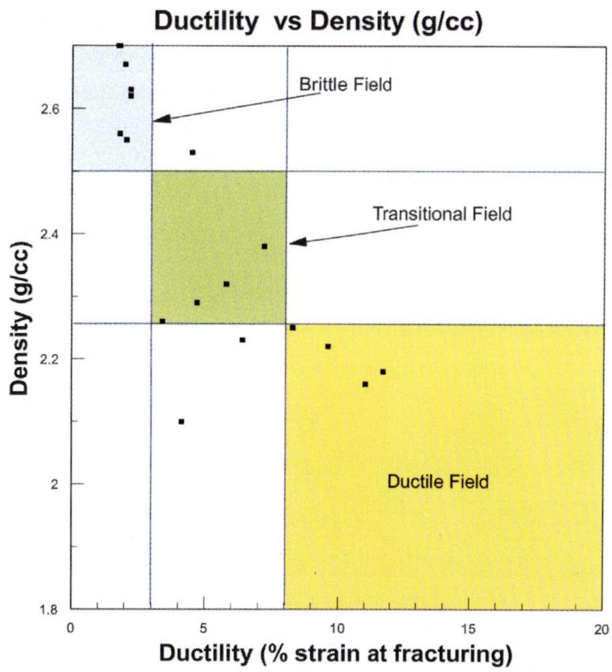

Fig. 5. Mechanical properties of mudrocks. Relationship between ductility (% strain at fracturing) and bulk density (g/cm^3), at a constant confining pressure of 1000 kg/cm^2, for a population of Neogene mudrocks from Japan. Mudrocks with densities in excess of ~2.5 g/cm^3 exhibit brittle behaviour and fracture at strains of less than 3%. Data from Hoshino et al. (1972).

when it can accommodate strains of 8–10% without fracturing and brittle when strain is less than 3% before fracturing (Fig. 5). Rock ductility is a function of lithology, confining pressure, pore fluid pressure, temperature, differential stress and strain rate (Davis and Reynolds, 1996).

Because the matrix permeability of buried and compacted mudrocks is extremely low, it is the fracture permeability that primarily controls seal capacity of these rocks. More relevant definitions are offered by Ingram and Urai (1999) who describe a ductile mudrock as one that can deform without dilatancy and the creation of fracture permeability, and a brittle mudrock as one that dilates and develops fracture permeability.

Experimental studies have shown that, for most lithologies, both rock strength and ductility increase with rising confining pressure (Handin et al., 1963; Gretener, 1981). This suggests that sedimentary rocks, including mudstones, increase in ductility during burial because confining pressure increases with depth. However, this inference results from treating the compaction of mudrocks as only a mechanical process and ignores the effects of chemical compaction in the deeper part of sedimentary basins (>2–3 km, 70–100°C) (Bjørlykke, 1999).

Bulk density (ρ_b) is one measured parameter which can be used to indicate the compaction state of mudrocks. However, mudrock density is also a function of matrix mineralogy, porosity, applied load, temperature and pore fluid pressure. Experimental results (using a population of Neogene mudstones from Japan) in Hoshino et al. (1972) have indicated that, for a constant confining pressure, ductility decreases as density increases, thereby supporting the view that compaction and diagenetic changes alter the picture of ductility at increased burial depths in the sedimentary column (Fig. 5). Shales with densities less than ~2.2 g/cm^3 exhibit ductile behaviour; shales with densities ~2.2–2.5 g/cm^3 are transitional and prob-

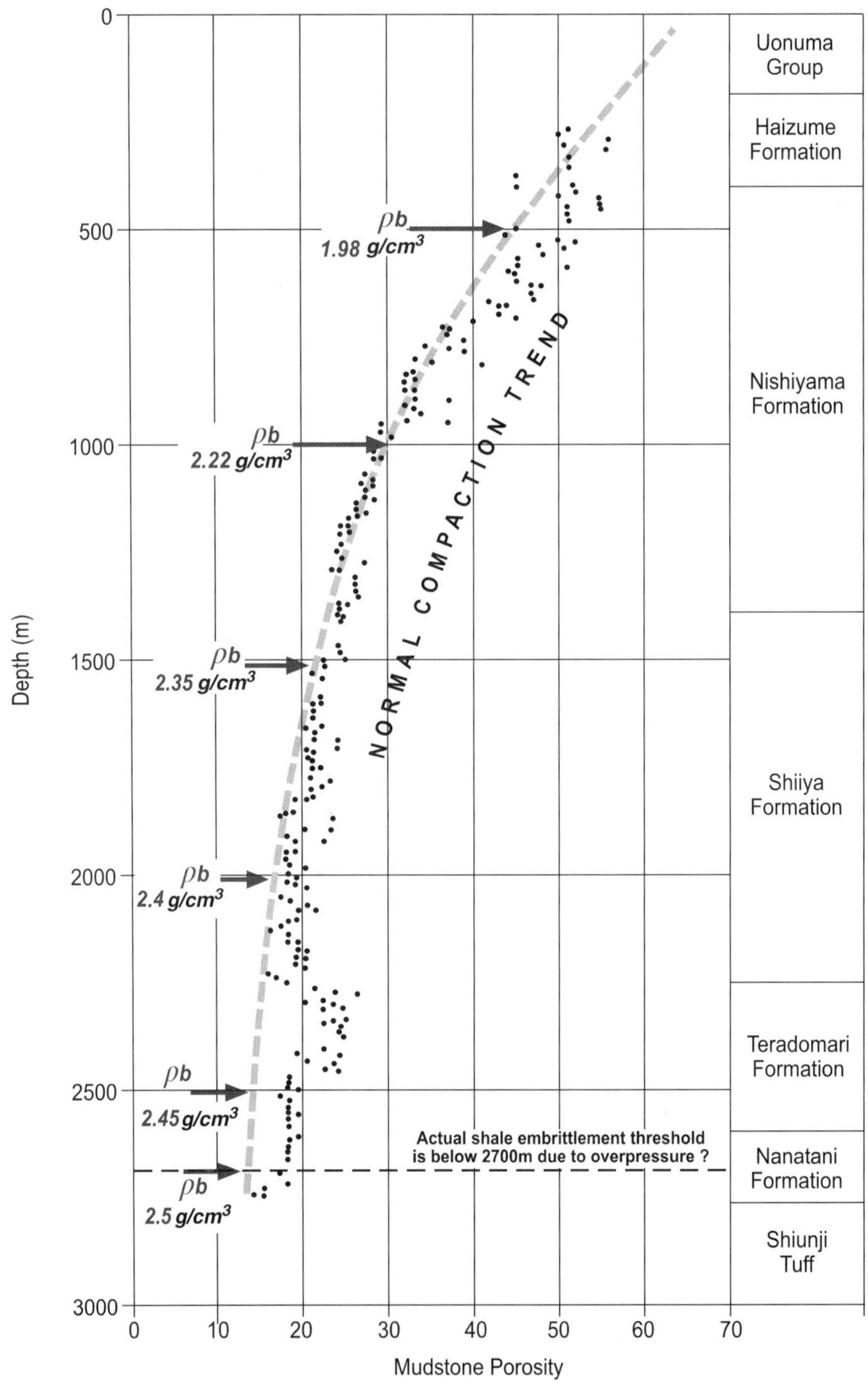

Fig. 6. Porosity–depth behaviour for a suite of Neogene mudrocks from Japan (after Magara, 1968). The organically rich, overpressured, mudrocks of the Teradomari and Nanatani formations form the cap-rock to hydrocarbon accumulations in the underlying fractured volcanic reservoirs of the Miocene Shiunji Tuff Formation and equivalents. A general increase in shale density with depth (normal compaction trend) is observed down to 2200 m. Below this depth mudstone porosity deviates from the 'normal compaction' trend, due to overpressure. (The relative enrichment of the Teradomari Formation in low density organic matter may also contribute to the deviation from the 'normal' trend.) Under conditions of 'normal compaction', embrittlement ($\rho_b = 2.5$ g/cm^3) of these Neogene claystones would occur at 2700 m; however, actual shale embrittlement probably occurs below a burial depth of 3000 m, due to the presence of overpressures.

ably exhibit a wide range of mechanical behaviour; shales with densities in excess of 2.5 g/cm^3 manifest brittle behaviour by fracturing at strains of less than 3%. If mudrock density can be used as a proxy for ductility then the onset of shale embrittlement can be estimated when the density/depth or porosity/depth behaviour of the claystone is known (Fig. 6). This collection of Neogene mudrocks presented in Magara (1968) indicates the potential role of overpressure in defining the shale embrittlement threshold for any given basin. Geothermal gradient and palaeotemperature history also exert a critical control on diagenesis and hence the rheological evolution of mudrocks.

With respect to cap-rock integrity during basin inversion the key factors are the timing and magnitude of the deformation and the mechanical behaviour (brittle vs ductile) of the cap-rock at the time of deformation. Brittle shales are more likely to rupture and leak than ductile shales and evaporites which may exhibit plastic flow under the applied deformation. However, even evaporitic rocks which serve as extremely efficient ductile seals, when overburden burial exceeds 1000 m, can manifest brittle behaviour at shallow depths (Downey, 1994).

Bolton et al. (1998) have shown experimentally that although elevated pore-fluid pressures reduce effective stress and enhance shear deformation, it is the consolidation state at the onset of shear that is the crucial factor with respect to the deformation style and resulting permeability. Timing of overpressure is particularly relevant in this regard as it can change the consolidation state of a mudstone cap-rock with respect to effective stress. Ingram and Urai (1999) have indicated that claystone cap-rocks that have undergone substantial uplift are prone to the formation of dilatant fractures as they are likely to be over-consolidated and have anomalous strength. However, in the context of exhumed basins, the mechanical behaviour of a mudstone cap-rock will be dependent upon whether or not embrittlement has been achieved prior to exhumation.

Physical processes which occur during exhumation

A number of physical processes may impact cap-rocks during exhumation, including erosion, tectonic deformation, shear failure, hydrofracturing due to disequilibrium pore pressure conditions and a changing hydrodynamic regime. Each of these processes must be examined in the context of the evolutionary changes that may be occurring in the cap-rock, in the petroleum system and at the basin scale. For example both the shear strength and tensile strength of mudrocks increase through burial and compaction (Fig. 7), hydrocarbons may migrate into the trap,

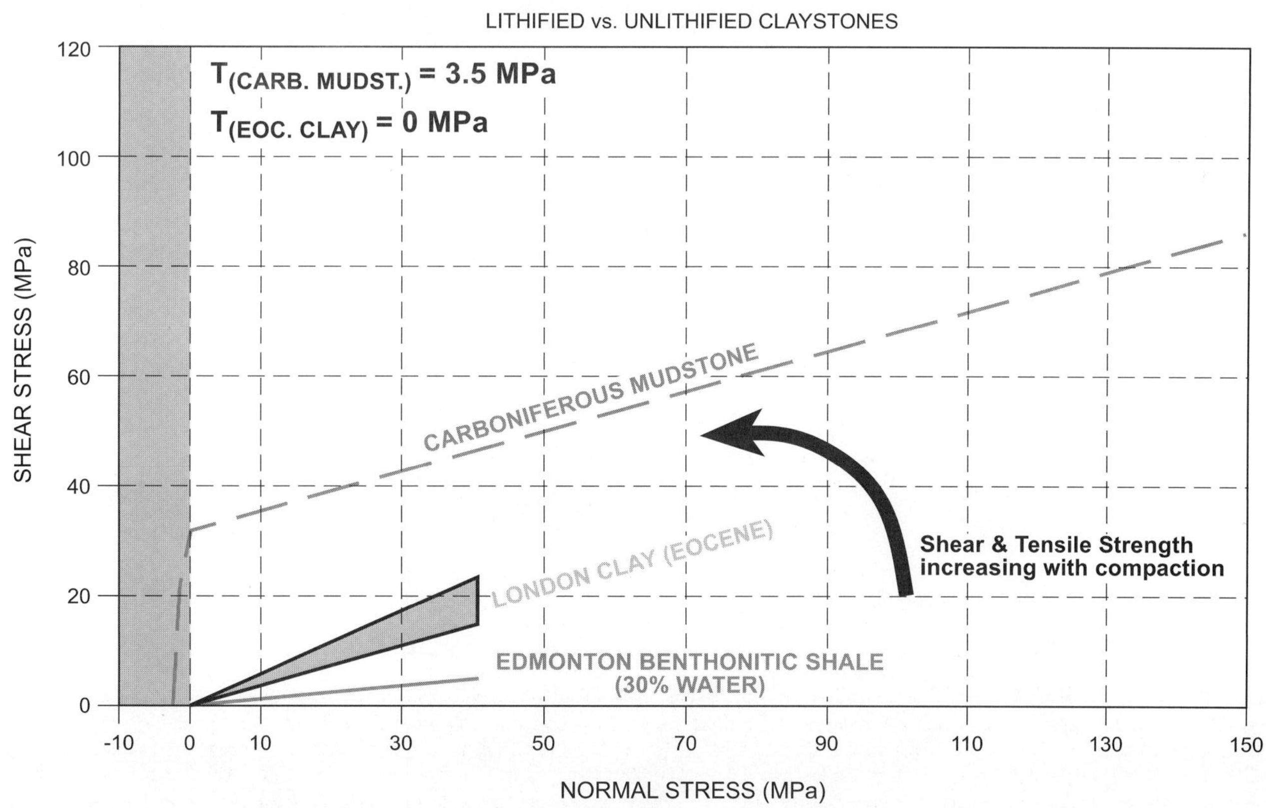

Fig. 7. Mohr–Coulomb failure envelopes for claystones at different levels of compaction. Both shear strength and tensile strength increase with mechanical and chemical compaction. After Cartwright and Lonergan (1996).

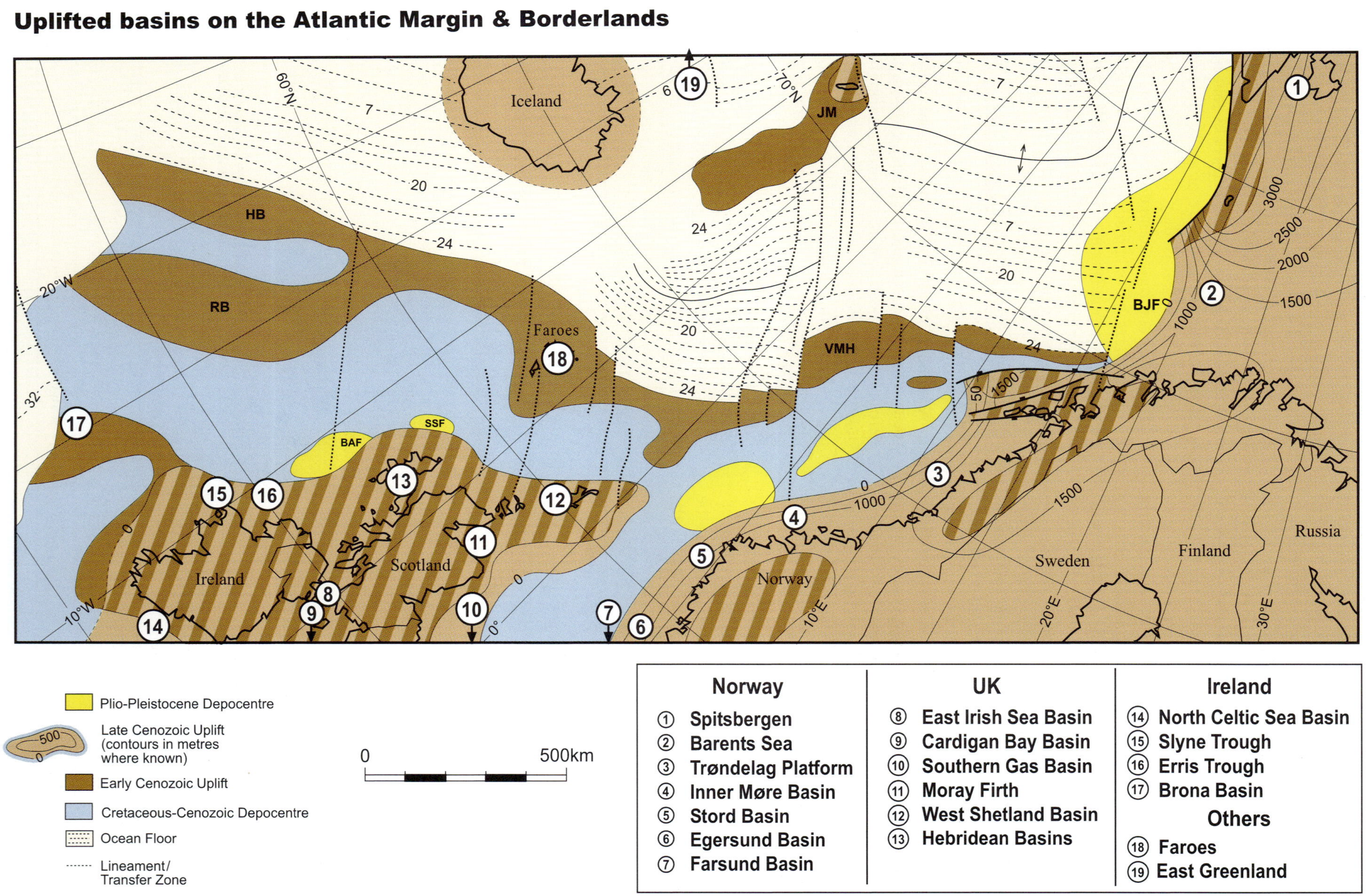

Fig. 8. Uplifted basins of the northeastern Atlantic Margin and borderlands. Many of these basins are characterised by Cenozoic exhumation events which have occurred after the initial migration of hydrocarbons into traps.

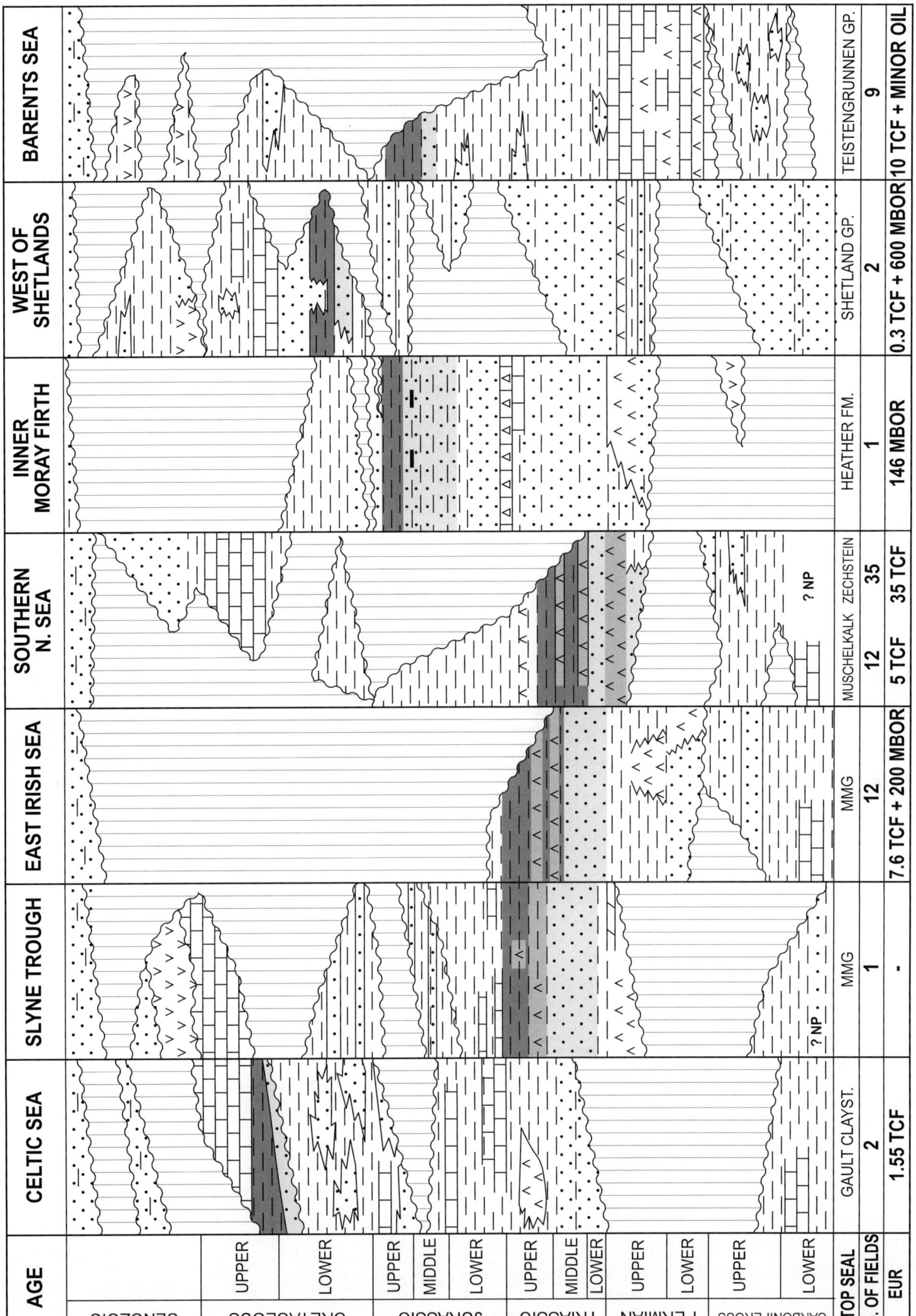

Fig. 9. Regional seals to hydrocarbon accumulations in exhumed basins of the Atlantic Margin and borderlands. Shales and evaporites are the most common cap-rock lithologies in this setting. *MMG* = Mercia Mudstone Group. (MBOR = million barrels oil recoverable.)

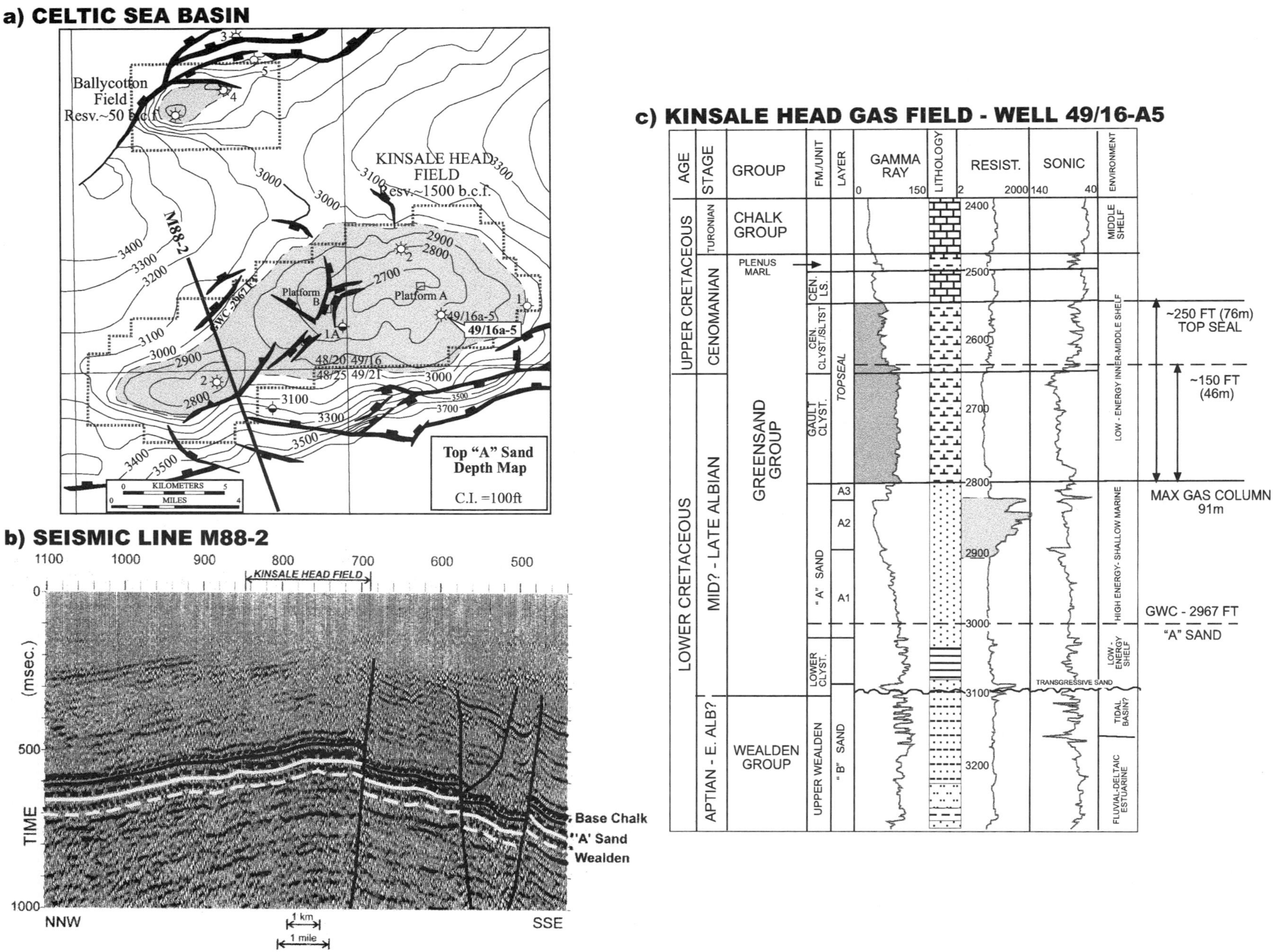
a) CELTIC SEA BASIN
Ballycotton Field Resv.~50 b.c.f.
KINSALE HEAD FIELD Resv.~1500 b.c.f.
M88-2
Platform B
Platform A
49/16a-5
48/20 49/16
48/25 49/21
Top "A" Sand Depth Map
C.I. =100ft
KILOMETERS
MILES
b) SEISMIC LINE M88-2
KINSALE HEAD FIELD
TIME (msec.)
NNW
SSE
1 km
1 mile
Base Chalk
'A' Sand
Wealden
c) KINSALE HEAD GAS FIELD - WELL 49/16-A5
AGE
STAGE
GROUP
FM./UNIT
LAYER
GAMMA RAY
LITHOLOGY
RESIST.
SONIC
ENVIRONMENT
UPPER CRETACEOUS
LOWER CRETACEOUS
TURONIAN
CENOMANIAN
MID? - LATE ALBIAN
APTIAN - E. ALB?
CHALK GROUP
PLENUS MARL
GREENSAND GROUP
WEALDEN GROUP
CEN. LS.
CEN. CLYST./SLTST
GAULT CLYST.
"A" SAND
LOWER CLYST.
UPPER WEALDEN
TOPSEAL
A3
A2
A1
"B" SAND
MIDDLE SHELF
LOW - ENERGY INNER-MIDDLE SHELF
HIGH ENERGY- SHALLOW MARINE
LOW - ENERGY SHELF
TIDAL BASIN?
FLUVIAL-DELTAIC ESTUARINE
TRANSGRESSIVE SAND
~250 FT (76m) TOP SEAL
~150 FT (46m)
MAX GAS COLUMN 91m
GWC - 2967 FT
"A" SAND

thereby locally increasing overpressure in the trap, and hydraulic head may be developed due to the uplift of the basin margins.

Fluids in the subsurface may manifest static behaviour (hydrostatic condition) or dynamic behaviour (hydrodynamic condition). Hubbert (1953) demonstrated that under hydrodynamic conditions accumulations of oil or gas will invariably exhibit inclined oil– or gas–water interfaces and in such cases the computation of hydrocarbon columns, based on the assumption of hydrostatic conditions, will be spurious. Water flow in sedimentary basins is driven by geographic variations in water potential which can change considerably in nature and distribution during the evolution of a basin (Wells, 1988). The early compaction history of an extensional basin, characterised by continual subsidence, will result in up-dip water flow to the basin margins. The driving force of this system is the relatively high water potential in the basin centre generated by water released through the processes of mechanical compaction and clay mineral transformation. This water-potential system changes if significant topographic relief forms adjacent to the basin, due to isostatic effects or some other mechanism. The earlier hydraulic system of the basin is then reversed. Elevated water tables along the basin margin create a hydraulic head which drives water flow inward towards the basin centre, provided upward discharge is possible there.

A significant consequence of basin inversion is a change in hydrodynamic conditions within the basin. The pattern of exhumation convolved with the pre-existing basin morphology may result in the outcrop of key aquifers, a redistribution of recharge and discharge areas and a change in the direction of gravity driven fluid flow within the basin. The existence of regional, topographically driven groundwater flow-systems has been documented for several exhumed sedimentary basins (Wells, 1988; Deming et al., 1992; Bredehoeft et al., 1994; Deming, 1994; Cramer et al., 1999).

Hydrodynamic effects on seal capacity may, for all practical purposes, be ignored except in those basins which manifest clear evidence of hydraulic gradients. In these basins hydrodynamic flow may modify seal retention capacity by either increasing or decreasing the driving pressure against the seal (Allen and Allen, 1990). When the hydrodynamic force has an upward vector, it adds to the buoyancy force thus reducing the hydrocarbon column heights the seal can support. In the case of a downward vector it reduces the buoyancy force on the seal and permits the retention of an increased hydrocarbon column.

Empirical observations and discussion — exhumed Atlantic Margin and borderland basins

Many Atlantic Margin and borderland basins are characterised by exhumation during the Cenozoic (Fig. 8). Although prior exhumation events may have occurred during the evolution of these basins, in terms of top-seal assessment, Cenozoic exhumation is most critical as it generally occurs in these basins after the initial migration of hydrocarbons into traps.

Shale and evaporite cap-rocks form the main regional seals to hydrocarbon accumulations in exhumed basins of the Atlantic Margin and borderlands (Fig. 9). Shale cap-rocks of Jurassic–Cretaceous age are prevalent in the Celtic Sea, Inner Moray Firth, West of Shetlands and Barents Sea basins and mixed evaporite and shale seals of Triassic age are encountered in three basins (Slyne Trough, East Irish Sea and Southern North Sea). The prodigious Zechstein evaporite seal is cap-rock to an estimated ultimate recovery of >150 TCF in the southern Permian Basin, including the giant Groningen accumulation (97 TCF) and approximately 35 TCF contained in 35 accumulations in the UK sector of the Southern North Sea Basin (Glennie, 1997) (Fig. 9).

There are a number of hydrocarbon trapping and top-seal configurations observed in these exhumed basins. In the Celtic Sea Basin, a maximum gas column of 91 m in the Kinsale Head Gas Field is retained by 46 m of Gault Claystone in a compressional anticlinal flexure (Fig. 10). This basin centred accumulation experienced approximately 900 m of exhumation during the early Cenozoic and has been overprinted by a compressional deformation which is poorly constrained but probably post-Paleocene in age (Murdoch et al., 1995). Local evidence suggests that the maximum burial depth of the Gault Claystone was in the range of 1700–1800 m which may not have been sufficient to achieve shale embrittlement prior to the applied deformation associated with exhumation and compressional inversion. Under this scenario, it

Fig. 10. Kinsale Head Gas Field, a compressional inversion structure in the exhumed Celtic Sea Basin. (a) Depth structure map on top main reservoir, indicating that the gas–water contact (GWC at −2967 ft.) is coincident with the spill point to the north of the structure. (b) NNW–SSE seismic line, illustrating reverse faulting on the southern limb of the anticline. (c) Type log for the 'A' Sand/Gault Claystone reservoir/top-seal couplet at the Kinsale Head Field. A maximum gas column of 91 m, in the Greensand reservoir, is retained in situ, without apparent leakage, by 46 m of Gault Claystone. It is postulated that Cenozoic exhumation of the Gault Claystone, in the Kinsale Head area, occurred prior to shale embrittlement, thereby inhibiting the formation of dilatant fractures. Data from Taber et al. (1995).

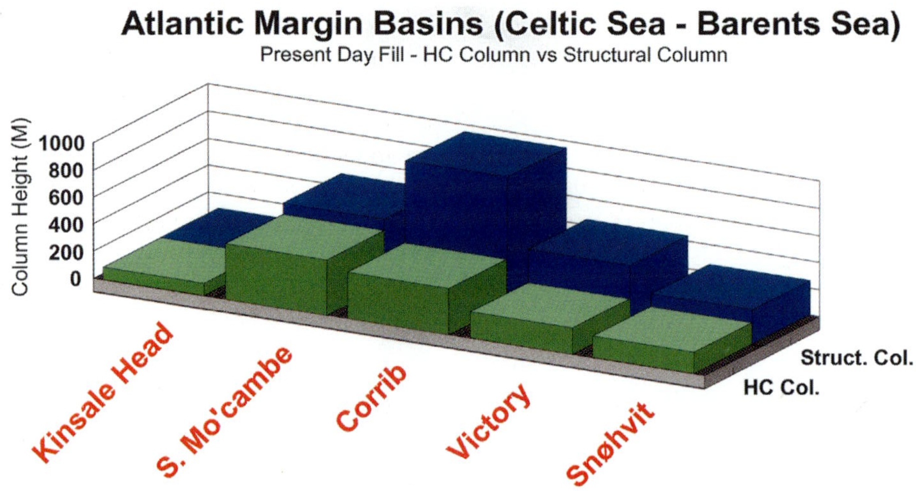

Fig. 11. Underfilled traps in exhumed Atlantic Margin basins. Of the 5 major gas accumulations from the Celtic Sea to the Barents Sea (Kinsale Head, South Morecambe, Corrib, Victory, Snøhvit) only the Kinsale Head Field is full to structural spill point, present-day.

is postulated that fracture development was inhibited as the Gault Claystone responded by plastic flow to Cenozoic deformation. This hypothesis is consistent with the experimental results of Bolton et al. (1998) which suggests that underconsolidated clayey sediments, undergoing shear, deform by bulk volume loss which reduces permeability, and result in weakly developed deformation fabrics which have little impact on the hydrological properties of the claystone.

All four generic leakage mechanisms (tectonic breaching, capillary leakage, hydraulic leakage and diffusion) operate in both continually subsiding basins and exhumed basin settings. However, the critical aspect of trap leakage in exhumed basin settings is a lower probability of trap replenishment due to the 'switching off' of hydrocarbon generation during regional uplift. In such cases, top-seal failure (induced by tectonic breaching or hydraulic leakage) *during exhumation* may result in the catastrophic loss of a pre-existing hydrocarbon fill whereas *post-exhumation* these traps can only be replenished from a curtailed hydrocarbon budget, which consists primarily of re-migrating oil and gas. This is consistent with the observation that a number of hydrocarbon accumulations in exhumed basin settings along the Atlantic Margin are characterised by underfilled traps (Fig. 11). Empirical observation also indicates that hydrocarbon accumulations in exhumed basins are characterised by hydrostatically pressured or modestly overpressured reservoirs, whereas significantly overpressured reservoirs are common in basins which have experienced relatively continual subsidence (Fig. 12). This suggests a close causal relationship between regional uplift, hydrocarbon re-migration and dissipation of overpressures.

Exhumation may also have positive implications for the capillary and hydraulic retention capacity of mudrock seals. Increased mechanical compaction

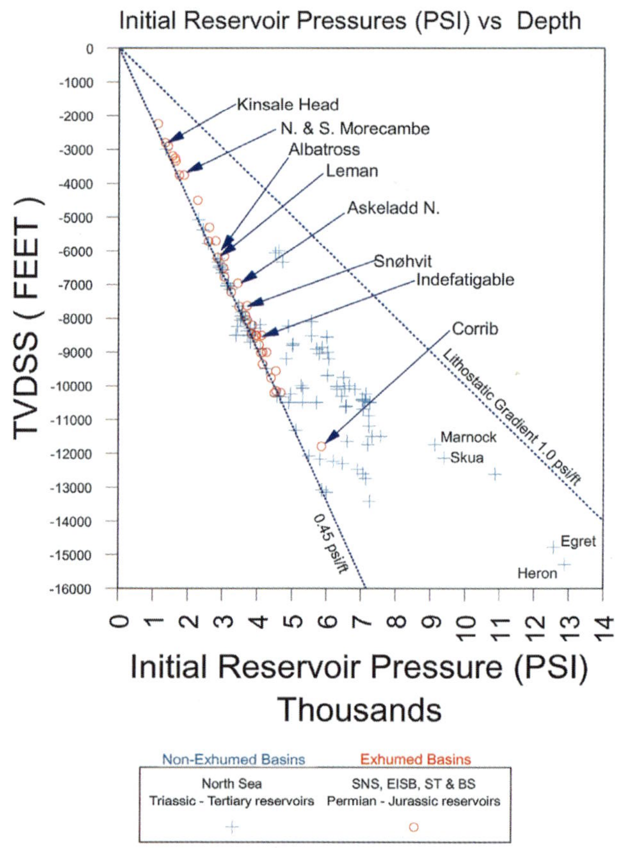

Fig. 12. Comparison of initial reservoir pressures versus depth for hydrocarbon accumulations in exhumed basins versus non-exhumed basins. Accumulations in exhumed basins are characterised by hydrostatically pressured or modestly overpressured reservoirs, whereas significantly overpressured reservoirs are common in basins which have experienced relatively continual subsidence. *SNS* = Southern North Sea Basin; *EISB* = East Irish Sea Basin; *ST* = Slyne Trough; *BS* = Barents Sea Basin (TVDSS = true vertical depth sub-sealevel). Data compiled essentially from Spencer et al. (1986), Abbotts (1991), Pooler and Amory (1999).

due to burial results in reduced interconnected pore throat sizes and increasing shear strength and tensile strength for a claystone rock. For example, when a

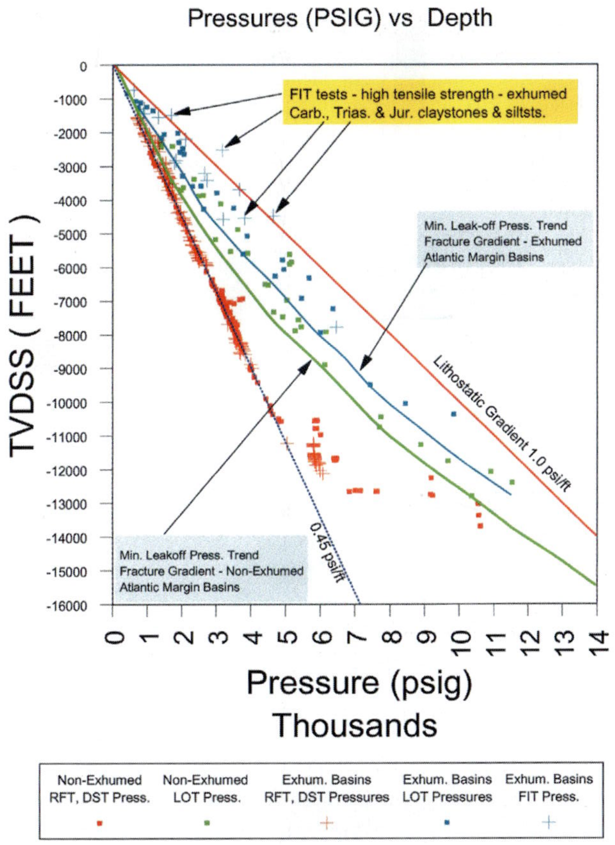

Fig. 13. Formation pressures (RFTs and DSTs), Formation integrity tests (FITs) and Leak-off tests (LOTs) for basins offshore Ireland. Data indicate that, for any given burial depth, the minimum horizontal stress or fracture pressure (defined by the lower envelope of LOT pressures) is higher in exhumed basins than in those characterised by continual subsidence.

claystone is exhumed it retains the tensile strength of its maximum burial depth and consequently, a higher pore fluid pressure will be required to induce hydrofracturing than for a claystone in a continually subsiding basin at the same depth. Leak-off tests (LOTs) and Formation Integrity tests (FITs) from a sub-set of the Atlantic Margin basins support these observations (Fig. 13). These data indicate that, for any given burial depth, the minimum horizontal stress or fracture pressure (defined by the lower envelope of LOT pressures) is higher in exhumed basins than in those basins characterised by continual subsidence. Critically, there are a number of FITs performed on exhumed claystones of Carboniferous, Triassic and Jurassic age which indicate that these seal rocks have very high tensile strengths, appropriate to their maximum burial depth, prior to exhumation. This suggests that *post-exhumation* top-seal integrity is relatively high in many of the Atlantic Margin and borderland basins under low differential stresses. However, the anomalously high shear strength of exhumed mudstones may result in the development of dilatant shear fractures (under low confining pressures) if shale embrittlement has been achieved prior to exhumation. The absence of seismic chimneys across major gas accumulations in many of these basins (e.g. Celtic Sea, EISB, SNS) suggests that dynamic leakage through the top-seal is not occurring present day and that pore fluid pressures are below the top-seal capillary and hydraulic leakage thresholds for these accumulations.

Conclusions

A number of effective regional cap-rocks are recognised in the petroleum systems of exhumed Atlantic Margin basins. However, the observation of underfilled traps, close to hydrostatically pressured reservoirs and breached traps with residual oil and gas shows, suggests that top-seal behaviour exercises a critical control on hydrocarbon distribution and re-distribution in exhumed basin settings. The following are the principal conclusions of this review.

(1) Lithofacies is a major control on top-seal efficiency in exhumed basins settings. The juxtaposition of a halite directly above the hydrocarbon bearing reservoir offers the most favourable condition for the retention of hydrocarbons in exhumed traps. However, the empirical evidence from the Atlantic Margin suggests that mudrocks can form efficient top-seals in exhumed basins under certain conditions.

(2) The behaviour of any cap-rock lithology during exhumation is dependent upon the physical and mechanical characteristics of the cap-rock at the time of exhumation and the timing and conditions of the associated deformation relative to the timing of hydrocarbon emplacement.

(3) Mudrocks may form effective cap-rocks in exhumed basins when the deformation associated with exhumation occurs prior to embrittlement and the cap-rock exhibits ductile behaviour. Where exhumation occurs post-embrittlement the shale cap-rock will facilitate hydrocarbon leakage through the development of extensive fracture networks.

(4) *Syn-exhumation* top-seal efficiency (fluid retention capacity) is a major exploration risk in exhumed basin settings, though *post-exhumation* top-seal integrity in these basins may be relatively high. This suggests that a major exploration risk factor in exhumed basin settings pertains to the limited hydrocarbon budget available post-regional uplift and the efficiency of the re-migration process.

Acknowledgements

The authors would like to thank Graham Yielding and Robert Hunsdale for their helpful reviews of the manuscript. We would also like to thank John Kipps

(Statoil (UK) Ltd.) who draughted most of the diagrams in this paper.

References

Abbotts, I.L. (Editor), 1991. United Kingdom Oil and Gas Fields, 25 Years Commemorative Volume. Geological Society, London, Memoir 14.

Allen, P.A. and Allen, J.R., 1990. Basin analysis: principles and applications. Blackwell Scientific Publications, Oxford, 449 pp.

Berg, R.R., 1975. Capillary pressures in stratigraphic traps. Am. Assoc. Pet. Geol. Bull., 59 (6): 939–956.

Bjørlykke, K., 1999. Principal aspects of compaction and fluid flow in mudstones. In: A.C. Alpin, A.J. Fleet and J.H.S. Macquaker (Editors), Muds and Mudstones: Physical and Fluid Flow Properties. Geol. Soc. London, Spec. Publ., 158: 73–78.

Bolton, A.J., Maltman, A.J. and Clennell, M.B., 1998. The importance of overpressure timing and permeability evolution in fine-grained sediments undergoing shear. J. Struct. Geol., 20 (8): 1013–1022.

Bredehoeft, J.D., Wesley, J.B. and Fouch, T.D., 1994. Simulations of the origin of fluid pressure, fracture generation and the movement of fluids in the Uinta Basin, Utah. Am. Assoc. Pet. Geol. Bull., 78: 1729–1747.

Capuano, R.M., 1993. Evidence of fluid flow in microfractures in geopressured shales. Am. Assoc. Pet. Geol. Bull., 77: 1303–1314.

Cartwright, J.A. and Lonergan, L., 1996. Volumetric contraction during the compaction of mudrocks: a mechanism for the development of regional-scale polygonal fault systems. Basin Res., 8: 183–193.

Clayton, C.J. and Hay, S.J., 1994. Gas migration mechanisms from accumulation to surface. Bull. Geol. Soc. Denm., 41: 12–23.

Cowan, G., Burley, S.D., Hoey, N., Holloway, P., Bermingham, P., Beveridge, N., Hamborg, M. and Sylta, Ø., 1999. Oil and gas migration in the Sherwood Sandstone of the East Irish Sea Basin. In: A.J. Fleet and S.A.R. Boldy (Editors), Petroleum Geology of Northwest Europe: Proceedings of the 5th Conference. Geological Society, London, pp. 41–61.

Cramer, B., Poelchau, H.S., Gerling, P., Lopatin, N.V. and Littke, R., 1999. Methane released from groundwater: the source of natural gas accumulations in northern West Siberia. Mar. Pet. Geol., 16: 225–244.

Davis, G.H. and Reynolds, S.J., 1996. Structural Geology of Rocks and Regions. Wiley, Chichester, 2nd edition, 650 pp.

Deming, D., 1994. Fluid flow and heat transport in the upper continental crust. In: J. Parnell (Editor), Geofluids: Origin, Migration and Evolution of Fluids in Sedimentary Basins. Geol. Soc. London, Spec. Publ., 78: 27–42.

Deming, D., Sass, J.H., Lachenbruch, A.H. and Derito, R.F., 1992. Heat flow and subsurface temperature as evidence for basin-scale groundwater flow. Geol. Soc. Am. Bull., 104: 528–542.

Dewhurst, D.N., Yang, Y. and Alpin, A.C., 1999. Permeability and fluid flow in natural mudstones. In: A.C. Alpin, A.J. Fleet and J.H.S. Macquaker (Editors), Muds and Mudstones: Physical and Fluid Flow Properties. Geol. Soc. London, Spec. Publ., 158: 23–43.

Doré, A.G. and Jensen, L.N., 1996. The impact of late Cenozoic uplift and erosion on hydrocarbon exploration: offshore Norway and some other uplifted basins. Global Planet. Change, 12: 415–436.

Doré, A.G., Lundin, E.R., Jensen, L.N., Birkeland, Ø., Eliassen, P.E. and Fichler, C., 1999. Principal tectonic events in the evolution of the northwest European Atlantic margin. In: A.J. Fleet and S.A.R. Boldy (Editors), Petroleum Geology of Northwest Europe: Proceedings of the 5th Conference. Geological Society, London, pp. 41–61.

Downey, M.W., 1984. Evaluating seals for hydrocarbon accumulations. Am. Assoc. Pet. Geol. Bull., 68: 1752–1763.

Downey, M.W., 1994. Hydrocarbon seal rocks. In: L.B. Magoon and W.G. Dow (Editors), The Petroleum System — from Source to Trap. Am. Assoc. Pet. Geol. Mem., 60: 159–164.

Gabrielsen, R.H. and Kløvjan, O.S., 1997. Late Jurassic–early Cretaceous caprocks of the southwestern Barents Sea: fracture systems and rock mechanical properties. In: P. Møller-Pedersen and A.G. Koestler (Editors), Hydrocarbon Seals: Importance for Exploration and Production. Norwegian Petroleum Society (NPF), Special Publication 7. Elsevier, Amsterdam, pp. 73–89.

Glennie, K.W., 1997. History of exploration in the southern North Sea. In: K. Ziegler, P. Turner and S.R. Daines (Editors), Petroleum Geology of the Southern North Sea: Future Potential. Geol. Soc., Spec. Publ., 123: 5–16.

Gretener, P.E., 1981. Pore pressure: fundamentals, general ramifications and implications for structural geology (revised edition). Am. Assoc. Pet. Geol., Educ. Course Note Ser., 4: 15–33.

Grunau, H.R., 1987. A worldwide look at the cap-rock problem. J. Pet. Geol., 10 (3): 243–266.

Hall, D.M., Duff, B.A., Elias, M. and Gytri, S.R., 1997. Pre-cretaceous top-seal integrity in the greater Ekofisk area. In: P. Møller-Pedersen and A.G. Koestler (Editors), Hydrocarbon Seals: Importance for Exploration and Production. Norwegian Petroleum Society (NPF), Special Publication 7. Elsevier, Amsterdam, pp. 231–242.

Handin, J.H., Hager Jr., R.V., Friedman, M. and Feather, J.N., 1963. Experimental deformation of sedimentary rocks under confining pressure: pore pressure tests. Am. Assoc. Pet. Geol. Bull., 47: 717–755.

Hoshino, K., Koide, H., Inami, K., Iwamura, S. and Mitsui, S., 1972. Mechanical properties of Japanese Tertiary sedimentary rocks under high confining pressures. Geol. Surv. Jpn., Rep., 244, 200 pp.

Hubbert, M.K., 1953. Entrapment of petroleum under hydrodynamic conditions. Am. Assoc. Pet. Geol. Bull., 37 (8): 1954–2026.

Hubbert, M.K. and Rubey, W.W., 1959. Role of pore fluid pressures in the mechanics of overthrust faulting. Geol. Soc. Am. Bull., 70: 115–205.

Ingram, G.M. and Urai, J.L., 1999. Top-seal leakage through faults and fractures: the role of mudrock properties. In: A.C., Alpin, A.J. Fleet and J.H.S. Macquaker (Editors), Muds and Mudstones: Physical and Fluid Flow Properties. Geol. Soc. London, Spec. Publ., 158: 125–135.

Kettel, D., 1997. The dynamics of gas flow through rock salt in the scope of time. In: P. Møller-Pedersen and A.G. Koestler (Editors), Hydrocarbon Seals: Importance for Exploration and Production. Norwegian Petroleum Society (NPF), Special Publication 7. Elsevier, Amsterdam, pp. 175–186.

Knipe, R.J., Fisher, Q.J., Jones, G., McAllister, E., Needham, T., Bolton, A., Davies, R., Edwards, E., Harris, S.D., Henson, D., Li, A., Odling, N., Pecher, R., Porter, J.R., Allin, J. and White, E., 2000. Quantification and prediction of fault seal parameters: the importance of the geohistory. In: Extended Abstracts, Hydrocarbon Seal Quantification, Norwegian Petroleum Society (NPF) Conference, Clarion Hotel, Stavanger, 16–18 October.

Kontorovitch, A.E., Mandel'baum, V.S., Surkov, V.S., Trofimuk, A.A. and Zolotov, A.N., 1990. Lena-Tuguska Upper Proterozoic–Palaeozoic petroleum superprovince. In: J. Brooks (Editor), Classic Petroleum Provinces. Geol. Soc. London, Spec. Publ., 50: 203–219.

Krooss, B.M., Leythaeuser, D. and Schaefer, R.G., 1992. The quantification of diffusive hydrocarbon losses through cap rocks of natural gas reservoirs — a reevaluation. Am. Assoc. Pet. Geol. Bull., 76 (3): 403–406.

Leythaeuser, D., Schaefer, R.G. and Yukler, A., 1982. Role of diffusion in primary migration of hydrocarbons. Am. Assoc. Pet. Geol. Bull., 66: 408–429.

Magara, K., 1968. Compaction and migration of fluids in Miocene mudstones, Nagaoka Plain, Japan. Am. Assoc. Pet. Geol. Bull., 52: 2466–2501.

Montel, F., Caillet, G., Pucheu, A. and Caltagirone, J.P., 1993. Diffusion model for predicting reservoir gas losses. Mar. Pet. Geol., 10: 51–57.

Murdoch, L.M., Musgrove, F.W. and Perry, J.S., 1995. Tertiary uplift and inversion history in the North Celtic Sea Basin and its influence on source rock maturity. In: P.F. Croker and P.M. Shannon (Editors), The Petroleum Geology of Ireland's Offshore Basins. Geol. Soc. London, Spec. Publ., 93: 297–319.

Osborne, M.J. and Swarbrick, R.E., 1997. Mechanisms for generating overpressure in sedimentary basins: a reevaluation. Am. Assoc. Pet. Geol. Bull., 81 (6): 1023–1041.

Palciauskas, V.V. and Domenico, P.A., 1980. Microfracture development in compacting sediments: relation to hydrocarbon maturation kinetics. Am. Assoc. Pet. Geol. Bull., 64: 927–937.

Pedersen, T. and Bjørlykke, K., 1994. Fluid flow in sedimentary basins: model of pore water flow in a vertical fracture. Basin Res., 6: 1–16.

Pooler, J. and Amory, M., 1999. A subsurface perspective on ETAP — an integrated development of seven Central North Sea fields. In: A.J. Fleet and S.A.R. Boldy (Editors), Petroleum Geology of Northwest Europe: Proceedings of the 5th Conference. Geological Society, London, pp. 993–1006.

Sclater, J.G. and Christie, P.A.F., 1980. Continental stretching: an explanation of the post-mid-Cretaceous subsidence of the Central North Sea basin. J. Geophys. Res., 85: 3711–3739.

Schloemer, S. and Krooss, B.M., 1997. Experimental characterisation of the hydrocarbon sealing efficiency of cap rocks. Mar. Pet. Geol., 14 (5): 565–580.

Schowalter, T.T., 1979. Mechanics of secondary hydrocarbon migration trapping. Am. Assoc. Pet. Geol. Bull., 63: 723–760.

Secor, D., 1965. Role of fluid pressure in jointing. Am. J. Sci., 263: 633–646.

Seedhouse, J.K. and Racey, A., 1997. Sealing capacity of the Mercia Mudstone Group in the East Irish Sea Basin: implications for petroleum exploration. J. Pet. Geol., 20: 261–286.

Sibson, R.H., 1995. Selective fault reactivation during basin inversion: potential for fluid redistribution through fault-valve action. In: J.G. Buchanan and P.G. Buchanan (Editors), Basin Inversion. Geol. Soc. London, Spec. Publ., 88: 3–19.

Spain, D.R. and Conrad, P.C., 1997. Quantitative analysis of top-seal capacity: offshore Netherlands, Southern North Sea. Geol. Mijnbouw, 76: 217–226.

Spencer, A.M., Campbell, C.J., Hanslien, S.H., Holter, E., Nelson, P.H.H., Nysæther, E. and Ormaasen, E.G. (Editors), 1986. Habitat of Hydrocarbons on the Norwegian Continental Shelf. Graham and Trotman, London.

Taber, D.R., Vickers, M.K. and Winn Jr., R.D., 1995. The definition of the Albian 'A' Sand reservoir fairway and aspects of associated gas accumulations in the North Celtic Sea Basin. In: P.F. Croker and P.M. Shannon (Editors), The Petroleum Geology of Ireland's Offshore Basins. Geol. Soc. London, Spec. Publ., 93: 227–244.

Watts, N.L., 1987. Theoretical aspects of cap-rock and fault seals for single- and two-phase hydrocarbon columns. Mar. Pet. Geol., 4 (11): 274–307.

Wells, P.R.A., 1988. Hydrodynamic trapping in the Cretaceous Nahr Umr Lower Sand of the North Area, offshore Qatar. J. Pet. Technol., March.

D.V. CORCORAN *Statoil Exploration (Ireland) Ltd., Statoil House, 6, George's Dock, IFSC, Dublin 1, Ireland*
E-mail: dermot.corcoran@statoil.com

A.G. DORÉ *Statoil (UK) Ltd., 11a, Regent Street, London SW1Y 4ST, UK*

Empirical estimation of fault rock properties

Susanne Sperrevik, Paul A. Gillespie, Quentin J. Fisher, Trond Halvorsen and Rob J. Knipe

Faults in clastic sequences are often significant barriers to single-phase fluid flow and can act as absolute barriers to the flow of non-wetting phases over geological time. Knowledge of the fault rock flow properties, as well as the width of the fault zone is required in order to conduct fluid flow simulations in faulted reservoirs. In this paper we present an equation for estimating fault zone thickness from fault throw based on outcrop data from Sinai and Northumberland. These data show that the throw/thickness relationship is dependent on lithology, and can be related to the clay content of the fault zone. The permeability and threshold pressures of fault rocks are dependent on factors such as the mineralogical composition of the faulted rock, the effective stress conditions and the time–temperature history of the reservoir prior to, during and following deformation. A strong power law relationship is established between threshold pressure and permeability, which is insensitive to the faulting mechanisms. The permeability and the threshold pressures of both the host rocks and the fault rocks can be represented by functions which are dependent on the clay content and the maximum burial depth (i.e. time–temperature history), whereas for the fault rocks the depth (i.e. effective stress conditions) at the time of deformation also needs to be taken into account. The database from which these empirical relationships were derived contains core measurements from faults and their associated host rocks in siliciclastic sequences from the North Sea. Many types of fault rock are contained within the database (disaggregation zones, cataclastic faults, phyllosilicate-framework faults and clay smears) and these have experienced a wide range in their maximum burial depths (2000–4500 m). In reservoir simulation the sealing effect of the faults can be represented as transmissibility modifiers for each grid cell, calculated from knowledge of fault rock permeability, the width of the fault zone, the grid block permeabilities and the geometry of the simulation grid. We have applied the technique to a number of North Sea reservoirs, using the new equation for calculating fault rock permeability. However, even if the new equation produced lower permeabilities than previously published relationships, in all cases the transmissibility modifiers generated by this technique proved consistently too high (1–2 orders of magnitude) in order to produce good history matches. In order to further improve the model, and to get better history match, we think that it is important to include capillary effects, relative transmissibility multipliers, the new equation for calculating fault zone width and to better constrain the clay content of the fault zone. However, better methods are still required for capturing complex fault geometries in the reservoir model.

Introduction

Knowledge of the fluid flow properties of faults is important in hydrocarbon exploration, appraisal and development, and the properties change both along strike, along dip, and also through geologic time (Knott, 1993). In periods when faults are active, they may be favourable pathways for fluid flow, whereas cementation may close them for fluid flow in quiet periods (e.g. Knipe, 1992). Faults can also act as barriers to fluid flow either by juxtaposing reservoir units against impermeable lithologies (Smith, 1966, 1980; Downey, 1984; Watts, 1987), or by creating fault rock with lower permeability and higher capillary entry pressures than the undeformed reservoir (Weber et al., 1978; Gibson, 1994).

The capillary properties of rocks are controlled by the size of the pore throats, and are in theory independent of the thickness of the rock itself. As long as the fault rock is continuous, and none of the interconnected pore throats are too large, the fault will seal. However, it is more likely that a fault seal will be continuous, without any interconnected large pores if the fault rock is not very thin. The thickness and distribution of fault rocks are dependent of the thickness and properties of the rocks involved in the faulting, as well as the fault architecture. Many faults show a very complex geometry with lenses, horses and duplexes (Childs et al., 1997; Gabrielsen and Clausen, 2001). If sand lenses occur in the fault zone, this may lead to increased communication across the fault. On the other hand, if more than one fault plane exists, and each plain can sustain a certain pressure this may lead to increased sealing capacity. Damage zones around faults in clastic rocks also normally have reduced permeability compared to the undeformed rocks (Antonellini and Aydin, 1994, 1995; Antonellini et al., 1994). Complex fault architecture and wide damage

Hydrocarbon Seal Quantification edited by A.G. Koestler and R. Hunsdale. NPF Special Publication 11, pp. 109–125, Published by Elsevier Science B.V., Amsterdam.

zones will additionally reduce the overall permeability and increase the tortuosity, thereby reducing the average flow.

The manner in which fluid flow is best modelled depends upon the time-scale over which flow occurs. For very slow flow rates (hydrocarbon migration) capillary and gravitational forces control two-phase fluid flow. Therefore, the capillary entry pressures must be constrained for hydrocarbon migration modelling and seal studies. During production, as flow rates increase, the viscous forces dominate and permeability becomes an increasingly important parameter. Where there are multiple hydrocarbon phases, relative permeabilities and the capillary threshold pressures must also be considered. The capillary threshold pressure is especially important in the late stages of the production phase when the hydrocarbon column height decreases and faults that earlier were open to fluid flow can start to act as barriers.

As the permeability and capillary entry pressure of large faults in reservoirs cannot be measured directly, methods are required for prediction of the fluid flow properties of fault rocks on the basis of reservoir properties that can be estimated. Empirical equations exist that relate permeability to the clay content of the fault rocks based on core-plug measurements. As we show in this paper, the error in the existing equations can be as much as four orders of magnitude. The main focus of this work has been to develop improved equations for estimating fault rock permeability and fault rock capillary entry pressure. We show that the petrophysical properties of fault rocks are highly dependent on clay content (Gibson, 1994; Yielding et al., 1997), and the geohistory (Knipe, 1992; Fisher and Knipe, 1998). In this study we have therefore investigated the variation in permeability and capillary entry pressure due to variations in clay content, maximum burial depth (which we use as a proxy for time–temperature history) and the depth (a proxy for effective stress conditions) at the time of faulting. A very consistent dataset has been used in which all samples are from the same kind of geological setting and measurements were performed with consistent methods on all samples.

In addition we present a new equation for estimating fault zone thickness from fault throw and clay content based on outcrop data from Sinai (Knott et al., 1996) and new data from Northumberland.

An overview of different parameters referred to in this paper is given in the Notation.

Classification of fault rocks

Faults in porous siliclastic rocks often have fault rocks that have lower porosity, lower permeability and higher capillary entry pressure than their surrounding host rocks (e.g. Antonellini and Aydin, 1994; Fulljames et al., 1997; Fisher and Knipe, 1998). The changes in properties occur because porous media tend to deform by porosity collapse, grain size reduction and the mixing of phyllosilicates with framework grains and grain fragments in the fault zone. In addition, we often see increased mineralization effects in the fault zone, and many fault rocks experience enhanced quartz cementation or grain contact quartz dissolution following initial deformation (Fisher and Knipe, 1998). The clay content of a fault zone, V_f can be estimated using a variety of algorithms. For continuous clay smear the most appropriate criteria are the Clay Smear Potential, CSP (Bouvier et al., 1989) or the Shale Smear Factor, SSF (Lindsay et al., 1993), whereas for fault rocks resulting from mixing of lithologies V_f can be modelled using the Shale Gouge Ratio, SGR (Yielding et al., 1997; our Eq. 1).

$$V_f \approx \mathrm{SGR} = \frac{1}{T}\sum V_m \Delta z \qquad (1)$$

In this study the fault rock types are classified according to the classification used by Fisher and Knipe (1998). A brief summary is given below.

Faults in clean sandstones

The development of faults in clean sandstones depends mainly on the relationship between the initial porosity of the sand and the stress conditions at the time of deformation (Dunn et al., 1973; Engelder, 1974; Mandl et al., 1977) (Fig. 1). Sands and sandstones that deformed under low mean effective stress conditions compared to their hydrostatic yield strength tend to dilate during deformation. The rock will compact if the mean effective stress conditions at the time of deformation are high compared to the hydrostatic yield strength of the sand (Engelder, 1974; Mandl et al., 1977; Sperrevik et al., 2000). Faults in clean sandstones, with throws in the scale of millimetres to centimetres without a discrete slip plane are variously referred to as deformation bands, shear bands or granulation seams (Aydin and Johnson, 1978; Nelson, 1985; Antonellini and Aydin, 1994). In this study, we distinguish between disaggregation zones in which the material deformed by particulate flow and grain reorientation without significant grain fracturing, and cataclasites that have experienced grain breakage in addition to pore collapse by reorientation of grains. Our term cataclasite is used in agreement with the term cataclastic slip band described by Fowles and Burley (1994), and the definition of deformation bands given by Aydin and Johnson (1978, 1983). In this paper we have not used

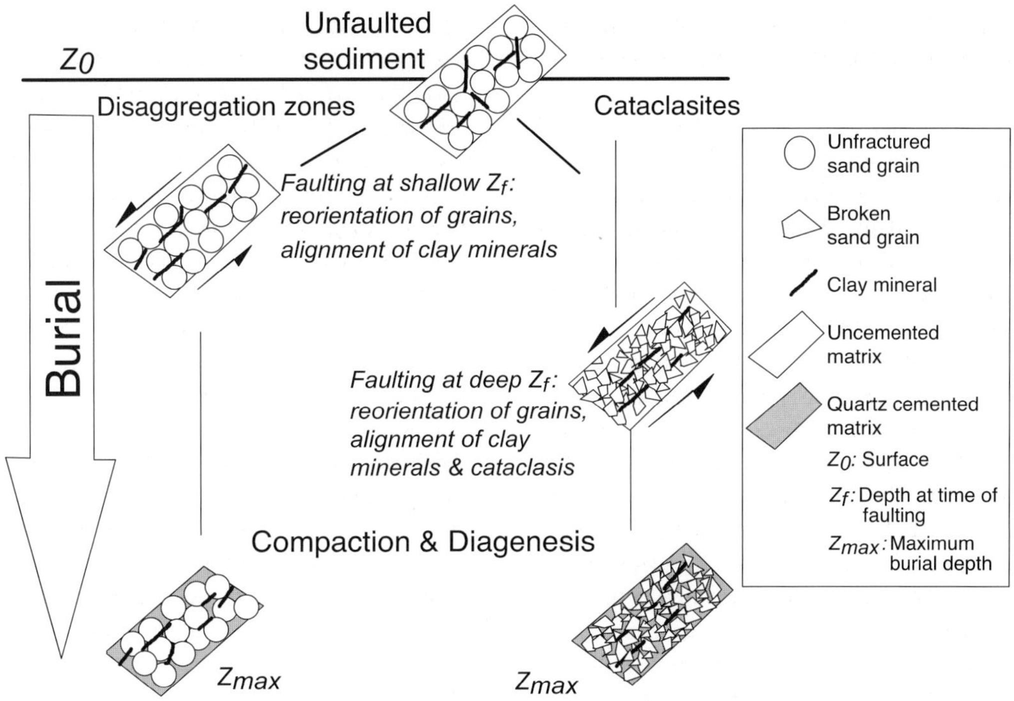

Fig. 1. Diagram illustrating the processes controlling fault zone properties.

the term deformation band as it has also often been used about small faults in sandstones without grain breakage (Antonellini et al., 1994; Fossen and Hesthammer, 1997; Swierczewska and Tokarski, 1998).

Experiments have shown that cataclastic faults in porous sand/sandstones deform first by pore collapse and then, given sufficient effective stress, by grain breakage (Mandl et al., 1977) (Fig. 1). Hence cataclastic faults are formed under higher mean effective stress conditions than disaggregation zones. Porosity, pore throat radii and permeability can be further reduced in cataclasites because they experience enhanced quartz cementation due to the increased area of grain–grain contacts (Fig. 1). In Middle Jurassic sandstones of the North Sea, pervasive quartz cementation begins at temperatures of about 90°C (e.g. Bjørlykke, 1983; Bjørlykke and Egeberg, 1993), which for a thermal gradient of 30°C/km corresponds to a depth of about 3000 m. If deformation occurred long before the rock reaches this temperature, the newly formed fracture surfaces become polluted and hence quartz cementation will be inhibited (Rimstidt and Barnes, 1980; Fisher and Knipe, 1998; Fisher et al., 2000).

Disaggregation zones are defined in this paper as faults in clean sandstones with throws on the scales of millimetres to centimetres, in which little or no grain breakage has occurred. Experiments have shown that these features develop under relatively low effective stress (Fig. 1), and the grains deform by particulate flow and eventually pore collapse (Mandl et al., 1977). As a consequence of their stress sensitivity, disaggregation zones generally form at depths < 500 m. Disaggregation zones develop in clean sandstones ($V_m < 0.15$), and they usually have similar petrophysical properties to their host sediment and therefore do not tend to represent significant barriers to fluid flow (Fisher and Knipe, 1998).

Phyllosilicate-framework fault rocks

Fault rocks that have evolved from impure sandstones where V_m is between 0.15 and 0.4 are called phyllosilicate-framework fault rocks (Fisher and Knipe, 1998). Mixing and alignment of clay minerals in the fault zone produce this type of fault rock, and the permeability is controlled by the pore characteristics and distribution of inter-detrital clay mineral framework structures (Knipe, 1992; Fisher and Knipe, 1998). During burial, the porosity and permeability of phyllosilicate-framework fault rocks tend to decrease and their capillary pressures increase as a result of both chemical and mechanical compaction (Fisher and Knipe, 1998).

Clay smears

Clay smears form by the deformation of clay-rich sequences ($V_m > 0.40$). The most common way for clay smears to form is by shear, as clay from the clay bed is dragged into the fault zone (Weber et al., 1978; Lehner and Pilaar, 1997; Sperrevik et al., 2000). Clay smears formed by the abrasion of clay-rich sequences over rough fault surfaces, or injection of mobile clays into fault zones have also been described (Knott, 1993; Lindsay et al., 1993).

Cementation seals

In addition to the fault rock types described above, fault seals can also form due to the precipitation of new phases (Hindle, 1989; Knipe, 1992). Common examples of minerals forming this type of fault rock in the North Sea are calcite, siderite and barite (Fisher et al., 2000). Some of the faulted samples in the database are carbonate cemented, but pervasive carbonate cementation tends to be a very localised phenomenon in the reservoirs examined. We have therefore not considered data obtained from carbonate cemented intervals as these obscure the trends that are created by the closed system diagenetic processes (e.g. quartz cementation and grain-contact quartz dissolution) that are common to all of the reservoirs examined.

Fault zone thickness

Under conditions of Darcy flow the volumetric flow rate, q, across a fault is given by:

$$q = \frac{k_f}{F} A \frac{\Delta P}{\mu} \quad (2)$$

where k_f is the fault permeability, F is the fault thickness, A is the cross-sectional area of the fault, ΔP is the pressure drop across the fault and μ is the fluid viscosity (Manzocchi et al., 1999). The fault zone thickness is therefore important in controlling the rate of flow across faults if viscous forces dominate.

The determination of fault zone thickness is strongly dependent on the criteria for defining the boundaries of the fault zone (Evans, 1990; Childs et al., 1997). In this study the fault zone thickness, F, is defined as the zone of the fault in which there is fault rock, i.e. rock with a significantly altered permeability formed principally by frictional processes. This definition of the fault zone broadly corresponds to the "fault core" of Caine et al. (1996) which is defined as the zone "where most of the displacement is accommodated". Fault zone thickness, F, is usually predicted from the local throw (T) on the fault surface and there is generally thought to be a linear relationship between F and T (Hull, 1988; Evans, 1990; Knott et al., 1996; Walsh et al., 1998). However, empirical plots of F against T show a wide scatter, which is thought to be the result of: (1) problems in defining the width of the fault zone; (2) variations in fault mechanism (e.g. normal, strike-slip and thrust); (3) variations in lithology; (4) complexity of fault zones, including splaying and pod development (Blenkinsop, 1989; Walsh et al., 1998); (5) dip variations; (6) variation in the stress conditions at the time of deformation.

Some information on F can be gained from core and image log data, but usually only very small faults are represented. Outcrop data must therefore be used to constrain the thicknesses of large-scale faults.

In this study we have tried to investigate if there is any relationship between the fault zone thickness, F, and the fault throw (T) and lithology using outcrop data from onshore fields that provide close analogues to the studied North Sea reservoirs.

Data have been collated from normal faults in sandstone/shale sequences from Sinai (Knott et al., 1996) and from Northumberland, U.K. (this publication, Fig. 2). The fault zone definition used by Knott et al. (1996) is consistent with the one that we have used. The Sinai data are from tectonic normal faults in Nubian sandstones, with an average dip of 63° (Moustafa, 1996). The Northumberland data are from Hartley Steps and Crag Point, which are coastal exposures of tectonic normal faults in Carboniferous sandstones, shales and coals. The sequence was lithified before faulting, and the average fault dip is 67°. In both datasets, the faults were divided according to the lithology of the footwall and hangingwall sequences. The fault zone thickness versus throw data for the different localities plot in the same fields for a given lithology. However, as noted by Knott et al. (1996), the faults that juxtapose sandstone and shale tend to have lower thickness/throw ratios than faults that juxtapose sandstone and sandstone. The regression lines (Fig. 2) show that the faults with sandstone/sandstone juxtapositions are on average about three times thicker than the sandstone/shale faults for a given displacement. The smaller thickness of faults in fine-grained lithologies has also been confirmed by observation of small faults in core from the Gullfaks Field, northern North Sea (Hesthammer and Fossen, 2000). The faults with shale/shale juxtaposition fall in a similar locus to the sandstone/sandstone faults. The faults with shale/shale juxtaposition are characterised by small interconnected fault strands separated by more intact shale. The fault zone thickness measured in the field was taken across this zone and may well be an overestimate of the thickness of fault rock. However, shale/shale juxtapositions are of little importance in reservoir modelling and so will not be considered further here.

The reduced thickness/throw ratios of faults that are associated with shales suggest that when shale enters fault zones, the fault zone becomes weaker than the surrounding rock and the deformation becomes localised (Knott et al., 1996). This is consistent with laboratory studies that have shown that the frictional

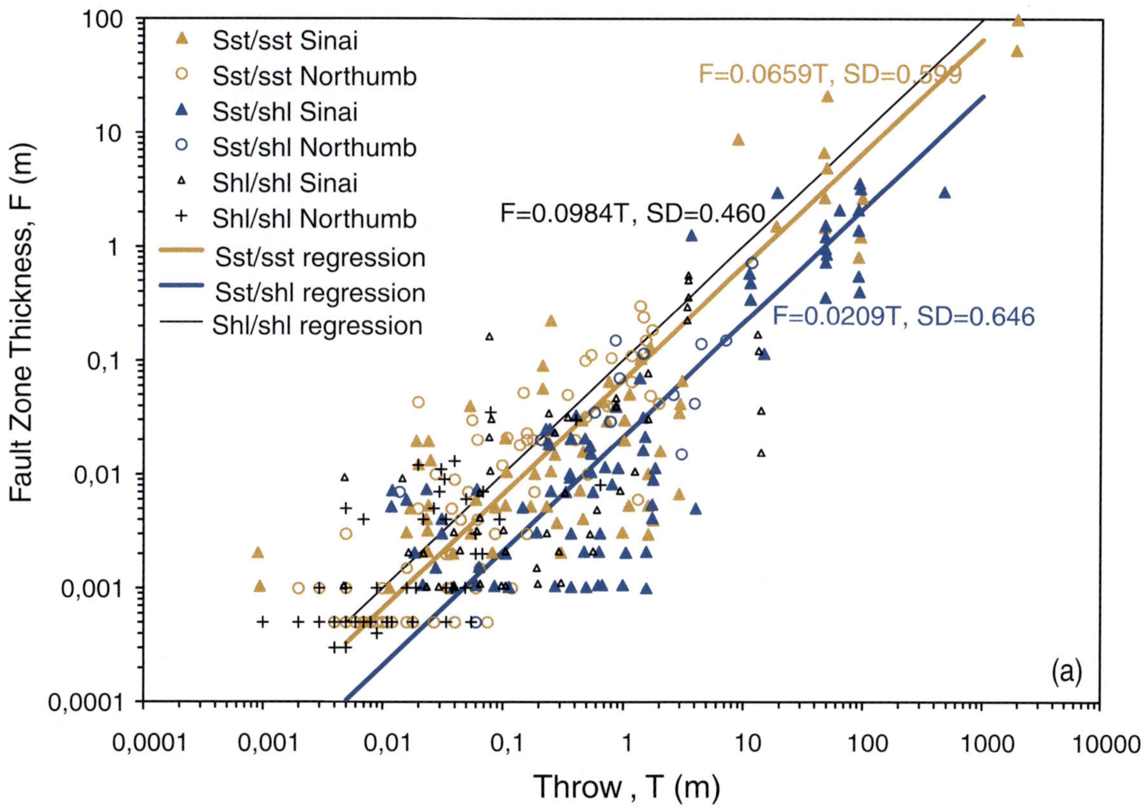

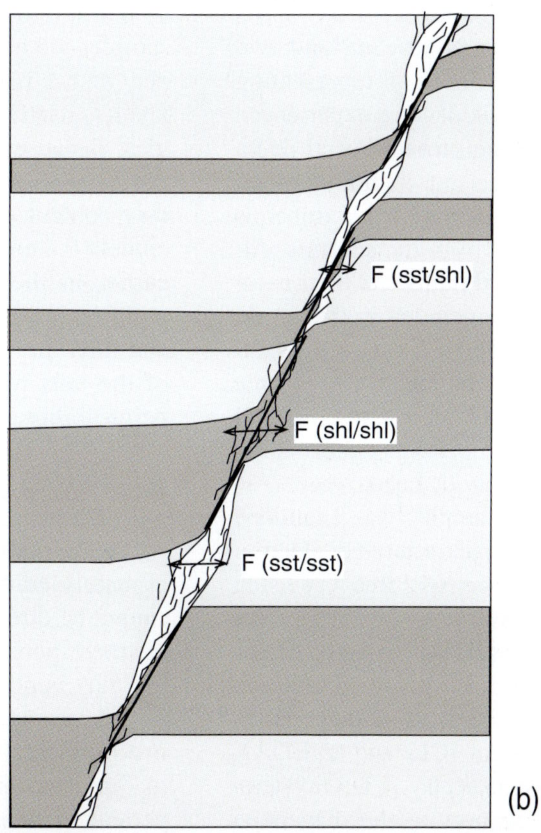

Fig. 2. (a) Plot of fault zone thickness (F) against local fault throw (T) for tectonic normal faults from Sinai (after Knott et al., 1996) and Northumberland. Data are subdivided according to the lithologies either side of the fault (e.g. sst/shl = sandstone/shale). Least squares regressions lines were produced assuming that thickness is directly proportional to throw, with a log-normal distribution of errors. Equations of the regressions lines are given together with the standard deviation (*SD*) of $\log_{10}(F/T)$. Student's *t*-test (Till, 1974) showed that the distinction between the ratios of thickness to throw for the sst/sst and sst/shl datasets is significant at the 95% confidence limit. (b) Idealized fault zone in clastic sequence showing variations in thickness and permeability (thickness exaggerated).

strength of shaley fault rocks is substantially less than that of non-shaley fault rocks (Morrow et al., 1984).

The results in Fig. 2 show that fault zone thickness, F, is a function of the local throw, T, and amount of shale in the fault zone, which can be estimated using the Shale Gouge Ratio. An empirical equation was derived by assigning the sandstone/sandstone regression line an SGR of zero, and the sandstone/shale an SGR of 0.5:

$$F = T\left(0.06V_f^2 - 0.12V_f + 0.0659\right) \quad (3)$$

By including the effect of lithology, this equation reduces some of the error in predicting fault zone thickness from throw, and we suggest that this relationship is used when calculating fault zone transmissibility multipliers in reservoir models. However, the relationship could be better constrained with datasets that for each fault record also V_f in addition to F and T. Investigation of rock properties (porosity and strength) at the time of faulting, and also the effective stress conditions at that time should improve determination of F.

Fluid flow properties in core samples

The data presented in this paper are from seven different fields in the northern North Sea and two fields from the Mid Norwegian Shelf, all representing clastic reservoirs. None of the fields have experienced major uplift and erosion, so the present burial depth is the maximum burial depth (z_{max}). Faults in which there has been significant cementation by minerals other than quartz were omitted from the analysis. Altogether data from ~100 normal faults and their associated unfaulted material were included in this work. Petrographical analyses were carried out on all samples to constrain the burial history, and to classify the fault rock type. All petrophysical and petrographical analyses were carried out by the Rock Deformation Research Group at the University of Leeds.

The microstructure of the samples was examined by secondary electron (SE), back-scattered electron (BSE) and cathode luminescence (CL) imagery using a CAMSCAN CS44 high-performance scanning electron microscope. This is equipped with a high-resolution solid state four quadrant back-scattered electron (BSE) detector, a secondary electron detector (SE), a cathode luminescence detector (CL) and an EDAX energy dispersive X-ray spectrometry (EDS) system. The timing of faulting in respect to the diagenetic history was established by identifying the authigenic phases that were deformed and those that grew after faulting. In many situations this places only broad limitations on the depth of deformation because most diagenetic alteration occurred during shallow (eodiagenesis) and deep burial (mesodiagenesis), whereas most faulting occurred between these depths.

The faults we have investigated generally have a thickness of 1–2 mm, and the throw is in the range of millimetres to centimetres. Permeability measurements were made on 27 mm diameter plugs using water permeametry, as described in Fisher and Knipe (1998). The permeability was measured for both the faults and their associated unfaulted material at a confining pressure of ~0.8 MPa. Additionally, capillary entry pressure measurements were made using mercury injection porosimetry. Petrographical analyses, including measurement of clay content in the host rock, were carried out using SEM (Scanning Electron Microscopy). For the cataclasites, disaggregation zones and phyllosilicate-framework rocks, the clay content of the fault rock is assumed to be similar to the clay content of the neighbouring host rocks as the fault throws are very small (mm–cm scale). Some control measurements of the clay content of the fault rock itself were performed using SEM, which showed that the clay content of the fault rock is equivalent to the clay content of the host rock. However, this assumption is invalid for clay smears where clay is sheared and incorporated into the fault zone from one or multiple discrete clay layers. In the clay smeared samples we have investigated, the composition of the clay smear is similar to the composition of the clay layer(s) itself, and is not a result of mixing of lithologies. Because the clay content of the clay smears has not been measured, they were assigned an arbitrary clay content of 0.7 (i.e. average value between clay content 0.4 and 1.0; see the description of clay smears earlier in the paper). In the following sections we will first present the results for the host rock samples, and then the results for the faulted rocks. The study of the host rocks allows the underlying controls on permeability to be established.

Fluid flow properties of unfaulted host rocks

The permeability of host rocks in porous media is mainly controlled by the effective porosity which cannot be directly measured (David et al., 1994). The effective porosity is generally reduced with increasing clay content, as clay minerals tend to block the pore throats (Fig. 3). If compaction is not inhibited by overpressure, bulk porosity will decrease by increased loading (mechanical compaction) and increased temperature (chemical compaction, Bjørlykke and Egeberg, 1993). Both temperature and confining stress generally increase with increasing depth. One should therefore expect the permeability to decrease with increasing clay content and increasing maximum burial depth.

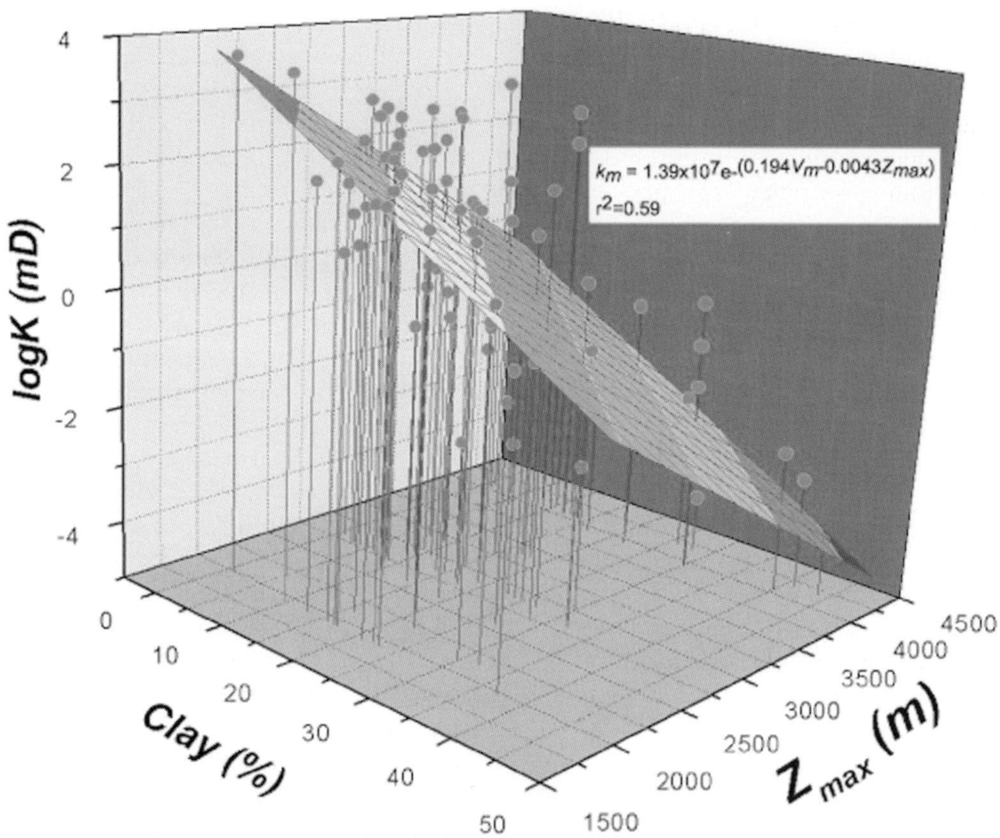

Fig. 3. 3D plot showing the relationship between measured clay content, permeability, and maximum burial depth for host rocks. The plane represents the exponential least squares regression (Eq. 4).

Fig. 3 shows that there is a clear negative exponential relationship between the clay content, V_m and the permeability, k_m for a given depth interval. The same relationship can be seen between k_m and the maximum burial depth, z_{max} for a given V_m. A least squares planar regression yields:

$$k_m = 1.39 \times 10^7 e^{-(0.194 V_m - 0.0043 Z_{max})} \quad (4)$$

Hence k_m is exponentially related to both V_m and z_{max}. This expression is only calculated for host rock samples with $V_m < 0.4$ as our database did not contain more clay-rich examples. The host rock data also show a clear relationship between the k_m and the Hg–Air capillary entry pressure, P_m (Fig. 4, Eq. 5).

$$P_m = 29.91 k_m^{-0.4022} \quad (5)$$

The general power-law form of this relationship is supported by data from Ibrahim et al. (1970) and Harper and Lundin (1997) and the results of a laboratory stick-and-ball model (Harper and Lundin, 1997). The P_m measurements can be converted to capillary pressures for hydrocarbon systems (P_{chw}) and to hydrocarbon column heights (H) using the relationships:

$$P_{chw} = \frac{\gamma_{hw} \cos\theta_{hw} P_{cma}}{\gamma_{ma} \cos\theta_{ma}} \quad (6)$$

(Purcell, 1949), and

$$H = \frac{P_{chw}}{0.433(\rho_w - \rho_h)} \quad (7)$$

(Leverett, 1941; Smith, 1966)

Published values for subsurface hydrocarbon–water interfacial tensions range from 5 to 35 dynes/cm for oil–water systems and from 70 to 30 dynes/cm for gas–water systems (Berg, 1975; Schowalter, 1979) and are related to the composition of the hydrocarbons, as well as temperature and pressure.

Fluid flow properties of fault rocks

Various methods have been used for relating known host rock properties to the fault rock permeability, k_f. For example, k_f has been related to host rock porosity (Fulljames et al., 1997). However, most commonly k_f is estimated from SGR, which is the estimated average clay content of the fault zone:

$$\log k_f = -4\mathrm{SGR} - \frac{1}{4}\log(D) \times (1 - \mathrm{SGR})^5 \quad (8)$$

(Manzocchi et al., 1999) and

$$K_f = 10^{(-5\mathrm{SGR})} \quad (9)$$

(in the FAPS software, Badleys Earth Sciences).

The main difference between these equations is

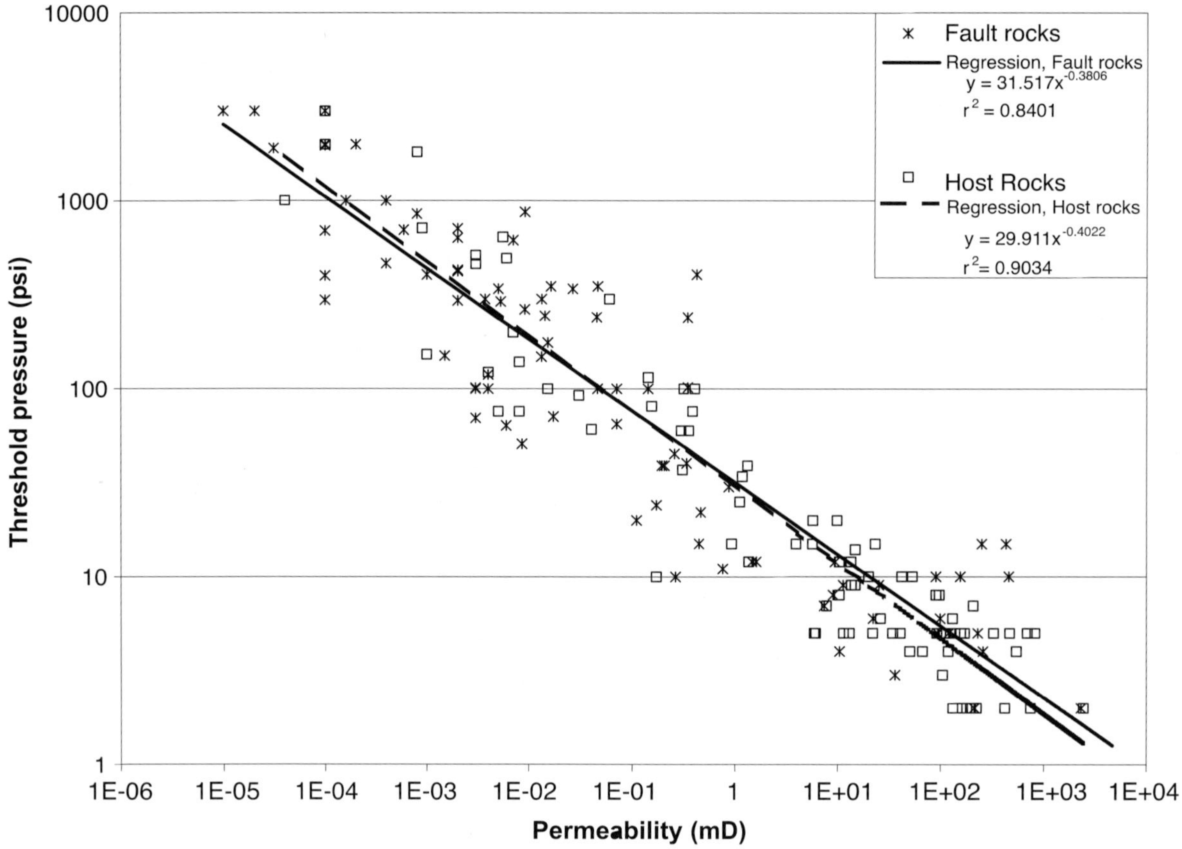

Fig. 4. Plot showing the relationship between measured permeability and measured Hg–Air threshold pressure for host rocks and for fault rocks. There is no significant difference between the data for fault rocks and their host rocks. Equations of least-squares regression given in legend.

that Eq. 8 includes a displacement term (D), which attempts to capture the effect noted by Antonellini and Aydin (1994) that in clean sandstones, there is a great permeability reduction at displacements of greater than ~1 m. However, this permeability reduction is very localised and only constitutes a very small part of the total fault zone thickness. Our opinion is that if low permeabilities are applied to the full thickness of fault zones the permeability reduction of the fault zone will be exaggerated. As Fig. 5 shows, neither Eq. 8 nor Eq. 9 are able to predict $k_f > 10$ mD. Both of the equations give a poor representation of the permeability variation measured on faults in cores (Fig. 5), and the difference between the calculated and the measured values of k_f can be as much as four orders of magnitude. A principal reason is probably that the equations are based on relatively few data. In addition the properties were not measured with one consistent method, and the data are from variable fault rocks occurring in various geological settings including both normal and strike-slip faults.

In the rest of this paper we will show how we have derived new empirical equations for estimating fault rock permeability and capillary entry pressures of fault rocks. We will also describe how they have been tested and implemented.

Data analysis

The fluid flow properties of fault rocks are fundamentally controlled by the ensemble properties of the pore network, which are related to the size of the pore throat radii and the tortuosity. These geometric properties are in turn related to the grain mineralogy, the grain size, grain sorting, and the extent of later diagenetic alteration. It is not possible at this stage to make a predictive theoretical model based on these fundamental properties of the fault rock. Instead, we have derived empirical relationships for estimating fault rock permeability and capillary threshold pressure.

Empirical equation for estimating fault rock permeability

Our data from faults in cores show that there is a clear relationship between fault rock clay content, V_f and fault rock permeability, k_f. There is a general decrease in k_f with increasing V_f (Fig. 5). However, the relationship is clearly affected both by the deformation mechanism (reflected by fault rock type) and the maximum burial depth, z_{max}.

For a given fault rock type the relationship between fault rock clay content (V_f) and fault rock permeability (k_f) changes with maximum burial depth (z_{max}). Clearly, k_f diminishes with increasing z_{max} (Fig. 6).

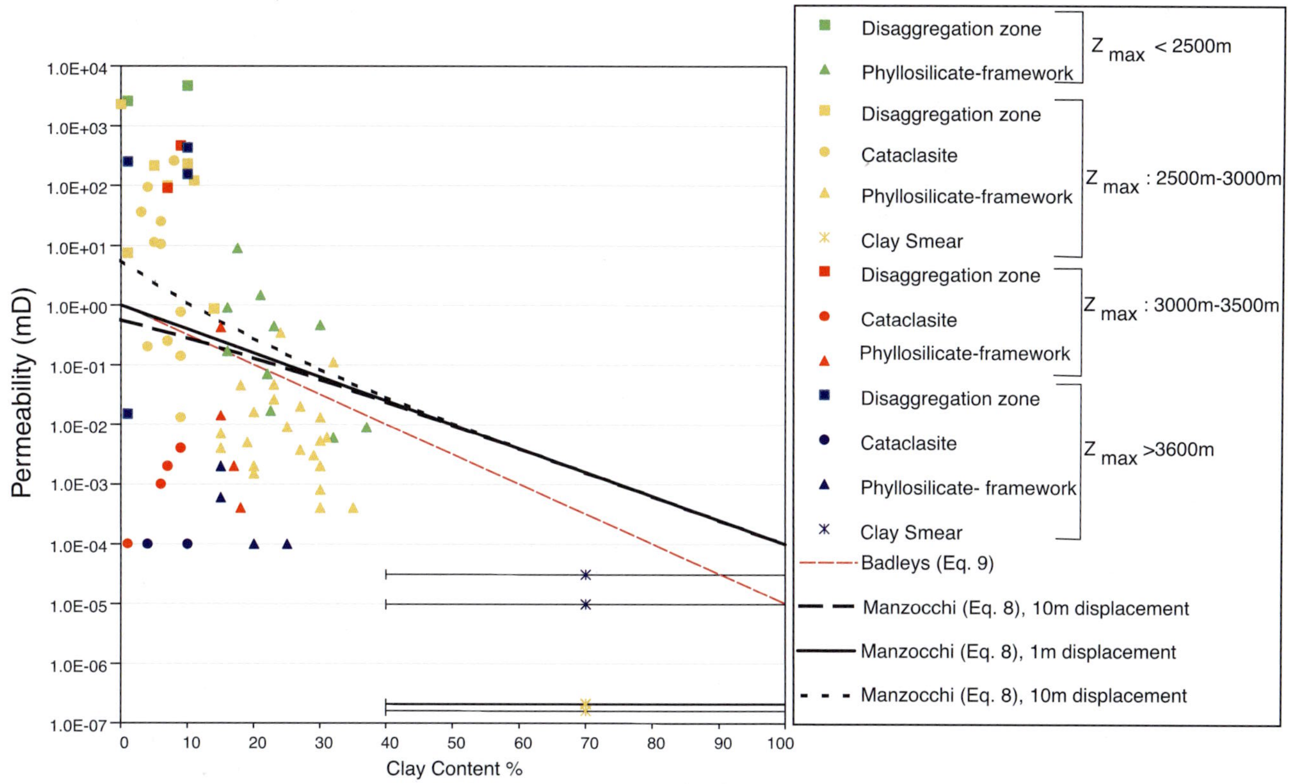

Fig. 5. Plot showing fault rock permeability versus fault rock clay content measured on faults in cores. The data points are grouped according to maximum burial depth (different colours represent different depths), and the different fault rocks have been given different geometric symbols. The permeability generally decreases with increasing fault rock clay content and increasing maximum burial depth. When fault rock clay content is $\lesssim 15\%$ the permeability is dependent on whether or not the fault zone has experienced cataclasis. The lines represent earlier published relationships (Eqs. 8 and 9) between fault permeability and fault rock clay content.

Increasing z_{max} corresponds to increasing overburden stress and temperature. The negative exponential decrease of k_f with z_{max} is comparable to the decrease observed for host rocks (Fig. 3).

For a given maximum depth of burial (z_{max}), the relationship between fault rock clay content (V_f) and fault rock permeability (k_f) varies with fault rock type. Disaggregation zones show the highest k_f, cataclasites and phyllosilicate framework fault rocks are intermediate and the clay smears show the lowest k_f (Figs. 5 and 7).

For a given fault rock clay content (V_f), the fault rock permeability (k_f) is highly dependent on deformation mechanism:

- Where $V_f > 0.15$ (phyllosilicate-framework faults and clay smears) the k_f–V_f relationship is more or less negative exponential (Fig. 7). However, k_f is also dependent on the maximum burial depth, z_{max}, and best fit approximations demonstrate that k_f decreases systematically with increasing z_{max} (Fig. 6). This effect is probably related to compaction of the fault rock after deformation.
- Where $V_f < 0.15$, the relationship between V_f and k_f is not exponential. However, results fall into two different categories: disaggregation zones with high k_f, and cataclasites with lower k_f (Figs. 6 and 7).

The disaggregation zones form when faulting occurs relatively shallowly (<500 m), whereas the cataclasites form at deeper levels (>500 m) and may experience quartz cementation if the temperature is above 90ºC. It is therefore reasonable to assume that the difference in permeability between disaggregation zones and cataclasites are related to differences in deformation processes. Whether or not a clean sandstone (of a certain strength and porosity) experiences cataclasis is mainly controlled by the effective stress at time of faulting, which in turn can be linked to the depth at time of deformation, z_f.

We have assumed that fluid flow properties of fault rocks are mainly controlled by fault zone clay content (V_f), maximum burial depth (z_{max}) and depth at time of deformation (z_f). The different samples were analysed and classified according to these parameters. The fault rock type reflects the deformation mechanism, and hence for the clean sandstone faults the fault rock type also reflects the depth at time of deformation, z_f. An equation was devised that fits the measured fault rock permeability data:

$$k_f = a_1 \exp\left\{-\left[a_2 V_f + a_3 z_{max} + (a_4 z_f - a_5)(1 - V_f)^7\right]\right\} \tag{10}$$

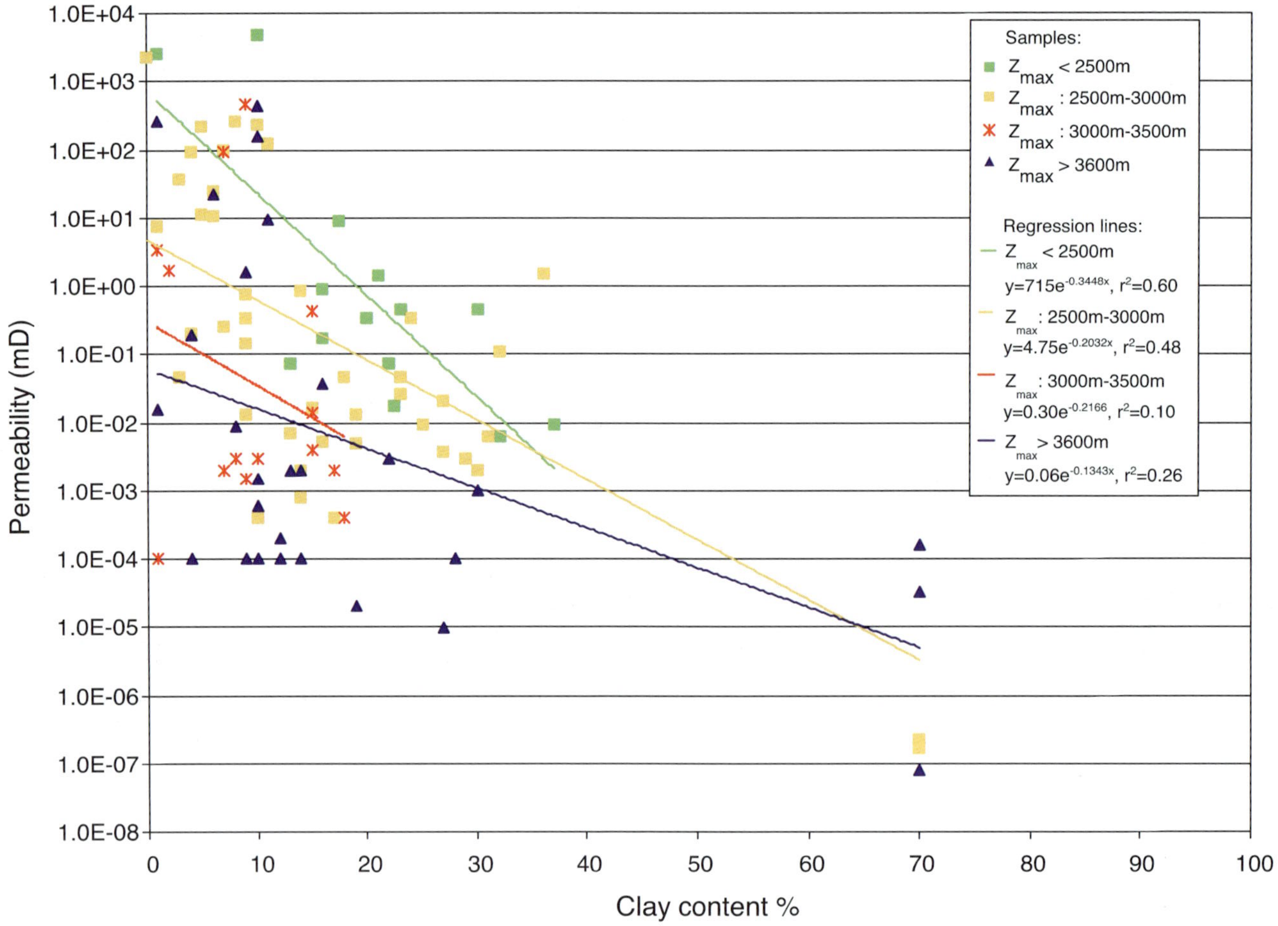

Fig. 6. Plot showing fault rock permeability against fault rock clay content measured on faults in cores. The data are grouped according to maximum burial depth. Exponential least-squares regression lines are shown for each group of data, showing a systematic decrease in permeability both with fault rock clay content and with depth.

where the constants are: $a_1 = 80000$; $a_2 = 19.4$; $a_3 = 0.00403$; $a_4 = 0.0055$; $a_5 = 12.5$.

Fig. 7 shows the fault rock permeability, k_f, calculated using Eq. 10 with different maximum burial depths, z_{max}. The different curves on the individual plots represent calculated k_f using different values for the depth at time of deformation, z_f. As seen from Fig. 7 the suite of curves shifts down with increasing z_{max}. This reflects permeability reduction caused by compaction. Depending on the assumed z_f, Eq. 10 predicts high or low permeabilities in clean sands ($V_f < 0.15$). For the more clay-rich sediments ($V_f > 0.15$), k_f is more controlled by z_{max} than by z_f.

Empirical equation for estimating fault rock capillary threshold pressure. The fault rock data show a clear relationship between the measured Hg–Air fault rock threshold pressure, P_f and the fault rock permeability, k_f (Fig. 4). A power law regression between k_f and P_f yields:

$$P_f = 31.838 \times k_f^{-0.3848} \tag{11}$$

This relationship is statistically indistinguishable from the relationship found for the host rocks (cf. Eq. 5). It is remarkable that the relationship between P and k seems to be entirely insensitive to whether or not the rock is faulted and so is insensitive to grain size and texture. The relationship may instead be a reflection of the general percolation properties of porous media. By combining Eq. 10 and Eq. 11 we get an equation for calculating the Hg–Air fault rock threshold pressure P_f. Fig. 8 shows P_f estimated from Eq. 8 compared to the P_f measured on faults in cores (points).

Generally P_f increases with increasing V_f, but as for k_f, there is a distinction between the properties of disaggregation zones and cataclasites (Fig. 8). The relationship between P_f and V_f is more or less the inverse of the relationship between k_f and V_f.

Testing the fluid flow property equations

After the equations were derived based on the measurements shown in Fig. 4 to Fig. 8, we obtained access to new data (burial depth, clay content, fault rock type, permeability and Hg–Air threshold pressure) from 36 more fault rocks (Fig. 9). The faults were

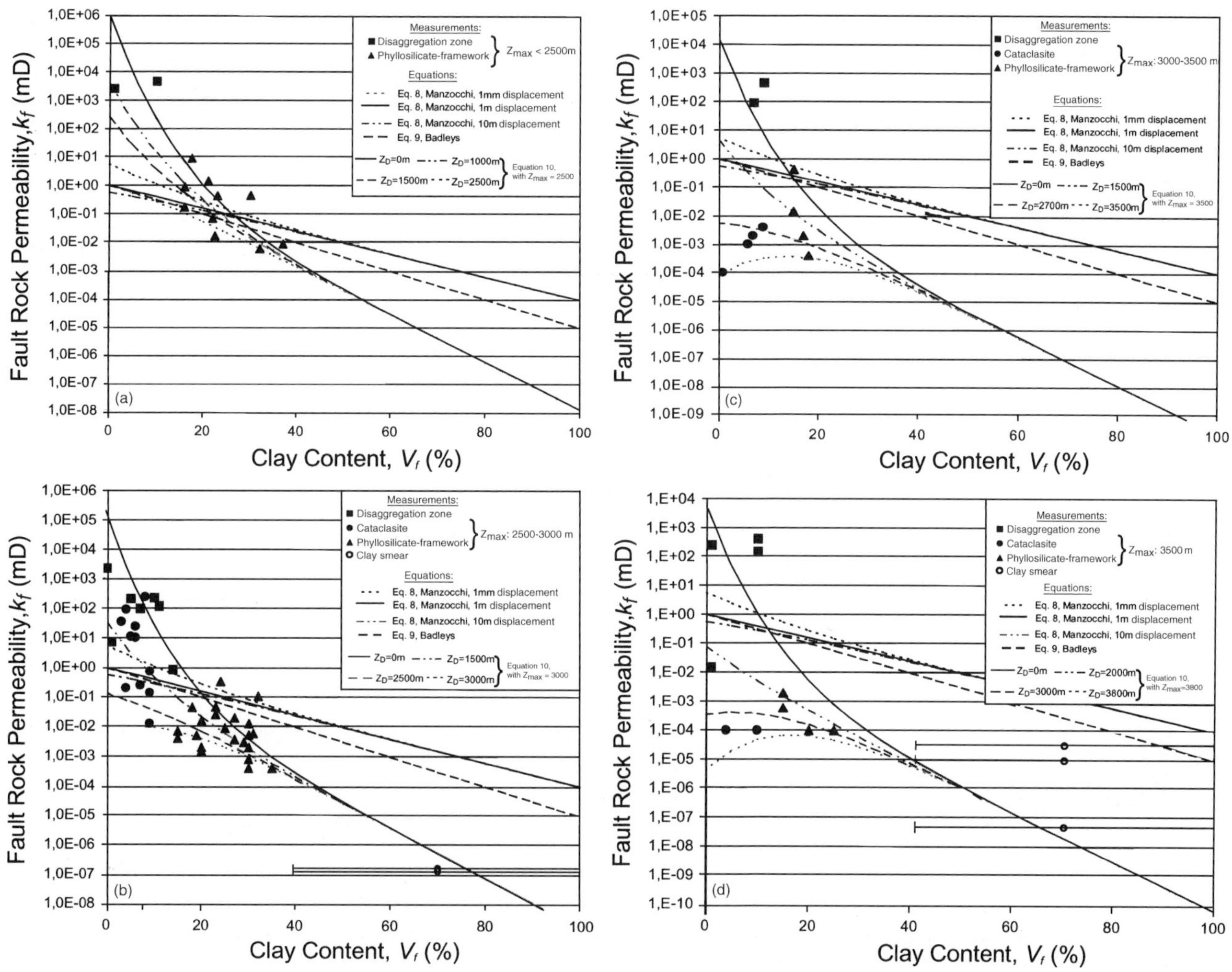

Fig. 7. Fault rock permeability vs. fault rock clay content measured from faults in core (points), and calculated from Eq. 10 (thin lines). The various lines represent the calculated fault rock permeability for a given z_{max} using different depths at time of faulting. The lines representing Eqs. 8 and 9 are also shown; (a) $z_{max} = 2500$ m, (b) $z_{max} = 3000$ m, (c) $z_{max} = 3500$ m, (d) $z_{max} = 3800$ m.

also from Middle Jurassic reservoirs in the North Sea and were analysed using the same methods and equipment as the other samples discussed in this study. The measured permeability values were compared with those calculated using Eqs. 8, 9 and 10 (Fig. 9a). We have also compared the measured Hg–Air threshold pressures with those calculated using Eq. 10 combined with Eq. 11 (Fig. 9b). It is apparent that the new equations provide permeability and Hg–Air threshold pressure estimates that are in good agreement with the measured values, when setting $Z_{max} = 3500$ m which corresponds to the maximum burial depth of most of the new samples. The equations also provide a good representation of the variation in measured permeability and threshold pressure.

Implementation

A reservoir simulation model is built from cells that are arranged in layers. The size of the individual cells may vary, but 100 m × 100 m is typical for the horizontal dimensions. Each of the cells have values assigned to them, representing different parameters such as permeability, net/gross, clay content (the so-called Vshale parameter), porosity, etc. Faults are typically represented either by modifying the fluid flow properties within a grid cell (i.e if there are faults which are less than the grid size), or by a split in the simulation grid. In the latter case it is possible to use the method from Manzocchi et al. (1999) to implement the fluid flow properties of faults as transmissibility multipliers. A Norsk Hydro in-house program, FAULTMULT (Rivenæs and Dart, 2002), based on Manzocchi et al. (1999), allows us to calculate transmissibility multipliers cellwise for the entire fault, and the technique therefore makes it possible to represent the variation in permeability and fault zone thickness along the fault.

Transmissibility multipliers are a function of the permeability in the host rock grid cells on each side of the fault, the grid size of the cells, the permeability of the fault rock, and the width of the fault zone. In

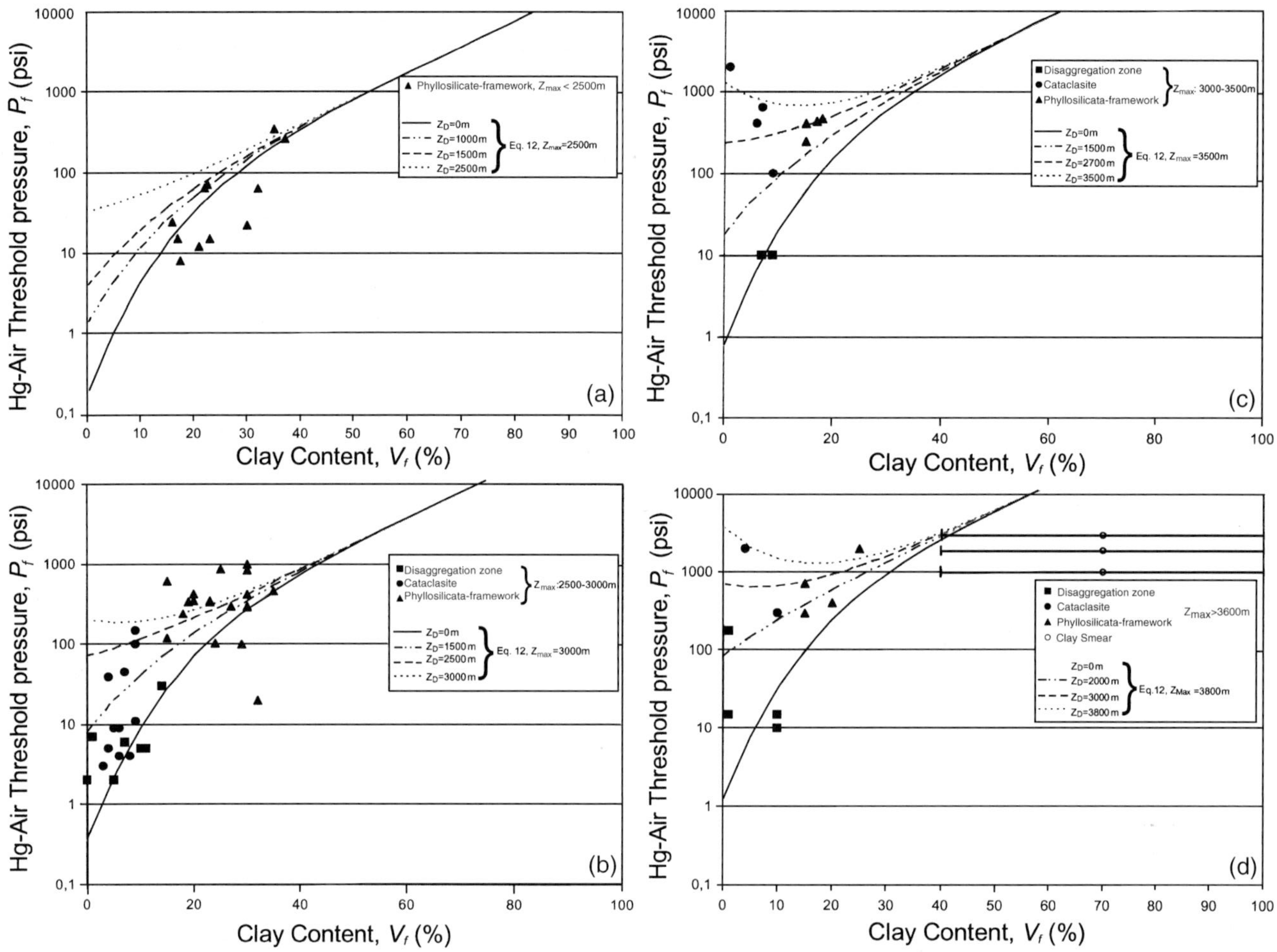

Fig. 8. (a) Threshold pressure vs. fault rock clay content measured from faults in core (points), and calculated from Eq. 10 combined with Eq. 11 (lines). The various lines represent the calculated fault rock threshold pressure for a given z_{max} using different depths at time of faulting: (a) $z_{max} = 2500$ m, (b) $z_{max} = 3000$ m, (c) $z_{max} = 3500$ m, (d) $z_{max} = 3800$ m.

FAULTMULT the properties and size of the host rock cells are read directly from the reservoir simulation grid, whereas the fault rock permeability and the width of the fault zone needs to be assigned or calculated. The local thickness of the fault zone is calculated from the local throw taken from the grid offset, using the empirical relationship between throw and fault rock thickness published by Manzocchi et al. (1999). The fault rock permeability can be calculated in FAULTMULT using Eq. 8 or Eq. 9, and as a part of this study Eq. 10 was also implemented. To estimate the fault rock permeability from Eq. 10 input about fault rock clay content, maximum burial depth and depth at time of deformation are required.

The Shale Gouge Ratio method (our Eq. 1, Yielding et al., 1997) is used to estimate the proportion of shale in the fault zone. This method requires host rock clay content and fault throw as input, which both can be read directly from the simulation grid.

In the cases where we have used Eq. 10 for calculating fault rock permeability, the reservoirs are thought to be presently at their maximum burial depth, and so z_{max} was also read directly from the reservoir simulation grid. The depth at the time of deformation is more difficult to determine, but it was estimated from knowledge about the geological history of the different fields and from petrographic analysis of fault rocks compared to their associated host rocks (fault rock type, mineral dissolution/precipitation). By combining this information with the geometry of the simulation grid and the grid block permeabilities, the sealing effect of the faults was represented as transmissibility modifiers for each grid cell.

We have applied the technique to a number of North Sea reservoirs. However, in all cases the transmissibility modifiers generated by this technique proved consistently too high (1–2 orders of magnitude) in order to produce good history matches (Rivenæs and Dart, 2002). This is despite the fact that the equation derived here (Eq. 10) tends to give lower k_f (and thereby lower transmissibility multipliers) than existing formulae (Eqs. 8 and 9). In the following discussion we will focus on how the representation of faults in fluid flow models can be improved.

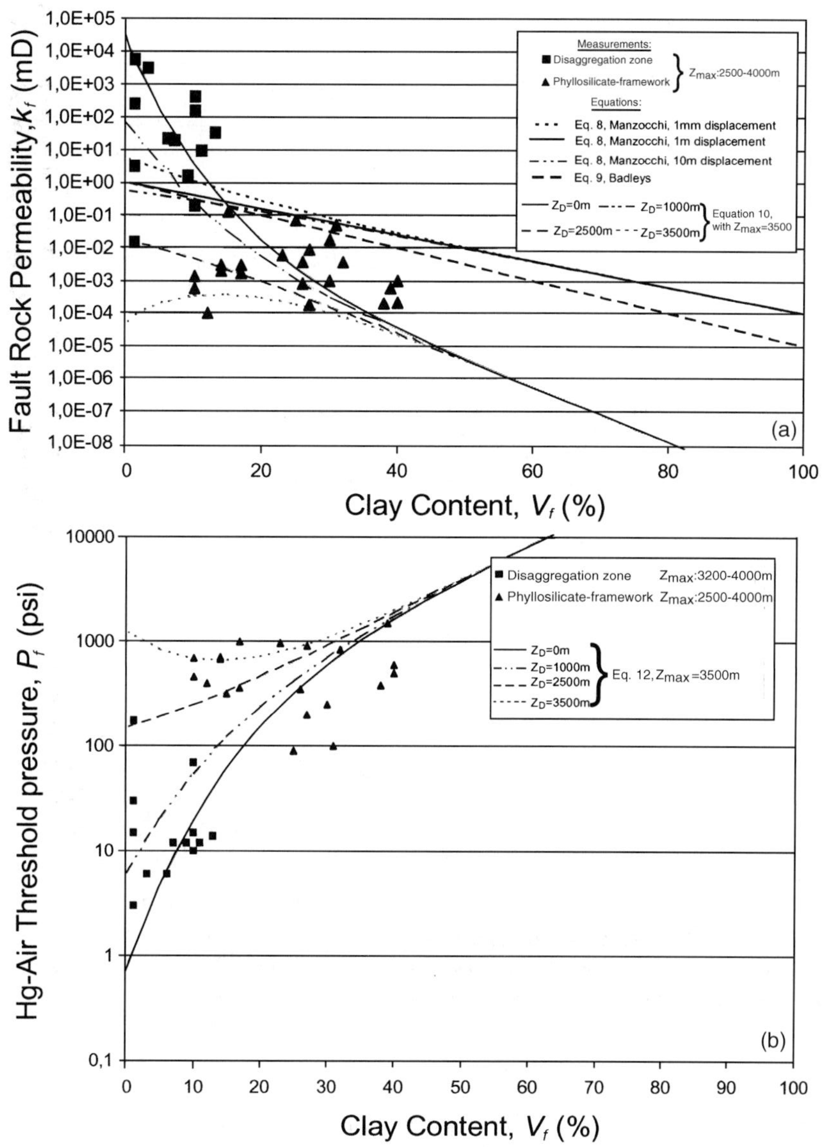

Fig. 9. (a) Comparison of permeability measurements from control samples and permeability estimated from Eqs. 8, 9 and 10. (b) Comparison of Hg–Air threshold pressure measurements from control samples and Hg–Air threshold pressure estimated from Eq. 10 combined with Eq. 11.

Discussion

The fluid flow properties of faults are dependent on many different factors such as the lithologies involved and their mineralogy (Knipe, 1992; Lindsay et al., 1993; Gibson, 1994; Gibson, 1996), the grain sorting (Antonellini and Aydin, 1994), the relationship between the deforming stress and the initial rock porosity (Engelder, 1974; Mandl et al., 1977; Aydin and Johnson, 1978; Antonellini et al., 1994), and cementation (Hindle, 1989; Fisher et al., 2000).

In this study we have derived empirical equations for estimating fault rock permeability and capillary threshold pressure as a function of clay content, maximum burial depth and depth at time of deformation. The equations are based on data from fault rocks in cores from sand/shale sequences. Although many other important parameters are also important for the fault rock permeability, including for example cementation by other minerals than quartz, and grain sorting, this approach has considerably reduced errors in prediction compared with existing methods.

Uncertainties related to the new equations

The estimates of z_f represent a major uncertainty when constraining the fault rock permeability and threshold pressure equations. The burial depth at time of faulting is not possible to measure directly, and has in this study been estimated from petrographical analysis. Structural reconstruction is another method, which can be helpful for estimating z_f. With respect to z_f, the equations have been made to fit the shallowest buried disaggregation zones when z_f is between 0 and 500 m and to fit the most quartz cemented cataclasites when z_f is more than about 3600 m. However, it is difficult to constrain the exact depths at which cataclasis starts. Antonellini et al. (1994) studied deformation bands in Arches National Park, Utah. They suggested that disaggregation zones were formed at depths of <100–200 m, whereas cataclasites developed at depths of ~1000–2000 m. The effective confining stress at which deformation changes from independent particulate flow to cataclasis depends on a number of factors including grain-size, void ratio, stress path and time (Lade et al., 1996). In other words, it is not possible to relate the change in deformation mechanisms precisely to a particular effective stress level or burial depth. However, rock mechanical experiments have shown that extensive grain crushing can occur in coarse sands at confining pressures as low as 2 MPa (Lee and Farhoomand, 1967), which is equivalent to burial depth of ~170 m under hydrostatic conditions. It would be expected that the onset of grain crushing would occur at higher effective stresses in finer material. For example, Lade et al. (1996) noticed a rapid increase in grain fracturing at confining stresses of >5 MPa (i.e. ~400 m burial) during deformation experiments on the finer Dense Cumbria Sand.

The new equations presented here are mainly based on measurements of samples with $V_f < 0.4$. The predicted values for $V_f > 0.4$ are therefore not validated, and the equations should be improved when measurements of permeability and threshold pressure on samples with high clay content is available.

The permeability measurements were made at pressures far lower than those in the subsurface. Therefore it is necessary to adjust the permeability measurements for the effect of stress relaxation of the samples. The ratio of the permeability in the subsurface (k) to the permeability at zero effective mean stress (k_0) is given by:

$$k/k_0 = \exp(-\lambda\sigma_m) \qquad (12)$$

where σ_m is the effective mean stress and λ describes the compressibility of the fault rock (David et al., 1994; Evans et al., 1997; Zhang et al., 2001). The value of λ has been determined experimentally for natural fault rocks and is 0.0045–0.014 MPa^{-1} for clay-free fault zones and 0.012–0.055 MPa^{-1} for clay-rich fault zones (Morrow et al., 1984; David et al., 1994) and so the stress dependency is higher in the clay-rich fault zones. Using fluid pressures and in situ stresses from one North Sea field we find that with $\lambda = 0.03$ MPa^{-1}, $k/k_0 = 0.51$, implying that the measured permeability should be reduced by a factor of two in the more shale-rich fault zones. In non-shaley fault zones the required reduction is less significant (e.g. if $\lambda = 0.01$, $k/k_0 = 0.80$).

Improved implementation of fault effects in the reservoir simulator

Using the transmissibility modifier method for assigning fluid flow properties to faults seems to give faults that are too open with respect to two-phase fluid flow. In order to improve the model, and to get better history match we think that capillary entry pressure effects (estimated using Eq. 10 in combination with Eq. 11, Rivenæs and Dart, 2002) and relative transmissibility multipliers (Manzocchi et al., 2000) have to be incorporated. However, there is also several other ways in which the existing method can be improved.

When estimating the proportion of shale in the fault zone using the Shale Gouge Ratio method (Eq. 1) it is common to derive the V_m of the host rocks from wireline logs (Vshale parameter). The Vshale parameter is derived from wireline logs and it often includes (some) mica, but does usually not include kaolin. Therefore the Vshale parameter does not correspond directly to the clay content of the host rock, and should be calibrated against core measurements.

When calculating the transmissibility modifier, the local thickness of the fault zone is estimated from the local throw taken from the grid offset using an empirical relationship between throw and fault rock thickness derived from outcrop analogues. Existing throw/thickness relationships do not consider variations related to lithology, but outcrop data from Sinai and Northumberland show that the throw/thickness relationship is dependent on lithology, and can be related to clay content (Fig. 2). We have shown that for a given throw a fault zone with sandstone/sandstone juxtaposition is wider than a fault zone where sandstone is juxtaposed against shale. In the future, the lithology-dependent relationship between throw and fault zone thickness (Eq. 3) should be used when assigning fault zone thickness in the reservoir model. However, the main problem regarding fault thickness estimations in reservoir simulation models is that faults in reservoir models are very simplified, and it is impossible to include complex fault architecture in the reservoir model. The only way to include the effects of fault architecture at present, is by capturing these effects by upscaling.

Another important aspect is that the SGR algorithm (Eq. 1) does not always capture the effects of clay smears. To incorporate the effects of clay smears other algorithms should be considered, for example the SSF (Lindsay et al., 1993) or the CSP (Bouvier et al., 1989), when estimating the clay content of the fault zones.

Also, both coal and limestone generally have low permeability/high entry pressure, and we know both from field exposures and cores (Lindsay et al., 1993; R.B. Færseth, pers. commun., 2000) that both coal, and limestone when faulted at shallow depths are able to smear along faults. We therefore suggest that the effects of coal layers and in some cases limestone beds, should be included when calculating SGR, SSF and CSP.

Upscaling

Fault zone thickness

Measurements of fault zone thickness used for deriving Eq. 3 were taken at discrete points on faults. However, the simulation model required thicknesses upscaled to values appropriate for the grid dimensions. The best fits in Fig. 2 go through the centre of the log permeability values and therefore represent geometric average values. Assuming flow straight through the fault, the thickness can be upscaled using the area-weighted harmonic average thickness (Manzocchi et al., 1999), which makes the upscaled thickness lower than the geometric mean. However, faults tend to have a much higher permeability parallel to the fault zone than perpendicular to the fault zone (Antonellini and Aydin, 1994; Evans et al., 1997; Zhang et al., 2001). Therefore once fluids enter the fault zone they may tend to flow diagonally across it or they may flow with a tortuous flow path (Zhang et al., 2001). The path length through the fault will therefore be longer than the thickness of the fault and this extra path length should be brought into the upscaling.

Permeability

The empirical relationships describing permeability apply to core plug scale measurements and so upscaling must be considered before the permeabilities can be applied to the reservoir simulator. If it is assumed that the permeability does not vary through the fault from one side to the other, then the upscaled permeability is given by the area-weighted arithmetic average (Manzocchi et al., 1999). However, field observations of faults with throws of more than a metre show that the fault rock is very heterogeneous, and varies both along and through the fault zone (e.g. Childs et al., 1997; Foxford et al., 1998). Flow through such a complex medium is best upscaled with a geometric average. Therefore no extra transformation of Eq. 10 is required for upscaling.

Thickness/permeability correlation

In the relationships derived here, both the permeability and the thickness are functions of SGR and the permeability is positively correlated with the thickness. Hence as the fault zone gets thinner it also has lower permeability and so upscaling on the assumptions of uncorrelated thickness and permeability will tend to overestimate the flow rate through the fault. A possible upscaling scheme for this situation is given in Manzocchi et al. (1999, their eq. 17).

In summary various factors in the upscaling tend to produce simulations that increase the sealing capacity of the faults: (1) diagonal flow of fluids; (2) heterogeneous nature of the fault rocks through the fault zone; and (3) positive correlation between the thickness and the permeability of the fault rocks

Conclusions

(1) Permeability and capillary threshold pressure of fault rocks in clastic sequences are sensitive to shale content of the faults zone, the depth at time of faulting and the maximum burial depth.

(2) There is a strong relationship between capillary threshold pressure and permeability that is the same for fault rocks and for host rocks. The relationship is therefore independent of the faulting process.

(3) Fault zone thickness is sensitive to the local throw on the fault plane and on the amount of shale in the faulted sequence.

(4) Because of sensitivity of both fault zone permeability and thickness to the shale content, there is thought to be a positive correlation between fault zone thickness and permeability.

(5) Empirical equations have been derived for the prediction of fault zone thickness, permeability and capillary threshold pressures that are applicable to tectonic normal faults in the North Sea clastic reservoirs.

(6) The equations may have validity for other normally faulted reservoirs in clastic sequences, although where there has between an unusual diagenetic or tectonic history the equations may be invalid. Therefore, the relationships should always be tested against or modified in the light of properties of fault rocks from the field in question.

(7) After consideration of upscaling and in situ stress effects, the empirical relationships can be applied in the reservoir simulator by use of transmissibility multipliers.

Notation

Parameters referred to in this chapter

SGR	Shale Gouge Ratio (fraction)
T	Fault throw (m)
$V_c l$	Host rock clay content (fraction)
Δz	Thickness of reservoir zone (m)
q	Volumetric flow rate
F	Fault zone thickness (m)
A	Cross-sectional area of the fault
ΔP	Pressure drop across the fault

μ Fluid viscosity
D Fault displacement (m)
k_f Fault rock permeability (mD)
k_m Host rock permeability (mD)
z_{max} Maximum burial depth (m)
z_f Burial depth at time of faulting (m)
V_m Host rock clay content (fraction)
V_f Fault rock clay content (fraction)
P_{chw} Hydrocarbon capillary pressure (psi)
P_m Mercury–air threshold pressure (psi), host rocks
P_f Mercury–air threshold pressure (psi), fault rocks
γ_{hw} Interfacial tension of hydrocarbon and water (dynes/cm)
γ_{ma} Interfacial tension of mercury and air (dynes/cm)
θ_{hw} Contact angle of hydrocarbon and air against the rock
θ_{ma} Contact angle of mercury and air against the rock
H Hydrocarbon column height (feet)
ρ_w Water density (g/cm^3)
ρ_h Hydrocarbon density (g/cm^3)

Acknowledgements

The authors would like to thank Peter Keller, Andreas Koestler, Chris Dart, Linn Arnesen, Arnd Wilhelms and Roy Gabrielsen for carefully reading and commenting on earlier versions of the manuscript. We will also thank Jan C. Rivenæs for implementing and testing the new permeability equation in FAULT-MULT. Norsk Hydro is thanked for financing the study, and for allowing us to publish the results.

References

Antonellini, M. and Aydin, A., 1994. Effect of faulting on fluid flow in porous sandstones: petrophysical properties. Am. Assoc. Pet. Geol. Bull., 78 (3): 355–377.
Antonellini, M. and Aydin, A., 1995. Effect of faulting on fluid flow in porous sandstones: geometric properties. Am. Assoc. Pet. Geol. Bull., 79 (5): 642–671.
Antonellini, M., Aydin, A., Pollard, D.D. and D'Onfro, P., 1994. Petrophysical study of faults in sandstone using petrographic image analysis and X-ray computerized tomography. Pure Appl. Geophys., 143 (1/2/3): 181–201.
Aydin, A. and Johnson, A.M., 1978. Development of faults as zones of deformation bands and as slip surfaces in sandstone. Pure Appl. Geophys., 116: 931–942.
Aydin, A. and Johnson, A.M., 1983. Analysis of faulting in porous sandstones. J. Struct. Geol., 5 (1): 19–31.
Berg, R.R., 1975. Capillary pressure in stratigraphic traps. Am. Assoc. Pet. Geol. Bull., 59: 939–956.
Bjørlykke, K., 1983. Diagenetic reactions in sandstones. In: A. Parker and B. Sellwood (Editors), NATO Advanced Study Institute on Sediment Diagenesis. D. Reidel, Dordrecht, pp. 169–213.
Bjørlykke, K. and Egeberg, P.K., 1993. Quartz cementation in sedimentary basins. Am. Assoc. Pet. Geol. Bull., 77: 1538–1548.
Blenkinsop, T.G., 1989. Thickness–displacement relationships for deformation zones: discussion. J. Struct. Geol., 11: 1051–1053.
Bouvier, J.D., Kaars-Sijpesteijn, C.H., Kluesner, D.F., Onyejekwe, C.C. and Van der Pal, R.C., 1989. Three-dimensional seismic interpretation and fault sealing investigations, Nun River Field, Nigeria. Am. Assoc. Pet. Geol. Bull., 73: 1397–1414.
Caine, J.S., Evans, J.P. and Forster, C.B., 1996. Fault zone architecture and permeability structure. Geology, 24: 1025–1028.
Childs, C., Walsh, J.J. and Watterson, J., 1997. Complexity in fault zone structure and implications for fault seal prediction. In: P. Møller-Pedersen and A.G. Koestler (Editors), Hydrocarbon Seals: Importance for Exploration and Production. Norwegian Petroleum Society (NPF), Special Publication 7. Elsevier, Amsterdam, pp. 61–72.
David, C., Wong, T.-F., Zhu, W. and Zhang, J., 1994. Laboratory measurement of compaction induced permeability change in porous rocks: implications for the generation and maintenance of pore pressure excess in the crust. PAGEOPH, 143: 425–456.
Downey, M.W., 1984. Evaluating fault seals for hydrocarbon accumulations. Am. Assoc. Pet. Geol. Bull., 68 (11): 1752–1763.
Dunn, D.E., LaFountain, L.J. and Jackson, R.E., 1973. Porosity dependence and mechanism of brittle fracture in sandstone. J. Geophys. Res., 78 (14): 2403–2417.
Engelder, J.T., 1974. Cataclasis and the generation of fault gouge. Geol. Soc. Am. Bull., 85: 1515–1522.
Evans, J.P., 1990. Thickness–displacement relationships for fault zones. J. Struct. Geol., 12 (8): 1061–1065.
Evans, J.P., Forster, C.B. and Goddard, J.V., 1997. Permeability of fault related rocks, and implications for hydraulic structure of fault zones. J. Struct. Geol., 19: 1393–1404.
Fisher, Q.J. and Knipe, R.J., 1998. Fault sealing processes in siliclastic sediments. In: G. Jones, Q.J. Fisher and R.J. Knipe (Editors), Faulting, Fault Sealing and Fluid Flow in Hydrocarbon Reservoirs. Geol. Soc. London Spec. Publ., 147: 117–134.
Fisher, Q.J., Knipe, R.J. and Worden, R.H., 2000. The microstructure of deformed and undeformed sandstones from the North Sea: its implications for the origin of quartz cement. In: R. Worden and S. Morad (Editors), Quartz Cementation in Sandstones. Int. Assoc. Sedimentol. Spec. Publ., 29: 129–146.
Fossen, H. and Hesthammer, J., 1997. Geometric analysis and scaling relations of deformation bands in porous sandstone. J. Struct. Geol., 19 (12): 1479–1493.
Fowles, J. and Burley, S., 1994. Textural and permeability characterization of faulted, high porosity sandstones. Mar. Pet. Geol., 11 (5): 608–623.
Foxford, K.A., Walsh, J.J., Watterson, J., Garden, I.R., Guscott, S.C. and Burley, S.D., 1998. Structure and content of the Moab fault zone, Utah, USA, and its implications for fault seal prediction. In: G. Jones, Q.J. Fisher and R.J. Knipe (Editors), Faulting, Fault Sealing and Fluid Flow in Hydrocarbon Reservoirs. Geol. Soc. London Spec. Publ., 147: 87–103.
Fulljames, J.R., Zijerveld, L.J.J. and Franssen, R.C.M.W., 1997. Fault seal processes: systematic analysis of fault seals over geological and production time scales. In: P. Møller-Pedersen and A.G. Koestler (Editors), Hydrocarbon Seals: Importance for Exploration and Production. Norwegian Petroleum Society (NPF), Special Publication 7. Elsevier, Amsterdam, pp. 51–59.
Gabrielsen, R.H. and Clausen, J.A., 2001. Horses and duplexes in extensional regimes: A scale modelling contribution. In: H.A. Koyi and N.S. Mancktelow (Editors), Tectonic Modeling: A Volume in Honor of Hans Ramberg. Geol. Soc. Am. Mem., 193: 219–233.
Gibson, R.G., 1994. Fault- zone seals in siliclastic strata of the Columbus Basin, Offshore Trinidad. Am. Assoc. Pet. Geol. Bull., 78 (9): 1372–1385.
Gibson, R.G., 1996. Compositional limitations on rock types contributing to zones of sealing phyllosilicate-rich fault gouge. In:

American Association of Petroleum Geologists 1996 annual convention. Annual Meeting Abstracts — American Association of Petroleum Geologists and Society of Economic Paleontologists and Mineralogists 5. American Association of Petroleum Geologists and Society of Economic Paleontologists and Mineralogists, Tulsa, OK, 52.

Harper, T.R. and Lundin, E.R., 1997. Fault seal analysis: reducing our dependence on empiricism. In: P. Møller-Pedersen and A.G. Koestler (Editors), Hydrocarbon Seals: Importance for Exploration and Production. Norwegian Petroleum Society (NPF), Special Publication 7. Elsevier, Amsterdam, pp. 149–165.

Hesthammer, J. and Fossen, H., 2000. Uncertainties associated with fault sealing analysis. Pet. Geosci., 6: 37–45.

Hindle, A.D., 1989. Downthrown traps of the NW Witch Ground Graben, UK North Sea. J. Pet. Geol., 12 (4): 405–418.

Hull, J., 1988. Thickness–displacement relationships for deformation zones. J. Struct. Geol., 10: 431–435.

Ibrahim, M.A., Tek, M.R. and Katz, D.L., 1970. Threshold pressure in gas storage. American Gas Association, Arlington, VA, 309 pp.

Knipe, R.J., 1992. Faulting processes and fault seal. In: R.M. Larsen, H. Brekke, B.T. Larsen and E. Talleraas (Editors), Structural and Tectonic Modelling and its Application to Petroleum Geology. Norwegian Petroleum Society (NPF), Special Publication 1. Elsevier, Amsterdam, pp. 325–342.

Knott, S.D., 1993. Fault seal analysis in the North Sea. Am. Assoc. Pet. Geol. Bull., 77 (5): 778–792.

Knott, S.D., Beach, A., Brockbank, P.J., Lawson, J., Brown, J.L., McCallum, J.E. and Welbon, A.I., 1996. Spatial and mechanical controls on normal fault populations. J. Struct. Geol., 18: 359–372.

Lade, P.V., Yamamuro, J.A. and Bopp, P.A., 1996. Significance of particle crushing in granular materials. J. Geotech. Eng., 122: 309–316.

Lee, K.L. and Farhoomand, I., 1967. Compressibility and crushing of granular soil in anisotropic triaxial compression. Can. Geotech. J., 4: 68–99.

Lehner, F.K. and Pilaar, W.F., 1997. The emplacement of clay smears in synsedimentary normal faults: inferences from field observations near Frechen, Germany. In: P. Møller-Pedersen and A.G. Koestler (Editors), Hydrocarbon Seals: Importance for Exploration and Production. Norwegian Petroleum Society (NPF), Special Publication 7. Elsevier, Amsterdam, pp. 39–50.

Leverett, M.C., 1941. Capillary behaviour in porous solids. AIME Pet. Trans., 142: 152–169.

Lindsay, N.G., Murphy, F.C., Walsh, J.J. and Watterson, J., 1993. Outcrop studies of shale smears on fault surfaces. In: S.T. Flint and A.D. Bryant (Editors), The Geological Modelling of Hydrocarbon Reservoirs and Outcrop. Int. Assoc. Sedimentol. Spec. Publ., 15: 113–123.

Mandl, G., de Jong, L.N.J. and Maltha, A., 1977. Shear zones in granular material. Rock Mech., 9: 95–144.

Manzocchi, T., Heath, A.E., Walsh, J. and Childs, C., 2000. Fault-rock capillary pressure: extending fault seal concepts to production simulation. In: Hydrocarbon Seal Quantification. Extended Abstracts, NPF, Stavanger, pp. 51–54.

Manzocchi, T., Walsh, J.J., Nell, P. and Yielding, G., 1999. Fault transmissibility multipliers for flow simulation models. Pet. Geosci., 5: 53–63.

Morrow, C.A., Sih, L.Q. and Byerlee, J.D., 1984. Permeability of fault gouge under confining pressure and shear stress. J. Geophys. Res., 89 (B5): 3193–3200.

Moustafa, A.R., 1996. Internal structure and deformation of an accommodation zone in the northern part of the Suez rift. J. Struct. Geol., 18: 93–107.

Nelson, R.A., 1985. Geologic Analysis of Naturally Fractured Reservoirs. Gulf Publishing, Houston, TX, 320 pp.

Purcell, W.R., 1949. Capillary pressures — their measurement using mercury and the calculation of permeability therefrom. AIME Pet. Trans., 186: 39–48.

Rimstidt, J.D. and Barnes, H.L., 1980. The kinetics of silica–water reactions. Geochim. Cosmochim. Acta, 44: 1683–1699.

Rivenæs, J.C. and Dart, C., 2002. Reservoir compartmentalisation by water-saturated faults — Is evaluation possible with today's tools? In: A.G. Koestler and R. Hunsdale (Editors), Hydrocarbon Seal Quantification. Norwegian Petroleum Society (NPF), Special Publication 11. Elsevier, Amsterdam, pp. 173–186 (this volume).

Schowalter, T.T., 1979. Mechanics of secondary hydrocarbon migration and entrapment. Am. Assoc. Pet. Geol. Bull., 63 (5): 723–760.

Smith, D.A., 1966. Theoretical considerations of sealing and non-sealing faults. Am. Assoc. Pet. Geol. Bull., 50 (2): 363–374.

Smith, D.A., 1980. Sealing and nonsealing faults in Lousiana Gulf Coast Salt Basin. Am. Assoc. Pet. Geol. Bull., 64 (2): 145–172.

Sperrevik, S., Færseth, R.B. and Gabrielsen, R.G., 2000. Experiments on clay smear formation along faults. Pet. Geosci., 6: 113–123.

Swierczewska, A. and Tokarski, A.K., 1998. Deformation bands and the history of folding in the Magura Nappe, western outer Carpathians (Poland). In: K. Decker, R. Lillie and C. Tomek (Editors), PANCARDI; The Lithospheric Structure and Evolution of the Pannonian/Carpathian/Dinaride Region. Tectonophysics, 297: 73–90.

Till, R., 1974. Statistical Methods for the Earth Scientist. An Introduction. MacMillan Education Ltd., London, 154 pp.

Walsh, J.J., Watterson, J., Heath, A.E. and Childs, C., 1998. Representation and scaling of faults in fluid flow models. Pet. Geosci., 4: 241–251.

Watts, N.L., 1987. Theoretical aspects of cap-rock and fault seals for single and two phase hydrocarbon columns. Mar. Pet. Geol., 4: 274–307.

Weber, K.J., Mandl, G., Pilaar, W.F., Lehner, F. and Precious, R.G., 1978. The role of faults in hydrocarbon migration and trapping in Nigerian growth fault structures. In: Offshore Technology Conference 10, Paper OTC 3356, Houston, TX, pp. 2643–2652.

Yielding, G., Freeman, B. and Needham, D.T., 1997. Quantitative fault seal prediction. Am. Assoc. Pet. Geol. Bull., 81 (6): 897–917.

Zhang, S., Tullis, T.E. and Scruggs, V.J., 2001. Implications of permeability and its anisotropy in a mica gouge for pore pressures in fault zones. Tectonophysics, 335: 37–50.

S. SPERREVIK *Department of Geology, University of Bergen, Allegt. 41, N-5007 Bergen, Norway*
Present address: Norsk Hydro Research Centre, P.O. Box 7190, N-5020 Bergen, Norway;
E-mail: susanne.sperrevik@hydro.com

P.A. GILLESPIE *Norsk Hydro Research Centre, P.O. Box 7190, N-5020 Bergen, Norway*

Q.J. FISHER *Rock Deformation Research Group, Earth Sciences Department, University of Leeds, Leeds LS2 9JT, UK*

T. HALVORSEN *Department of Geology, University of Bergen, Allegt. 41, N-5007 Bergen, Norway*

R.J. KNIPE *Rock Deformation Research Group, Earth Sciences Department, University of Leeds, Leeds LS2 9JT, UK*

A method for including the capillary properties of faults in hydrocarbon migration models

C. Childs, Ø. Sylta, S. Moriya, J.J. Walsh and T. Manzocchi

One of the main mechanisms for generating fault seal is by shale smearing or shaley gouge formation in fault zones. Fault sealing potential in clastic sequences is related to the percentage of shale within the part of the sequence that has moved past a point on the fault surface, termed the Shale Gouge Ratio (SGR). Here a method is described for incorporating the capillary effects of fault seal due to the presence of shaley fault rock into migration modelling based on ray-tracing methods. Ray-tracing methods of hydrocarbon migration modelling assume that hydrocarbons migrate along the top of a permeable carrier interval and upwards along the steepest dip governed by buoyancy. We describe a new method in which the effects of faults on migration are accounted for by depressing the top carrier surface by an amount equal to the hydrocarbon column height that can be supported at the fault. The first step is to calculate the distribution of SGR for a representative sequence containing a single carrier interval for the entire range of fault offsets in a study area. From a knowledge of the throw variations along the length of a given fault trace, the distribution of SGR values over the fault surface can be derived. The SGR values are then converted to supportable hydrocarbon column heights, based either on a direct calibration between SGR and column height or indirectly via a calibration between SGR and fault rock capillary threshold pressure. The method accounts for fault throws both less than and greater than the carrier interval thickness. In the first case the minimum capillary threshold pressures of the fault rock determine the onset of hydrocarbon leakage. In the second case hydrocarbon leakage will occur along the fault surface with the migration pathway governed by the spatial distribution of fault rock capillary threshold pressures over the fault surface. The method identifies those points on faults that are likely to determine the paths of major migration arteries and should therefore help identify those areas where detailed structural mapping and fault seal analysis is required. Varying the relationship between SGR and fault seal capacity, by comparing modelling results with actual hydrocarbon distributions, allows examination of the sensitivity of modelling results to fault seal calibration parameters. The method is demonstrated for an area from the North Viking Graben. Of the many uncertainties attached to this type of migration modelling, the paucity of property data for within-plane flow is identified as the most significant.

Introduction

Faults play a key role in controlling both hydrocarbon migration pathways and the distribution and magnitude of hydrocarbon accumulations in extensional basins. Faults modify migration routes and trap hydrocarbons by offsetting carrier units or, where carrier units are juxtaposed across faults, by providing zones of fault rocks with high capillary threshold pressures (Fig. 1). Direct observation of fault zone threshold pressures in the subsurface is not possible, but in recent years empirical methods for risking the sealing potential of faults have been devised (Weber, 1987; Bouvier et al., 1989; Jev et al., 1993; Gibson, 1994; Fristad et al., 1997; Yielding et al., 1997). These methods, in combination with outcrop and laboratory studies provide a means of deriving estimates of the sealing potential of faults offsetting a particular sequence and therefore of fault zone threshold pressures. These methods are generally applied to individual fault-bounded hydrocarbon accumulations, but have not been incorporated into hydrocarbon migration models at basin scales.

Here a method is described for incorporating the results of fault seal potential studies and predicted fault rock capillary properties into a hydrocarbon migration modelling system based on ray tracing. The modelling incorporates the geometric effects of faults and both across- and within-fault hydrocarbon migration (Fig. 1). The method provides a first-order fault seal analysis of each of the mapped faults within an area allowing identification of the main migration arteries within a faulted carrier interval and the identification of fault seal-dependent traps. The method is intended to complement existing fault seal analysis practice in providing the regional migration framework in which the individual accumulations occur. The fault sealing mechanism considered is incorporation of shale/clay into fault zones but the modelling approach can be modified to incorporate other sealing processes.

Hydrocarbon Seal Quantification edited by A.G. Koestler and R. Hunsdale. NPF Special Publication 11, pp. 127–139, Published by Elsevier Science B.V., Amsterdam.

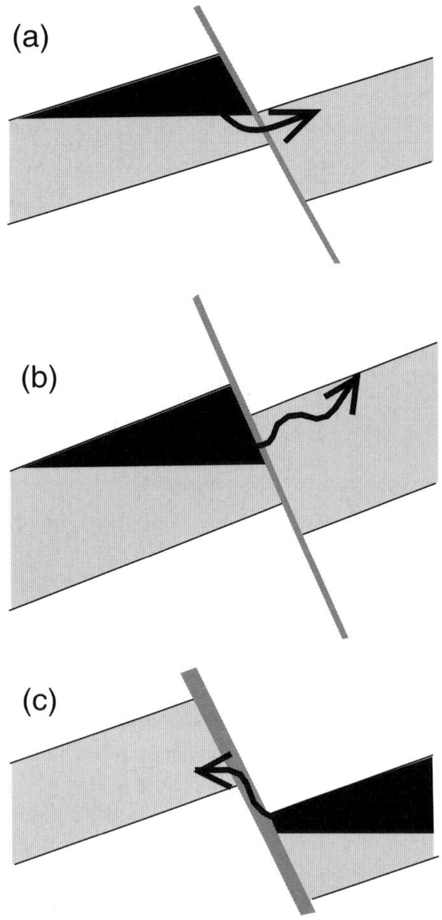

Fig. 1. Fault seal types considered. (a) Juxtaposition seal where the carrier interval is faulted against sealing lithologies and hydrocarbon exits the trap from the highest structural spill-point. (b) Membrane seal where the capillary properties of the fault retard migration and the leak point is determined by the distribution of fault rock capillary threshold pressure. (c) Juxtaposition seal where leakage occurs along the fault zone and is controlled by fault rock capillary threshold pressure.

Semi migration modelling

Semi is a computational modelling system designed to investigate migration and entrapment of oil and gas from source rocks to traps, and thereby provide explorationists with quantitative estimates of the amounts of hydrocarbons in (undrilled) traps, and the most likely phases to expect in them (Sylta, 1991; Krokstad and Sylta, 1996). In order to achieve this, Semi has in-built functions for simulating the generation and expulsion of oil and gas from source rocks using a grid-based method. The primary input to the software is a depth map of the top-carrier unit (Fig. 2a).

Oil migration is modelled by the ray-tracing method where oil migrates beneath the top surface of the carrier interval (Sylta, 1991). Upward migration due to buoyancy forces occurs as stringers or rivulets of hydrocarbons, which follow flow paths normal to structure contours (Fig. 2a). The modelled area naturally subdivides into catchment (drainage) areas, associated with each culmination of the top carrier surface (Fig. 2b). Once filled to capacity, hydrocarbons spill from a trap to structurally higher traps. The spill-paths between traps are seen as zones of high flow rates (Fig. 2c). The flow-rate maps are typically used to assess hydrocarbon migration through time. Oil spills in preference to gas, but the liquid phase is modelled to carry solution gas during migration. The gas phase may carry large quantities of condensates during migration when pore pressures are sufficient. The result of modelling is an estimate of the spatial distribution, level of fill and hydrocarbon composition of each accumulation in the modelled area (Fig. 2d). The sensitivity of migration patterns and spill paths can be analysed from flow pathways and drainage areas during migration.

Although the base of the carrier interval is not explicitly represented in the migration modelling, the carrier interval is assigned a thickness, defined either as a constant value or from an isopach map. The carrier thickness is used in volumetric calculations and is incorporated implicitly in all of the fault seal cases described below.

Fault seal calibration

One of the main mechanisms for generating fault seal in clastic sequences with interbedded course- and fine-grained units is by incorporation of clay and shale into fault zones (Weber et al., 1978; Lindsay et al., 1993; Yielding et al., 1997). Outcrop observations show that shale or clay may be incorporated into fault zones by 'dragging' of individual beds into the fault (Fig. 3a), or by the concentration of the shale fraction in 'dirty' sands within the fault zone (Knipe, 1993; Gibson, 1998). These observations provide a rationale for a proposed relationship between the sealing capacity of faults and the proportion of shale within the faulted sequence (Fristad et al., 1997; Fulljames et al., 1997). Quantification of the relationship between the amount of shale in a sequence and fault seal capacity involves calibrating some function of the proportion of shale within a faulted sequence against fault sealing behaviour observed from known hydrocarbon accumulations. The most commonly applied function of the proportion of shale within a faulted sequence is simply the percentage shale within that part of the sequence that has moved past a point on the fault surface, often referred to as the Shale Gouge Ratio (Fig. 3b; Fristad et al., 1997; Yielding et al., 1997). Published calibrations of SGR against fault seal potential (Gibson, 1994; Fristad et al., 1997) indicate that faults form effective hydrocarbon seals when

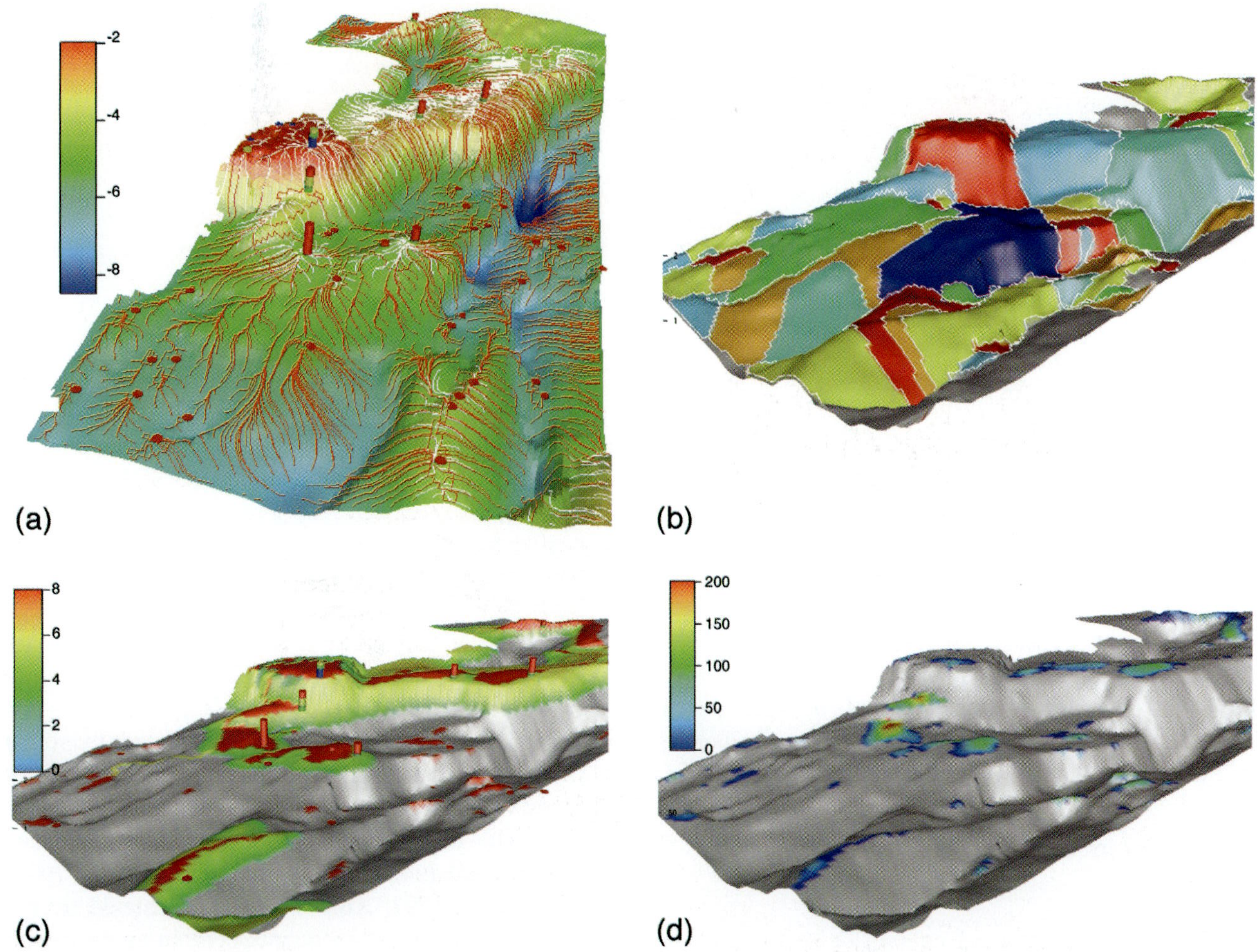

Fig. 2. (a) Depth map (scale in km) with simulated hydrocarbon migration flow paths plotted as red lines. Cylinders show trapped hydrocarbon volumes: blue = free capacity, green = oil, yellow = solution gas, red = free gas. Units for depth map are kilometres. (b) Hydrocarbon migration catchment (drainage) areas define the area sourcing each trap. (c) Oil flow-rate map. Warm colours show areas with high flow-rates, cool colours show low flow-rates. Trapped oil and gas shown as red and blue, respectively. Grey areas contain no oil-flow areas. Topography is depth of carrier at migration time. Units are Sm^3/km per million years. (d) Modelled oil column thickness. Grey regions contain no oil column. Units are m. (a) is viewed from the south and (b–d) are viewed from the southeast.

SGR is greater than ca. 15–20% (Fig. 4a), although this value is expected to vary between basins depending on shale lithology, depth of burial at the time of faulting, etc. Published calibrations also indicate that fault seal capacity may increase with increasing SGR; again this relationship will vary between basins. A relationship between SGR and fault seal potential is supported by published laboratory data, as demonstrated by the negative correlation between the percentage phyllosilicate of a fault rock sample and its effective pore-throat radius (Fig. 4b; Gibson, 1998).

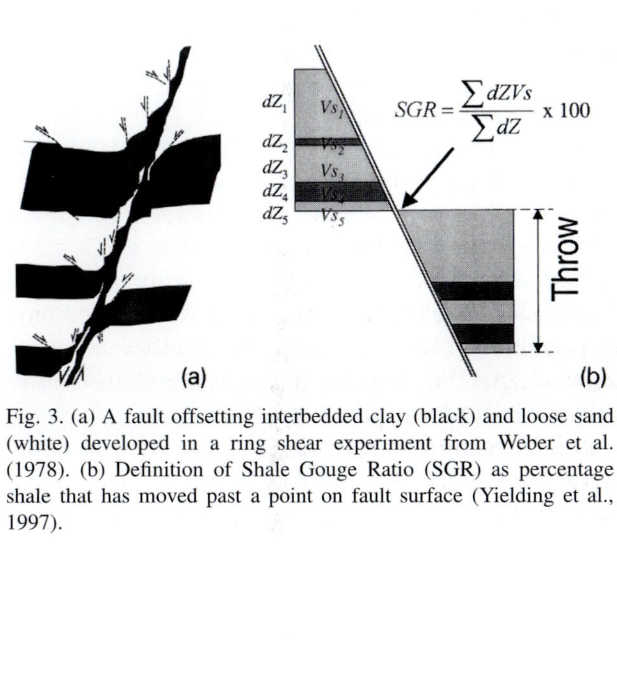

Fig. 3. (a) A fault offsetting interbedded clay (black) and loose sand (white) developed in a ring shear experiment from Weber et al. (1978). (b) Definition of Shale Gouge Ratio (SGR) as percentage shale that has moved past a point on fault surface (Yielding et al., 1997).

Other faulting processes which can potentially give rise to sealing faults include cataclasis of sandstones and diagenetic effects, e.g. quartz cementation. Although both of these processes result in fault rocks with high threshold pressures (Gibson, 1998), it remains to be demonstrated that they result in the areal continuity of high threshold pressures required to trap significant hydrocarbons (Weber and Daukoru, 1975; Gibson, 1998; Yielding, 2002). There are to date no published calibration data on which to base relationships between fault sealing capacity due to either

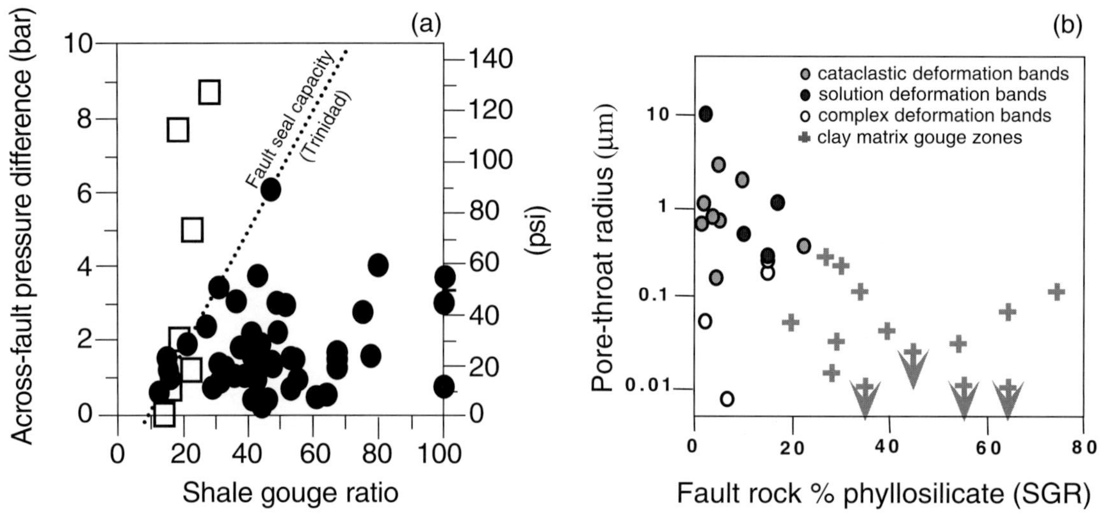

Fig. 4. (a) Published calibrations of SGR against fault seal potential. Squares are data from the North Sea (Fristad et al., 1997) and dots are from offshore Trinidad (Gibson, 1994). (b) Correlation between percentage phyllosilicate of fault rock and its effective pore-throat radius for a variety of fault rock types. Symbols with arrows are maximum values. From published laboratory data (Gibson, 1998).

cataclasis or diagenesis. Therefore we consider only the effects of shale/clay smearing in our modelling, but other potential relationships, e.g. between depth of burial and quartz cementation, could in principle be incorporated in the modelling process.

Method for incorporating fault-sealing properties

There are three ways of treating faults in Semi. If faults are not explicitly included, referred to below as the open fault case, a smooth grid across the faults is produced and the fault is effectively represented by a shear zone with a vertical thickness and capillary properties which are the same as those of the carrier interval. In this case, fault seal due to juxtaposition of the carrier with sealing lithologies is not modelled and fault related accumulations are due only to culminations in the top carrier surface. If faults are included in the modelling, without their fault rock capillary properties, then geometric, or juxtaposition, sealing is modelled and fault-related accumulations are determined by geometric across-fault spill-points. In the third method, described here, the fault rock capillary properties are included and each of the three processes shown in Fig. 1 is incorporated in the modelling.

The method presented here allows a first-order fault seal analysis to be applied routinely to each mapped fault. The method effectively collapses the results of 3D fault seal attribute analysis into a form suitable for input to Semi without the loss of structural validity. The method devised changes the elevation of the top carrier grid at mapped fault traces according to the heights of calculated potential trapped oil columns at each point on the fault trace. At present it can be applied only to single carrier systems, but will be extended to cope with multiple carrier intervals.

Fault sealing properties are calculated from the lithological sequence within the study area and fault offsets of the top carrier interval, calculated within Semi. The calculations are based on a sequence/throw juxtaposition diagram (Bentley and Barry, 1991; Childs et al., 1997; Knipe, 1997) constructed from a well log or lithological sequence containing the carrier interval (Fig. 5). The input sequence is best described by a 'Vshale' curve which expresses the amount of shale in the sequence (Fig. 5). The Vshale curve is offset by a notional fault with throws between zero and the thickness of the input sequence to give a triangular plot. SGR values are calculated at all points on the sequence–throw juxtaposition diagram. SGR values are converted to fault rock effective pore-throat radii (r), using data such as those in Fig. 4b, then to threshold pressures (P_t) and, for particular fluid densities, to potential trapped column heights (h) from the equation

$$2d \times \cos\phi / r = P_t = (\rho_w - \rho_{hc})gh$$

where ρ_w and ρ_{hc} are brine and hydrocarbon densities respectively, d is surface tension and ϕ is the contact angle with the solid interface (Schowalter, 1979). Alternatively, for individual calibration data sets, e.g. the Trinidad data in Fig. 4a, a direct conversion from SGR to threshold pressure can be made. When combining or comparing calibration data sets for different hydrocarbon types, the data should be converted to effective pore-throat radius.

In estimating oil column heights, two different geometries must be considered (Fig. 5). The first is where the carrier interval is self-juxtaposed and oil

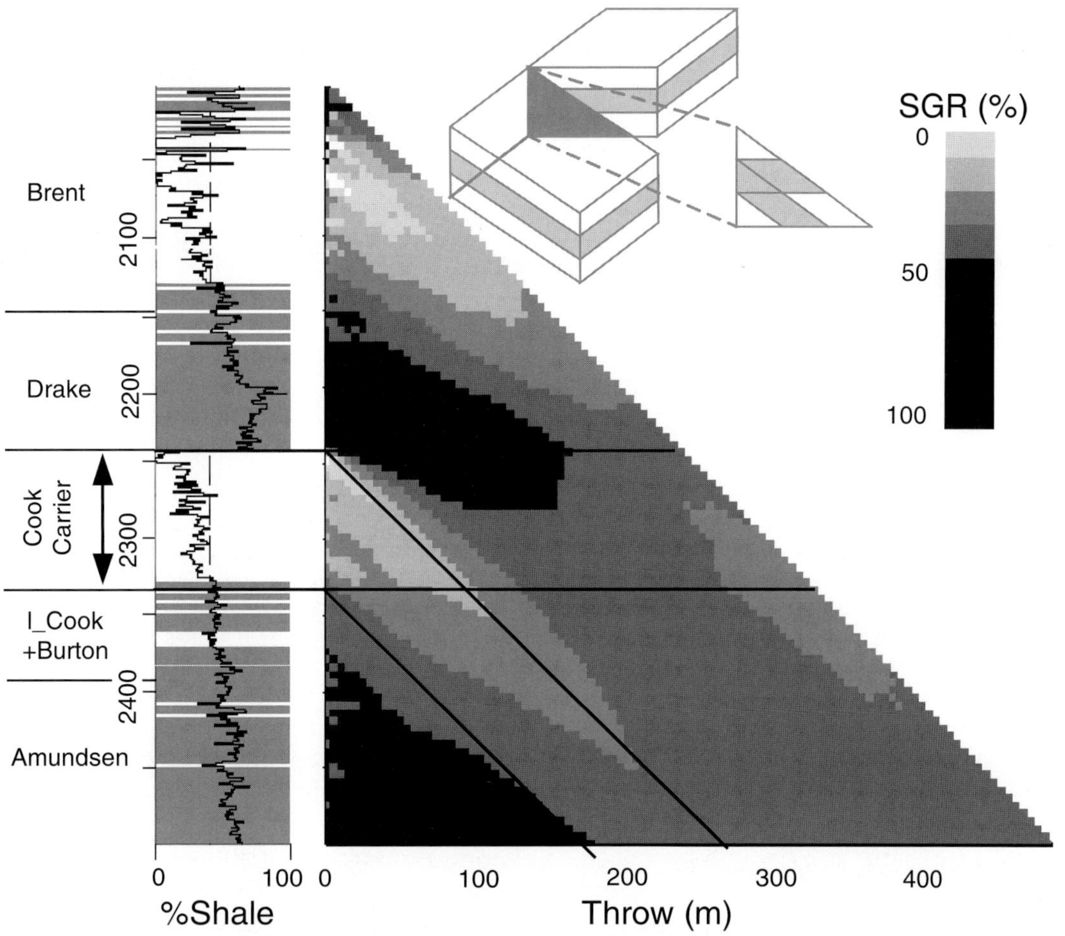

Fig. 5. Sequence–throw juxtaposition diagram constructed from a well in the Gullfaks Field. The diagram construction is illustrated in the inset and described in the text. In the inset, the stippled area represents the carrier interval. The Vshale curve for the well is shown on the left of the figure (TVD in metres) and filled grey where Vshale > 40. The sequence–throw diagram is contoured for SGR. The horizontal and diagonal lines are the top and base of the carrier interval on the upthrown and downthrown side of the fault, respectively.

leakage will occur across the fault (Fig. 1b). In this case the column height which can be supported by the fault at a particular throw value is determined by the threshold pressures of the fault zone separating the footwall and hangingwall carrier interval (Fig. 6i). The oil column height is therefore determined by the minimum buoyancy pressure required to intersect the fault zone threshold pressure curve (Fig. 6ii). Within Semi, the point on the fault trace is represented as a grid cell within the top carrier depth map (see Fig. 6iii). This grid cell is assigned an elevation, relative to the top carrier interval on the downthrown side of the fault, equal to the height of the potential oil column that can be trapped by a fault of this throw. The fault trace is therefore represented by a 'curtain' which hangs beneath the top of the carrier interval. Assignment of elevation allows direct application of the Semi ray-tracing methodology.

The second situation is where the fault throw exceeds the thickness of the carrier interval (self-separated, Fig. 7) and oil leakage occurs along the fault assuming that the rocks on either side of the carrier interval are completely sealing to hydrocarbons. In this case the height of the trapped oil column is governed by fault zone threshold pressures along that portion of the fault that connects the footwall and hangingwall carrier interval (Fig. 7i). The SGR profile for that part of the fault trace separating the reservoir in the footwall and hangingwall is converted to a series of threshold pressure values (Fig. 7ii). Each threshold pressure value is then converted to a potential trapped oil column height that is converted to oil column height relative to the top carrier (Fig. 7iii). A cross-section through the relevant portion of the fault can then be represented as a series of grid cell elevations (Fig. 7iv). The deepest elevation point on this cross-section represents the maximum oil column height that could be supported at this location on the fault trace if migration were limited to vertical flow up the fault. In map view, a fault trace separating the hangingwall and footwall carrier interval is represented as a grid of elevations of potential hydrocarbon column heights. This elevation grid represents sealing capacity over that part of the fault surface between

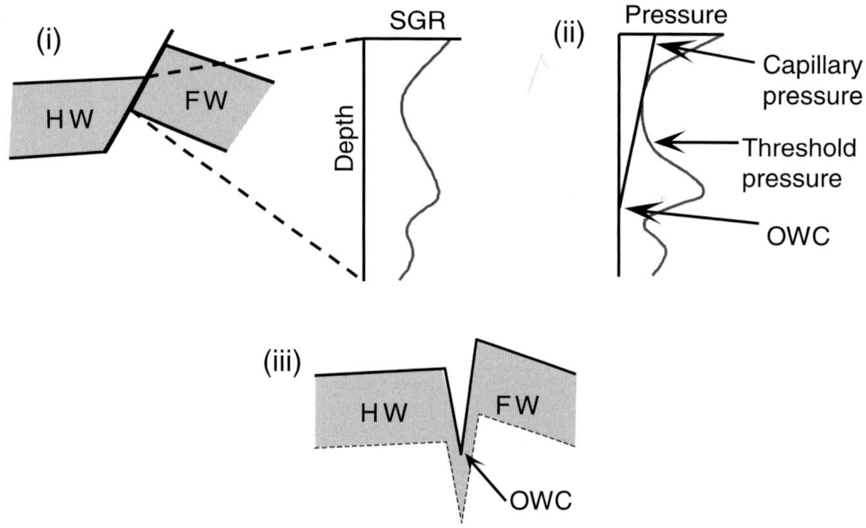

Fig. 6. Schematic diagram illustrating the method of calculation of oil column heights and their incorporation into Semi when the carrier interval (stippled) is juxtaposed against itself. The SGR versus depth curve for the relevant fault trace segment is converted to a threshold pressure curve (i). The depth of the oil–water contact is determined by the minimum capillary pressure required to intersect the fault threshold pressure curve (ii) and the Semi grid at the fault trace is assigned this depth (iii). Although not explicitly represented in the modelling software, for illustration purposes, the base of the carrier interval is shown by the dashed line in (iii).

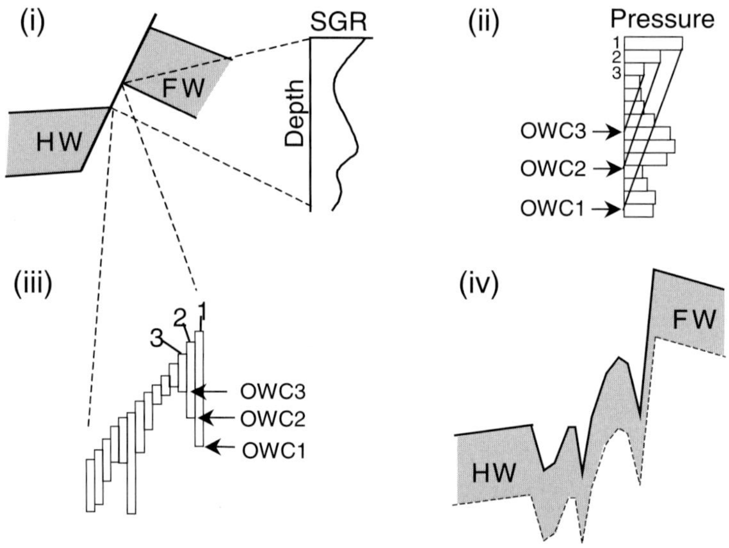

Fig. 7. Schematic diagram illustrating the method of calculation of oil column heights and their incorporation into Semi when the carrier interval is completely offset (i). The SGR versus depth curve for the relevant fault trace segment is converted to a threshold pressure curve and to a potential sealed column height at each point on the fault surface (ii). The column heights are added to the interpolated carrier bed elevation within the fault heave polygon (iii) providing potential oil–water contact depths in the Semi model (iv).

the top of the carrier interval on the hangingwall and footwall sides of the fault. The result of ray tracing across this elevation grid is equivalent to deriving the path of minimum resistance to flow across a grid of fault surface threshold pressures and is equivalent to modelling migration within the fault zone.

Available SGR calibration data are derived from juxtaposed reservoir intervals. Our approach to the estimation of fault sealing in the self-separated case assumes that SGR calibration data can be used to constrain the sealing capacity of faults within the plane of the fault, i.e. the capillary property distribution preventing along-fault flow. This approach assumes that fault rock capillary properties are isotropic at a point on a fault and ignores the textural and compositional anisotropy of fault zones. Given the absence of any data to constrain the effect and significance of anisotropy, our approach allows sensitivity studies to be performed to determine whether an anisotropy function is required to match observed hydrocarbon distributions.

The methodology outlined results in a single top carrier map which implicitly incorporates the capillary sealing properties of each mapped fault. The advantage of the methodology is that it can be easily applied to large areas with many faults, rather than focusing on a few critical faults, as is typically done in exploration work today.

Example of application of the method

The feasibility of the approach can be demonstrated by an example from the Gullfaks Field where several fault seal property cases are simulated without changing other parameters. The objective in choosing this study area is not to exactly match the Gullfaks data, but to demonstrate the practical application of the method and its use in sensitivity studies.

Semi model set up

Within the Gullfaks Field there are reservoir intervals in the Brent, Cook and Statfjord Formations. We have chosen to model the hydrocarbon distributions in the Cook Formation. Only oil accumulations are found in the Cook reservoir. Different oil–water contact (OWC) levels in fault-bounded blocks (A to E, Fig. 8a) indicate that in this field faults play an important role in controlling migration pathways and oil distribution.

Hydrocarbon migration in the Gullfaks area occurred from Tertiary to recent times. Relatively minor changes in carrier bed topography due to Tertiary subsidence have not been incorporated into the modelling. Faulting is of Late Jurassic age so that fault throws and seal capacities estimated from SGR did not change during hydrocarbon migration; structural backstripping and fault throw restoration was therefore not required. For larger faults within the study area there are juxtapositions of the Brent and Cook Formations (Fig. 8b) but, because of the single carrier interval modelling requirement, hydrocarbon migration between these reservoir intervals is not incorporated in the model. Nevertheless the model illustrates both the methodology and the importance of incorporating fault sealing properties in migration modelling.

The Gullfaks Field is strongly overpressured and close to hydraulic fracture gradient (Heum, 1996). Previous work indicates that overpressuring limits the caprock seal capacity to a maximum oil column height of 200 m (Karlsson, 1986). Within the Semi model we have assigned a 200 m seal capacity to the crest of the Gullfaks structure and applied a depth-dependent seal capacity to the deeper parts of the structure. Sufficient hydrocarbon is injected to fill the field and excess hydrocarbon leaks from the crests of the structural highs. Hydrocarbons within the Gullfaks Field are, on geochemical grounds, thought to be sourced both from the northwest and east (Petterson et al., 1990) and within our model hydrocarbon is injected into the two points labelled S1 and S2 in Fig. 8a.

Calibration of fault seal property

Fault seal capacity for the Gullfaks Field was calibrated by comparing SGR values, on those parts of mapped fault surfaces where reservoir rocks are juxtaposed, against the capillary pressure in the fault-bounded accumulation. Where a hydrocarbon-bearing reservoir is juxtaposed against a water-bearing reservoir this is equivalent to the across-fault pressure. Where a fault separates two juxtaposed hydrocarbon accumulations, the capillary pressure on the side with the deeper oil–water contact was used. Capillary pressure was estimated using densities of 0.8 and 1.0 for oil and water, respectively. The results of the calibration on the nine faults analysed (Fig. 9) are in broad agreement with those of Fristad et al. (1997) in that the onset of fault seal occurs at SGR values of 15–20%. A positive correlation between SGR and fault seal capacity is poorly constrained by the available data so three possible relations labelled Case 2 to 4 in Fig. 9 were tested in the migration modelling.

Incorporation of fault seal properties into Semi

As described in the previous section, the first step for incorporating fault seal properties into Semi migration modelling is to calculate SGR on fault surfaces. A single Vshale log was used to calculate the SGR values at fault grid nodes (Fig. 5) for all the faults in the model area, though there are in reality some lateral variations in the carrier lithology and thickness. The calculated SGR was converted to fault rock threshold pressure. The model was run for five different cases, an open fault case (Case 1), a case which incorporates geometric fault seal only (Case 1b) and three cases incorporating fault capillary properties as defined by the three SGR–threshold pressure relations in Fig. 9 (Cases 2, 3 and 4). The fault rock threshold pressure was converted to hydrocarbon column heights using a hydrocarbon density of 0.8, and this height was added to the carrier depth map providing the modified carrier surface (Fig. 10). Faults with high threshold pressures are represented by deep depressions at the fault zone.

Modelling results

The depressions at fault zones form barriers to flow, and hydrocarbons either migrate around or are trapped behind these barriers. Fig. 11 shows an example of these effects. In Case 1b (Fig. 11a), where

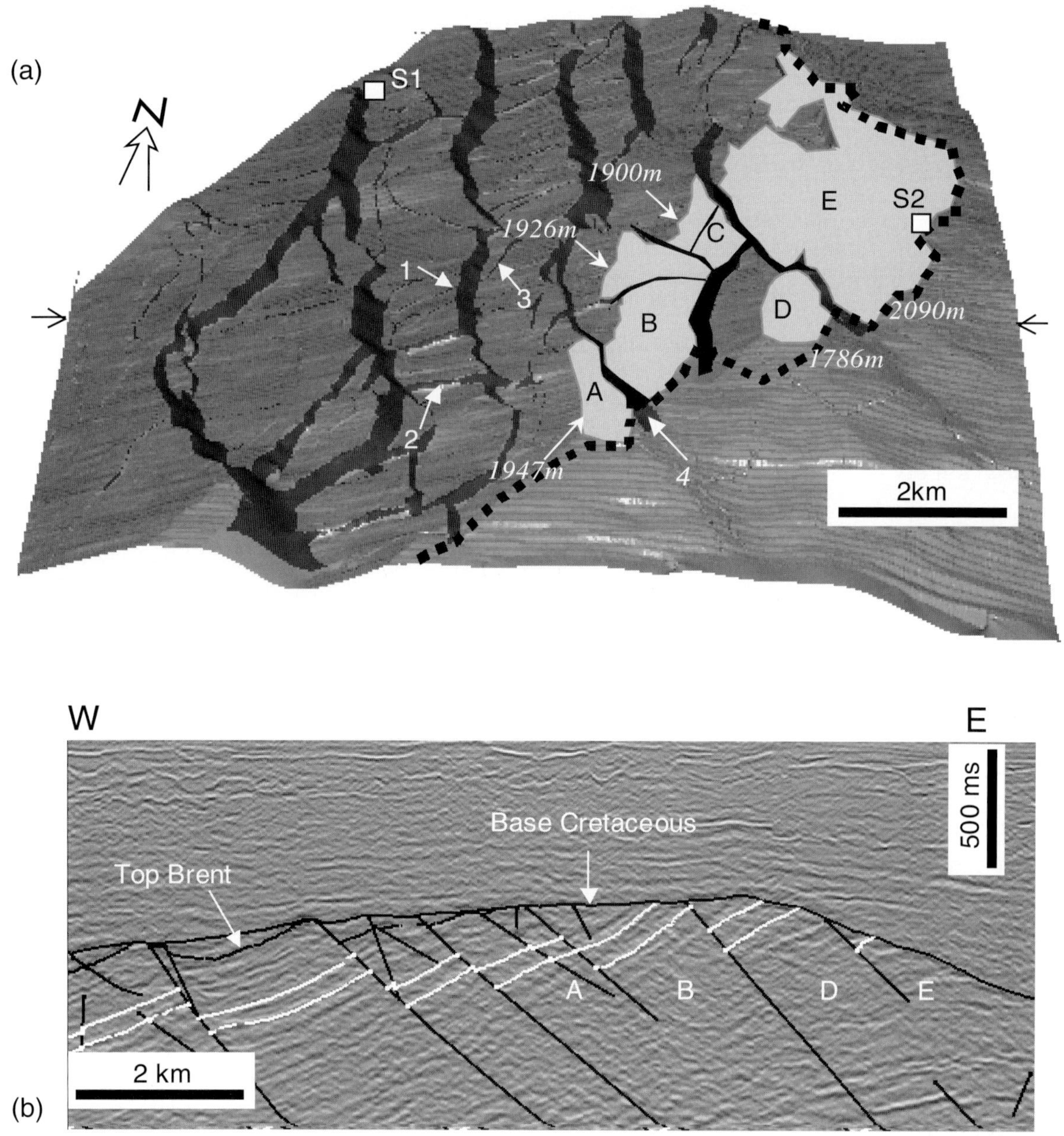

Fig. 8. (a) 3D view from the south of the top of the Cook Formation in the Gullfaks Field. Stippled areas labelled *A* to *E* are oil accumulations and oil–water contact (OWC) levels are shown. *S1* and *S2* are the locations of two hydrocarbon injection points used in the modelling. The carrier interval is eroded to the east of the bold dashed line providing a truncation seal on the eastern edge of Block E. Faults numbered *1* to *4* are discussed in the text. (b) Seismic line through the Gullfaks Field located between the arrows at the edge of (a). The fault blocks in which hydrocarbon accumulations *A*, *B*, *D* and *E* occur are labelled. The two seismically mapped horizons shown in white are the top and base of the Cook Formation.

only the geometric properties of faults are included, a small hydrocarbon accumulation occurs in the footwall closure of Fault 1. As more oil enters the accumulation from the north, excess oil spills to the south through the highest spill point (see pink arrows in Fig. 11). In Case 2 a larger hydrocarbon accumulation occurs due to the sealing capacity of the bounding fault to the south (Fault 2). Moreover, the excess oil spills to the east where the fault displacement is partitioned onto two fault strands (Faults 1 and 3) as the sealing capacity of the fault to the south is higher than that of either of the two strands. This example illustrates the potential sensitivity of hydrocarbon migration arteries to fault capillary properties.

Hydrocarbon distributions at the end of each of the five model runs are shown in Fig. 12. The oil distribution pattern and OWC levels are different in each case. In Case 1, the open fault case where all faults are effectively represented as shear zones, a single large oil accumulation is formed over Blocks A through E. The identical OWC at 1928 m is controlled by the caprock seal capacity of the crest.

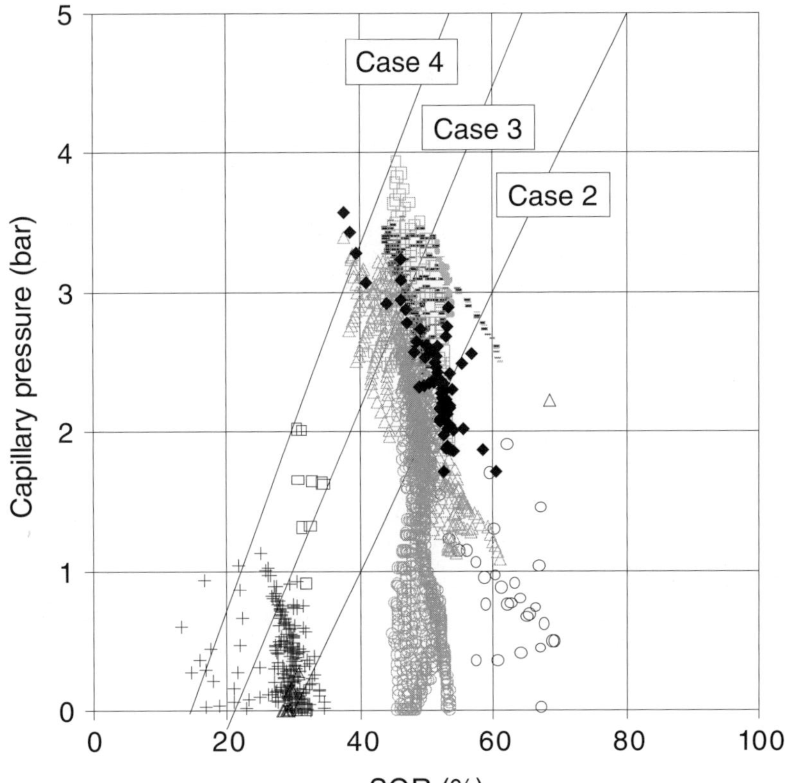

Fig. 9. Cross-plot of calculated SGR against capillary pressure for nine faults in the Gullfaks Field. The lines labelled *Case 2*, *3* and *4* are the relationships between SGR and fault rock threshold pressures used in modelling of the example case study. SGR values were calculated on the fault surface, using Vshale logs in wells adjacent to the faults. The filled diamonds are from an area of juxtaposition of the Cook and Brent reservoir formations on the fault separating the D and E blocks (see text).

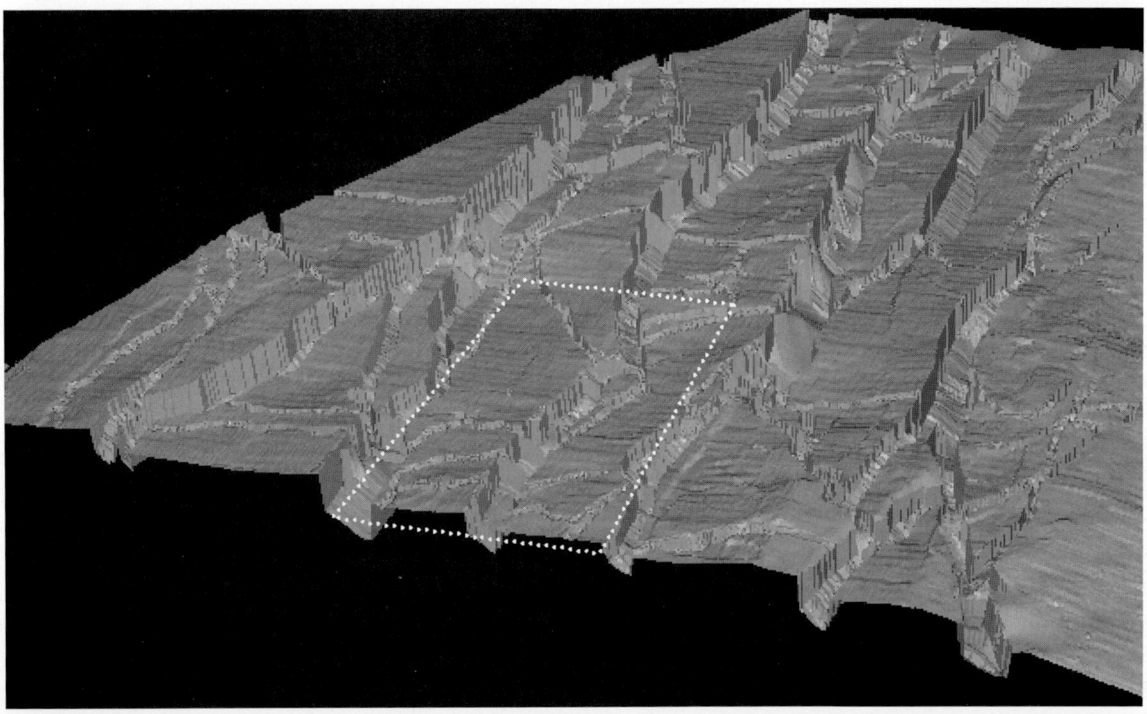

Fig. 10. 3D view, towards the north-northwest of modified top-carrier surface with depressions at fault zones as modelled in Case 4. The area outlined by the dashed line is shown in detail in Fig. 11. The cross-sectional view at the front of the picture is along the line of the seismic section in Fig. 8b.

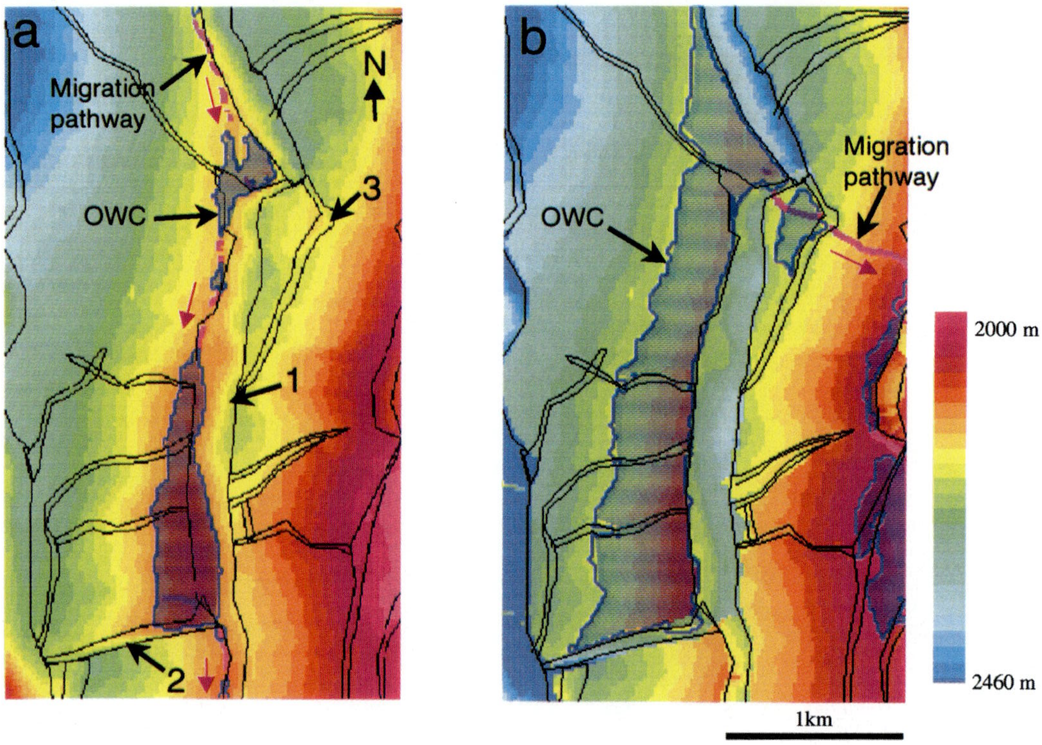

Fig. 11. Structure contour map of the top of the model carrier unit (depth in metres) showing the migration pathways and hydrocarbon accumulations in (a) Case 1b, incorporating only geometric sealing effects and (b) Case 2. The shaded areas outlined in blue are oil accumulations and the pink lines are major migration arteries. The narrow black lines outline fault polygons and depressions within polygons can be seen from the structure contours in (b). Faults *1* to *3* are discussed in text. Note the differences in elevation within the fault polygons in (a) and (b).

For reasons discussed below the open fault case is the only realisation for which oil occurs in Block D. In Case 1b, where only geometric fault sealing effects are incorporated, two different oil–water contacts are developed, but those in Blocks A to C are identical. The results of Case 2 are identical to Case 1b in terms of hydrocarbon distribution in Blocks A to E. However, the migration pathway by which oil reaches these blocks is different in the two cases as shown in Fig. 11. Given a slightly different up-dip structure, these two migration pathways could result in very different hydrocarbon distributions. The modelling results for Cases 2 and 3 are again similar; however, Blocks A and B have the same OWC in Case 2 but different OWCs in Case 3. The difference is because Fault 4 (Fig. 8a) is non-sealing in Case 2 while it is partially sealing in Case 3. In Case 4, fault seal capacities are very high compared with caprock seal capacities. Therefore the oil does not migrate to the structural highs through faults but leaks upward through the caprock.

There is no oil in Block D in the model realisations which incorporate fault seal. This block is bounded by large faults with throws everywhere greater than the thickness of the Cook Formation carrier interval, so that migration into this block in the model, would require flow along the faults between the upthrown and downthrown carrier interval. Migration along the faults does not occur because the sealing capacity of the faults exceeds that of the caprock. These faults juxtapose Brent and Cook reservoirs with a minimum SGR of ~40 in the area of juxtaposition; SGR calibration points derived from this area of juxtaposition are shown in Fig. 9. The oil–water contact in the Brent in the E block is ~160 m deeper than that in the Cook of the D block suggesting that the seal capacity of this fault has been reached and migration of hydrocarbon into the D block is from the Brent in the B block. This illustrates that accurate modelling of the oil distribution within this fault block therefore requires the incorporation of multiple carrier units.

These models illustrate the very different hydrocarbon distributions and migration routes which can occur by varying the sealing properties of faults in migration modelling. In this particular study, Case 3 provides the closest fit to the present-day hydrocarbon distribution and reproduces differences in hydrocarbon column heights across faults. The absolute elevations of fluid contacts is not matched but a closer fit to the data could be achieved by varying the SGR versus fault rock threshold pressure relationship.

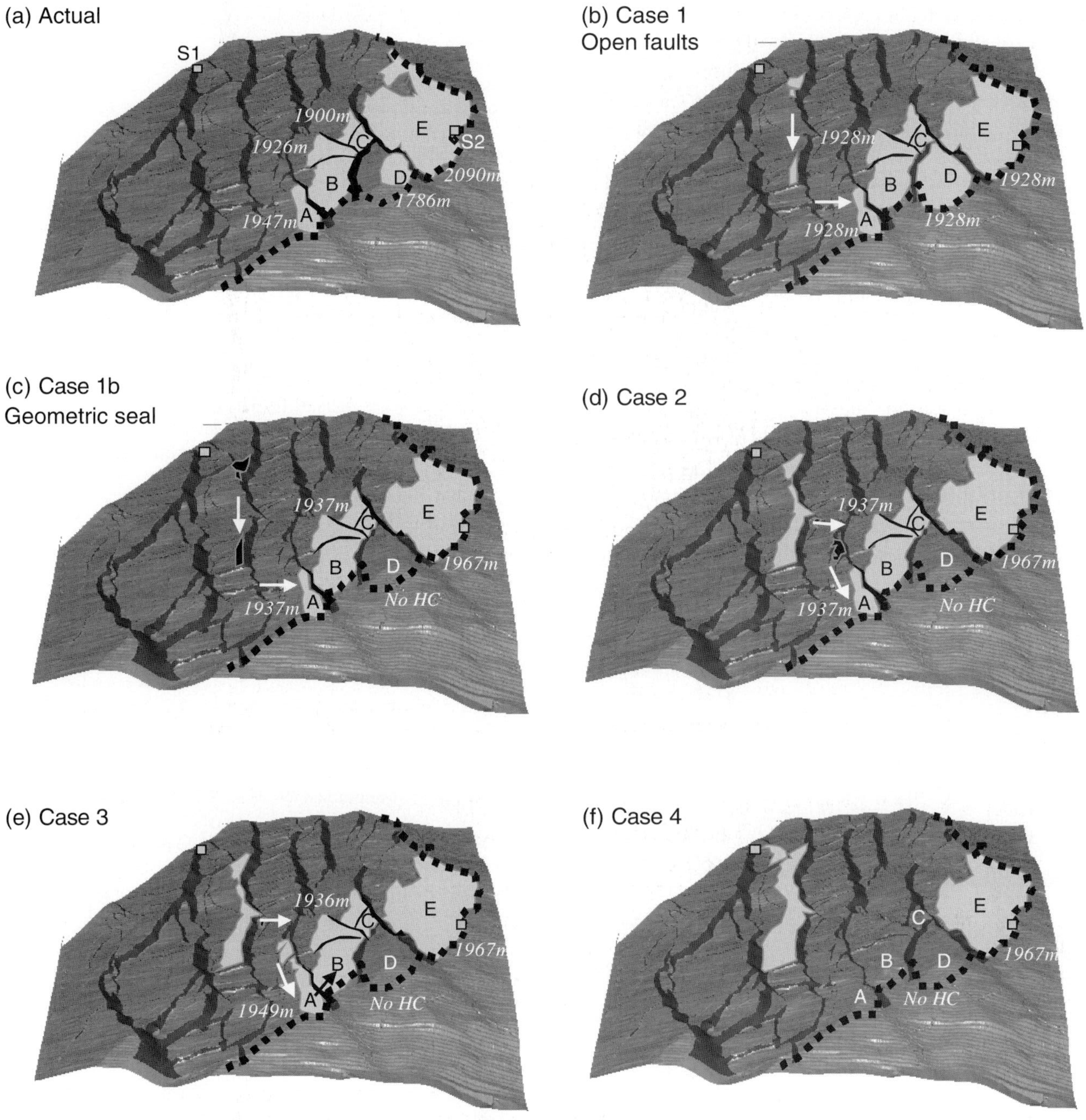

Fig. 12. (a) Known hydrocarbon distribution and modelled hydrocarbon distributions for (b) open faults, (c) incorporation of the geometric sealing effects, and (d–f) three different SGR versus fault rock threshold pressures (see Fig. 9). Stippled areas are oil accumulations with OWC levels in white. *S1* and *S2* are hydrocarbon injection points and arrows are migration pathways. Fault-bounded blocks are labelled *A* to *E*.

Discussion and conclusions

The main purpose of this paper has been to demonstrate a methodology for incorporating fault rock capillary properties into ray-tracing migration models. The principal use of this methodology is in risking exploration prospects that rely on fault seal. The basic approach may also be incorporated into other types of basin-scale hydrocarbon flow models.

The modelling results are strongly dependent on the lithological sequence used and the SGR/fault rock threshold pressure relationship. Sensitivity analysis on the effects of these two parameters can be conducted to establish their impact on hydrocarbon distribution. However, in order to reduce the uncertainty in predictions, more research into the SGR/fault rock threshold pressure relationship is required. Improvements in the lithological definition to allow lateral variations in stratigraphy will also make the method more robust.

An advantage of the method is that fault seal analysis can be routinely applied to all the seismically mapped faults within a basin allowing a quick appraisal of the likely significance of each fault. It can therefore be used as a screening method to identify those faults which, due to their geometry and sealing capacity, impact both migration pathways and sites of hydrocarbon accumulation. In highlighting those faults or parts of faults which are critical to hydrocarbon distributions, the method complements existing, more rigorous techniques that are used to assess the sealing potential of individual faults in exploration risk analysis, e.g. Fristad et al. (1997).

The approach also provides a stringent test of fault seal calibration data sets in that it requires not only that the sealing capacities of individual faults are matched in isolation, but demands that the spatial distribution of accumulations is also replicated in the modelling. The method can also be used to refine fault seal calibration data sets. In constructing a fault seal calibration database it is often necessary to decide whether a fault-bounded culmination is not filled because the sealing capacity of the fault has been reached or the culmination was not charged. Incorporating fault sealing capacity into migration modelling provides a means of refining fault seal calibration data sets by minimising this source of uncertainty. When extended to multiple carrier systems, we envisage that the method will provide a powerful tool for determining the filling histories of stacked, faulted reservoirs.

Acknowledgements

The licencees of PL050/PL050B and the Norwegian Petroleum Directorate (NPD) provided the Gullfaks data set. This work was largely funded by the EC Joule III Programme (Contract No. JOF3-CT95-0014). We thank Peter Hennings for his helpful review of the manuscript.

References

Bentley, M.R. and Barry, J.J., 1991. Representation of fault sealing in a reservoir simulation: Cormorant block IV UK North Sea. In: 66th Annual Technology Conference and Exhibition of the Society of Petroleum Engineers, Dallas, TX, pp. 119–126.

Bouvier, J.D., Kaars-Sijpesteijn, C.H., Kluesner, D.F., Onyejekwe, C.C. and Van der Pal, R.C., 1989. Three-dimensional seismic interpretation and fault sealing investigations, Nun River Field, Nigeria. Am. Assoc. Pet. Geol. Bull., 73: 1397–1414.

Childs, C., Walsh, J.J. and Watterson, J., 1997. Complexity in fault zones and implications for fault seal prediction. In: P. Møller-Pedersen and A.G. Koestler (Editors), Hydrocarbon Seals: Importance for Exploration and Production. Norwegian Petroleum Society (NPF), Special Publication 7. Elsevier, Amsterdam, pp. 51–59.

Fristad, T., Groth, A., Yielding, G. and Freeman, B., 1997. Quantitative fault seal prediction — a case study from Oseberg Syd. In: P. Møller-Pedersen and A.G. Koestler (Editors), Hydrocarbon Seals: Importance for Exploration and Production. Norwegian Petroleum Society (NPF), Special Publication 7. Elsevier, Amsterdam, pp. 51–59.

Fulljames, J.R., Zijerveld, L.J.J. and Franssen, R.C.M.W., 1997. Fault seal processes: systematic analysis of fault seals over geological and production time scales. In: P. Møller-Pedersen and A.G. Koestler (Editors), Hydrocarbon Seals: Importance for Exploration and Production. Norwegian Petroleum Society (NPF), Special Publication 7. Elsevier, Amsterdam, pp. 51–59.

Gibson, R.G., 1994. Fault zones seals in siliciclastic strata of the Columbus Basin, offshore Trinidad. Am. Assoc. Pet. Geol. Bull., 78: 1372–1385.

Gibson, R.G., 1998. Physical character and fluid-flow properties of sandstone-derived fault zones. In: M.P. Coward, T.S. Daltaban and H. Johnson (Editors), Structural Geology in Reservoir Characterisation. Geol. Soc. Spec. Publ., 127: 83–97.

Heum, O.R., 1996. A fluid dynamic classification of hydrocarbon entrapment. Pet. Geosci., 2: 145–158.

Jev, B.I., Kaars-Sijpesteijn, C.H., Peters, M.P.A.M., Watts, N.L. and Wilkie, J.T., 1993. Akaso Field, Nigeria: Use of integrated 3-D seismic, fault slicing, clay smearing, and RFT pressure data on fault trapping and dynamic leakage. Am. Assoc. Pet. Geol. Bull., 77: 1389–1404.

Karlsson, W., 1986. The Snorre, Statfjord and Gullfaks Oil Fields and the habitat of hydrocarbons on the Tampen Spur, offshore Norway. In: A.M. Spencer, C.J. Campbell, S.H. Hanslien, E. Holter, P.H.H. Nelson, E. Nysæther and E.G. Ormaasen (Editors), Habitat of Hydrocarbons on the Norwegian Continental Shelf. Norwegian Petroleum Society, Graham and Trotman, London, pp. 81–197.

Knipe, R.J., 1993. The influence of fault zone processes and diagenesis on fluid flow. In: A.D. Horbury and A.G. Robinson (Editors), Diagenesis and Basin Development. Am. Assoc. Pet. Geol., Stud. Geol., 36: 135–151.

Knipe, R.J., 1997. Juxtaposition and seal diagrams to help analyze fault seals in hydrocarbon reservoirs. Am. Assoc. Pet. Geol. Bull., 81: 187–195.

Krokstad, W. and Sylta, Ø., 1996. Risk assessment using volumetrics from secondary migration modelling: assessing uncertainties in source rock yields and trapped hydrocarbons. In: A.G. Doré and R. Sinding-Larsen (Editors), Quantification and Prediction of Petroleum Resources. Norwegian Petroleum Society (NPF), Special Publication 6. Elsevier, Amsterdam, pp. 219–235.

Lindsay, N.G., Murphy, F.C., Walsh, J.J. and Watterson, J., 1993. Outcrop studies of shale smears on fault surfaces. In: S. Flint and A.D. Bryant (Editors), The Geological Modelling of Hydrocarbon Reservoirs and Outcrop Analogues. Int. Assoc. Sedimentol. Spec. Publ., 15: 113–123.

Petterson, O., Storli, A., Ljosland, E. and Massie, I., 1990. The Gullfaks Field: geology and reservoir development. In: A.T. Buller et al. (Editors), North Sea Oil and Gas Reserves, II. The Norwegian Institute of Technology, Graham and Trotman, London, pp. 67–90.

Schowalter, T.T., 1979. Mechanics of secondary migration and entrapment. Am. Assoc. Pet. Geol. Bull., 63: 723–760.

Sylta, Ø., 1991. Modelling of secondary migration and entrapment of a multicomponent hydrocarbon mixture using equation of state and ray-tracing modelling techniques. In: W.A. England and A.J. Fleet (Editors), Petroleum Migration. Geol. Soc. Spec. Publ., 59: 111–122.

Weber, K.J., 1987. Hydrocarbon distribution patterns in Nigerian growth fault structures controlled by structural style and stratigraphy. J. Pet. Sci. Eng., 1: 91–104.

Weber, K.J. and Daukoru, E., 1975. Petroleum geology of the Niger Delta. In: 9th World Petroleum Congress Proceedings 2: Geology. Applied Science Publishers, London, pp. 209–221.

Weber, K.J., Mandl, G., Pilaar, W.F., Lehner, F. and Precious, R.G., 1978. The role of faults in hydrocarbon migration and trapping in Nigerian growth fault structures. In: 10th Annual Offshore Technology Conference, Society of Petroleum Engineers, Houston, TX, Vol. 4, pp. 2643–2653.

Yielding, G., 2002. Shale Gouge Ratio — calibration by geohistory. In: A.G. Koestler and R. Hunsdale (Editors), Hydrocarbon Seal Quantification. Norwegian Petroleum Society (NPF), Special Publication 11. Elsevier, Amsterdam, pp. 1–15 (this volume).

Yielding, G., Freeman, B. and Needham, D.T., 1997. Quantitative fault seal prediction. Am. Assoc. Pet. Geol. Bull., 81: 897–917.

C. CHILDS *Fault Analysis Group, Department of Geology, University College Dublin, Dublin, Ireland*
E-mail: fault@fag.ucd.ie

Ø. SYLTA *SINTEF Petroleumsforskning A/S, Trondheim, Norway*

S. MORIYA *Fault Analysis Group, Department of Geology, University College Dublin, Dublin, Ireland*

J.J. WALSH *Fault Analysis Group, Department of Geology, University College Dublin, Dublin, Ireland*

T. MANZOCCHI *Fault Analysis Group, Department of Geology, University College Dublin, Dublin, Ireland*

Quantitative fault seal assessment in hydrocarbon-compartmentalised structures using fluid pressure data

Dominique Grauls, Frederic Pascaud and Thierry Rives

Faults act as barriers or conduits to horizontal and vertical fluid flow. As a consequence of this time-dependent behaviour, their sealing efficiency cannot be easily assessed. Quantitative and predictive assessment of fault seals is, however, a key point, as the charging, trapping and preservation of hydrocarbon accumulations at depth are frequently fault-controlled. Fluid pressure data (P), recorded in a compartmentalised structure, constitutes the only available in-situ data for assessing the relative sealing contribution of faults at field scale. The results of this approach, supported by case studies from various geographical areas, are summarised as follows. (1) The fault lateral transmissibility or fault entry pressure seems to be controlled by four main factors: (a) the fault throw — also directly related to the thickness of the damaged zone; (b) the nature of the fault gouge or damaged zone — high entry pressure values are linked to the presence of clay smear or carbonate cementation, low values to cataclastic shear zones; (c) the hydrodynamism or the differential hydraulic potential on both sides of the fault; (d) the position of hydrocarbon kitchen with regards to the trap (downthrown or upthrown side of the fault). (2) A quantitative relationship between sealing efficiency and fault throw can be applied to the prediction of hydrocarbon potential in fault-related traps, down-dip compartments within a given structure or/and in different structures, provided that the geostructural histories are similar. (3) Vertical transfer becomes dominant as soon as the lateral reservoir connectivity through the damage zone is lost. In absence of lateral transmissibility, the fluid pressure (P) builds up inside the confined compartment, during later burial, until it reaches a hydrofracture threshold close to the in situ minimum stress (σ_{3_f}). The fault vertical transmissibility is therefore dependent on the minimum effective stress value or the difference between σ_{3_f} and P. Hydrocarbon migration will therefore occur under hydraulic fracture regime through fault or fracture reactivation processes. (4) This quantitative approach could be also implemented in both basin and reservoir modelling, to account for the role of faults in reservoir compartmentalisation.

Introduction

The general concepts and methodology presented hereafter are based on previous work conducted on capillary behaviour (Leverett, 1941; Purcell, 1949), on cap-rock sealing capacity and trapping (Hubbert, 1954; Berg, 1975; Schowalter, 1979; Chiarelli, 1992; Sales, 1993; Dahlberg, 1994; Deming, 1994; Heum, 1996), and on qualitative and quantitative fault seal assessment (Smith, 1980; du Rouchet, 1984; Watts, 1987; Gibson, 1994).

Fluid movements across a fault zone vary as a function of time and depth and are controlled by the lateral fault entry pressure (P_e). The P_e value depends on the following parameters:

- petrophysical characteristics of reservoir rocks on both sides of the fault zone;
- characteristics of the fault damage zone: clay smear, cementation, thickness and dip ... (Knipe, 1992);
- pressure regime (P_w) in the water zones, on both sides of the fault;
- buoyancy pressure (P_b) induced by the nature and thickness of hydrocarbon columns in both compartments (Gussow, 1954).

Position of drainage area which conditions the direction of hydrocarbon flow (Q_{hc})

Fault-related vertical fluid flow is, in addition to the above parameters, mainly controlled by in situ stress conditions, as hydrofracture-related vertical leakage occurs at minimum effective stress value close to zero ($\sigma'_3 = \sigma_3 - P_f = 0$), where σ_3 and P_f correspond to minimum principal stress and fracture initiation/reopening pressure respectively).

The importance of, and the conditions required for hydraulic fracturing have been described by Secor (1965), du Rouchet (1978), Meissner (1978), Watts (1987), Grauls and Cassignol (1992), Caillet et al. (1994). Fault-related mechanisms under such critical fluid pressure regime have also been emphasised by Sibson (1981), Gaarenstroom et al. (1993), Grauls and Baleix (1994), Finkbeiner et al. (1999), Flemings et al. (1999).

Hydrocarbon Seal Quantification edited by A.G. Koestler and R. Hunsdale. NPF Special Publication 11, pp. 141–156, Published by Elsevier Science B.V., Amsterdam.

These different concepts will be applied to selected case studies from the Caspian Sea, South China Sea, Gulf of Mexico and the North Sea.

Dynamic components of fault-related lateral trapping

Assuming similar reservoir characteristics on both sides of the fault, and hydrocarbon charging (Q_{hc}) from compartment 'A', the general relationship between the dynamic parameters is given in Fig. 1. Pressure profiles at A and B are reported as a function of depth on the right hand side of Fig. 1. At present day, in 'steady state' conditions, the trapping potential of the fault or the difference between the hydrocarbon–water contacts (HWC) can be directly related to the hydraulic pressure differential, δP_w, observed in A and B, and the fault entry pressure P_e (or capillary resistance to hydrocarbon flow). The fluid pressure profiles obtained from RFT or DST (see Fig. 1), combined with a good knowledge of HWC and hydrocarbon densities of compartments A and B, allow the entry pressure value (P_e) to be quantitatively assessed at larger scale. The value obtained provides a fault lateral seal assessment at field scale but cannot be compared, in any case, to capillary pressure values obtained from core samples.

With respect to Fig. 1, the fault lateral entry pressure value (P_e), expressed in MPa, is given by the following equation:

$$P_e = \delta P_w + \delta h\,(\rho_\omega - \rho_h)\,g \qquad (1)$$

where: $\delta P_w = P_w(A) - P_w(B)$ or differential hydraulic pressure regime (in MPa); δh = difference between hydrocarbon–water contacts (in metres); ρ_ω, ρ_h = density of water and hydrocarbon respectively (unitless); $g = 0.009806$ MPa/m.

Hydrostatic systems (pseudo-static conditions)

In this case, the position of water contacts is fully controlled by the P_e value (Fig. 2), which in turn depends on the characteristics of the fault gouge (facies, structure, thickness ...). The P_e value can be directly assessed from pressure plots by evaluating the pressure difference between the two hydrocarbon legs. Eq. 1 can be simplified as follows:

$$P_e = \delta h(\rho_\omega - \rho_h)g \qquad (2)$$

Cataclastic or shear bands in sand reservoirs

This type of shear discontinuity is often linked to small displacements occurring within sand-prone facies. Grain size reduction or grain rearrangement along shear displacements results in the formation of

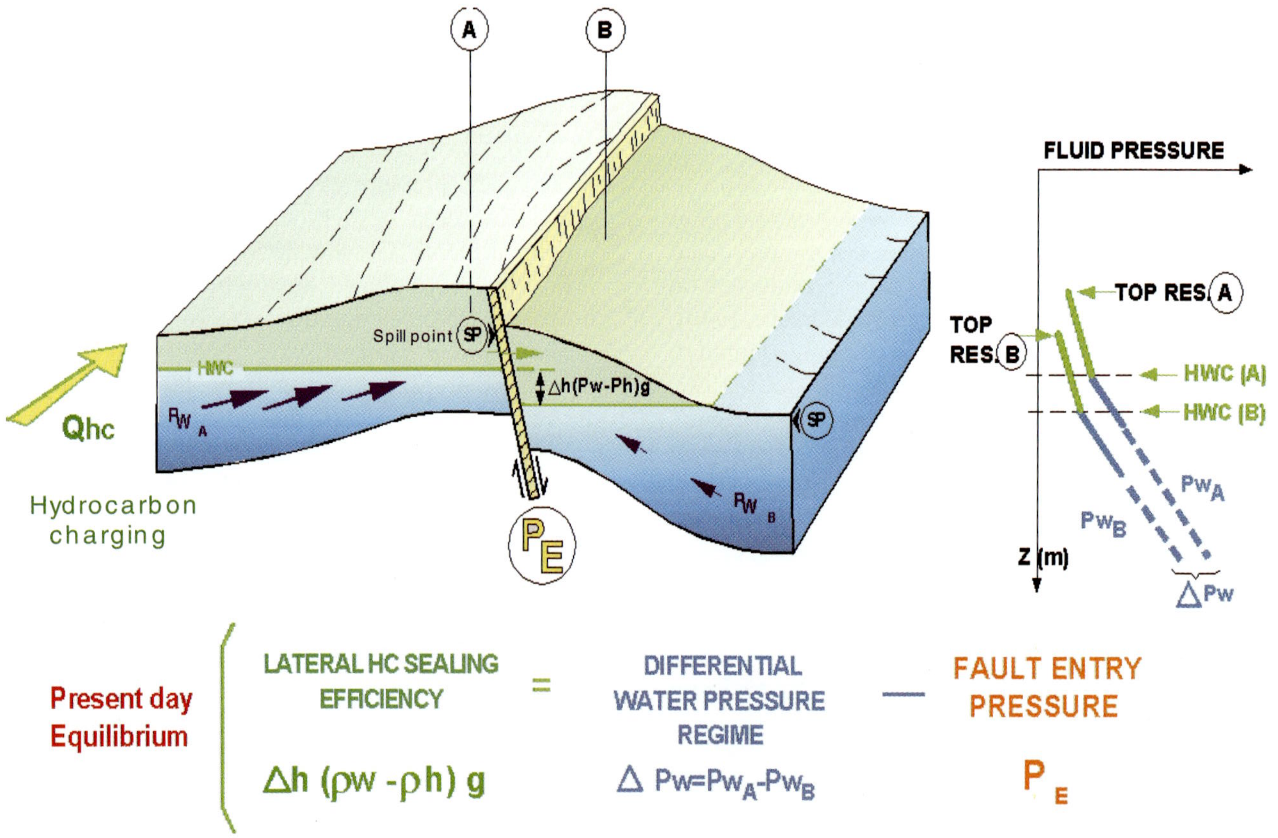

Fig. 1. Main lateral fault seal parameters: Q_{hc} = hydrocarbon charge; P_e = fault entry pressure; δP_w = hydraulic pressure differential; δ_h = difference in water tables; ρ_ω and ρ_h = water and hydrocarbon densities; SP = spill point; HWC = hydrocarbon water contact; Z = depth.

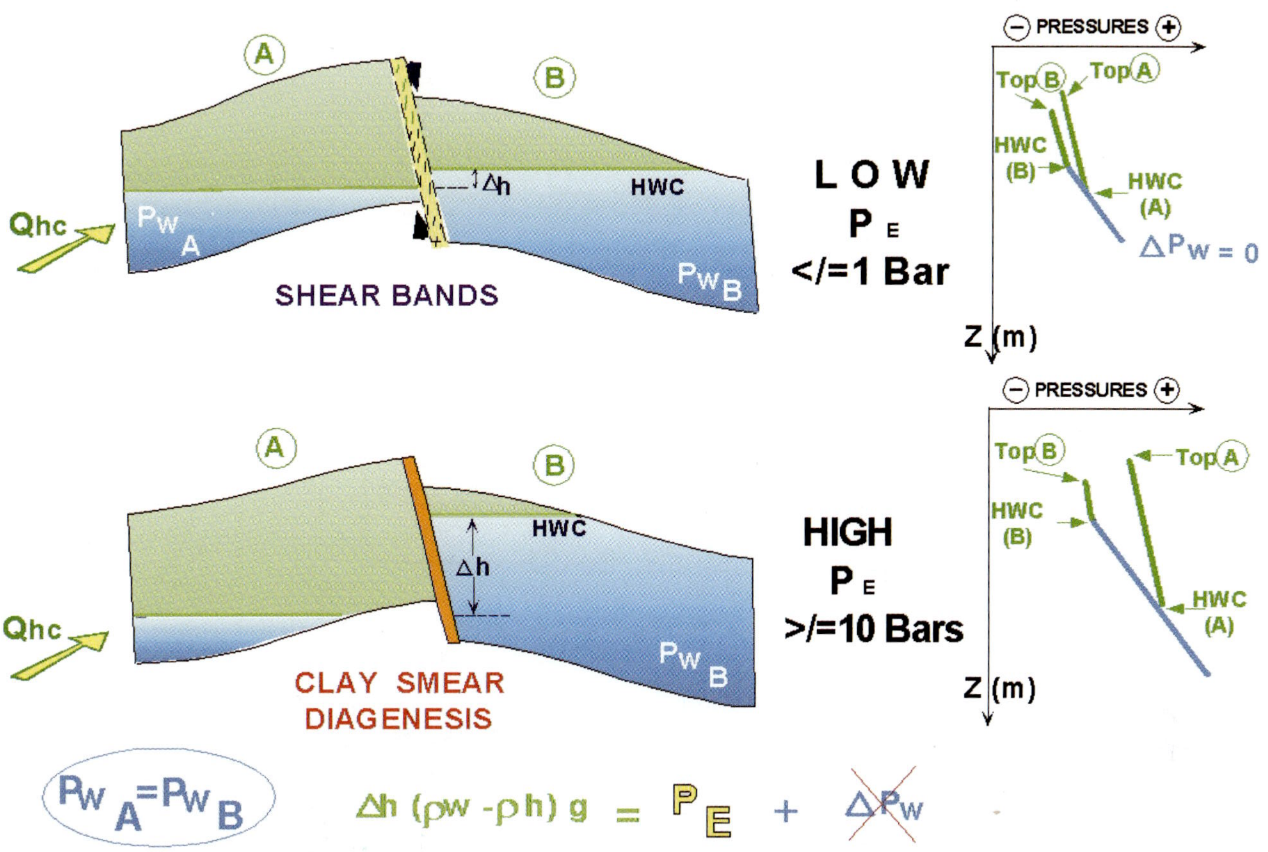

Fig. 2. Predominance of capillary sealing effect (no hydrodynamism): low sealing efficiency values (≤ 1 bar) are related to cataclastic subseismic faults; higher sealing efficiency values (≥ 10 bar) are linked to clay smear effect, or/and diagenesis (Q_{hc} = hydrocarbon charge; P_e = fault entry pressure; δP_w = hydraulic pressure differential; δh = difference in water tables; ρ_w and ρ_h = water and hydrocarbon densities; *SP* = spill point; *HWC* = hydrocarbon water contact; *Z* = depth).

a capillary barrier within the reservoir zone. The P_e value remains very low and the mean value does not exceed 1 bar or 0.1 MPa (Fig. 2). Local variations of HWC, less than 20 m, observed in apparently unfaulted reservoirs can be therefore accounted for (see Bahar field case in Caspian Sea for instance).

Clay smear or/and diagenesis

Fluid flow through the fault zone may become more restricted due to the presence of clayey sediment or diagenetic cement within the fault gouge. Therefore entry pressure (P_e) increases up to mean values close to 10 bar or 1 MPa, or even exceeds this value. This type of seal is of great importance for assessing field reserves, as significant hydrocarbon columns in excess of 100 m can be trapped against faults in downthrown compartments (Fig. 2). This type of trap is typically observed in deltaic environments (South China Sea, offshore Brunei, Gulf of Mexico) and as well in North Sea-type settings. They constitute interesting targets in intensive exploration, especially down flank of main structures.

Hydrodynamic systems (dynamic conditions)

Assuming that the fault entry pressure value (P_e) is negligible with respect to the hydrodynamism, two different situations can be considered (Fig. 3).

(1) *'Water drive' effect*. The pressure potential in the water leg, more active in compartment A (P_{w_A}), shows hydrocarbon flow Qhc being flushed across the fault zone and trapped in compartment B where $P_{w_B} < P_{w_A}$. The pressure differential is therefore positive with respect to hydrocarbon flow and typical mean δP_w values vary from 2 to 20 bar (top Fig. 3). The 'water drive' effect therefore constitutes a major risk in the exploration of compartments controlled by faults with low P_e values. A good example of this situation has been observed in SEI 330 block in GOM (Losh et al., 2002).

$$\delta P_w - \delta h(\rho_\omega - \rho_h)g = 0 \tag{3}$$

(2) *'Pressure seal' effect*. The differential pressure (δP_w) on both sides of the fault zone is negative (bottom of Fig. 3) and water flow counteracts the hydrocarbon charge Q_{hc}. The hydrocarbons therefore

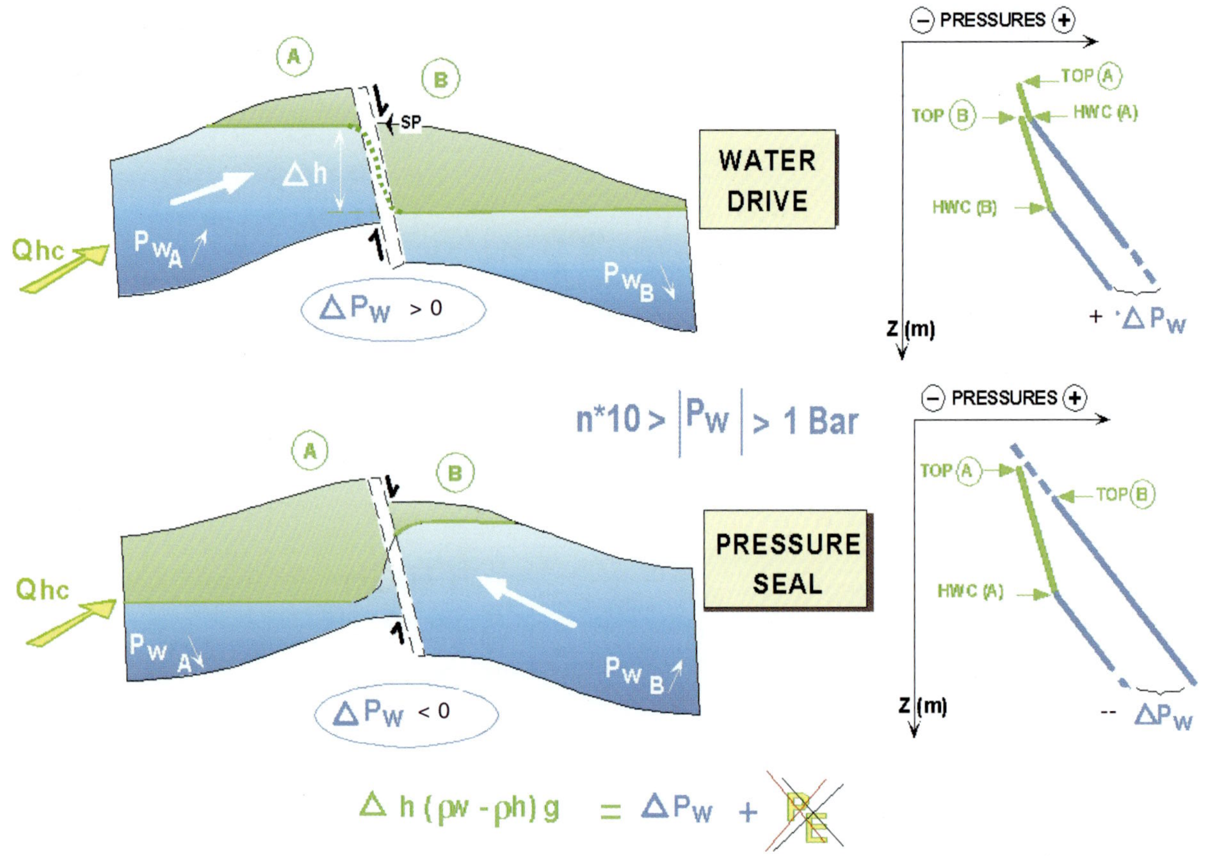

Fig. 3. Predominance of pressure sealing (fault entry pressure > 0): negative differential pressure ($-\delta P_w$) tends to flush hydrocarbon (Q_{hc}) through the fault zone up to internal spill point (water drive effect). In opposition, positive differential pressure values ($+\delta P_w$) counteract the hydrocarbon flow (Q_{hc}) and tends to trap hydrocarbons (pressure seal effect).

tend to be trapped in compartment A.

$$-\delta P_w + \delta h(\rho_\omega - \rho_h)g = 0 \quad (4)$$

Thus, dynamic trapping induced by hydrodynamism is efficient both at prospect and at basin scale.

Dynamic components of fault-related vertical trapping

The presence of clay smear or/and diagenetic cement, or an increase in vertical fault displacement, can completely restrict fluid flow across the fault plane. This results in a rapid build-up of the internal pressure with burial and hydrocarbon charging up to a critical pressure value at top seal. The process is very similar to the 'pressure cell' model introduced by Mann and Mackenzie (1990). As shown in Fig. 4, a vertical-potential leakage can occur at leak-points 1 and 2 (LP_1 and LP_2) whenever the fluid pressure value at these points slighly exceeds the fracture reopening pressure (P_f), which is in turn close to the in situ minimum stress (σ_3). At steady state conditions, the equilibrium is described by the following general equation:

$$P_f = \sigma_3 = P_w + h\,(\rho_\omega - \rho_h)\,g \quad (5)$$

where: P_f corresponds to the fracture reopening pressure in MPa; σ_3 to the in-situ minimum principal stress; P_w to the hydraulic pressure value at leak-point depth LP_1 or LP_2; $h(\rho_\omega - \rho_h)g$ corresponds to the buoyancy effect with respect to hydrocarbon columns.

The hydraulic leakage process can occur at any depth where the reservoir's highest structural point intersects the fault plane. This occurs especially in contexts where recent active burial and compartmentalisation lead to the development of high pressures (HP) at depths exceeding 3000 m (HP fields in offshore Norway, Elgin and Franklin in UK central North Sea . . .). Hydrocarbon charging associated with hydraulic pressure increase, or tectonic stress relaxation favour transient fluid escape related to fracture reopening and vertical fracture permeability increase.

The maximum P_f value at top seal fluctuates between values of σ_3 and σ_{3_f} corresponding to the regional minimum principal stress and to the stress normal to fault plane, respectively. Fracture or fault permeability dramatically increases as soon as P_f within the fault plane reaches σ_{3_f}. The differential between σ_3 and σ_{3_f} varies from 1 to 5 MPa depending on depth and the dip of fracture or fault. This process of vertical leakage can lead to massive hydrocarbon

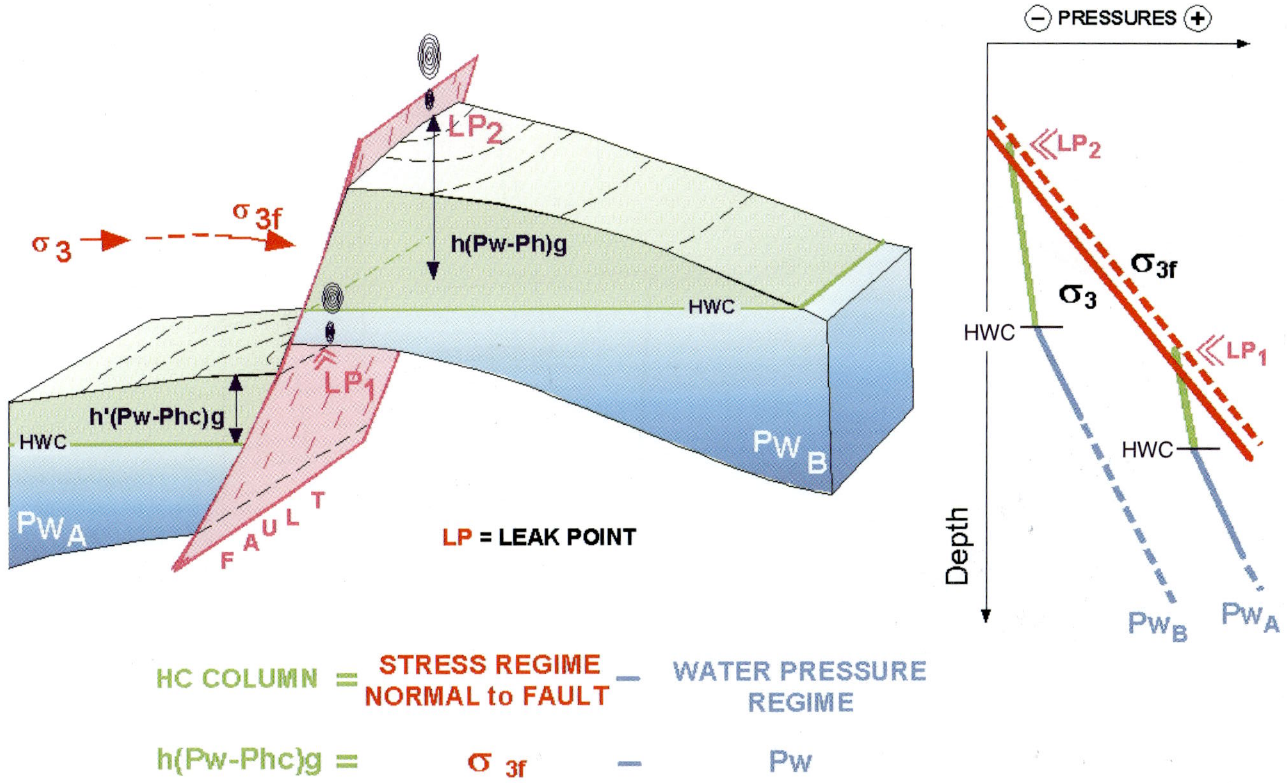

Fig. 4. Main vertical fault seal parameters: hydrocarbon columns trapped against the fault depend on the stress regime normal to the fault and the hydraulic pressure gradient in the water legs; transient leakage occurs at leak points LP_1 and LP_2 (Q_{hc} = hydrocarbon charge; δP_w = hydraulic pressure differential; δh = difference in water tables; ρ_ω and ρ_h = water and hydrocarbon densities; *HWC* = hydrocarbon water contact; Z = depth; σ_3 = regional minimum stress regime; σ_{3_f} = normal to fault stress regime).

dysmigration, and therefore constitutes a major risk in exploration (Watts, 1987). However, the presence or absence of hydrocarbons in a given structure must be considered with respect to the timing of charging and to the quantitative balance between 'fill' and 'leak'. For a three-phase petroleum system, vertical leakage tends to concentrate the liquid phase with respect to the gas vapour phase. The magnitude of the minimum principal stress can be reasonably approximated as a function of depth by making use of leak-off-test data from different geographical areas (Breckels and Van Eekelen, 1982), or by combining both pressure and leak-off data from various tectonic regimes (Grauls, 1997, 1999).

Case studies

Case studies from different geographical areas (Caspian Sea, South China Sea, Gulf of Mexico, and the North Sea) have been selected to illustrate these fault seal concepts.

Caspian Sea (Bahar field)

The Bahar field located in the Caspian Sea provides a good illustration of low capillary sealing efficiency related to cataclastic and subseismic fault zones.

Geological setting

The Bahar field in the Caspian Sea is located on a NNW–SSE compressive structure resulting from present-day active compressive tectonics. As reported by Narimanov et al. (1998) and displayed on the structural map and cross-section in Fig. 5, the structure is bounded on the western flank by reverse faults and compartmentalised by E–W and N–S subseismic faults into seven main compartments. The isobath map (left-hand side of Fig. 5) of the top of the Upper Balakhany 'X' reservoir unit shows that the position of the different gas and oil contacts is affected by these vertical discontinuities.

Interpretation of pressure profiles

Information relative to fluid pressure data is from Abdullaev et al. (1998) and the pressure values at Upper (UBX), Lower (LBX) Balakhany and Periviv (P) reservoir units are displayed as a function of depth in Fig. 6. The following points must be emphasised.

(1) A consistent, hydrostatic pressure regime has been observed in the water zone of the different compartments. As a result of this static pressure regime,

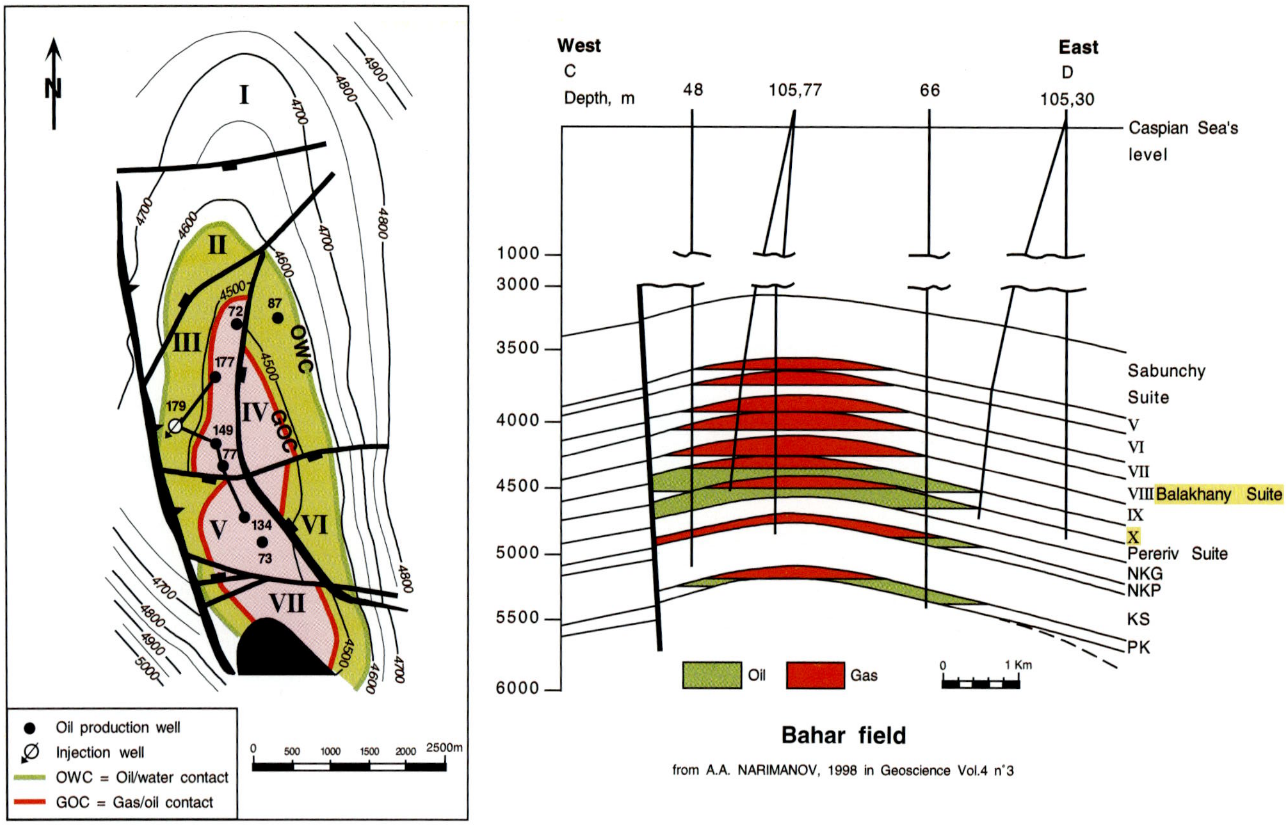

Fig. 5. Bahar field in Caspian Sea: structural map at top Upper Balakhany X and E–W cross-section (adapted from Narimanov et al., 1998).

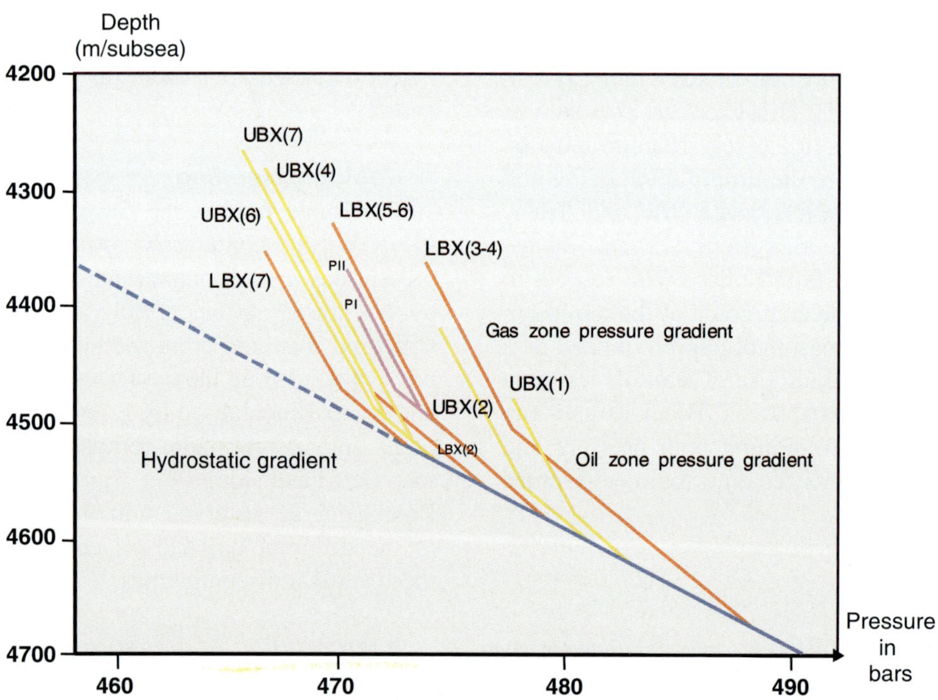

Fig. 6. Bahar field in Caspian Sea: pressure profiles as a function of depth in UBX, LBX, P reservoirs (adapted from Abdullaev et al., 1998).

an excellent overlay of pressure gradient can be clearly observed through the water legs (see Fig. 6). This indicates good lateral drainage efficiency at each reservoir unit over the entire structure.

(2) As a consequence of this pseudo-static pressure regime, the lateral sealing efficiency of the faults is directly accounted for by the pressure difference in the hydrocarbon legs of two adjacent compartments.

(3) The capillary sealing efficiency of the E–W and N–S barriers is low, with values typically ranging from 0.5 to 2 bar and increasing with fault throw (relative increase in clay smear). However, such values are significant enough to account for certain differences in water contacts: 20 to 100 m respectively.

This case provides a good illustration of a hydrostatic situation with a low lateral fault seal. A quick-look quantitative estimate of lateral fault entry pressure is directly obtained graphically by the pressure difference between hydrocarbon legs on both sides of fault.

South China Sea, offshore Brunei

Geological setting

This case study is located in a Tertiary basin, offshore Brunei. The Mid-Miocene and Pleistocene sedimentary section is sand–shale dominated and characterised by the presence of multiple sand layers that allow good lateral drainage and normal hydrostatic pressure regimes down to 4000 m in depth. As shown in a structural diagram in Fig. 7, the trapping of major hydrocarbon columns is controlled by a SW and NE dipping fault system. These synsedimentary faults are 5 to 10 km long, N–S oriented, with fault throws ranging from 100 to 200 m. The interest is focused on the 3500–3900 m 'F' fault interval where downthrown and upthrown compartments have been tested by wells BM-1 and AM-1, respectively (see Fig. 8). In both compartments the trapping of major HC columns can only be accounted for by efficient lateral fault seal.

Interpretation of pressure profiles

A quick-look on well pressure profiles in zones I and II provides the following conclusions:

(1) the existence of a hydraulic pressure differential between the two compartments ($\delta P_w = 3$ bar) indicates dynamic conditions. This pressure differential is graphically assessed from the water pressure legs in both compartments (left side in Fig. 8).

(2) However this slight pressure potential difference ($\delta P_w = 3$ bar) in the BM-1 to AM-1 water leg is not enough to allow hydrocarbons to be flushed across the fault zone. Hydrocarbon columns exceeding 150 m developed in compartment BM-1.

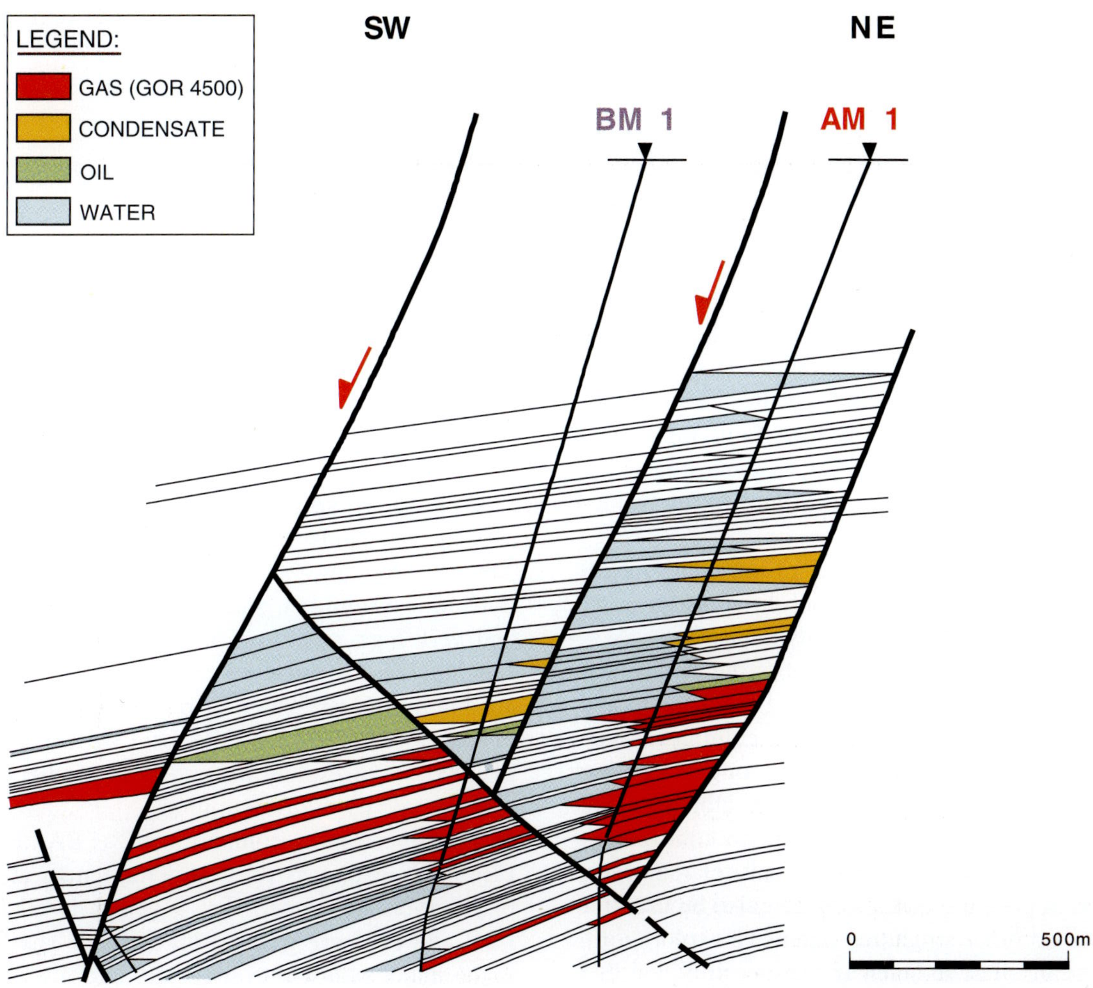

Fig. 7. South China Sea, offshore Brunei: field compartmentalisation and fault-related hydrocarbon accumulations.

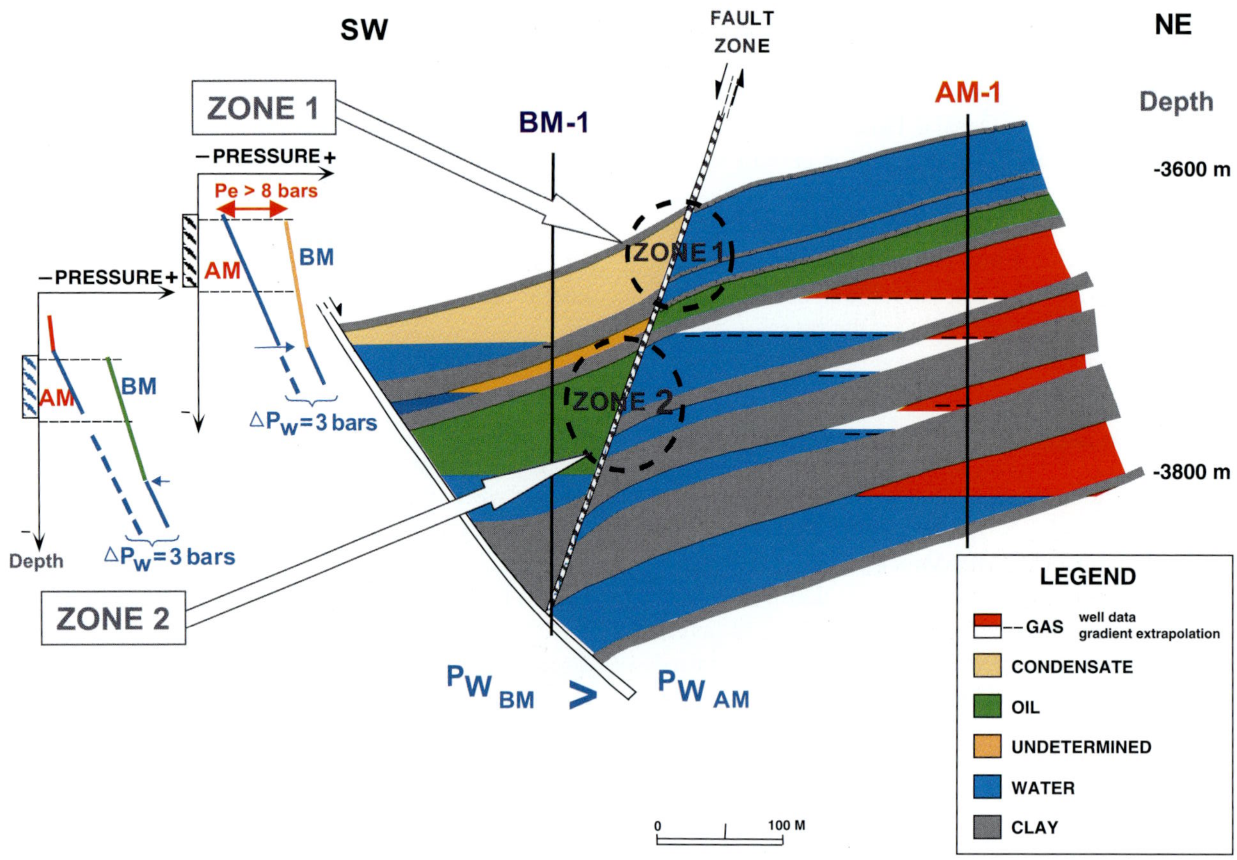

Fig. 8. South China Sea, offshore Brunei: zoom along the fault zone showing the relationship between pressure profiles and fault-related hydrocarbon accumulations in zones I and II (δP_w = hydraulic pressure differential).

(3) The lateral entry pressure value of the fault (P_e) is estimated to be close to 8 bar and corresponds to the differential pressure at the top of zones I or II. This P_e value is obtained by adding δP_w to the buoyancy pressure related to the hydrocarbon column.

(4) The 'clay smear' process which developed in such a sand–shale dominated context is responsible for the higher sealing efficiency values. Fault entry pressure values (P_e) increase with fault throw and thickness of damaged zone. Values close to 10 bar or 1 MPa are commonly linked to fault throw values ranging from 150 to 200 m.

This case is good example of efficient lateral fault seal due to clay smear in sub-hydrostatic water pressure conditions (slight hydrodynamism).

Despite a high sand-to-shale ratio, the clay smear appears to be a very efficient sealing mechanism and well-suited to the trapping of an HC column exceeding 200–250 m for oil and 70–100 m for gas. Linking these results to direct hydrocarbon indicators (seismic amplitude anomalies) should be a good contribution in the exploration of down-dip fault-related traps. This type of pressure quick-look can also be used for prognoses on other structures (intensive exploration), as well as in other sedimentary basins provided that the geohistory is similar.

Gulf of Mexico: South Eugene Island (SEI) block 330-331

This case study is a good example of a dynamic charging system where water-drive processes and vertical hydraulic fracture leakage are responsible for present-day hydrocarbon compartmentalisation. The data and diagrams used for illustrating this case study are taken from Losh et al. (2002).

Geological setting

This case study, located in SEI block 330 Gulf of Mexico, provides a good illustration of a present-day dynamic system. A sand–shale dominated setting and very high rate of sedimentation prevails in this specific area affected by salt diapirism. Reservoirs are highly overpressured at relatively shallow depth. Pressure values close to 5400 psi (385 bar) at a depth of 2000 m correspond to a pressure gradient of 1.92 specific-gravity-equivalent mud.

The three main compartments (FBA, FBB and FBC) are displayed on a structural map at the top of reservoir OI 1 (see Fig. 9). FBA is oil bearing, FBB and FBC are gas and oil bearing. The 'F' fault is the major fault controlling the fluid distribution in structure SEI (Losh et al., 2002).

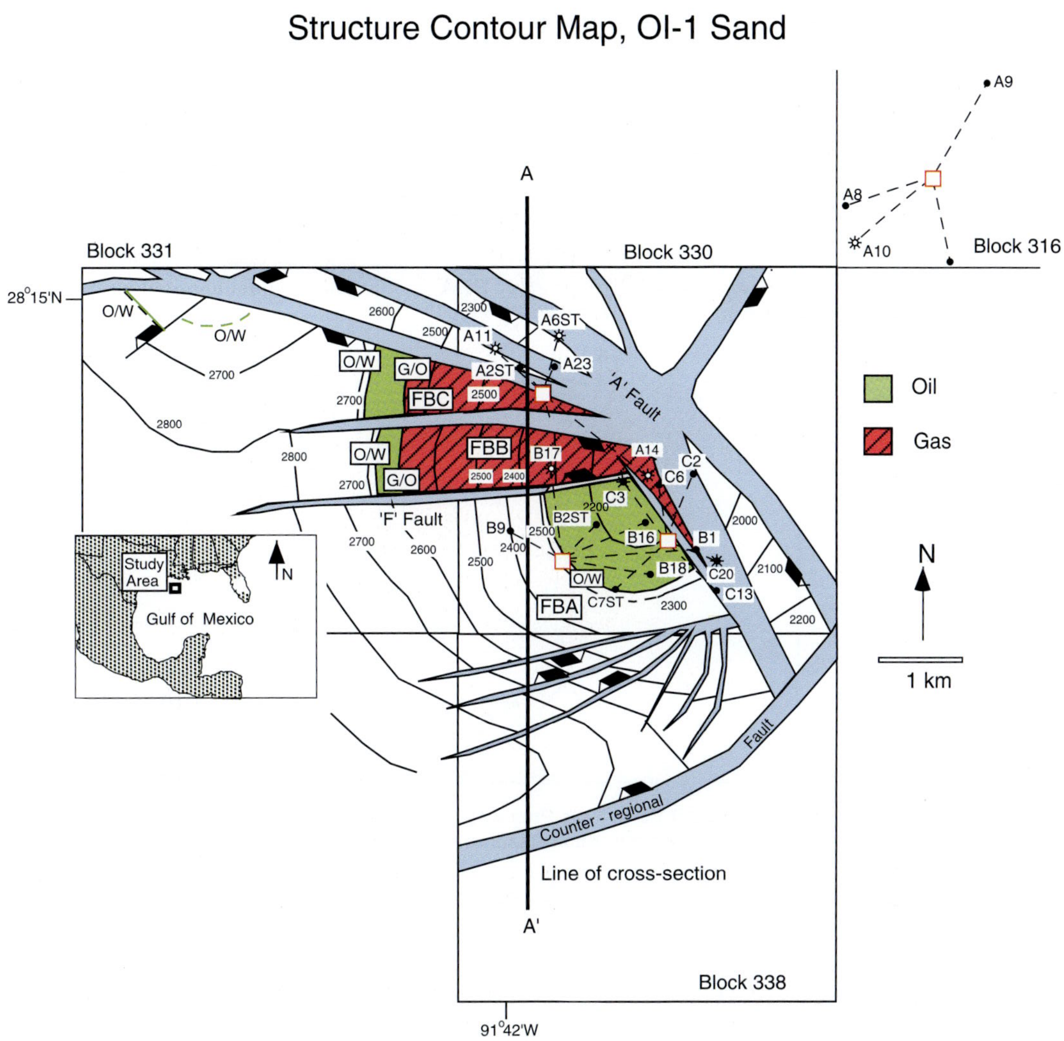

Fig. 9. Gulf of Mexico, South Eugen Island SEI block 330: structure contour map at OI 1 sand (from Losh et al., 2002).

Interpretation of pressure profiles

Pressure data taken from different compartments in the OI 1 sand reservoirs are plotted as a function of depth through the 2000–2800 m depth interval. Results are displayed on Fig. 10. The interpretation of these data leads to the following conclusions.

(1) Pressure profiles of FBB and FBC compartments are very similar, indicating a very low sealing efficiency of the fault separating these two compartments.

(2) The pressure difference in the water legs of compartments FBA and FBC is high and estimated at 60 bar (850 psi). This strong pressure differential ('water drive' effect) flushes hydrocarbons from FBA to FBC across the 'F' fault.

(3) The pressure value close to hydraulic fracture at the top of the structure (pressure gradient of 1.92 at 2000 m) suggests that vertical leakage may be present updip, in the 'A' fault area.

(4) The lateral sealing capacity of the 'F' fault is very low and very likely to be less than 50 psi or 3 to 4 bar according to the differential pressure observed at 2100 m (see Fig. 10). The main hydrocarbon trapping mechanism and present-day distribution is essentially dynamic ('water drive' effect).

(5) The present-day hydrocarbon distribution across this field can be accounted for in three steps (Losh et al., 2002): (1) homogeneous gas condensate charging phase, (2) gas-washed oil flushes the gas across the leaky 'F' fault, (3) the FBA compartment is totally filled by oil (Fig. 11).

This case shows that a quick-look interpretation of pressure data can lead to a better understanding of migration processes in dynamic hydrocarbon systems. A dynamic overpressured setting is responsible for the present-day hydrocarbon distribution across the 'F' fault.

North Sea, UK

In a restricted area (less than 50 km^2) this North Sea gas field represents an excellent summary of

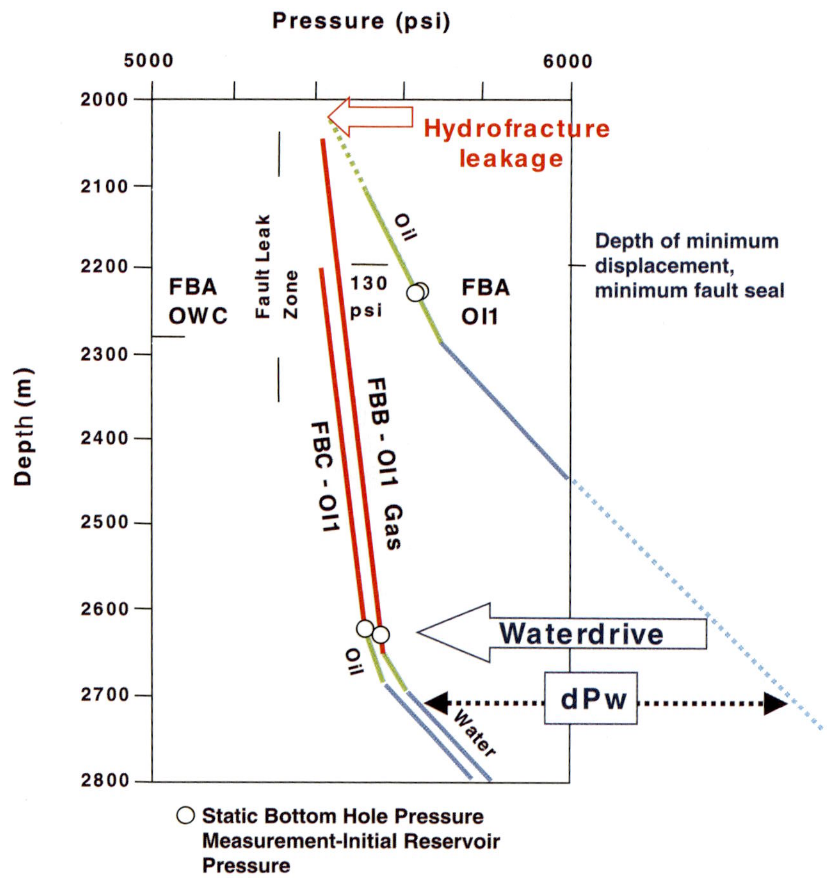

Fig. 10. Gulf of Mexico, South Eugen Island SEI block 330: pressure profiles versus depth, blocks FBA FBB and FBC (from Losh et al., 2002).

the different fault seal mechanisms involved in fluid compartmentalisation: lateral sealing (static and dynamic sealing) and vertical sealing (hydraulic fracture leakage).

Geological setting

This gas field is located on the UK side of the Viking Graben. A structural map at the top of the Jurassic reservoir (Fig. 12) and an E–W cross-section (Fig. 13) show the impact of fault orientation, throws, free water tables and field compartmentalisation on the spatial distribution of hydrocarbons.

Hydrocarbons have been charged since Miocene times from a kitchen located to the northeast and charging is still active today. The deformation of the Jurassic series terminated at the end of the Cretaceous and the three intersecting fault trends segregated the series into 6 main compartments (see Figs. 12 and 13).

The major compartments are limited by N–S major faults with considerable throws, up to 300 m. Intermediate-size compartments are controlled by N–S to N20° fault system with throw values ranging from 150 to 50 m. The low throw and subseismic faults trend NW–SE and NE–SW in a conjugate system and constitute a third-order barrier with throws ranging from 50 to less than 10 m.

Interpretation of pressure data

The pressure records have been gathered from ten wells (see Fig. 12). These data were analysed with respect to fault throw. The vertical and lateral evolution of pressure regimes in different compartments is displayed in Figs. 14 and 15 and leads to the following conclusions.

(1) The lateral efficiency of a fault seal with respect to hydrocarbon cross flow must be considered in a relative way and appears to be related to the importance of fault throw as pointed out by Knipe in 1992 as well as to the thickness of damaged zone (Walsh et al., 1998).

(2) The general westward overpressure decrease across the field indicates a strong hydraulic potential or 'water drive effect' from compartment 6 to 1 (see Figs. 14 and 15). Pressure regimes evolve within a short lateral distance (10 km) from a hydraulic fracture regime in the eastern panel of the field (value of 250 to 300 bar in compartment 6) to hydrostatic pressure conditions westward (overpressure from 0 to 15 bar in compartments 1 and 2). This 'water drive effect' controlled the migration of hydrocarbons generated by the deep kitchen which developed further to the east.

(3) The knowledge of hydraulic pressure in water

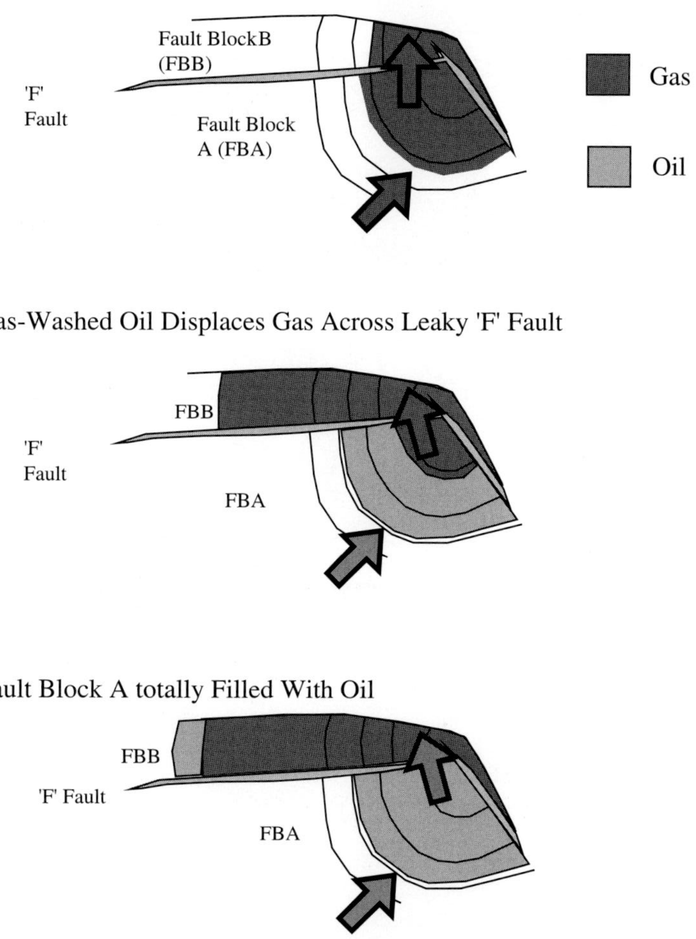

Fig. 11. Gulf of Mexico, South Eugen Island SEI block 330: filling history for OI 1 sand (from Losh et al., 2002).

legs and buoyancy pressures related to hydrocarbon columns in different compartments made it possible (1) to quantitatively assess the fault entry pressure at prospect or field scale by combining the well information relative to the overpressure in water legs, the hydrocarbon heights, and fluid densities, and (2) to establish a ranking of fault barriers in terms of lateral entry pressure at field scale (see legend of Fig. 14). Such results related to fault sealing efficiency can be used as input in both numerical basin simulations and reservoir modelling.

(4) An empirical relationship between fault sealing efficiency or fault entry pressure and fault throw (see Fig. 16) can be used as a predictive tool and applied to other fields in an intensive exploration context. As Fig. 16 shows, the fault seal efficiency (E) expressed in bar increases with fault throw (R). A log–log linear relationship can be established ($\log E \simeq \log R - 1$). Typical fault seal efficiency values of 1 and 10 bar correspond to fault throw values of approximately 10 and 100 m, respectively. The diagram shows that the increase in lateral connectivity or lateral flow across adjacent compartments is linked to the decrease in fault throw. This relation, however, does not apply to the fault limiting compartments 6 and 5. The lateral sealing efficiency of this major fault is too high. Diagenetic precipitation likely associated with significant vertical fluid flow would account for this high lateral sealing efficiency value. As the lateral connectivity between 5 and 6 was lost, the pressure built up during recent burial until it reached the hydraulic fracture threshold, which is close to the normal to fault stress regime. As seen in Fig. 15, present-day pressure values at the top of compartment 6 reached 600 bar, which is close to the least principal stress value S_3 at 3400 m/msl. Compartment 6 evolved in closed system. The only possibility for the hydrocarbon fluids is to escape vertically from compartments 6 to 5 under hydraulic fracture pressure regime (see Figs. 14–16) and to charge compartment 5 transiently. The domain of application of these empirical relationships must be restricted, however, to sedimentary basins or settings with similar geohistory. A fault seal model obtained in the North Sea context cannot be expected to provide satisfactory results in the Niger Delta for instance.

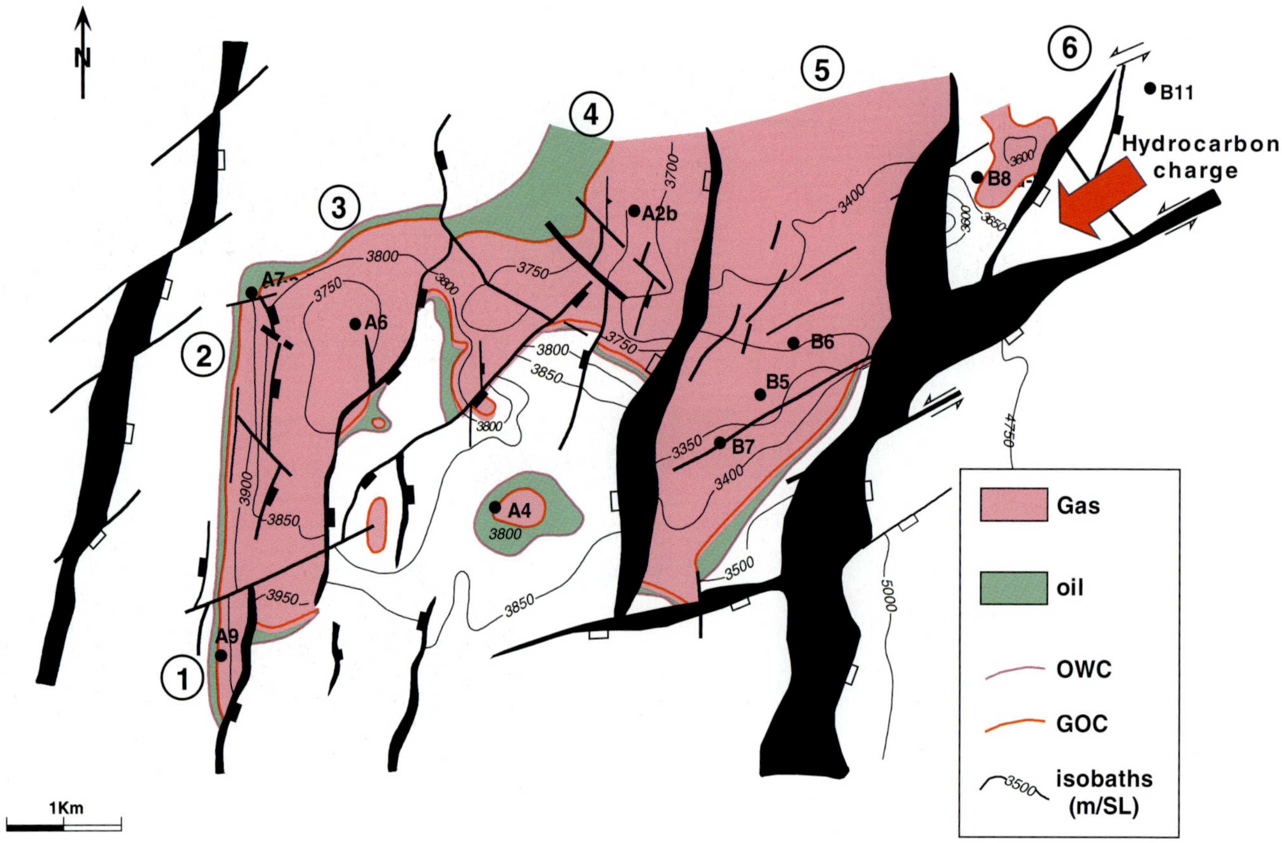

Fig. 12. North Sea gas field, structural map at top Jurassic play: field compartmentalisation (*1* to *6*) and fault-controlled hydrocarbon accumulations.

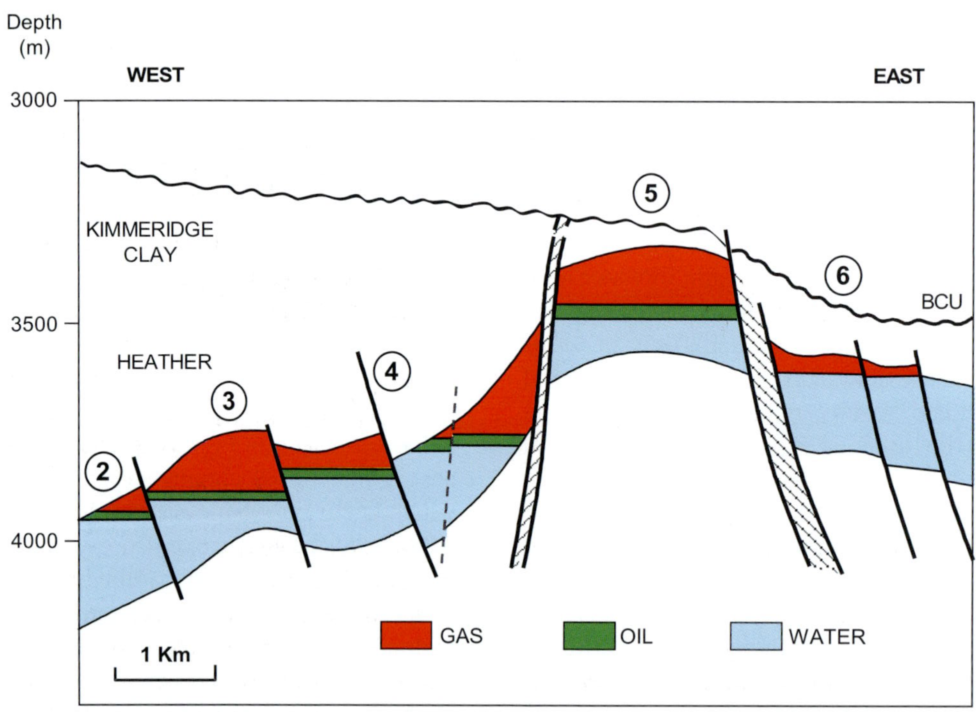

Fig. 13. North Sea gas field, E–W cross section through compartments 6 to 2: relationship between fault throws and free water contact.

(5) The overpressure distribution across the field, the fault seal efficiency, and hydrocarbon columns are consistent with the degree of maturity of oils from the different compartments. As shown in Fig. 17 the least mature hydrocarbons (0.7% R_o maturity index) were encountered in the western compartments, and

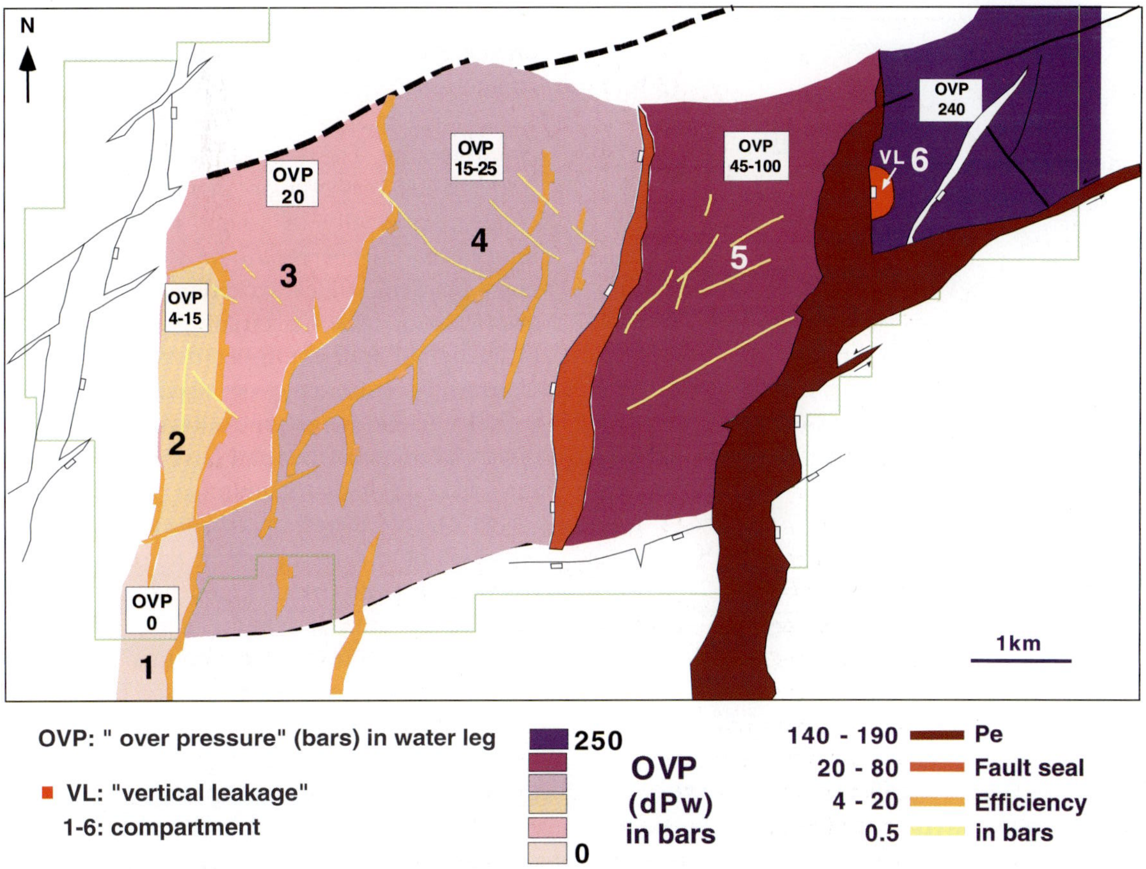

Fig. 14. North Sea gas field, overpressure distribution map view (*OVP* = overpressure value; *VL* = vertical leakage).

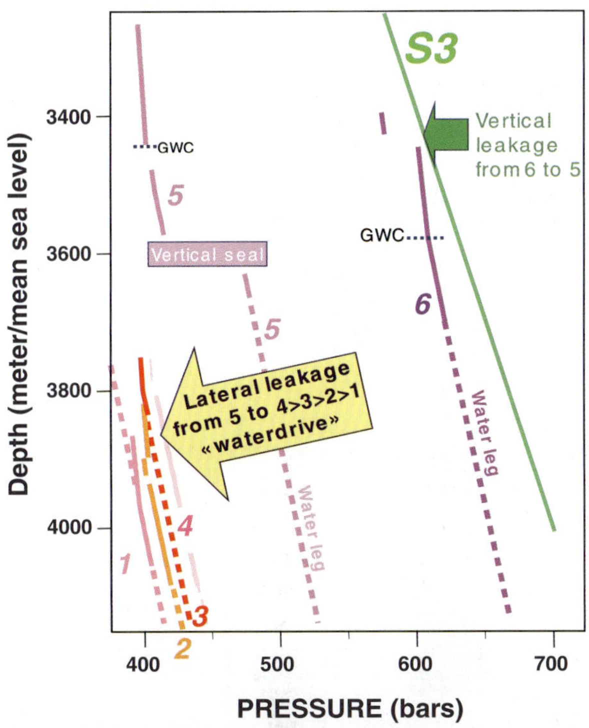

Fig. 15. North Sea gas field, pressure profiles versus depth mean sea level for different compartments: evidence of westwards water drive across the field and vertical leakage at compartment 6 (σ_3 = in-situ far field minimum stress).

the most mature (1% R_o maturity index) in compartments 5 and 6, to the east. As the hydrocarbons were generated deeper from the eastern side of the structure, this is consistent with the dynamic migration scheme of this area.

Summary and implications

The fluid pressure records, when supported by a good structural frame and fault network definition, appear to be a very cost-effective and powerful tool for assessing the fault seal efficiency at prospect scale. The general seal concepts already applied to the cap-rock efficiency of conventional traps seem to be well adapted to fault seal evaluation at prospect or compartment scale.

The general conclusions obtained from this conceptual approach and case study applications are summarised in Fig. 18. The following points must be emphasised.

Fault efficiency must be considered both with respect to the lateral fluid flow across the fault damaged zone, and to vertical fluid transfer.

• *Lateral fault seal or entry pressure* seems to be controlled, in both case studies, by four main parameters:

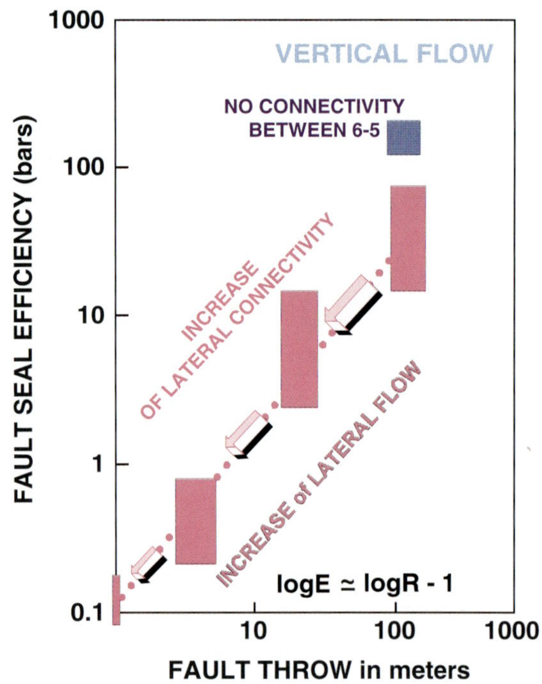

Fig. 16. North Sea gas field, fault seal efficiency versus fault throw in log–log scale.

(1) The *fault throw*, directly linked to the thickness of the faulted or *'damaged' zone*. A 'log–log' relationship between throw and lateral efficiency was defined, for instance, in a North Sea structure (see Fig. 16). This provides a quantitative fault seal approach that can be applied in a predictive way to structures of equivalent geohistory.

(2) The *nature of fault gouge* also controls the fault entry pressure. The 'clay smear' seems to be the most frequent and very efficient in a sand–shale dominated context, especially in young Tertiary basins where diagenesis did not contribute to fault sealing. Fault entry pressure values exceeding 10 bar are common and lead to, in some circumstances, a positive re-evaluation of fault-controlled downdip compartments. Conversely, cataclastic or 'shear bands', linked to subseismic faults, do not allow lateral sealing efficiency to exceed 1 bar and have no strong impact on the hydrocarbon distribution (see Figs. 2 and 5–8).

(3) The existence of a *hydraulic pressure differential* on both sides of a fault favours migration and trapping of hydrocarbons, depending on the position of drainage area and fault lateral efficiency. Good knowledge of the pressure regime in water legs is therefore of prime importance (see Figs. 3, 9, 10, 14 and 15). The water pressure gradient is directly linked to the degree of confinement, which is itself related to the amount of fault throw.

(4) The *position of the hydrocarbon kitchen* or the *origin of hydrocarbon source* with respect to down- or upthrown compartments is, as well, a key parameter controlling the filling of fault-related traps (Figs. 1–3).

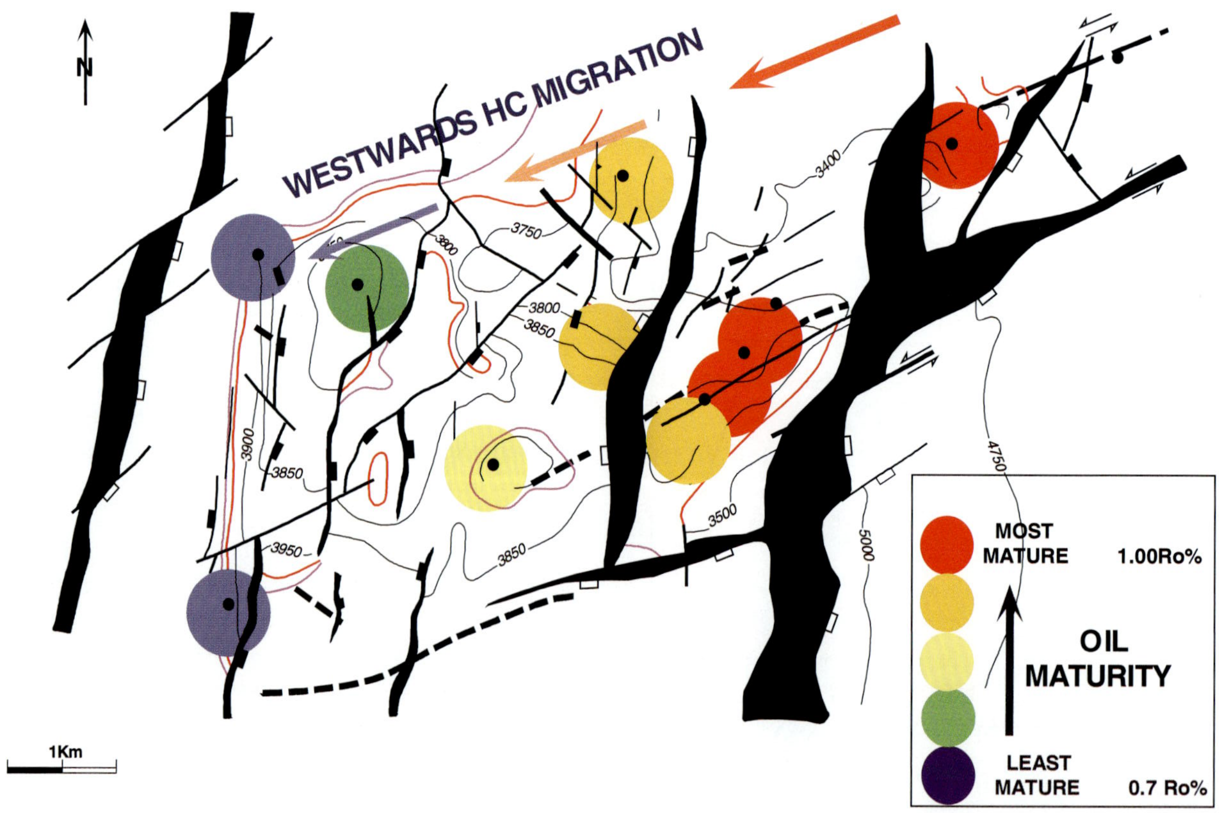

Fig. 17. North Sea gas field, distribution of hydrocarbon maturities is consistent with westwards migration, overpressure and fault seal efficiency model at field scale.

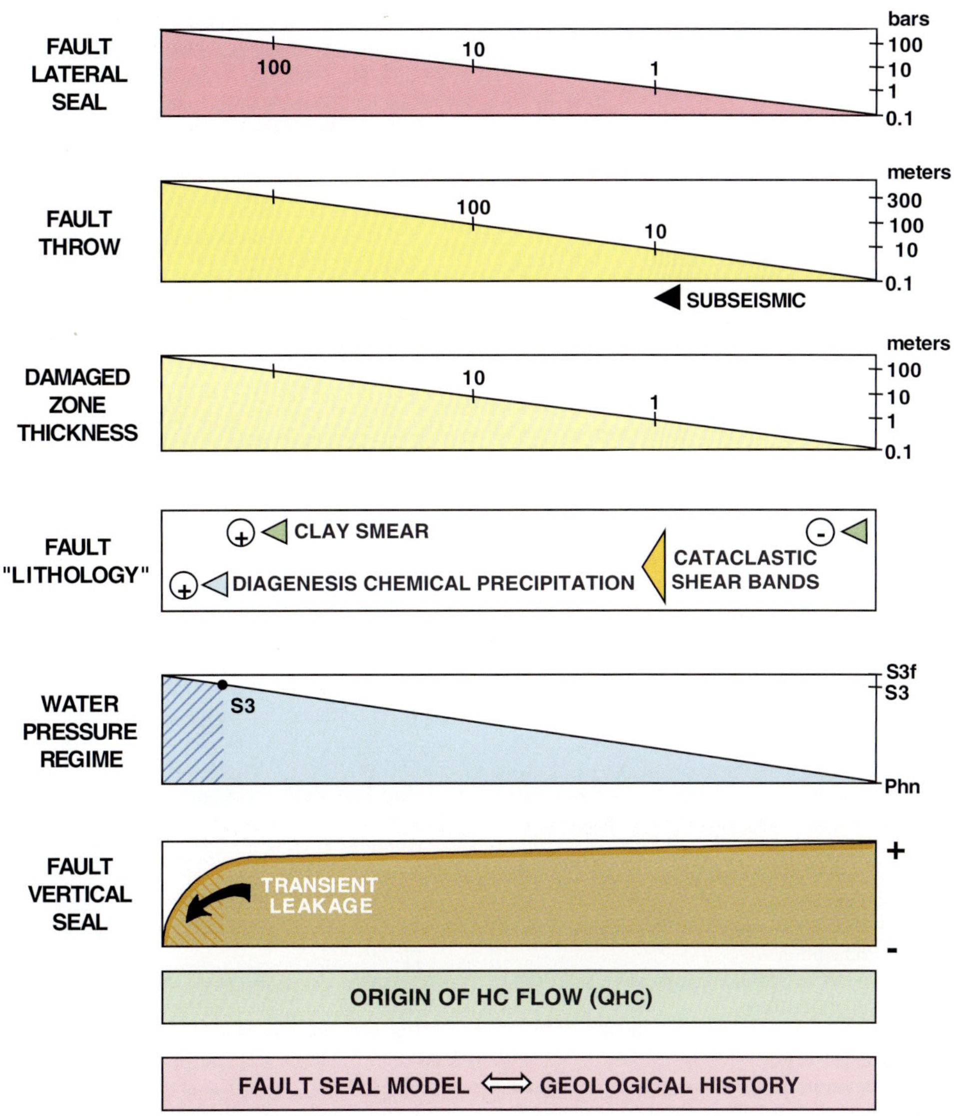

Fig. 18. Fault seal parameters summary: relationship between fault lateral sealing, fault throw, damage zone thickness, fault gouge, water pressure regime, fault vertical sealing, origin of hydrocarbons (P_{hn} = normal hydrostatic pressure; σ_3 = regional least principal stress; σ_{3_f} = normal to fault stress regime; Q_{hc} = hydrocarbon flow).

• *Vertical fault entry pressure* becomes predominant as soon as the fault throw is too important, i.e. exceeding the sand reservoir thickness, or if the nature and thickness of the damaged zone no longer allow transverse fluid transfers. At similar burial rates, the drainage efficiency linked to the compartment size becomes critical. As the fluid cannot escape laterally across the fault zone, the pressure builds up very rapidly up to a given threshold close to the in-situ stress normal to the fault (see Figs. 4, 14 and 15). The fault permeability will exhibit chaotic behaviour, which is a function of effective stress normal to the fault and time. This mechanism favours the escape of gas-prone hydrocarbons from the highest structural point along the fault plane (LP or leak-point area).

• *Implications* from the above conclusions will have a direct impact on the following points.

(1) *The evaluation of compartmentalised structures*. A qualitative assessment of hydrocarbon structures would result at an early stage of exploration in a better understanding of fault-related hydrocarbon migration mechanisms.

(2) *The development strategy in a compartmented prospective area*. The possibility for assessing the fault efficiency gives access to a predictive approach that should apply for re-evaluating some neglected exploration areas where fault-related traps are present. The sealing efficiency versus fault throw relationship can be applied in a predictive way only if geostructural histories are similar. Relationships developed in compartmented North Sea Jurassic structures cannot be applied, for instance, in offshore Tertiary contexts and vice versa.

(3) *Modelling in exploration and production*. The quantitative relationship between fault throw and efficiency can be integrated in basin modelling packages,

as well as in reservoir simulation models in order to account for the contribution of fault semi-barrier potential during depletion.

• *Additional work* is still needed and should focus on the following aspects: (a) relation between static pressure model and results of Shale Gouge Ratio approach; (b) relation between pressure data and long-term production test data; (c) relation between fault or fracture permeability in shaly cap rocks and in-situ effective stress regime; (d) fault lateral transmissivity at production time scale.

References

Abdullaev, T., Fält, L.M., Akhundov, A., Graas, G.W., Kvamme, T., Flolo, L.H., Mehmandarov, K., Narimanov, A.A., Olsen, T.S., Seljekog, G., Skontorp, O., Sultanzade, T., Tank, N. and Valieva, E., 1998. A reservoir model for the main Pliocene reservoirs of the Bahar field in the Caspian Sea, Azerbaidjan. Geoscience, 4 (3): 259–270.

Berg, R., 1975. Capillary pressure in stratigraphical traps. Am. Assoc. Pet. Geol. Bull., 59: 939–956.

Breckels, I.M. and Van Eekelen, H.A.M., 1982. Relationship between horizontal stress and depth in sedimentary basins J. Pet. Technol., 34: 2191–2198.

Caillet, G., Deboaisne, G., Mathis, B. and Roux, C., 1994. The present day stress regime in some deep structures of quadrant 25, offshore Norway. Bull. Elf Aquitaine, 18: 382–390.

Chiarelli, A., 1992. La migration et le piègeage des hydrocarbures — concepts et principes d'application. Internal report.

Dahlberg, E., 1994. Applied Hydrodynamics in Petroleum Exploration. Springer, Berlin, 2nd edition.

Deming, D., 1994. Factors necessary to define a pressure seal. Am. Assoc. Pet. Geol. Bull., 78: 1005–1009.

du Rouchet, J., 1978. Stress fields, a key to oil migration. Am. Assoc. Pet. Geol. Bull., 65: 74–85.

du Rouchet, J., 1984. Migration in fracture networks, an alternative of the supply of the giant tar accumulations of Alberta in Canada. J. Pet. Geol., 7: 381–402.

Finkbeiner, T., Zoback, M. and Stump, P.B., 1999. *In situ* stress, pore pressure, and hydrocarbon migration in the South Eugene Island field, Gulf of Mexico. In: A. Mitchell and D. Grands (Editors), Overpressures in Petroleum Exploration Elf Special Publication, pp. 150–157.

Flemings, P., Siahaan, V., Hicks, P.J. and Stump, P.B., 1999. Secondary migration via fracture permeability in shales: illuminating the relationship between pressure, stress, and column height. In: A. Mitchell and D. Grands (Editors), Overpressures in Petroleum Exploration Elf Special Publication, pp. 139–145.

Gaarenstroom, L., Tromp, R.A., de Jong, M.C. and Brandenburg, A.M., 1993. Overpressures in the central North Sea: implications for trap integrity and drilling safety. In: J.R. Parker (Editor), Petroleum Geology of Northwest Europe: Proceedings of the 4th Conference. The Geological Society, London, pp. 1305–1314.

Gibson, R., 1994. Fault zone seals in siliciclastic strata of the Columbus basin, offshore Trinidad. Am. Assoc. Pet. Geol. Bull., 78 (9): 1372–1385.

Grauls, D., 1997. Minimum principal stress assessment as a control of overpressures in sedimentary basins. Geofluids II march 1997.

Grauls, D., 1999. Overpressures: causal mechanisms, conventional and hydro-mechanical approaches in oil and gas science technology. Euroconference on Pore Pressure, Scale Effect and the Deformation of Rocks, Aussois, France, IFP Special Issue Vol. 54(6), pp. 667–678.

Grauls, D.J. and Baleix, J.M., 1994. Role of overpressures and in situ stresses in fault controlled hydrocarbon migration. Mar. Pet. Geol., 11: 734–742.

Grauls, D.J. and Cassignol, C., 1992. Identification of a zone of fluid pressure-induced fracture from log and seismic data — a case history. First Break, 11 (2): 59–68.

Gussow, W., 1954. Differential entrapment of oil and gas: a fundamental principle. Am. Assoc. Pet. Geol. Bull., 38: 816–853.

Heum, O., 1996. A fluid dynamic classification of hydrocarbon entrapment. Pet. Geosci., 2: 145–158.

Hubbert, M., 1954. Entrapment of petroleum under hydrodynamic conditions. Am. Assoc. Pet. Geol. Bull., 37: 1954–2026.

Knipe, R.J., 1992. Faulting processes and fault seal. In: R.M. Larsen, H. Brekke, B.T. Larsen and E. Talleraas (Editors), Structural and Tectonic Modelling and its Application to Petroleum Geology. Norwegian Petroleum Society (NPF), Special Publication 1. Elsevier, Amsterdam, pp. 325–342.

Leverett, M., 1941. Capillary behaviour in porous solids. AIME Petrol. Trans., 142: 152–159.

Losh, S., Walter, L., Muelbroek, P., Martini, A., Cathles, L. and Whelan, J., 2002. Reservoir fluids and their implication into South Eugene Island B330 reservoirs, offshore Louisiana. Am. Assoc. Pet. Geol. Bull., submitted.

Mann, D.M. and Mackenzie, A.S., 1990. Prediction of pore fluid pressures in sedimentary basins. Mar. Pet. Geol., 7: 55–65.

Meissner, 1978. Petroleum geology of the Bakken shales formation, Williston basin North Dakota and Montana. Proceedings of 1978 Williston basin symposium, Montana Geological Society, Billings, pp. 207–227.

Narimanov, A.A., Akperov, N.A. and Abdullaev, T., 1998. The Bahar oil and gas-condensate gas field in the South Caspian Basin. Geoscience, 4 (3): 253–258.

Purcell, W., 1949. Capillary pressure — their measurements using mercury and the calculation of permeability therefrom. AIME Petrol. Trans., 186: 39–48.

Sales, J., 1993. Closure versus seal capacity — a fundamental control on the distribution of oil and gas. In: A.G. Doré, J.H. Augustson, C. Hermanrud, D.J. Stewart and Ø. Sylta (Editors), Basin Modelling: Advances and Applications. Norwegian Petroleum Society (NPF), Special Publication 3. Elsevier, Amsterdam, pp. 399–414.

Schowalter, T., 1979. Mechanics of secondary migration and entrapment. Am. Assoc. Pet. Geol. Bull., 63: 723–760.

Secor, D., 1965. Role of fluid pressure in jointing. Am. J. Sci., 263: 663–669.

Sibson, R., 1981. Controls on low stress hydro-fracture dilatancy in thrust, wrench and normal fault terrains. Nature, 289: 665–667.

Smith, D.A., 1980. Sealing and non-sealing faults in Louisiana Gulf Coast salt basin. Am. Assoc. Pet. Geol. Bull., 64: 145–172.

Walsh, J.J., Watterson, J., Heath, A.E. and Childs, C., 1998. Representation and scaling of faults in fluid flow models. Pet. Geosci., 4: 241–251.

Watts, N., 1987. Theoretical aspects of cap-rock and fault seals for single and two phase 'columns'. Mar. Pet. Geol., 4: 274–307.

D. GRAULS *Total Fina Elf e.p., Subsurface and Petrophysics, Avenue Larribau, 64018 Pau, France*
E-mail: dominique.grauls@totalfinaelf.com

F. PASCAUD *GEOGRAPH, rue Cail, 75010 Paris, France*

T. RIVES *Total Fina Elf e.p., Structural Geology Department, Avenue Larribau, 64018 Pau, France*
Present address: Total Fina Elf Nederland, Den Haag, The Netherlands

Havana — a fault modeling tool

Knut Hollund, Petter Mostad, Bjørn Fredrik Nielsen, Lars Holden, Jon Gjerde, Maria Grazia Contursi, Andrew J. McCann, Chris Townsend and Einar Sverdrup

Improved knowledge on faults and hydrocarbon seal put pressure on geologists and reservoir engineers doing reservoir modeling. All geo-knowledge must be built into the reservoir models to assure that it is taken into account in the decision processes. The need for advanced modeling tools is increasing. This paper describes the development of a fault modeling tool, the methodology behind it and examples of fault modeling studies. The general focus is on the uncertainty related to faults. The tool can be used for sensitivity analysis of fault effects, including studies of the flow effects of all faults scales, adding faults to simulation grids, and studies of the geometric uncertainty of the faults. The work started out as a development of a tool for stochastic modeling of sub-seismic scale faults. The faults can be added to a flow simulation grid as both displacement and seal. The current tool has been designed to operate together with the Eclipse flow simulator and the IRAP RMS program package. IRAP RMS is the main tool for visualizing output and Eclipse is used to examine the effect of the faults on hydrocarbon recovery. The techniques for modeling of fault seal, outputting results in a format that Eclipse can directly utilize, and the possibility for displacing simulation grids has proved useful also to seismic scale faults. This has led to further development, more detailed fault models and improvements of the general fault modeling capabilities. Examples of fault modeling, including three field examples, Statfjord, Heidrun and Sleipner, are presented to illustrate ways of including fault modeling as part of the reservoir modeling workflow.

Introduction

Faults of widely varying sizes are present in most geological rock formations. In petroleum reservoirs, faults are likely to influence fluid flow patterns and they contribute significantly to defining the size and the shape of the reservoir. Whether the focus is on exploration, field development or well planning for a producing field, it is important that all available knowledge is taken into account in the decision processes. Possible effect of faults on hydrocarbon recovery and in-place volumes must be estimated.

Fault properties are usually highly uncertain. In order to examine the possible outcomes it might be necessary to run a significant number of sensitivities. A fault sensitivity study typically involves the whole reservoir modeling workflow. The faults are modeled by geologists, but to study their effect on hydrocarbon recovery, they usually are built into a flow simulation model and examined by reservoir engineers. To make this feasible the modeling procedures must be efficient. Fault modeling tools are called for.

The fault modeling tool Havana has been developed at the Norwegian Computing Center (NCC) in cooperation with Statoil, Norsk Hydro and formerly Saga Petroleum. The first version was released in 1993 as the result of a fault modeling research project. At that time NCC had been involved in fault modeling research since 1989. Since the first release, several fault modeling features have been developed and implemented within the Havana code. The first model with a detailed parametric representation of each fault (PFM) was introduced in 1998 and a combined uncertainty model for faults and horizons was released in 2001. Development is still ongoing.

Havana was originally designed as a tool for stochastic modeling of sub-seismic faults. As such it generates elliptical fault objects, which can be added to a flow simulation grid as both displacement and seal. The inputs to the seal calculations are fault thickness and permeability parameters and its output is transmissibility multipliers. These multipliers can be used directly in flow simulations.

The techniques in Havana for modeling of fault seal, outputting results in a format that Eclipse can directly utilize, and the possibility for displacing simulation grids has proved useful also to seismic scale faults. This has led to further development of Havana to improve its capabilities as a general fault modeling tool. A parametric fault model (PFM) has been implemented. This is a more complex fault format than the elliptical one.

Hydrocarbon Seal Quantification edited by A.G. Koestler and R. Hunsdale. NPF Special Publication 11, pp. 157–171, Published by Elsevier Science B.V., Amsterdam.

The current version of Havana has been designed to operate together with the Eclipse flow simulator (Eclipse, 1999) and the IRAP RMS program package (IRAP RMS, 2000). Havana has a focus on integration and interaction with an Eclipse reservoir model. Properties are modeled in terms of geological concepts, but the results will typically be explored by running flow simulation using Eclipse. IRAP RMS is the main tool for visualizing output from Havana. Havana has a close integration with IRAP RMS formats for faults and surfaces, making it possible to transport objects in both directions for visualization or as part of building a model in IRAP RMS.

Features

Havana features include: 3D modeling; modeling of both fault planes and reservoir; advanced stochastic models including fault truncation and fault interaction or use of fault displacement density for generation of realistic fault patterns; conditioning on well observations of the presence (or absence) of faults and their properties; conditioning on well observations of geological horizons; flexible stochastic models for both fault geometries and fault permeability properties (Fig. 1).

For a really useful fault modeling tool, more than an advanced fault model is needed. It is important that the fault modeling is part of the reservoir modeling workflow. To be of any value the modeled faults must show up in a geological model or a flow simulation model. Large variation in reservoirs and fault data call for flexibility. Faults generated or read into a Havana model may be manipulated or applied in several ways. Fault sets can be 'split' according to the sizes for different treatment. Reading several sets into Havana can also join fault sets. A flow simulation grid can be deformed according to the fault models and the transmissibility multipliers may be generated according to the simulated fault surface properties. Alternatively, geometry and seal effect of faults can be transformed into a change in permeability field values. Modeled faults may also be used to deform surfaces as part of building a geological model. Havana can also compute intersections of well paths with fault sets.

The need for a fault modeling tool for larger faults (near and above seismic resolution) led to the Havana PFM and more features. The displacement is allowed to vary by any amount along the fault surface in the PFM model, and strike changes and variable heights along the fault can be modeled. Fault permeability can now be modeled using shale-gouge ratio (Yielding et al., 1997), shale smear or clay smear potential (Lindsay et al., 1993) in addition to the absolute and displacement-dependent fault permeability models which were previously implemented. Several ways of modeling fault thickness are available. These range from simple Gaussian or lognormal distributions, to models of spatial variations along the fault surface, via methods where the uncertainty is linked to displacement or facies information.

Fault models

Faults vary hugely in size and the level of detail needed in order to examine their impact on fluid flow is variable. Consequently, Havana uses different fault models, differing in complexity from very simple to quite flexible. The two types currently implemented are the following.

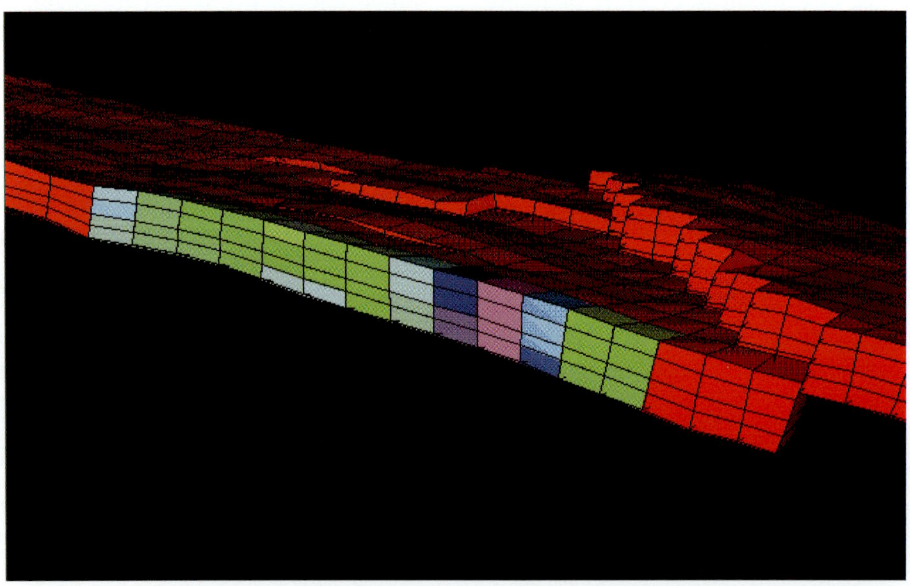

Fig. 1. Fault displacement and properties modeled using Havana.

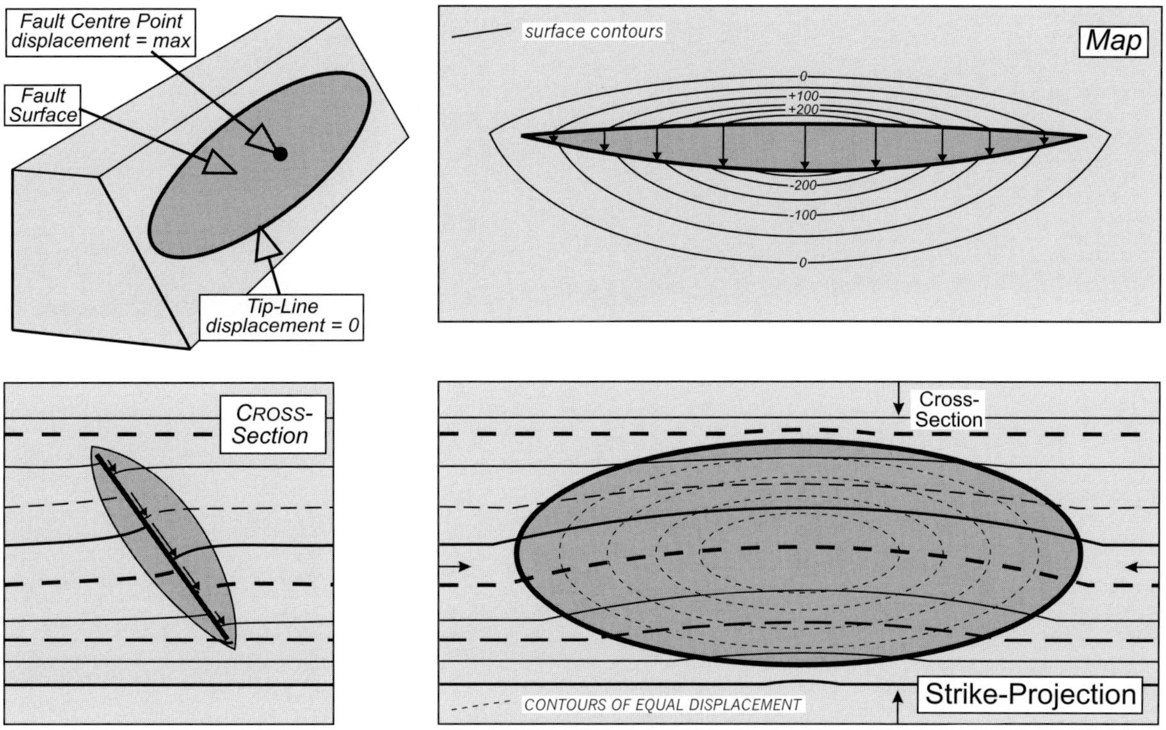

Fig. 2. Elliptic fault model.

(1) The elliptic fault model. The fault plane in this model is an ellipse (Fig. 2). Elliptic faults are typically used for modeling small (below seismic resolution) faults.

(2) The parametric fault model (PFM). The fault surface is here a sequence of bilinear planes. The planes meet at pillars (Fig. 3). PFM faults are typically used to model larger faults known or partially known from seismic data.

In addition Havana uses IRAP RMS and Eclipse fault data. As mentioned, Havana faults can be exported to IRAP RMS for visualization or further modeling. The other way round is also possible and relevant. For IRAP RMS faults the fault plane is a triangulated surface, i.e. it is very flexible. Havana reads, writes and uses IRAP RMS fault data including the fault surfaces. The fault surface can be used directly in the simulation of sub-seismic faults, extensions of IRAP RMS faults can be modeled in Havana and the IRAP RMS faults can be approximated by elliptic or PFM faults and used in the Havana modeling. For details of the IRAP RMS fault data consult IRAP RMS (2000).

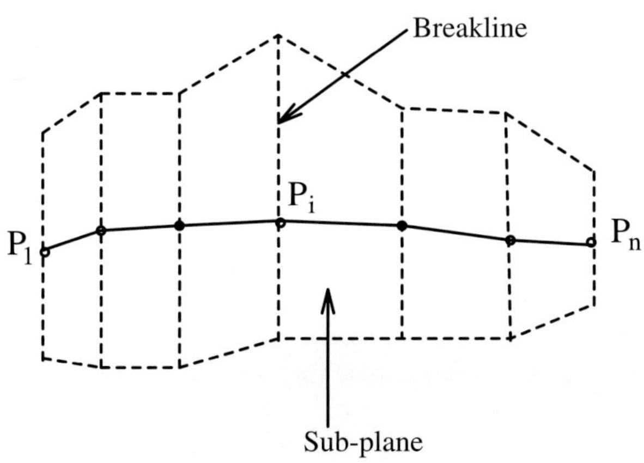

Fig. 3. Parametric fault model.

The Eclipse grid used by Havana is a corner-point grid with straight pillars. Such a grid is specified by the coordinates of the top and bottom of the pillars (called 'coord lines') and the vertical position of the grid cell corners for each coord line (called 'zcorn values'). For Eclipse a fault can be modeled by adjusting the zcorn values, causing non-neighboring cells to be connected and/or adding fault seal as transmissibility modifiers at cell boundaries. The main purpose of Havana can be said to be generation of Eclipse fault data, i.e. deform a grid and calculate transmissibility modifiers. Havana reads, writes and uses Eclipse data in order to accomplish this. In addition Havana can read faults previously specified in an Eclipse grid using the Eclipse codeword faults for further modeling of sealing properties. The data specified by the Eclipse codeword faults are simply a specification of a collection of grid cell connections.

Fault deformation operator

The feature that distinguishes Havana from many other fault modeling packages is that it models not only the fault surface, but also the deformation that

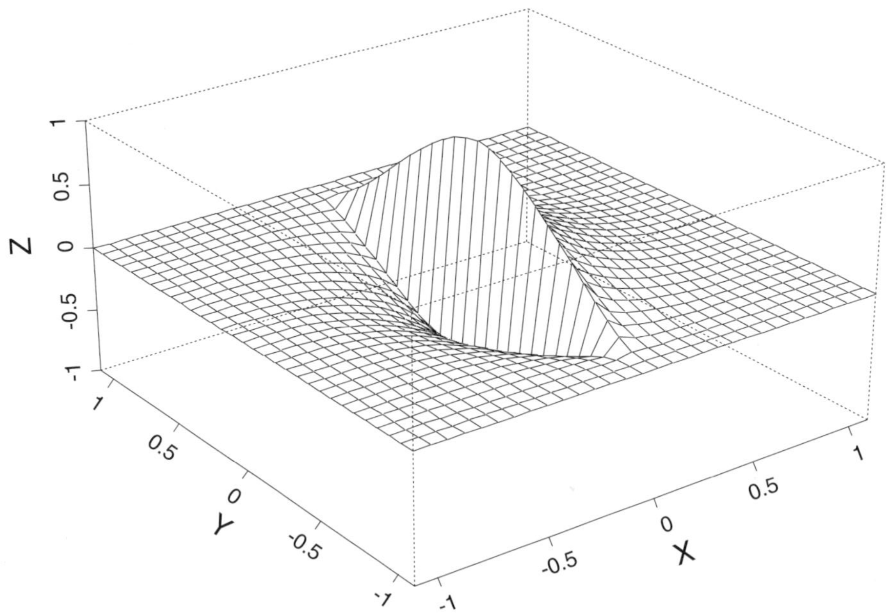

Fig. 4. Elliptic fault deformation operator.

created the fault (or at least an idealized version of this deformation). Inside the volume of deformation a deformation operator is defined, moving the reservoir in different directions on each side of the fault plane. The definition of the deformation operator is based on results presented in Barnett et al. (1987) and Walsh and Watterson (1989). The displacement is defined as a dip-slip displacement.

For elliptical faults, the volume of deformation extends in an ellipsoid around the fault plane (see Fig. 4). PFM faults have a triangulated region around the fault surface in which the deformation occurs.

The fault deformation operator makes it possible to model for example sub-seismic faults, and then study their 'geometric' effects on the reservoir, in addition to the flow effects of the fault surfaces. In heterogeneous reservoirs, the geometric effect may be quite important. The modeling of the faulting deformation for larger faults also makes it possible to easily and accurately generate correspondingly faulted Eclipse grids, where the faults appear as non-neighbor grid cell connections.

The deformation is determined by a mathematical formula that can be inverted. This makes it possible to both 'apply' and 'remove' faults. This is very useful from a modeling point of view as can be seen in the section 'Simulation of structural model'.

Fault truncation

Truncation is another part of the fault modeling that greatly improves the geological realism of modeled fault patterns (see Fig. 5). Both elliptic and PFM faults may be truncated against other fault planes. The truncation also applies to the model of the deformation. Thus one may model patterns of branching fault planes.

Impact on the flow properties

Both the thickness and permeability in the fault plane are modeled (see Fig. 6). These properties are important in determining the impact the faults have on the flow. The models for thickness and permeability are quite flexible, and there are also ways to visualize these properties along the fault planes before they are transformed into transmissibility multipliers. There are options for smear-gouge ratio and shale-smear factor. These may also be applied to faults on Eclipse format.

There are two ways in Havana to study the impact of faults on production and flow properties. The first is to add the faults to the Eclipse grid (Fig. 1). Faults are added to the Eclipse grid by first adding their displacement effect, and thereafter adding the transmissibility multipliers to represent the fault seal effect. Only the component of the displacement parallel to the coord lines is used for displacement, so if the grid is dipping in a different direction than the faults, then their total displacement appears to be reduced. The transmissibilities are computed so that this effect appears in the same cells as the discontinuity of the displacement.

The second way to study the impact of faults on production and flow properties is to add them directly to some gridded permeability field representing the reservoir (Fig. 9).

The choice between the two options may have a significant impact on the result. Generally, faults that

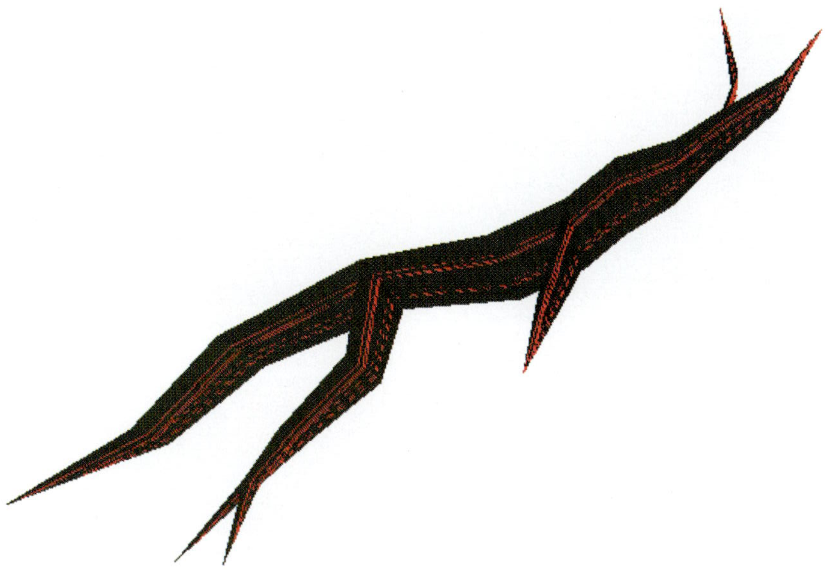

Fig. 5. Branching fault modeled using truncations.

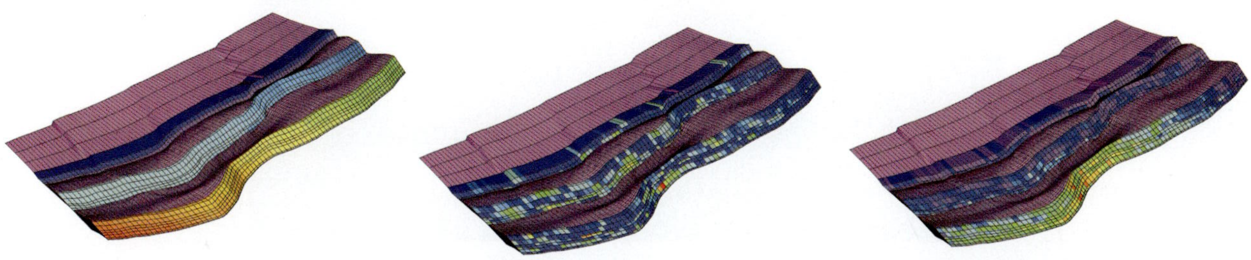

Fig. 6. Fault properties: displacement property (left) and two realizations of thickness property.

are larger than the Eclipse grid cells should be entered into the Eclipse grid. Both their displacement and their fault seal effect are then reasonably well represented. However, if the faults are equal to or smaller in size than the Eclipse grid blocks, they may very well disappear completely. Such faults should be added to the permeability.

Simulation

Simulation of sub-seismic faults

The first stochastic fault model implemented in Havana (reported in Munthe et al., 1993) did not easily generate realistic fault patterns compared to real fault systems. A study of a South Yorkshire fault map provided by the Fault Analysis Group, University of Liverpool, led to the current fault pattern model (reported in Munthe et al., 1994). The model is built on the following premises for sub-seismic faults: (1) the larger of these faults tend to 'repulse' each other, i.e. they are more separated in space than they would have been if they were located randomly and independently; and (2) the smaller faults tend to be located around the larger faults (Fig. 7).

To accommodate these effects, the elliptic fault model differentiates between the larger faults, which are called mother faults, and the smaller faults, which are called children faults. There is one stochastic model for mother faults, and one stochastic model for the children faults, given the mothers. The basic simulation algorithm is:

(1) Read in interpreted faults (or previously simulated faults).
(2) Simulate new mother faults, using a point process with interaction.
(3) Simulate children faults based on the realization of the mother faults.
(4) Truncate faults.

The mother faults are distributed throughout the reservoir according to a marked point process (Stoyan et al., 1987), i.e. the state-of-the-art geostatistical technique for modeling geological 'objects' like sand bodies, calcite deposits, faults and so on. The location of the faults as well as the properties of the faults are stochastic. The locations of mother and children faults are specified by intensity maps which should indicate regions more likely to have sub-seismic faults than other regions. Intensity maps could be derived from seismic attribute maps.

Fig. 7. A Havana fault pattern realization.

Fig. 8. A relay ramp structure (left) and its damage zone (right).

To accommodate that the larger faults are separated in space, interaction functions must be specified. It is important to include the interpreted faults when new faults are added, so that they may interact with each other.

Children faults are located in a neighborhood

around the mother faults. Their location is therefore based on the point process which governs the mother faults as well as the deviation from the mother location. Both interpreted and new faults will be given children faults.

The user can specify how the children faults should be distributed in the length direction, the height direction, and the reverse drag direction of their mother. Several parameters are available to control the distribution. It is possible, e.g. to have increased intensity at fault tips, within fault relay ramps and/or different fault intensity at footwall and hanging-wall sides of the mother fault (see Fig. 8).

The properties concerning the orientation of the faults like strike and dip, can be specified in several ways. Both variables are on a continuous scale and nothing indicate a need for asymmetric distributions, thus they are modeled using a simple Gaussian distribution with a given mean (expectation) and standard deviation. These parameters can be given as constants or trend maps. There is also a third option for the strike direction, where the user may specify a number of main directions. Each main direction is given by an expected value and standard deviation as well as the fraction of faults to be simulated for each main direction.

The strike and dip for the children faults are also assumed to follow Gaussian distributions. The expected value in these distributions is controlled by the strike or dip direction for the mother fault, while the standard deviation is given by the user. Parameters are available to control how the expected strike of the children faults relate to the strike of the mother fault. For example, a fraction of the children faults can be specified to have expected strike perpendicular to the strike of mother fault.

In Childs et al. (1989) and Heffer and Bevan (1990) the maximal displacement in fault populations are reported to follow a fractal distribution. Relationship between the displacement and the size parameters are reported in Barnett et al. (1987), Walsh and Watterson (1988) and Gillespie et al. (1992). For Havana the fault displacements are drawn from a fractal distribution with fractal dimension, minimum displacement and maximum displacement as specified by the user. Mother faults will however always have displacements that are larger than those of the children faults. The length of the principal axes for both mother and children faults are derived from the displacement. The length (l), height (h) and reverse drag (r) of the fault are assumed to approximately be related to the displacement (d)in the following way:

$$l = (d/c_1)^{1/p_1}, \quad c_1 l^{p_2} = d, \quad h = l/c_2, \quad c_3\sqrt{l \cdot h}$$

The parameters have to be given by the user, together with a variability around the values from these deterministic functions.

Conditioning

A fault realization may be conditioned to assure that individual faults observed in wells are reproduced by the simulations. A fault realization may also be conditioned so that none of the fault operators move the positions where seismic surfaces have been observed in the wells. Further, a fault realization may be conditioned so that seismic horizons are not moved outside the specified seismic resolution band.

The fact that a seismic horizon has been observed in some well does not imply that sub-seismic faulting never moved the observed point. Rather, it means that the point has been moved to its present location by all the faulting events that have taken place.

Havana models ‘pre-faulted’ or ‘non-faulted’ horizons corresponding to how the observed seismic horizons were before any sub-seismic faulting took place. These ‘pre-faulted’ horizons must be constructed so that when simulated faults are added to them, the seismic horizons will pass through the wells where they have been observed, and they will stay within the seismic uncertainty band. In Havana, the ‘pre-faulted’ horizons are constructed by adding an ‘adjustment operator’, a 3D kriging operator, to the seismic horizons. The operator moves points vertically, and is zero except for places where seismic horizons have been observed in the wells, or where the fault realization sends the seismic surfaces out of the seismic uncertainty band. At these points, special ‘conditioning points’ are computed. Around these points, the value of the adjustment operator is kriged with an exponential variogram, with ranges determined from the size of the largest simulated faults.

Now, conceptually, to make a stochastic reservoir description, one should first simulate a sub-seismic fault system, then apply the resulting adjustment operator to the seismic horizons to get ‘pre-faulted’ horizons, then simulate facies and petrophysics between these horizons, and finally applying the sub-seismic fault system to the result. This is clearly impractical, but fortunately, it is possible to make a shortcut. Almost the same result as the above is obtained by first simulating facies and petrophysics realizations between the seismic horizons, and then apply the adjustment operator. In fact, the adjustment operator of Havana is viewed as a part of the fault system realization, and is stored together with it.

Simulation of structural model

There is a considerable uncertainty in the large seismic faults modeled by the parametric format. The uncertainty is in both the exact position and the displacement. This uncertainty is important for both volume calculations and positioning of wells, as well as in a full reservoir characterization framework.

A common stochastic model for the horizons and faults has been established. The model for the horizons may include depth conversion. Each horizon is described by a time horizon or an expected horizon in depth. A linear velocity model is assumed. All horizons are modeled as correlated Gaussian fields, and erosion is handled by truncation of the Gaussian fields. Parameters in the velocities and trend surfaces are conditioned using Bayesian conditioning (see Abrahamsen, 1992). The model handles a large number of horizons and well observations simultaneously. Observations of deviating wells influence position and uncertainty in position above and below the observations. The inversion of the fault operator is necessary in order to condition on the well observations.

Realizations of the structural model are generated by the following workflow:

(1) Generate the horizons based on time horizons, trend maps, the velocity model, prior model for all parameters and all well observations of horizons.
(2) Apply the inverse fault operator.
(3) Smooth the horizons in the neighborhood of the fault planes.
(4) Add Gaussian field (noise) for maintaining well observations and variability in the support area of the smoother.
(5) Apply the fault operator.
(6) Calculate footwalls, hanging walls and branch lines for all faults.

Fault modeling examples

Only a few papers describing applications of Havana have been published (Damsleth et al., 1998; England and Townsend, 1998), but Havana is used as part of the reservoir modeling workflow for both Norsk Hydro and Statoil and have been used for several fault studies. The program is also in use at Conoco and British Gas.

Statoil has used Havana for fault studies on several fields, including Åsgard and Norne (small seismic faults (elliptical) and fault seal calculations), Statfjord, Sleipner, Heidrun, Gullfaks (sub-seismic fault modeling) and Huldra. A few examples are presented below.

Modeling a relay ramp structure

Fig. 8 shows two examples that demonstrate subseismic realizations as created by Havana. The first example shows two single, sub-parallel Havana faults (mothers) that exhibit a relay zone in between (left display). The typical character of a relay zone is that of a complexly deformed (faulted) area. The children faults in this example are those structures that are located in the damage zones of the mother faults and within the relay ramp structure (right display). The realization of 300 children faults as shown here, was modeled in one single step by specifying the intensity field of the relay structure, the damage zone geometry, strike and dip variations, number of (children) faults, as well as fault population rules. As will be demonstrated below, the introduction of such structures will have impact on the permeability field of the reservoir.

Small faults are commonly included in the flow simulation grid without displacement, but by modifying the grid permeability. Fig. 9 displays a part of an Eclipse grid where the effect on the permeability field by the relay ramp structures in Fig. 8 has been implemented. The introduction of fault plane effects depends on parameters such as fault rock permeability and thickness, as specified by the user. Because the faults in this example have been given lower permeability values than the surrounding reservoir rocks, the permeability is reduced in the area close to the relay ramp structure.

Streamlines for investigation of fault effect on fluid flow

As a quick alternative to a full Eclipse flow simulation, the effect of sub-seismic faults can be examined by streamline calculations. Fig. 10 shows a realization of a fault model that is introduced to a reservoir that already contains large, interpreted faults. In the new fault model both new 'mother' faults (50) and 'children' faults (950) are introduced according to fault population and distribution relationships defined by the geologist.

By changing fault parameters such as the number of faults, size and throw variations, strike and dip, as well as repulsion and truncation rules, new fault realizations can be generated rapidly. The effect of introducing the sub-seismic fault structures on the hydrocarbon recovery of the reservoir will be shown below.

Figs. 11–14 demonstrate how the effect of subseismic faults on fluid flow can be investigated by the use of the streamline module within IRAP RMS. The method includes dynamic information from wells,

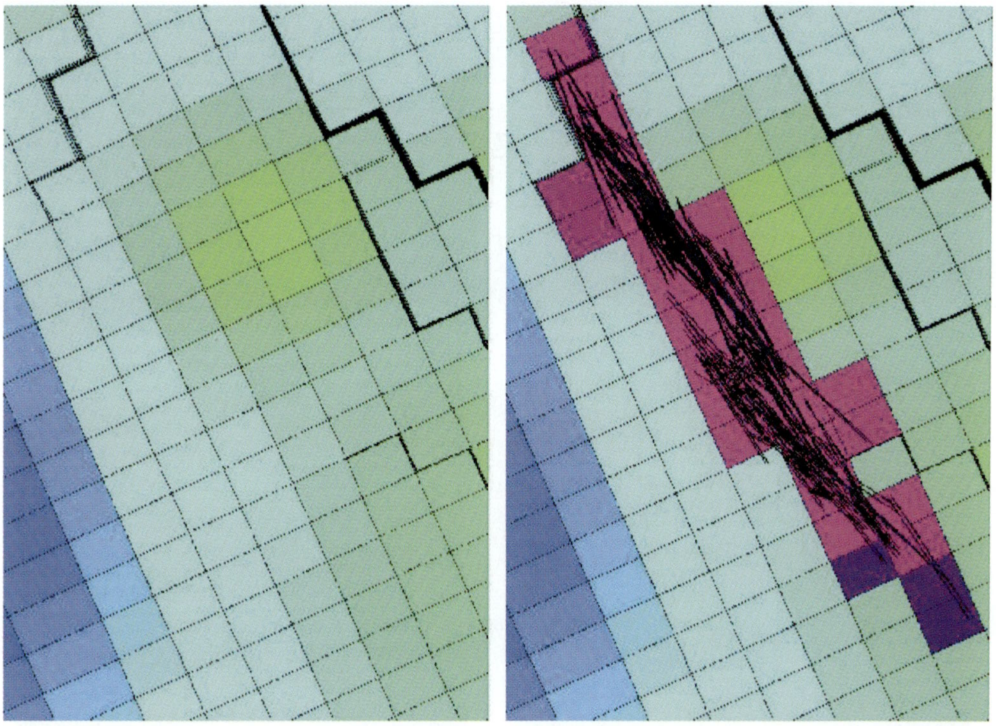

Fig. 9. A part of an Eclipse grid color coded by the permeability values. The original permeability field is shown to the left, while the modified permeability field is shown to the right.

Fig. 10. A realization of sub-seismic faults (right display) generated into a model with interpreted faults (left display).

and represents a quick and visually important step in validating fault models for continued modeling. In the example below modified permeability fields have been obtained from the sub-seismic fault pattern shown in Fig. 10, as well as transmissibility information for the large faults as derived from Havana.

Statfjord

Havana formed a crucial role in building the first geometrically realistic model of the Statfjord East Flank. This structure comprises three main slumped fault blocks sitting at the top of a major rotated fault block. This rotated block forms the main part

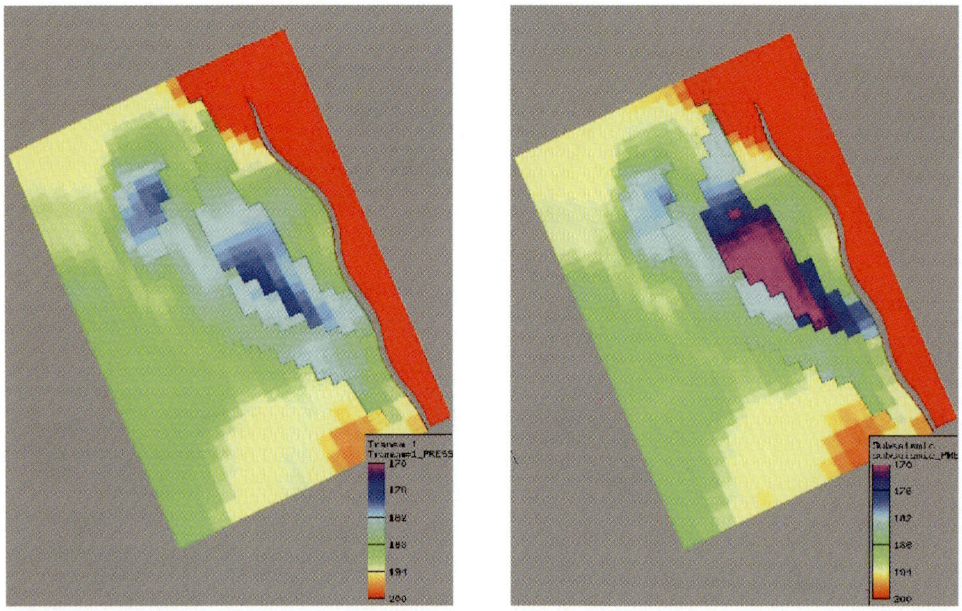

Fig. 11. Reservoir pressure after 100 days of production/injection. The original model to the left and the model with sub-seismic faults to the right.

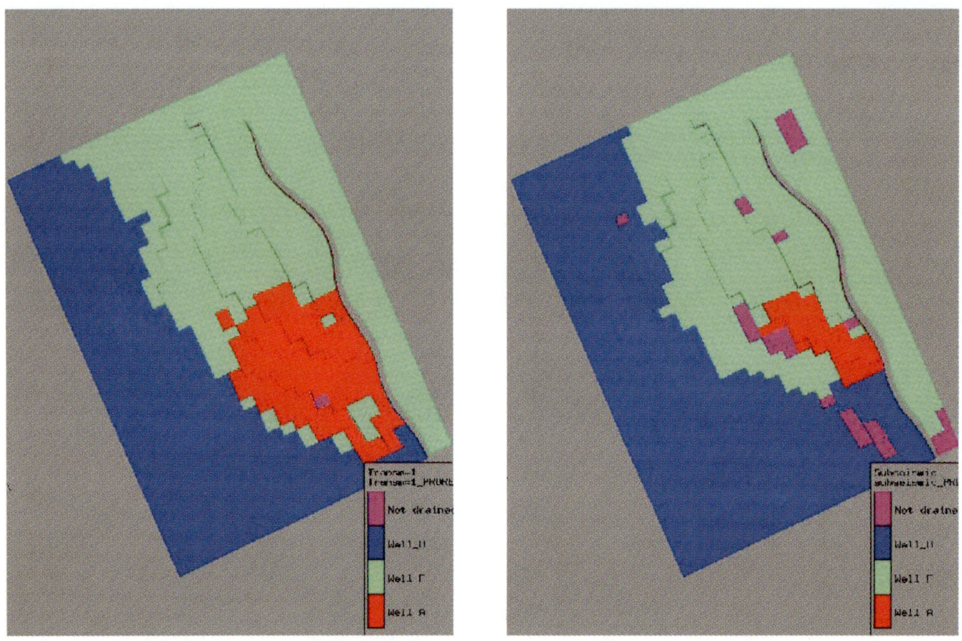

Fig. 12. Production regimes. The original model to the left and the model with sub-seismic faults to the right.

of the Statfjord Field. The seismic data quality over the East Flank is generally very poor with only the Base Cretaceous unconformity and the basal slope failure to the slump blocks visible. The three main slump faults can usually be mapped on seismic along with several smaller slump faults and it has been observed that these slump faults detach deeper in the stratigraphy the further east they lie.

Previous models have relied heavily on well data, and because there is little structural control from the seismic data this often led to a model where geological horizons became deeper eastwards. The slump faulting downthrows the stratigraphy to the east, but the actual layering dips westwards towards the east-dipping slump faults. Consequently these previous models were not able to realistically capture the known structural geometries.

The modeling process used relied upon the geological concepts outlined above in order to try and generate a realistic 3D framework model. The well data were not used until the final stage of modeling and even then not all the data could be utilized. The following modeling steps were followed (Fig. 15):

(1) A set of unfaulted and extrapolated geological

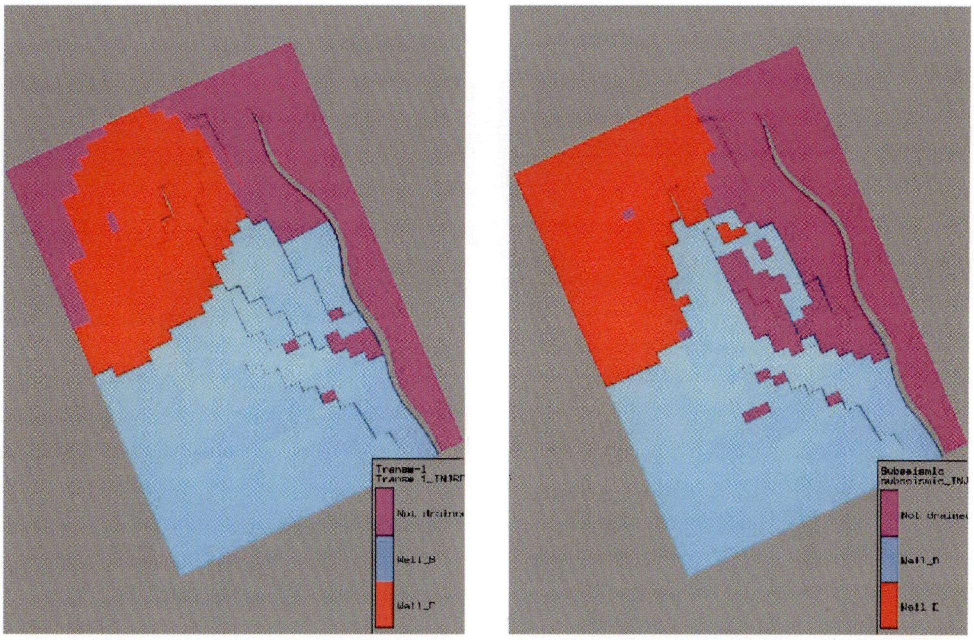

Fig. 13. Injection regimes. The original model to the left and the model with sub-seismic faults to the right.

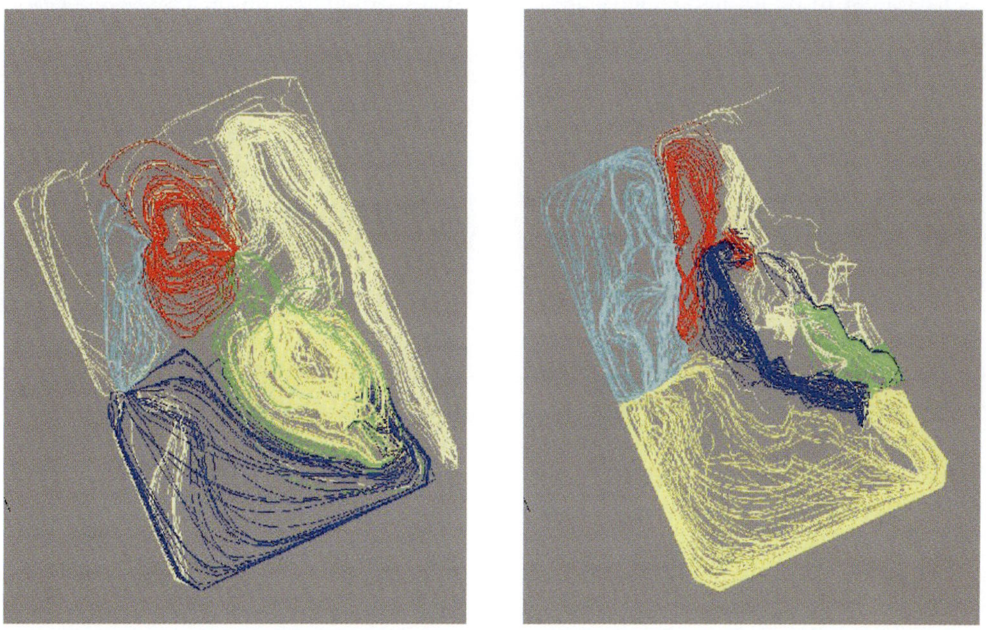

Fig. 14. Streamlines after 100 days of production/injection. The original model to the left and the model with sub-seismic faults to the right.

horizons was created. These were generated using an intra Ness horizon from the main field where it is well constrained and projecting it eastwards. The other horizons were then mapped using a combination of main field thickness data and well data from the East Flank.

(2) The three main slump faults were converted into Havana's PFM fault format. Two versions of these faults were generated, the first had zero displacements and the second had true displacement values that were extracted from the seismic data at 4 or 5 specific points along the length of each fault.

(3) The zero-displacement PFM faults were converted to the IRAP RMS fault format, using the extrapolated surfaces to generate fault lines for each horizon. These faults were then imported into IRAP RMS.

(4) A 3D grid was built in IRAP RMS using the extended surfaces and the zero-displacement faults. The faults were used to control the positioning of the coord lines of the grid so that later the displaced faults can be added without forming a zigzag structure. The number of grid blocks included in the model was varied in its different regions. This was to improve

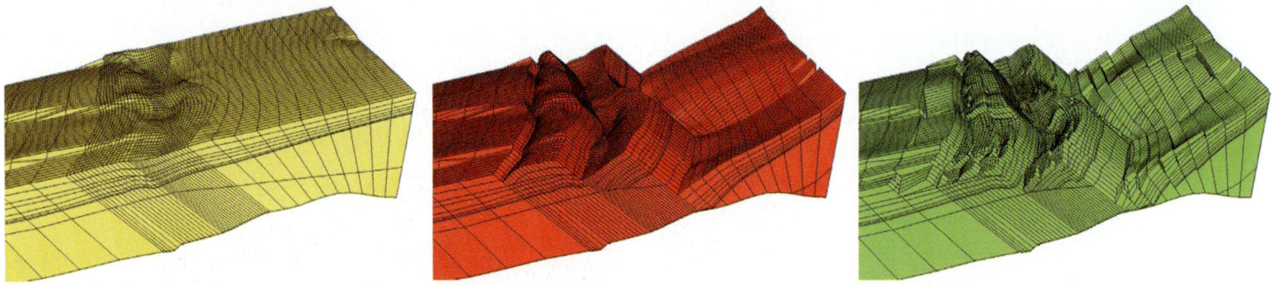

Fig. 15. Statfjord East Flank fault modeling. To the left 3D grid built around the fault framework (zero displacement), at the middle 3D grid displaced by major faults, to the right 3D grid displaced by smaller seismic faults.

the efficiency of the simulation where details were not required (i.e. few blocks were used on the main field and on the most easterly slump block) and in the two main slump blocks, where more detail was required, a finer grid size was used.

(5) This 3D grid was exported as an Eclipse format and Havana was used to displace it using the true displacement faults. The faults had the same location as their zero-displacement counterparts and therefore allowed the displacement to be located exactly along a single coord line. Havana has a function for displacing a selected number of layers within a 3D grid. This option was used to help include the effect of the detachment surfaces.

(6) The smaller faults were modeled as Havana elliptical faults and these were added to the grid as zigzag structures.

(7) The final grid was then adjusted to the wells wherever possible. In some cases this gave unsatisfactory results, in which case the wells had to be either relocated to give a more acceptable result or omitted altogether.

(8) The 3D grid was then truncated by the erosional Base Cretaceous unconformity and the main bounding fault to the east (not illustrated in the figures).

(9) The final step of the modeling was to use Havana to estimate the effects of fault seal and convert this to Eclipse transmissibility multipliers.

Heidrun

A detailed 3D reservoir model was built covering three fault blocks from within the Heidrun Field (H, I and J blocks). The main aim of the study was to examine the effects on production and well placement of the smaller intra-block faults. These smaller faults are only just visible on the seismic data and because of this there is considerable uncertainty to their precise geometry. When the dataset was given to two different seismic interpreters, two distinctly differing fault patterns emerged.

A 3D structural model was built using the main horizons and the block-bounding faults by 'conventional' modeling techniques in IRAP RMS. The intra-block faults were included in this model with zero displacement so that the locations of the displaced structures were pre-defined. They were initially described using the Havana PFM format and then converted to the IRAP RMS format as surfaces and fault lines. The intra-block faults were then added later as displacements using Havana (Fig. 16). The advantage to taking this approach was that the uncertainty in displacement could be assessed because the displacement values lie within the PFM fault format files, which could be easily edited. Moreover, the different fault geometries could also be assessed by generating different sets of PFM format faults which reflect the different interpreted fault patterns.

Havana was used to generate the fault-related transmissibility multipliers. These were modeled using stochastic techniques for the estimation of fault permeability and thickness. This allowed the uncertainties related to fault sealing to be assessed and compared to those related to fault geometry and displacement.

Sleipner

Havana fault modeling techniques have been applied to the Sleipner Field where a reservoir simulation model already existed. The Sleipner Field has possibly one of the most complex fault patterns of any of the fields in the North Sea. The segment of the field being studied had a number of small seismically mapped faults, which had not been included in the simulation grid. These faults were converted to Havana's elliptical fault format and added to the grid as a displacement operator. The smaller faults within the Sleipner Field often intersect, so the truncation option in Havana was used in an attempt to recreate the correct fault geometry. There was some uncertainty to the extent of the smaller seismic faults. The concern was that some of the faults might have been longer than they had actually been mapped. Therefore a second set of faults was generated whereby the lengths of the faults were increased by 500 m. This was

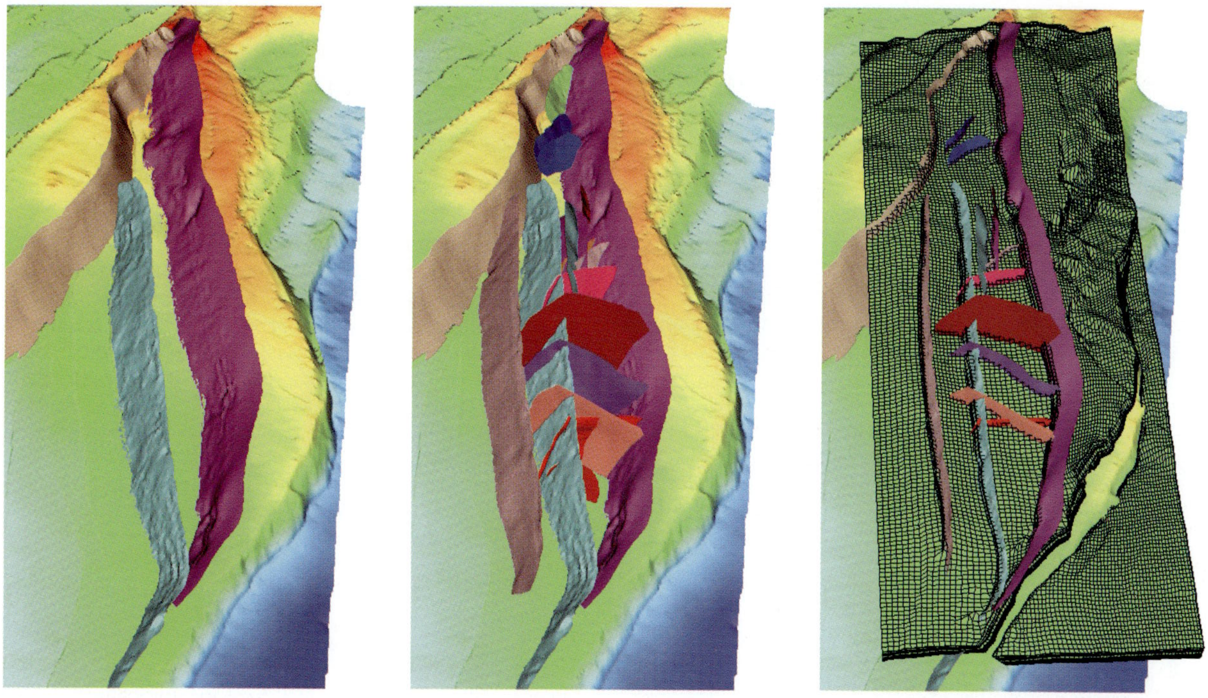

Fig. 16. A 3D fault model from the Heidrun Field showing the larger block bounding faults (left), the block bounding faults and the smaller block internal faults (middle), and 3D grid built on the structural model (right).

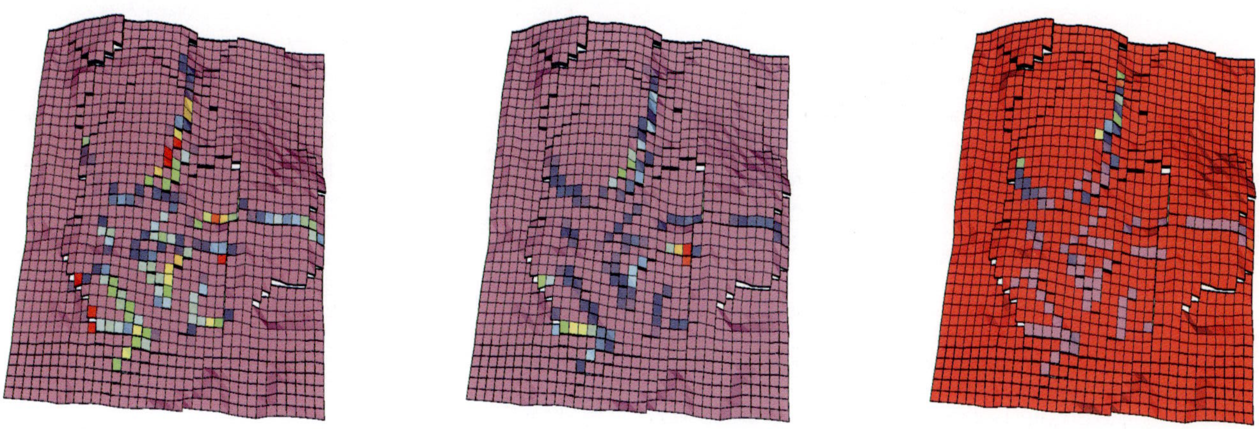

Fig. 17. Fault modeling of the Sleipner field. To the left displacement, at the middle thickness and to the right transmissibility multiplier.

carried out by simply editing the length parameter in the elliptical fault file.

Havana was used to account for the effects of fault seal using the SGR and shale smear algorithms. A comparison of the different techniques was used against known production data. The best history match was achieved when SGR and shale smear were both used in combination with the extended fault lengths. Part of the fault seal modeling in Havana includes calculating a number of fault properties (e.g. displacement is calculated from the 3D grid and thickness is modeled as a function of displacement). Some of these parameters are shown in Fig. 17.

Conclusions

Advanced fault models and flexible fault modeling tools are essential if the uncertainty related to fault geometry and fault sealing should be efficiently and consistently examined. For proper examination of the effect on fluid flow the fault models must be incorporated into the reservoir modeling workflow that results in flow simulation models.

The development of improved fault modeling techniques is still ongoing. Havana still is a research product; nevertheless, Havana already is a flexible tool that can be introduced at a number of stages in any 3D modeling process. Some of these options have been outlined in the examples presented. A sum-

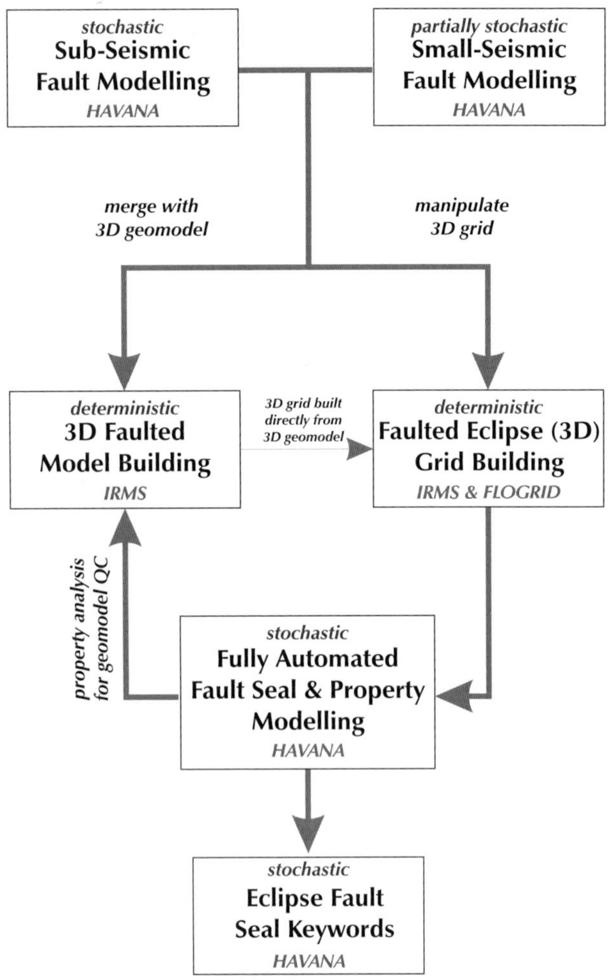

Fig. 18. Possible workflow for fault sensitivity studies. Sub-seismic faults (top, left) simulated using Havana can be incorporated in a geomodel (middle, left), the geomodel can be used for generation of a flow simulation model (middle, right), for quality control or well-planning purposes. Alternatively the sub-seismic faults can be added directly to the flow simulation model. Faults partially known from seismic data and modeled by Havana (top, right) can be included into the reservoir models similar to the sub-seismic faults. Some of the fault properties modeled in Havana can be read into the geomodel for quality control or further modeling, other parts of the seal modeling (typically larger faults) can be entered directly to the flow simulation grid (last box).

mary workflow diagram is presented in Fig. 18 which attempts to summarize how Havana fault modeling techniques can be used in combination with a number of other tools. The tools that have been specified are not necessarily unique in any sense whatsoever and Havana could also be used in combination with alternative commercial software products.

References

Abrahamsen, P., 1992. Bayesian kriging for seismic depth conversion of a multi-layer reservoir. In: A. Soares (Editor), Geostatistics Tróia '92, Vol. 1. Kluwer, Dordrecht, pp. 385–398.

Barnett, J.A.M., Mortimer, J., Rippon, J.H., Walsh, J.J. and Watterson, J., 1987. Displacement geometry n the volume containing a single normal fault. Am. Assoc. Pet. Geol. Bull., 71 (8): 925–937.

Childs, C., Walsh, J.J. and Watterson, J., 1989. A method for estimation of the density of fault displacements below the limits of seismic resolution in reservoir formations. In: A.T. Buller, E. Berg, O. Hjelmeland, J. Kleppe, O. Torsæter and J.O. Aasen (Editors), North Sea Oil and Gas Reservoirs, II. Graham and Trotman, London, pp. 309–318.

Damsleth, E., Sangolt, V. and Aamodt, G., 1998. Sub-seismic faults can seriously affect fluid flow in the Njord field off western norway — a stochastic fault modeling case study. In: Annual Technical Conference and Exhibition. Society of Petroleum Engineers, New Orleans, LA.

Eclipse, 1999. Reference Manual, Version 99a. Schlumberger GeoQuest, Oxfordshire.

England, W.A. and Townsend, C., 1998. The effects of faulting on production from a shallow marine reservoir — a study of the relative importance of fault parameters. In: Annual Technical Conference and Exhibition. Society of Petroleum Engineers, New Orleans, LA, pp. 489–500.

Gillespie, P.A., Walsh, J.J. and Watterson, J., 1992. Limitations of dimension and displacement data from single faults and the consequences for data analysis and interpretation. J. Struct. Geol., 14 (10): 1157–1172.

Heffer, K. and Bevan, T., 1990. Scaling relationship in natural fractures — data, theory and applications. In: SPE — Europec90. Society of Petroleum Engineers, The Hague, SPE 20981, pp. 367–376.

IRAP RMS, 2000. User Guide. ROXAR ASA, Stavanger.

Lindsay, N.G., Murphy, F.C., Walsh, J.J. and Watterson, J., 1993. Outcrop studies of shale smears on fault surfaces. In: S.S. Flint and I.D. Bryant (Editors), The Geological Modeling of Hydrocarbon Reservoirs and Outcrop Analogues. Int. Assoc. Sedimentol., Spec. Publ., 15: 113–123.

Munthe, K., Holden, L., Mostad, P. and Townsend, C., 1994. Modelling sub-seismic fault patterns using a marked point process. In: ECMOR IV, 4th European Conference on the Mathematics of Oil Recovery. Røros, SPE 49024, pp. 295–304.

Munthe, K.L., Omre, H., Holden, L., Damsleth, E., Heffer, K., Olsen, T.S. and Watterson, J., 1993. Subseismic faults in reservoir description and simulation. In: 68th Annual Technical Conference and Exhibition. Society of Petroleum Engineers, Houston, TX, SPE 26500.

K. HOLLUND — *Norwegian Computing Center, P.O. Box 114, Blindern, N-0314 Oslo, Norway*
E-mail: knut.hollund@nr.no

P. MOSTAD — *Norwegian Computing Center, P.O. Box 114, Blindern, N-0314 Oslo, Norway*

B.F. NIELSEN — *Norwegian Computing Center, P.O. Box 114, Blindern, N-0314 Oslo, Norway*

L. HOLDEN — *Norwegian Computing Center, P.O. Box 114, Blindern, N-0314 Oslo, Norway*

J. GJERDE — *Norwegian Computing Center, P.O. Box 114, Blindern, N-0314 Oslo, Norway*

M.G. CONTURSI — *Statoil's Research Centre, N-7005 Trondheim, Norway*

A.J. McCANN — *Norwegian Computing Center, P.O. Box 114, Blindern, N-0314 Oslo, Norway*

C. TOWNSEND — *Statoil's Research Centre, N-7005 Trondheim, Norway*

E. SVERDRUP — *Roxar Software Solutions A/S, P.O. Box 165, N-0212 Skøyen, Norway*

Stoyan, D., Kendall, W.S. and Mecke, J., 1987. Stochastic Geometry and its Applications. Wiley, New York.

Walsh, J.J. and Watterson, J., 1988. Analysis of relationship between displacements and dimensions of faults. J. Struct. Geol., 10 (3): 239–247.

Walsh, J.J. and Watterson, J., 1989. Displacement gradients on fault surfaces. J. Struct. Geol., 11: 307–316.

Yielding, G., Freeman, B. and Needham, D.T., 1997. Quantitative fault seal prediction. Am. Assoc. Pet. Geol. Bull., 81 (6): 897–917.

Reservoir compartmentalisation by water-saturated faults — Is evaluation possible with today's tools?

Jan C. Rivenæs and Chris Dart

A new method has recently become available for the prediction of fault seal properties for use in hydrocarbon reservoir production simulators as transmissibility modifiers. The method is based on a clay smear technique and empirical databases of absolute fault permeability and fault thickness. In this study we discuss the results of applying the method to the North Sea Oseberg and Brage fields. It was found that the method was unsuccessful and produced faults that were too open compared to what is known from production performance. The paper goes on to propose that high water saturations are likely to be present along fault zones in hydrocarbon reservoirs in sand/shale sequences. Such water saturations are not accounted for in the applied method. A simulation model was built for part of the Oseberg field where water saturations along a fault zone were represented explicitly. This model showed additional restriction of hydrocarbon flow over faults by the introduction of capillary entry pressure and relative permeability effects. Current limitations in computing power make it unrealistic to build full field reservoir simulations where fault volumes are represented by explicit properties. One way forward may be to produce detailed fault structure simulators with explicit fault properties and from these derive fault permeability formulae for use in full field simulation.

Introduction

A reservoir simulator is a numerical model that is applied to predict reservoir flow performance. This tool is quite important for estimating recovery factors, and is heavily applied in most oil companies. The reservoir geometry and properties are described by cells, where the shape of the cells is often an irregular six-sided polyhedron (so-called corner point grid). The centre of each cell is assigned physical properties (porosity, permeability, saturations, etc.). Reservoir compartmentalisation is described by the organisation of the cells (Fig. 1). Stratigraphic barriers are defined either explicitly by introducing tight and missing layers, or by assigning a factor that reduces or halts the flow between layers. This factor is termed a transmissibility multiplier (e.g. Manzocchi et al., 1999).

When modelling fault barriers, vertical splits are introduced into the grid to account for juxtaposition (Fig. 1). To describe flow restriction from one fault block to another, a transmissibility multiplier is applied to the fault plane. Fault zones are not modelled as volumes, and the explicit grid cell approach is in practice never used to represent the fault zones themselves. The reason is mainly technical. If we want to model the fault zone as a volume, the number of grid cells will increase so much that the model cannot be simulated with today's computer technology. Therefore, fault zones are mainly modelled implicitly as zero-volume planes.

The estimation of fault transmissibility multipliers has traditionally been in hands of the reservoir engineer, and is often assigned during the history-matching cycle, where the task is to match simulated to the observed reservoir performance. Quite often, constant values are applied for all faults or sets of faults. Hence, nothing or just very simplified discrimination, is applied for fault throw, fault direction, nature of juxtaposition, etc.

For many reservoirs, fault properties are important for recovery estimation and well planning. It is therefore necessary to increase focus on this subject, and fault analysis tools such as FAPS (e.g. Fristad et al., 1997) and the introduction of Shale Gouge Ratio (SGR) (Yielding et al., 1997) has given us important additional information. However, it is necessary that this information can be applied directly and digitally in the reservoir simulation model.

Although Knai and Knipe (1998) presented a method for incorporating fault properties based on triangle plots, Manzocchi et al. (1999) is possibly the first article that offers the reservoir engineer with a complete work flow for automatically assigning geo-

Hydrocarbon Seal Quantification edited by A.G. Koestler and R. Hunsdale. NPF Special Publication 11, pp. 173–186, Published by Elsevier Science B.V., Amsterdam.

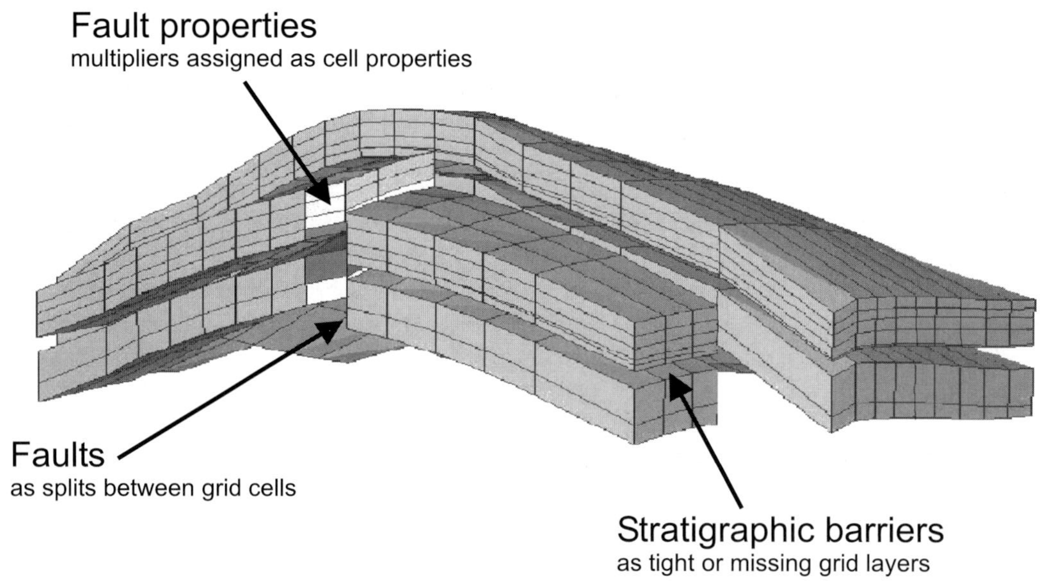

Fig. 1. Compartmentalisation of a gridded reservoir. Stratigraphic barriers are represented as grid layers or transmissibility modifiers from one layer to the next. Faults are represented as splits in the grid and transmissibility modifiers as properties of the cells adjacent to the faults.

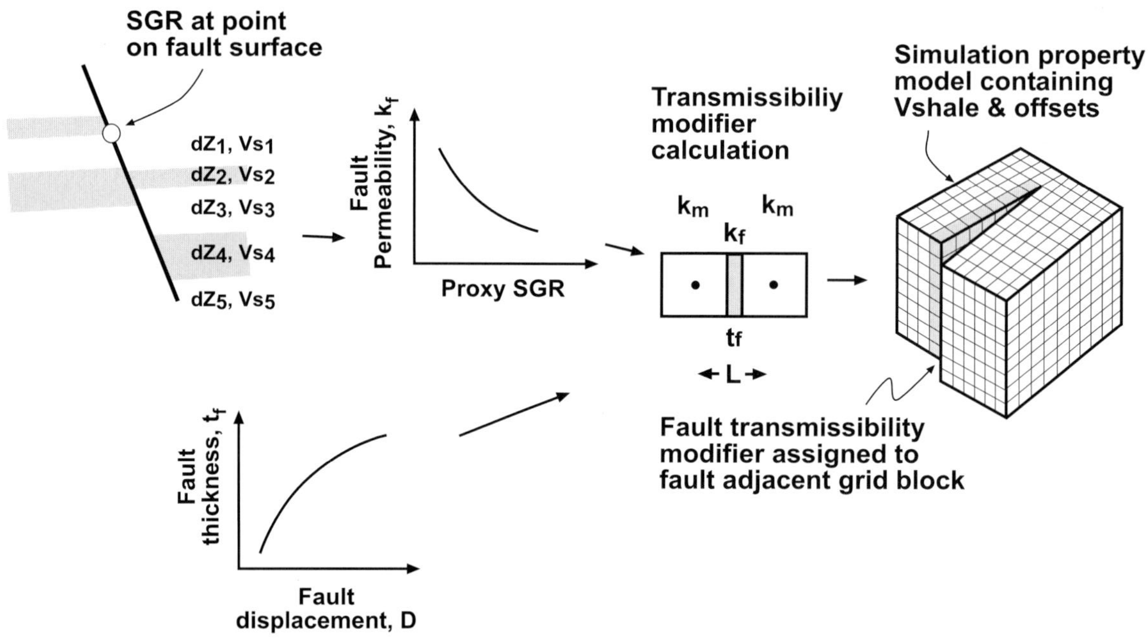

Fig. 2. Figure that outlines the workflow proposed by Manzocchi et al. (1999) for the assignment of fault transmissibility modifiers to offsets in a reservoir simulation model. Shale Gouge Ratio (SGR), a parameter that describes the clay content at a point on the fault (Yielding et al., 1997), is calculated by $\sum \Delta Z \cdot V_s/\text{throw}$, where ΔZ is the thickness of a particular layer in the slipped interval and V_s is the shale fraction of that layer. Empirical relationships are used to convert (i) fault clay content measured in plug samples (used as a proxy for SGR) to fault permeability (K_f), and (ii) fault displacement (D) to fault thickness (t_f). The transmissibility multiplier for the fault is derived from K_f and t_f, together with the permeability of the unfaulted matrix (K_m) and the distance between grid node centres in the simulation grid (L). A similar method is used in the FAULTMULT program (see Appendix A).

logically determined fault properties by interrogating a reservoir simulation model (Fig. 2). A prerequisite for the method is that the geometry of the simulation model represents the area and shape of faulted reservoir juxtapositions with a believable amount of confidence. The properties of the fault over the region of juxtaposed reservoir is then calculated as a transmissibility modifier. Empirical equations are used to relate fault displacement (offset) to fault thickness, and reservoir shale volume to fault permeability using the Shale Gouge Ratio technique (Yielding et al., 1997). Averaging methods are then used to incorporate an upscaling in the equations. Finally, fault permeability and fault thickness are combined with host rock permeability and grid cell spacing, to determine a transmissibility modifier value.

Experience using FAULTMULT

The Norsk Hydro in-house program FAULTMULT is an analogue to the commercial TransGen (marketed by Badleys), that applies the techniques of Manzocchi et al. (1999). FAULTMULT scans the reservoir simulation grid, automatically finds offsets (i.e. faults), and calculates SGR. Based on empirical relationships (see Sperrevik et al., 2002), the width of the fault zone and the permeability are estimated. This is translated to transmissibility multipliers that can be applied directly in the simulator (e.g. Eclipse). Technical details of FAULTMULT are given in Appendix A.

In the following sections, two case histories are described where FAULTMULT has been applied to currently producing fields.

Oseberg

The Oseberg field sits in an eastward tilted fault block located c. 140 km west of Bergen on the eastern margin of the Viking graben (Fig. 3). The reservoir is of Middle Jurassic age and comprises excellent reservoir quality shallow marine sandstones of the Oseberg and Tarbert Formations (Nipen, 1987; Helland-Hansen et al., 1992). Permeabilities range from 1 to 4 D, and porosity is around 24% (Johnsen et al., 1995). The Oseberg and Tarbert Formations are divided by a poorer quality Ness Formation (20–102 m thick), comprising isolated fluvial channel sandstones (Ryseth et al., 1998), which sits on thin (up to 16 m thick) progradational shoreface deposits of the Etive and Rannoch Formations. The reservoir is cut by extensional faults that formed both during deposition of the reservoir in the Middle Jurassic and also subsequently in the Late Jurassic (Færseth and Ravnås, 1998).

Oil migrated into the structure from a Draupne Formation source rock in the Viking Graben kitchen during the Late Cretaceous–Eocene (Johnsen et al., 1995). Gas migration began in the Eocene and probably continues today. Prior to production start-up in December 1988, a gas column of 380 m lay on an oil column of 210 m in the Oseberg field Alpha fault block. Present burial depth at the crest of the structure is 2120 m and negative effects of diagenesis on reservoir quality are minimal. The reservoirs in the Oseberg and Tarbert Formations are relatively thick (Oseberg 17–64 m; Tarbert up to 50 m, where not reduced in thickness by erosion at the base Cretaceous unconformity), relatively shale-poor and contain carbonate concretions. Vertical resolution of the seismic data over the field is ca. 10–20 m, and faults close to seismic resolution do not cause significant disruption of reservoir continuity.

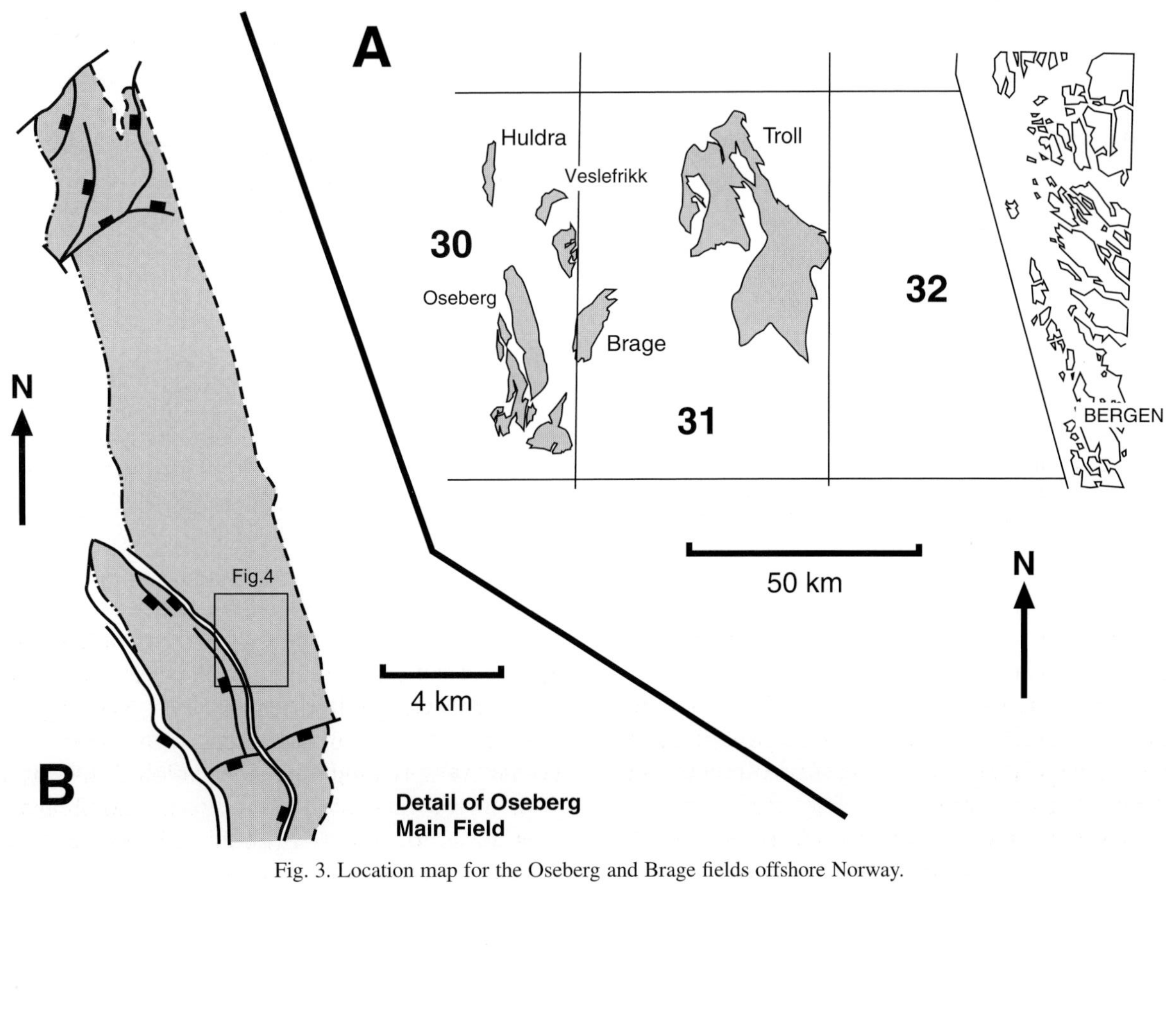

Fig. 3. Location map for the Oseberg and Brage fields offshore Norway.

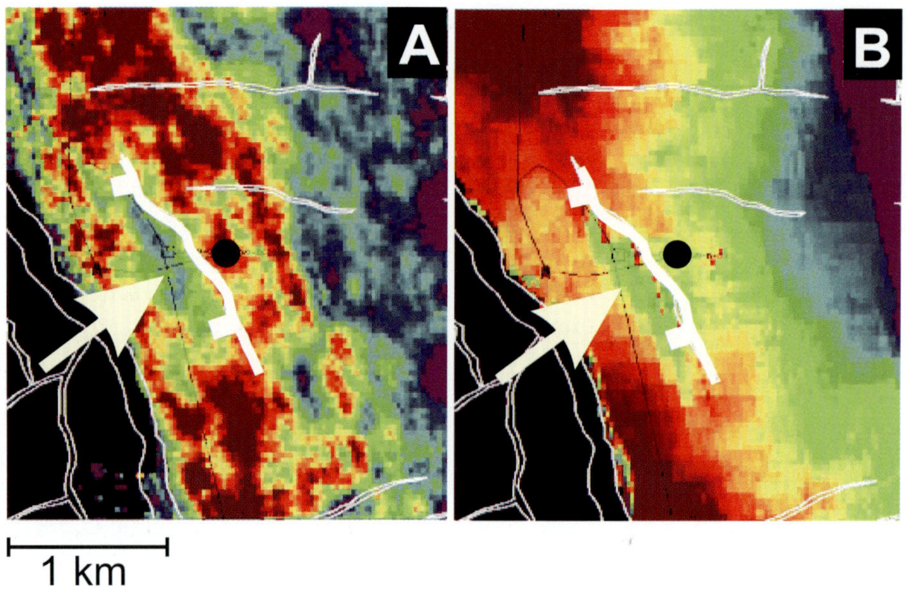

Fig. 4. Seismic response (A) around a small (ca. 10 m) fault in a clay-poor interval of the Oseberg reservoir indicates unswept oil west of the fault (see arrow). To reproduce the undrained oil pocket in the simulation model (B), a transmissibility multiplier of 0.0001 was applied. A is a 4D seismic technique, and is the difference in seismic amplitude at Base Brent derived from surveys acquired in 1997 and 1992 (blue = little change, red = large change). B is the acoustic impedance of synthetic seismic derived from fluid saturations in the reservoir simulator (green = high oil saturations).

No conclusive pressure differences were observed across faulted reservoir juxtapositions in the Oseberg field prior to production start-up. During the first ten years of production, pressure measurements (which had an uncertainty of <1 bar) indicated that faults did not operate as barriers to fluid flow (Johnsen et al., 1995). However, recent reservoir simulation and history matching results from the Oseberg Formation have shown that measurable across-fault pressure differences are now occurring under production due to a combination of pressure depletion, and gas and water injection. In one area a combination of seismic amplitude and reservoir simulation results has identified an unexploited pocket of oil that has been held back behind by a 10 m throw fault after a gas flood (Fig. 4). In another area premature water breakthrough occurred in a well due to redirection of injected water adjacent to a fault trace. In other areas of the field lateral pressure differences as high as 9 bar were reported over distances of ca. 2 km that contained at least one seismic mapped fault.

Studies of faults in cores (Norsk Hydro in-house work) show that the Oseberg field is characterised by relatively low fault intensities. Thin section analysis shows that the faults in the Oseberg field are associated with changes in packing, entrainment of clay and mica and clay smearing in the shaly parts of the sequence. Some cataclasis is also recorded, and calcite-cemented fractures occur in the vicinity of calcite-cemented concretions. Permeability reductions across individual cored faults are rarely greater than one order of magnitude. Cored faults are only rarely found in the Oseberg Formation, the Ness Formation showing most deformation features. The low clay content in the Oseberg Formation suggests that the incorporation of clay into the fault zones is not the primary sealing mechanism. We ran FAULTMULT by using fault zone formulae from both Manzocchi et al. (1999) and Sperrevik et al. (2002), but the transmissibility multipliers produced (ranging from 0.03 to 0.2) had to be reduced by a factor of two to three orders of magnitude (to 0.0001) to achieve a history match to the production data, and to match the seismic attributes (Fig. 4). Hence, we suggest that the method within FAULTMULT (covering SGR-based fault seal) does not capture all of the relevant processes.

Brage East

The Brage East field sits in a mildly faulted gentle dome structure located ca. 125 km west of Bergen on the eastern margin of the Viking graben (Fig. 3). The reservoir comprises bioturbated sandstones and shales of the Middle Jurassic Fensfjord Formation, which were deposited in marine shelf and shoreface environments (Ravnås and Bondevik, 1997; Ravnås et al., 1997). Reservoir quality is relatively modest with permeabilities ranging between 30 and 200 mD, although porosity lies around 27%. The reservoir is cut by extensional faults that formed both during deposition of the reservoir in the Middle Jurassic and subsequently in the Late Jurassic (Færseth, 1996; Færseth et al., 1997). Oil migrated into the structure probably during the Eocene (Johnsen et al., 1995).

Present burial depth at the crest of the structure is 2070 m. The reservoir is relatively thin (25–50 m), relatively shale-rich and contains carbonate nodules. Vertical resolution of the seismic data over the field is ca. 15–20 m, and faults close to seismic resolution cause significant disruption of reservoir continuity.

No across-fault pressure differences were present in the Brage field prior to production start-up in September 1993. However, significant across-fault pressure differences have occurred during production due to a combination of pressure depletion and water injection. A study combining RFT pressure data, seismic coherence data and acoustic borehole imaging conducted in 1997 showed that dynamic pressure barriers of up to 20 bars occurred at faults with throws up to 30 m and fault rock thicknesses of up to 50 cm (Dart et al., 1998). In other areas of the field lateral pressure differences as high as 160 bar are reported over distances of ca. 600–800 m.

Structural core logging, thin section analysis and borehole image response indicate that the faults are characterised by tighter grain packing, entrained micas and clays, and locally by calcite cementation (Aaland and Skjerven, 1998). Permeability measurements from plug samples across individual faults show permeability reductions relative to matrix values of up to one order of magnitude. The faults are clustered at seismically resolved offsets to form damage zones that are up to 30 m across for a fault with 150 m of throw. Fault intensity within the damage zones varies up to a maximum value of 100 planes per metre, and it is likely that the central portion of the damage zone forms an interconnected network of fault planes.

The relatively high clay content in the stratigraphy suggests that the incorporation of clay into the fault zones is likely to be the dominant sealing mechanism, and the field therefore appeared an ideal candidate for a FAULTMULT approach. Despite this, the transmissibility multipliers produced by the program had to be reduced by a factor of two to three orders of magnitude to produce a history match to the production data. Again, as in the Oseberg example, the method did not capture all of the relevant processes.

Why are FAULTMULT predictions wrong?

We suggest that the current method for the incorporation of geologically determined fault transmissibility modifiers (Manzocchi et al., 1999, FAULTMULT) does not work sufficiently well for the representation of cross-fault flow in reservoir simulation models. Furthermore, we propose that faults and fault damage zones in reservoirs are likely to be water-filled, and that capillary entry pressure and relative permeabilities are important factors for fault flow behaviour (see discussion section). These factors are not incorporated in the current method which is based solely on a treatment based on absolute permeabilities. It is well known that cap rock sealing is determined by the capillary entry pressure of the water-saturated cap-rock material (Schowalter, 1979; Watts, 1987), and static fault seal analysis for prospect definition also focuses on capillary entry pressure as the controlling mechanism (Yielding et al., 1997). Clay-smeared fault rocks are likely to have similar petrophysical properties to cap rocks, and it is probable that they also remain water-saturated as a reservoir is charged with hydrocarbons (Fig. 5). Also, outcrop studies show that damage zone haloes which occur around seismic mapped faults can form highly connected networks (e.g. Knipe et al., 1997; Foxford et al., 1998; Manzocchi et al., 1998). This network mesh offers the possibility of isolating fault-bounded lozenges of water-filled host rock during the charging process (Fig. 6).

Fault zone properties

Two factors have to be satisfied in order for significant volumes of hydrocarbons to flow across a water-saturated fault. First, the fault zones' capillary entry pressure must be overcome by the buoyancy pressure imposed by the hydrocarbon column. This results in the establishment of a narrow connected path of hydrocarbon from one side of the fault to the other. The second requirement is that the water saturation of the fault zone must be reduced (and thus the relative permeability of the fault zone with respect to the hydrocarbon increased) sufficiently to allow a significant volume of hydrocarbon to flow.

Fig. 5 shows what is typical when oil migrates into a reservoir. The initial water column is replaced by hydrocarbons as the hydrocarbon is trapped. As the rock is normally water-wet, this is a drainage process, where the capillary threshold pressure must be overridden in order to replace water with hydrocarbon. The threshold pressure is a function of permeability; the lower the permeability, the higher the threshold pressure (e.g. Ringrose and Corbett, 1994; Sperrevik et al., 2002).

If the buoyancy difference between hydrocarbon and water is less than the threshold pressure, the low-permeability fault zones may stay water-filled after migration of the hydrocarbon (Fig. 5B). When reservoir production imposes a differential pressure regime in the reservoir, flow across a fault will not be established until the pressure difference exceeds the fault rock threshold pressure (Fig. 5C,D).

Schematic relative permeability and capillary pressure curves for host rock and the fault zone rock are illustrated in Fig. 7. The imbibition curves (water dis-

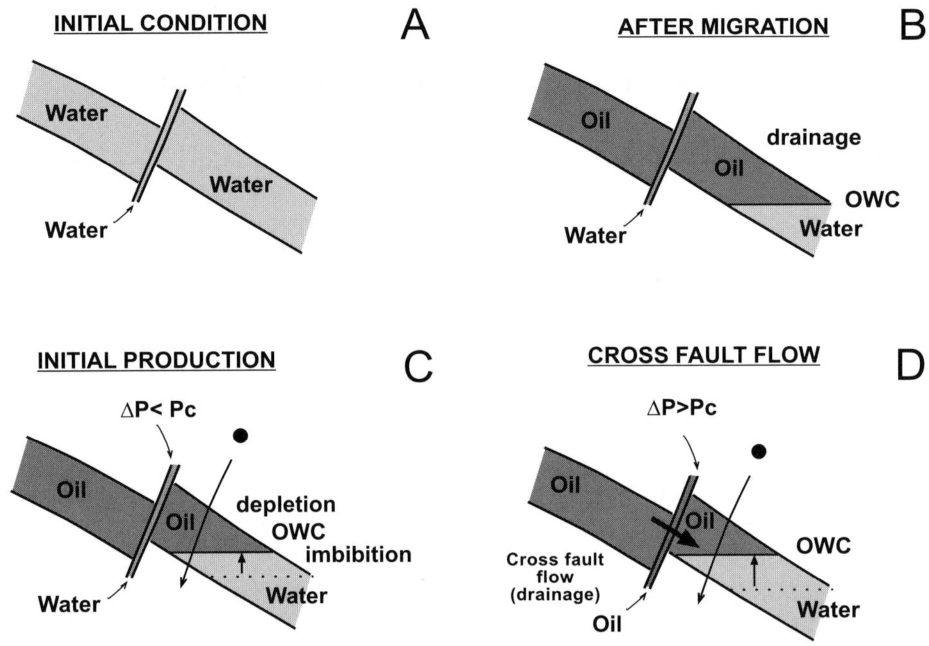

Fig. 5. Hypothesis showing how fault zones may be water filled after oil migration. See text for details.

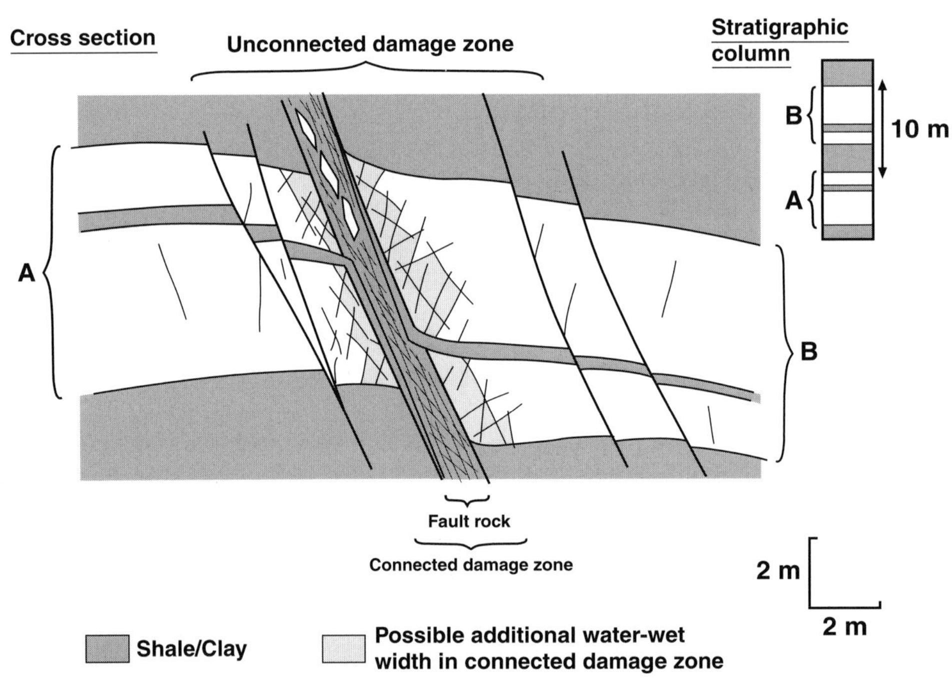

Fig. 6. Typical fault zone architecture from outcrop as compiled from faults in sand/shale sequences. Note that sandstone unit B has been downthrown in the hangingwall of the fault zone such that it now lies juxtaposed with sandstone unit A in the footwall (as a sand–sand juxtaposition). The dark grey layers within sandstone units are thin shale beds that are preferentially smeared into the fault zone (Lindsay et al., 1993). The connected damage zone peripheral to the fault rock is composed of a mesh of deformation bands. If these deformation bands have capillary entry pressures that are high enough, it is possible that the width of the preserved water-saturated zone may be more than the width of the clay-dominated fault rock (additional width indicated in light grey).

places oil, see Fig. 5C), and not the drainage curves, are most commonly used in reservoir simulations, and hysteresis (the difference between the drainage and imbibition curves) is often ignored.

We suggest that the fault rock may have relative permeability and capillary pressure curves that can be quite different from the host rock (Fig. 7, right part). Also, the hysteresis effect may be quite significant, and the threshold pressure for the drainage process may be substantial. The form of the relative permeability curves are often highly concave upwards, which implies that the sum of each phase permeability at intermediate water saturations is much lower than the absolute (one-phase) permeability.

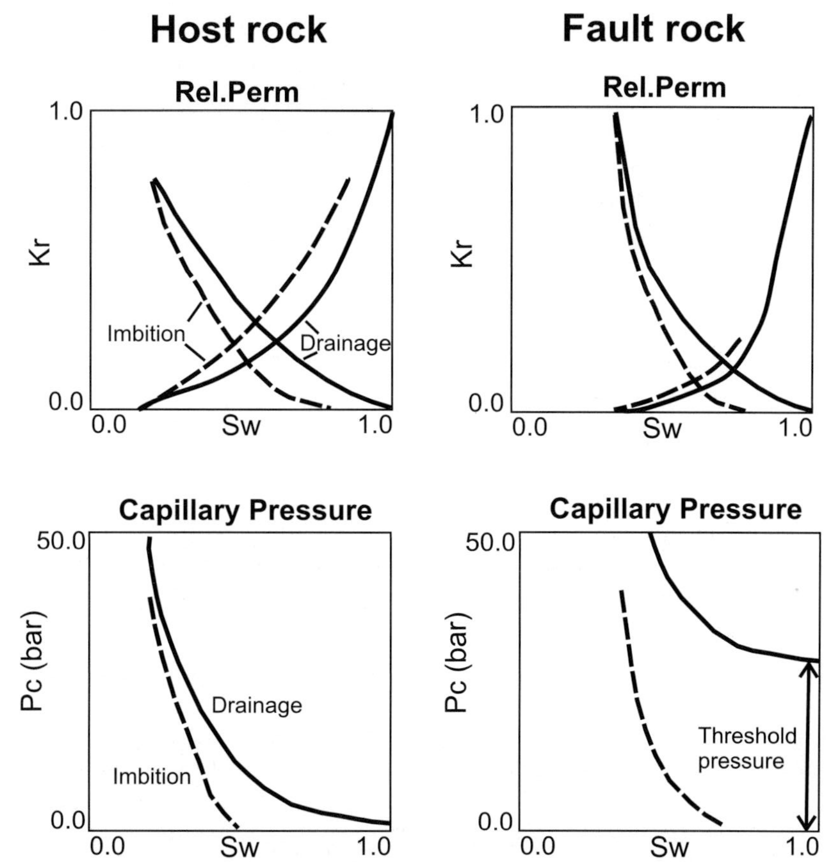

Fig. 7. Schematic diagram to show that fault rock relative permeabilities and capillary pressures will commonly be quite different from the corresponding host rock values. Both end-points and curvatures may differ significantly due to differences in permeability, lithology and rock wettability (e.g. Honarpour et al., 1986). This may have great impact on flow through a fault zone, but fault zone properties are commonly not modelled explicitly in reservoir simulators.

Synthetic example with water-filled fault zone

In order to determine the behaviour of water-filled fault zones, a small reservoir simulation model was built based on the Oseberg data. The grid model is shown in Fig. 8. The fault throw is around 10 m and the average horizontal permeability of the reservoir is ca. 3000 mD. The fault zone permeability was set explicitly to 0.001 mD.

We introduced separate sets of relative permeability and capillary pressures curves for the host rock and for the fault zone, and we activated the hysteresis option in the simulator in order to model the combined imbibition and drainage process that may occur

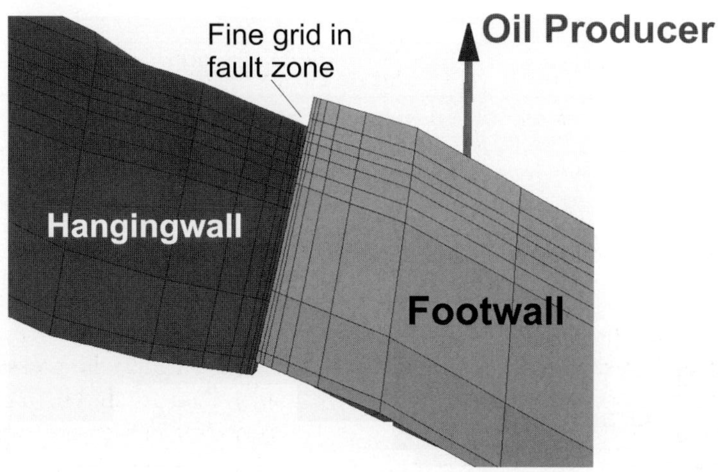

Fig. 8. The reservoir simulation model with a fine grid around the fault. Oil is produced from the footwall block while pressures in both blocks are monitored.

in different parts of the reservoir at the same time (Fig. 5C,D).

Simulation results

The results from the simulations are given in Fig. 9. The footwall pressure decreases rapidly (as expected) since the oil producer is placed there (Graph 1). Two curves are plotted for the hangingwall pressures. In case A, the fault zone has reduced absolute permeability, but the fluid properties (water saturation, relative permeabilities and capillary pressures) are inherited from the host rock. This will in effect be the same as a FAULTMULT model. In case B, the permeabilities are the same, but the fault zone is water-filled, and has relative permeability functions and capillary pressures according to Fig. 7. The drainage threshold pressure is 10 bars, sufficient to keep the water saturation within the fault zone if no external pressure differences are introduced. The choice of threshold pressure for the fault permeability is in accordance with Ringrose and Corbett (1994).

The difference in pressure development for the two cases is significant. If we zoom into the first year of pressure development (Graph 2, shaded area in Graph 1), we see that for case B the pressure in the hangingwall block is unaffected until the threshold pressure is exceeded. If a non-zero multiplier is applied as in case A, there will be an effect on pressure from day one of depletion. Hence, in order to 'break the seal', explicit fault rock and fluid properties (case B) must be modelled.

In Fig. 10, the fault-zone permeability in case A was multiplied by 0.1 (case C). We see now that the pressure history is much closer to case B. Thus, we found in this experiment that the effect of introducing unique fault-zone properties reduced the 'apparent' fault permeability by almost one order of magnitude.

Discussion

In this paper we have only considered the effects that water-saturated faults have on production from hydrocarbon reservoirs. However, as seen in the example described above, the introduction of this factor alone may produce a decrease in across-fault flow sufficient to explain the production behaviour of the Brage and Oseberg field examples.

Fault water saturation is not the only explanation why the original FAULTMULT simulations created faults that were too open. Some possible reasons are listed below, but see also Sperrevik et al. (2002) for a more detailed treatment.

(1) The empirical formulae used may over-estimate fault permeability and under-estimate the effective fault thickness (upscaled thickness of both central

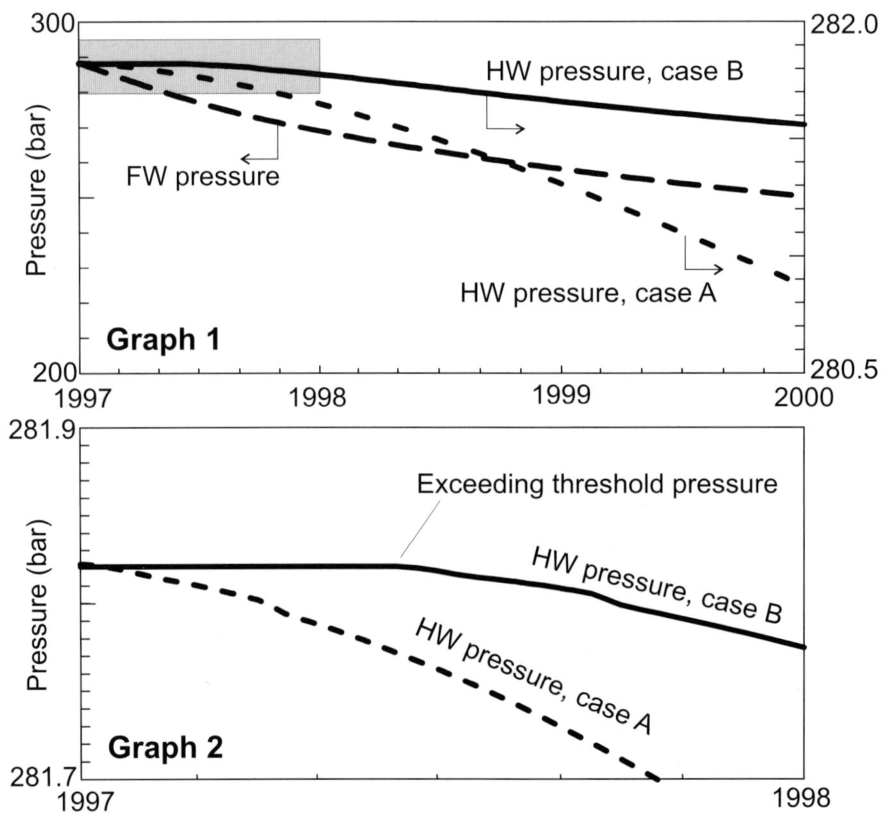

Fig. 9. In Graph 1, the footwall (FW) pressures decreases ca. 30 bar during 3 years, while the hangingwall (HW) pressures decreases much less. In Graph 2, the grey shading in Graph 1 is magnified in order to monitor the exact pressure history in the HW block. See text for further details.

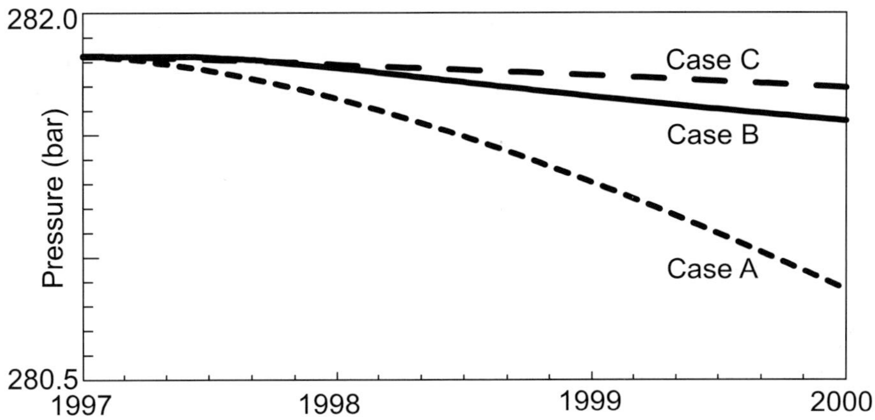

Fig. 10. The effect of using explicit relative permeability and capillary pressures (case B) is almost the same as reducing the fault zone permeability by a factor 10. See text for further explanation.

fault rock, plus thicknesses of additional peripheral damage zone fault rocks). In particular it should be noted that absolute permeability measurements need to be corrected for overburden effects.

(2) The SGR parameter may underestimate the amount of entrained clay in clay smeared faults. In fact shale SGR values reflect a proportional mixing of the faulted stratigraphy to generate the fault rock material and not the preferential entrainment of clays which is known to occur from the study of outcrop analogues (Lehner and Pilaar, 1997).

(3) Calcite or quartz overgrowth cementation may be an additional fault sealing process (Sverdrup and Bjørlykke, 1997) that is not considered in the technique described in this paper. It should also be noted that the absolute permeability of clay-bearing fault rocks will decrease with maximum burial depth due to compaction and diagenesis (see Krooss et al., 1998). Again this is not accounted for.

(4) The upscaling techniques may over-estimate the effective permeability and under-estimate the effective thickness of the fault zone at the scale of a typical grid block in a reservoir simulator (i.e. $100 \times 100 \times 0$ m). Both fault zone and damage zone structure (Fig. 6) is complex and still poorly understood at these scales. Outcrop analogue data suggest that damage zones are thicker in sandstone than in shale-rich stratigraphies (Beach et al., 1999; Sperrevik et al., 2002).

Modifications to the empirical formula for absolute fault permeability within the bounds suggested by Sperrevik et al. (2002) do reduce the values of the transmissibility multipliers generated by FAULT-MULT, but not sufficiently to produce the required history matches on the Brage and Oseberg fields. Therefore this is not the whole answer. Also abnormal levels of fault sealing due to enhanced diagenesis and compaction are unlikely at Oseberg and Brage due to the limited local development of calcite-cemented faults and relatively shallow maximum burial depths. The impact of using a different clay smear descriptor or different upscaling routine at Oseberg and Brage has not been investigated at the time of writing, and in the authors' opinion these factors could be a focus for further investigation.

If, however, for the sake of argument, we assume that the initial method used for predicting fault seal behaviour under production is seriously impaired because the effects of water-saturated faults are not included, it is interesting to speculate under which conditions water-saturated faults are most likely to occur and under what circumstances they should be addressed in the hydrocarbon industry. This is dealt with in the following subsections.

How common are water-saturated faults in hydrocarbon reservoirs?

Almost all porous rocks within a sedimentary basin, including porous rocks in passive fault zones, are initially water-filled (Schowalter, 1979). Oil migrates into the porous reservoir rock such as a sandstone and is trapped beneath the cap rock. The cap rock remains water-saturated, giving it a sufficiently high capillary entry pressure to trap a hydrocarbon column. At a first approximation clay-smeared fault zones in hydrocarbon reservoirs in sand/shale sequences can be considered as analogues to a shale cap rock (Watts, 1987). If clay-smeared faults are assumed to have the same properties as the cap rock then it follows that all faults in hydrocarbon reservoirs in sand/shale sequences will also be water-saturated. If this is the case then water-saturated clay-smeared faults in hydrocarbon reservoirs are universal and a reservoir simulation implementation should be available for all sand/shale reservoirs (i.e. a very significant percentage of the world's hydrocarbon reserves).

Fault zones in sand/shale sequences are, however, not identical to cap rocks. For example, fault zones may contain a significant sand component in the form of matrix blocks within damage zones, and entrained pods of sandstone within the fault rock (Childs et al., 1997; Fig. 6). As long as the migrating hydrocarbon is isolated from these sandy fault components they will remain water-soaked, but in general it is clear that the higher the sandstone component the less likely that the fault zones within the hydrocarbon column will be water-filled.

Dependence on depth of burial, structural style and in-situ stress

There are three main periods during the evolution of hydrocarbon reservoir that are critical for determining whether reservoir compartmentalising faults will be water-saturated:

- trap filling
- geological history following trap fill
- production history

The effective isolation of the water-filled fault zone from migrating hydrocarbons is dependent on encapsulating the fault zone material with a membrane with a sufficiently high capillary entry pressure to withstand the hydrocarbon column. The fault zone components with the highest capillary entry pressures are likely to be clays (Gibson, 1998). Compaction and diagenesis decreases clay pore size with burial and thus increase the capillary entry pressure (Krooss et al., 1998), and there is a marked increased in diagenetic effects at ca. 2–3 km depth (Bjørlykke, 1999). Therefore burial to depths greater than 2–3 km prior to hydrocarbon migration will enhance the probability of preserving water-saturated faults. Because compaction and diagenetic effects on the physical properties of the rock will be largely unrecoverable (Magara, 1976, 1996), fields that have been buried and then uplifted to shallower depths will have a higher probability of containing water-saturated faults.

It is now generally accepted that active faults are responsible for the transport of substantial quantities of water and other fluids during basin subsidence (Sibson et al., 1975; Hooper, 1991; Flemming et al., 1998), and that currently active faults are likely to contain episodically flowing water (Townend and Zoback, 2000). It is likely that reactivated faults transport fluids within a hydrocarbon column. However, these fluids are just as likely to be hydrocarbons as water. Activation of a fault that sits within a hydrocarbon column will significantly increase the chances of providing the hydrocarbon access to the fault zone by the creation of new fault splays. Once the hydrocarbon has entered the fault, along-fault transport parallel to the fault zone fabric, may be quite efficient (Arch and Maltman, 1990). The likelihood of slip subsequent to trap filling will be dependent on how the fault pattern is oriented relative to the evolving stress fields that are applied to the reservoir during the time from filling to the present day. Not only the orientations, but also the magnitudes of stress will play a role, as will the presence of abnormally high fluid pressures, which will tend to hold faults open in the tighter parts of the stratigraphy. Increasing burial, compaction and diagenesis make mudrocks more brittle (Dewhurst et al., 1999). This implies that hydrocarbon invasion of a water-filled fault due to reactivation and slip may be easier in rocks that are, or have been, buried to greater depths. A rock mechanics treatment (for example see Scholz, 1990) is necessary to evaluated the chances of reactivation and subsequent expulsion of water from fault zones for a particular field. Such a treatment is beyond the scope of this paper.

When fault zones in sand/shale sequences are examined at outcrop (e.g. Sverdrup and Bjørlykke, 1997; Kattenhorn et al., 2000) and cores (Kulander et al., 1990) they are often characterised by open joints in addition to typical fault rock types such as clay smears and deformation bands. The extent to which jointing exists in intact fault zones in the subsurface remains unclear, and the joints we observe at the surface may be related to exhumation, or unloading during recovery of core to the surface. In the sub-surface, open fractures in poorly cemented sedimentary rocks, typically with relatively low shear strengths, are expected to close rapidly due to high confining pressures (Bjørlykke and Høeg, 1997). High fluid pressures may be one mechanism by which joints can be held open in the subsurface (Engelder, 1993), given that the matrix rock has sufficient strength (i.e. has been cemented/compacted enough). Under such circumstances water-saturated faults within a hydrocarbon column may be more likely to be invaded by hydrocarbon during fault reactivation.

Production of hydrocarbons and the injection of water and gas can have a profound influence on the fluid pressure and magnitudes of stress in a reservoir. In certain formations fault slip may be induced, leading the creation of shear failure and subsidence (e.g. Ekofisk, Teufel et al., 1991). It is unclear at present how general this phenomenon is and how it applies to siliciclastic reservoirs. Again a rock mechanics treatment is necessary to evaluate the chances of fault reactivation under production. Coupled fluid/rock mechanics simulations are required to tackle these problems.

To summarise, water-filled faults in hydrocarbon reservoirs are most likely to occur in rocks that have been buried deeply and have not undergone fault reactivation after trap filling. Conversely, they are least likely to occur in rocks that have been deeply buried and have subsequently undergone fault reactivation after trap filling.

How routinely should water-saturated faults be implemented?

Where production history data exist for a particular field it is possible to build a simulation model and adjust the fault transmissibility modifiers until a history match is achieved. This approach is often used and described by Fulljames et al. (1997) who goes on to relate the derived transmissibility modifiers to fault throw in order to derive a predictive tool. This kind of approach may be sufficient for individual fields where sufficient production history data are available. However, given the (i) geological, (ii) fluid, and (iii) simulation grid geometry variation between fields it is unlikely that such a relationship could be applied to try and predict the performance of an as yet unproduced field. If our predictive tools are to be reliable then they have to be based on a correct physical understanding of the processes operating.

If, as we suggest, most compartmentalising faults in clastic hydrocarbon reservoirs are water-filled, then evaluation of the production performance of new fields should address the influence of this effect. For existing fields, where sufficient production history data exist, it may be possible to find situations that can be analysed and modelled to get a feel for the likely ranges of fault transmissibility multiplier values that are relevant. However, we would suggest that when simulation techniques are available that can routinely account for water-soaked faults, that these should also be applied to existing fields. In this scenario better process understanding would reduce the economic risk of increase oil recovery (IOR) efforts by producing more robust infill well locations and better production forecasting.

Where should we go from here?

Simulations show that the presence of water in the fault zone has significant impact on how the fault restricts flow. This may explain why the FAULTMULT approach on the Oseberg and Brage fields always gives too optimistic results.

It is difficult to implement water-saturated faults correctly in reservoir simulators without representing the fault zone as a volume, not just as split nodes as done today. As mentioned earlier, this may be difficult in terms of computer capacity. Hence, we need an approach that can help us achieve the future's solution with today's technology.

One approach is to make a suite of small prototype models, where the fault zones are represented explicitly (as in this paper). From that, one can estimate a fault permeability formula that incorporates the fluid effects indirectly. Combining this with a segment threshold pressure in the simulator, may be adequate for most cases.

Another approach is outlined by Manzocchi et al. (2002), who applied a directional relative permeability to the upstream cell. Their method appears promising, but published results from a real field example are needed to fully evaluate the usefulness of the technique.

Conclusions

(1) A method now exists for the derivation of fault transmissibility modifiers for use in reservoir simulation independent of history matching experience. Inputs are fault throw and Vshale, together with empirical relationships between (i) fault clay content and absolute fault permeability and (ii) fault throw and fault thickness.

(2) Application of this method to the mature Oseberg and Brage fields in the North Sea, where production history data exist, has shown that the fault transmissibility modifiers produced are two to three orders of magnitude too high and give faults that are too open to flow.

(3) In this paper it is proposed that the main reason why the method is unsuccessful is that the high water saturations along fault zones in hydrocarbon reservoirs are not accounted for by the method. An experimental simulation model is constructed to test this hypothesis. The results illustrate the importance of including capillary entry pressure and relative permeability for an adequate description of across-fault flow.

(4) High water saturations can be expected to occur if the capillary entry pressure of the faulted material is high enough to prevent hydrocarbon from entering the fault zone during trap filling.

(5) Preservation of faults with high water saturations are expected to be favoured by deep burial prior to trap filling and minimal fault reactivation subsequently and under production.

(6) Today, computer power limitations make it impractical to represent faults as explicit volumes with independent water saturations.

(7) One way forward is to simulate detailed models of fault structure to derive fault permeability formulae that can be applied in full field simulation.

Acknowledgements

The following have contributed significantly to the description of faulting in the cases described: R. Gabrielsen, K.J. Hersvik, R. Knipe, J. Korstgård, H. Rutledal, I. Skaar, O.P. Wennberg. P. Gillespie, S. Sperrevik and E. Sverdrup have been an invaluable aid in the discussions that form the foundations for this paper. Roy Gabrielsen is also thanked for a very constructive review.

Appendix A. Description of FAULTMULT

FAULTMULT is a Norsk Hydro in-house program designed for assigning transmissibility multipliers in reservoir simulation models, based on geological input. It is currently designed for interacting with the commercial simulator Eclipse.

FAULTMULT reads a parameter file, which holds all relevant information the program needs for the calculation, such as

- Names of input files (grid and property).
- Vshale parameter, or 1 − NTG (NTG = net-to-gross) if this should be applied for approximating Vshale.
- Formula that gives width of fault zone, to a function of throw.
- Formula that deduces fault zone permeability from throw, SGR, etc. FAULTMULT will evaluate the formula at run-time, which gives the user great flexibility on formula construction.
- Output options. The most important outputs are MULTX and ULTY, which can be applied directly in the simulator.

After parsing the input file, FAULTMULT reads the grid geometries. It scans every row and column in the grid looking for split nodes (faults). If a split node is found, throw is measured, fault zone width is estimated, and SGR is computed (Yielding et al., 1997; Fig. 2). This is applied to compute fault zone permeability.

The next step is to compute the transmissibility multiplier. This is illustrated in Fig. 11. First, the transmissibility T_1 is computed by assuming no fault zone present. The calculations follow the following formula:

$$T_1 = \left(\frac{\Delta X_i}{K_i} + \frac{\Delta X_{i+1}}{K_{i+1}} \right)^{-1}$$

Here, ΔX is length of grid cells (see Fig. 11) and K is cell permeability. Notice that the permeability on cell position $i+1$ is an arithmetic, area weighted average of the cells juxtaposed (interval A in Fig. 11). Next, the permeability in the cell under investigation is reduced by adding a imaginary fault zone (e.g. using formulas from Sperrevik et al., 2002). These are averaged harmonically. Transmissibility calculation is redone, resulting in T_2, and the final multiplier M is

$$M = \frac{T_2}{T_1}$$

Various output from FAULTMULT can be inspected graphically in commercial geoscientific software such as IRAP RMS.

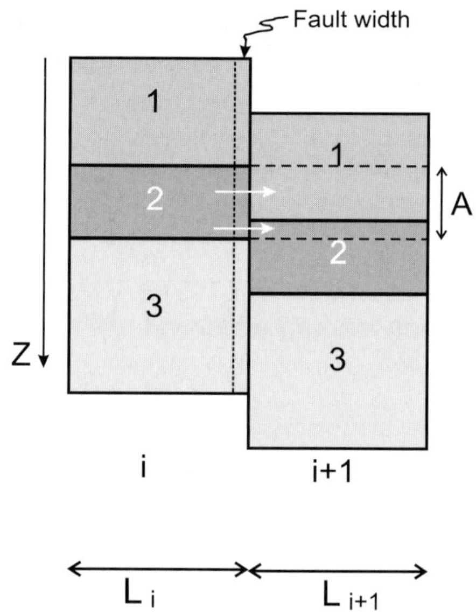

Fig. 11. Computation of the transmissibility multiplier is illustrated in this example. Two columns of cells with width L are divided by a split node (fault). From cell layer 2 in the left hand column, there is flow into both cell 1 and cell 2 in the right hand column. To estimate the multiplier, a transmissibility calculation between left cell 2 and the arithmetic average of permeabilities in the right column cells (interval A), gives the transmissibility value T_1. Next, a hypothetical fault zone with estimated width and permeability is averaged harmonically (since it is perpendicular to flow) into left cell 2. The width of this fault zone may be derived from throw, and the properties may be derived by using clay smear (SGR) technique. This gives transmissibility T_2. The multiplier assigned to upstream left cell is thus T_2/T_1.

References

Aaland, R.K. and Skjerven, J., 1998. Fault and fracture characteristics of a major fault zone in the northern North Sea: analysis of 3D seismic and oriented core in the Brage Field (Block 31/4). In: M.P. Coward, T.S. Daltaban and H. Johnsen (Editors), Structural Geology in Reservoir Characterisation. Geol. Soc. London Spec. Publ., 127: 209–229.

Arch, J. and Maltman, A.J., 1990. Anisotropic permeability and tortuosity in deformed wet sediments. J. Geophys. Res., 95: 9035–9047.

Beach, A., Welborn, A.I., Brockbank, P.J. and McCallum, J.E., 1999. Reservoir damage around faults: examples from the Suez rift. Pet. Geosci., 5: 109–116.

Bjørlykke, K., 1999. Principal aspects of compaction and fluid flow in mudstones. In: A.C. Aplin, A.J. Fleet and J.H.S. MacQuaker (Editors), Muds and Mudstones: Physical and Fluid Flow Properties. Geol. Soc. London Spec. Publ., 158: 73–78.

Bjørlykke, K. and Høeg, K., 1997. Effects of burial diagenesis on stresses, compaction and fluid flow in sedimentary basins. Mar. Pet. Geol., 14: 267–276.

Childs, C., Walsh, J.J. and Watterson, J., 1997. Complexity in fault zone structure and implications for fault seal prediction. In: P. Møller-Pedersen and A.G. Koestler (Editors), Hydrocarbon Seals: Importance for Exploration and Production. Norwegian Petroleum Society (NPF), Special Publication 7. Elsevier, Amsterdam, pp. 61–72.

Dart, C.J., Fjellbirkeland, H., Robson, A. and Wennberg, O.P., 1998. Structural interpretation verified using acoustic borehole images — Brage field, Norway (Extended Abstract). EAEG 60th Conference and Technical Exhibition, Leipzig, 2, pp. 3–11.

Dewhurst, D.N., Yang, Y. and Aplin, A.C., 1999. Permeability and fluid flow in natural mudstones. In: A.C. Aplin, A.J. Fleet and J.H.S. MacQuaker (Editors), Muds and Mudstones: Physical and Fluid Flow Properties. Geol. Soc. London Spec. Publ., 158: 23–43.

Engelder, T., 1993. Stress Regimes in the Lithosphere. Princeton University Press, Princeton, NJ.

Færseth, R.B., 1996. Interaction of Permo-Triassic and Jurassic extensional fault blocks during the development of the northern North Sea. J. Geol. Soc. London, 153: 931–944.

Færseth, R.B. and Ravnås, R., 1998. Evolution of the Oseberg Fault Block in the context of the northern North Sea structural framework. Mar. Pet. Geol., 15: 467–490.

Færseth, R.B., Knudsen, B.E., Liljedahl, T., Midtbøe, P.S. and Søderstrøm, B., 1997. Oblique rifting and sequential faulting in the Jurassic development of the northern North Sea. J. Struct. Geol., 19: 1285–1302.

Flemming, C.G., Couples, G.D. and Haszeldine, R.S., 1998. Thermal effects of fluid flow in steep fault zones. In: G. Jones, Q.J. Fisher and R.J. Knipe (Editors), Faulting, Fault Sealing and Fluid Flow in Hydrocarbon Reservoirs. Geol. Soc. London Spec. Publ., 147: 217–229.

Foxford, K.A., Walsh, J.J., Watterson, J., Garden, I.R., Guscott, S.C. and Burley, S.D., 1998. Structure and content of the Moab fault Zone, Utah, USA and its implications for fault seal prediction. In: G. Jones, Q.J. Fisher and R.J. Knipe (Editors), Faulting, Fault Sealing and Fluid Flow in Hydrocarbon Reservoirs. Geol. Soc. London Spec. Publ., 147: 87–103.

Fristad, T., Groth, A., Yielding, G. and Freeman, B., 1997. Quantitative fault seal prediction: a case study from Oseberg Syd. In: P. Møller-Pedersen and A.G. Koestler (Editors), Hydrocarbon Seals: Importance for Exploration and Production. Norwegian Petroleum Society (NPF), Special Publication 7. Elsevier, Amsterdam, pp. 107–124.

Fulljames, J.R., Zijerveld, L.J.J. and Franssen, R.C.M.W., 1997. Fault seal processes: Systematic analysis of fault seals over geological and production time scales. In: P. Møller-Pedersen and A.G. Koestler (Editors), Hydrocarbon Seals: Importance for Exploration and Production. Norwegian Petroleum Society (NPF), Special Publication 7. Elsevier, Amsterdam, pp. 51–59.

Gibson, R.G., 1998. Physical character and fluid-flow properties of sandstone-derived fault zones. In: M.P. Coward, T.S. Daltaban and H. Johnsen (Editors), Structural Geology in Reservoir Characterisation. Geol. Soc. London Spec. Publ., 127: 83–97.

Hansen, S., 1996. Quantification of net uplift and erosion on the Norwegian Shelf south of 66N from sonic transit times of shale. Nor. Geol. Tidsskr., 76: 245–252.

Helland-Hansen, W., Ashton, M., Lømo, L. and Steel, R.J., 1992. Advance and retreat of the Brent delta: Recent contributions to the depositional model. In: A.C. Morton, R.S. Haszeldine, M.R. Giles and S. Brown (Editors), Geology of the Brent Group. The Geological Society, London, pp. 109–127.

Honarpour, M., Koederitz, L. and Harvey, A.H., 1986. Relative Permeability of Petroleum Reservoirs. CRC Press, Boca Raton, FL, 143 pp.

Hooper, E.C.D., 1991. Fluid migration along growth faults in compacting sediments. J. Pet. Geol., 14: 161–180.

Johnsen, J.R., Rutledal, H. and Nilsen, D.E., 1995. Jurassic reservoirs; field examples from the Oseberg and Troll fields: Horda Platform area. In: S. Hanslien (Editor), Petroleum Exploration and Exploitation in Norway. Norsk Petroleums Forening (NPF), Special Publication 4. Graham and Trotman, London, pp. 199–234.

Kattenhorn, S.A., Aydin, A. and Pollard, D.D., 2000. Joints at high angles to normal fault strike: an explanation using 3-D numerical models of fault perturbed stress fields. J. Struct. Geol., 22: 1–23.

Knai, T.A. and Knipe, R.J., 1998. The impact of faults on fluid flow in the Heidrun field. In: G. Jones, Q.J. Fisher and R.J. Knipe (Editors), Faulting, Fault Sealing and Fluid Flow in Hydrocarbon Reservoirs. Geol. Soc. London Spec. Publ., 147: 269–282.

Knipe, R.J., Fisher, Q.J., Jones, G., Clenell, M.R., Farmer, A.B., Harrison, A., Kidd, B., McAllister, E., Porter, J.R. and White, E.A., 1997. Fault Seal Analysis: successful methodologies, application and future directions. In: P. Møller-Pedersen and A.G. Koestler (Editors), Hydrocarbon Seals: Importance for Exploration and Production. Norwegian Petroleum Society (NPF), Special Publication 7. Elsevier, Amsterdam, pp. 15–38.

Krooss, B.M., Schloemer, S. and Ehrlich, R., 1998. Experimental investigation of molecular transport and fluid flow in unfaulted and faulted pelitic rocks. In: G. Jones, Q.J. Fisher and R.J. Knipe (Editors), Faulting, Fault Sealing and Fluid Flow in Hydrocarbon Reservoirs. Geol. Soc. London Spec. Publ., 147: 135–146.

Kulander, B.R., Dean, S.L. and Ward, B.J., 1990. Fractured Core Analysis. American Association of Petroleum Geologists, Tulsa, OK.

Lehner, F.K. and Pilaar, W.F., 1997. The emplacement of clay smears in synsedimentary normal faults: inferences from field observations near Frechen, Germany. In: P. Møller-Pedersen and A.G. Koestler (Editors), Hydrocarbon Seals: Importance for Exploration and Production. Norwegian Petroleum Society (NPF), Special Publication 7. Elsevier, Amsterdam, pp. 39–50.

Lindsay, N.G., Murphy, F.C., Walsh, J.J. and Watterson, J., 1993. Outcrop studies of shale smears on fault surfaces. Spec. Publ. Int. Assoc. Sedimentol., 15: 113–123.

Magara, K., 1976. Thickness of removed sedimentary rocks, palaeopore pressure and palaeotemperature, southwestern part of the Western Canada Basin. Am. Assoc. Pet. Geol. Bull., 60: 554–565.

Manzocchi, T., Ringrose, P.S. and Underhill, J.R., 1998. Flow through fault systems in high-porosity sandstones. In: M.P. Coward, T.S. Daltaban and H. Johnsen (Editors), Structural Geology in Reservoir Characterisation. Geol. Soc. London Spec. Publ., 127: 65–82.

Manzocchi, T., Walsh, J.J., Nell, P. and Yielding, G., 1999. Fault transmissibility multipliers for flow simulation models. Pet. Geosci., 5: 53–63.

Manzocchi, T., Heath, A.E., Walsh, J.J. and Childs, C., 2002. The representation of two phase fault-rock properties in flow simulation models. Petrol. Geosci., in review.

Nipen, O., 1987. Oseberg. In: A.M. Spencer, E. Holter, C.J. Campbell, S.H. Hanslie, P.H.H. Nelson, E. Nysæther and E. Ormåsen (Editors), Geology of the Norwegian Oil and Gas Fields. Norwegian Petroleum Society, Stavanger, pp. 379–387.

Ravnås, R. and Bondevik, K., 1997. Architecture and controls on the Bathonian–Kimmeridgian shallow-marine synrift wedges of the Oseberg–Brage area, northern North Sea. Basin Res., 9: 197–226.

Ravnås, R., Bondevik, K., Helland-Hansen, W., Lømo, L., Ryseth, A. and Steel, R.J., 1997. Sedimentation history as an indicator of rift initiation and development: the Late Bajocian–Bathonian evolution of the Oseberg–Brage area, northern North Sea. Nor. Geol. Tidsskr., 77: 205–232.

Ringrose, P.S. and Corbett, P.W.M., 1994. Controls on two-phase flow in heterogeneous sandstones. In: J. Parnell (Editor), Geofluids: Origin, Migration and Evolution of Fluids in Sedimentary Basins. Geol. Soc. London Spec. Publ., 78: 141–150.

Ryseth, A., Fjellbirkeland, H., Osmundsen, I., Skålnes, A. and Zachariassen, E., 1998. High resolution stratigraphy and seismic attribute mapping of a fluvial reservoir unit: Ness Formation, Oseberg field. Am. Assoc. Pet. Geol. Bull., 82: 1627–1651.

Scholz, C.H., 1990. The Mechanics of Earthquakes and Faulting. Cambridge University Press.

Schowalter, T.T., 1979. Mechanics of secondary hydrocarbon migration and entrapment. Am. Assoc. Pet. Geol. Bull., 63: 723–760.

Sibson, R.H., Moore, J.M. and Rankin, A.H., 1975. Seismic pumping: a hydrothermal fluid transport mechanism. J. Geol. Soc. London, 131: 653–659.

Sperrevik, S., Gillespie, P.A., Fisher, Q.J., Halvorsen, T. and Knipe, R.J., 2002. Empirical estimation of fault rock properties. In: A.G. Koestler and R. Hunsdale (Editors), Hydrocarbon Seal Quantification. Norwegian Petroleum Society (NPF), Special Publication 11. Elsevier, Amsterdam, pp. 109–125 (this volume).

Sverdrup, E. and Bjørlykke, K., 1997. Fault properties and the development of cemented fault zones in sedimentary basins: field examples and predictive models. In: P. Møller-Pedersen and A.G. Koestler (Editors), Hydrocarbon Seals: Importance for Exploration and Production. Norwegian Petroleum Society (NPF), Special Publication 7. Elsevier, Amsterdam, pp. 91–106.

Teufel, L.W., Rhett, D.W. and Farrel, H.E., 1991. Effect of reservoir depletion and pore pressure drawdown on in situ stress and deformation in the Ekofisk Field, North Sea. In: J.C. Roegiers (Editor), Rock Mechanics as a Multidisciplinary Science. Proceedings of the 32nd US Symposium, Oklahoma, pp. 63–72.

Townend, J. and Zoback, M.D., 2000. How faulting keeps the crust strong. Geology, 28: 399–402.

Watts, N.L., 1987. Theoretical aspects of cap rock and fault seals for single- and two-phase hydrocarbon columns. Mar. Pet. Geol., 4: 274–307.

Yielding, G., Freeman, B. and Needham, D.T., 1997. Quantitative fault seal prediction. Am. Assoc. Pet. Geol. Bull., 81: 897–917.

J.C. RIVENÆS *Norsk Hydro ASA, P.O. Box 7190, N-5020 Bergen, Norway*
E-mail: jan.christen.rivenas@hydro.com

C. DART *Norsk Hydro ASA, P.O. Box 7190, N-5020 Bergen, Norway*

Geological implications of a large pressure difference across a small fault in the Viking Graben

C. Childs, T. Manzocchi, P.A.R. Nell, J.J. Walsh, J.A. Strand, A.E. Heath and T.H. Lygren

Two discrete pressure cells with a 128 bar pressure difference within a 100 m thick Tarbert reservoir interval are separated by a fault with a throw of ~50 m. Given that high permeability reservoir rocks are juxtaposed across the fault, the observed across-fault pressure difference cannot be maintained for reasonable fault rock permeabilities over geologically significant periods (>10,000 years). A high resolution flow model of a 60 km^2 area straddling the pressure compartments is used to investigate the main parameters controlling pressure compartmentalisation. Single-phase hydrodynamic flow modelling demonstrates that the observed pressure distribution requires low across-fault transmissibilities and relatively high hydrodynamic flow rates (10 m^3/day). The most significant contributor to the high flow rates is gas generation and migration into the high pressure cell. Low fault transmissibilities are attributed either to shale smearing or, less likely, to extensive quartz cementation of the fault rock. Our study shows that pressure compartmentalisation in the North Sea can be controlled by relatively small displacement faults and highlights the importance of high resolution 3-D geological models in understanding overpressure distribution.

Introduction

Faults are commonly invoked as flow barriers between discrete pressure blocks but the character of block boundary faults is rarely described in detail in the literature. In general, faults separating blocks with a large pressure difference have large (100s of metres) throw. Here, however, we report a large (128 bar) difference in overpressure across a fault with a low (~50 m) throw which juxtaposes high permeability reservoir rocks and would not generally be expected to support a large pressure difference. We have mapped as accurately as possible the fault pattern between adjacent blocks and apply known fault parameter ranges to constrain single-phase flow modelling of the area. The results of this modelling have implications for basinal flow rates and fault permeabilities.

The study area

The study area lies on the eastern flank of the Viking Graben within Norwegian sector Block 30 (Fig. 1) and includes the southern part of the Tune gas field (Fig. 2). The Tune reservoir sequence is the Middle Jurassic Tarbert Formation with additional reserves in the Triassic Statfjord Formation. The Tarbert reservoir interval has a constant thickness of ~100 m throughout the study area and across major faults, indicating that faulting post-dated deposition of the Tarbert reservoir section. The reservoir interval comprises two main high permeability intervals, the Tarbert 2b–c and the Tarbert 2e–g, separated by an 8 m thick mudstone interval, the Tarbert 2d (Fig. 3). A third high permeability sand, the Tarbert 1b, is ~10 m thick.

Within the Tune Field a Tarbert gas–oil contact at 3577 m and an oil–water contact at 3589 m were established in the 30/5-2 well. The 30/8-1S well is gas-bearing while the 30/8-3 well to the east of the Tune Field is water-bearing. The crest of the Tarbert accumulation is at 3400 m.

Fluid pressures

RFT data are available for four wells within the study area (Fig. 4). The Tarbert section within the Tune Field is at 150 bar overpressure (wells 30/5-2 and 30/8-1S), while the adjacent fault block to the east is at ca 22 bar overpressure (30/8-3 and 30/9-19a). Wells 30/5-2 and 30/8-1S have identical pressures within the gas while the 30/8-3 and 30/9-19a wells have a 2 bar pressure difference. These data indicate that there are two distinct pressure cells, each without significant internal lateral pressure gradients, separated by a permeability barrier which supports a 128 bar pressure difference.

Hydrocarbon Seal Quantification edited by A.G. Koestler and R. Hunsdale. NPF Special Publication 11, pp. 187–201, Published by Elsevier Science B.V., Amsterdam.

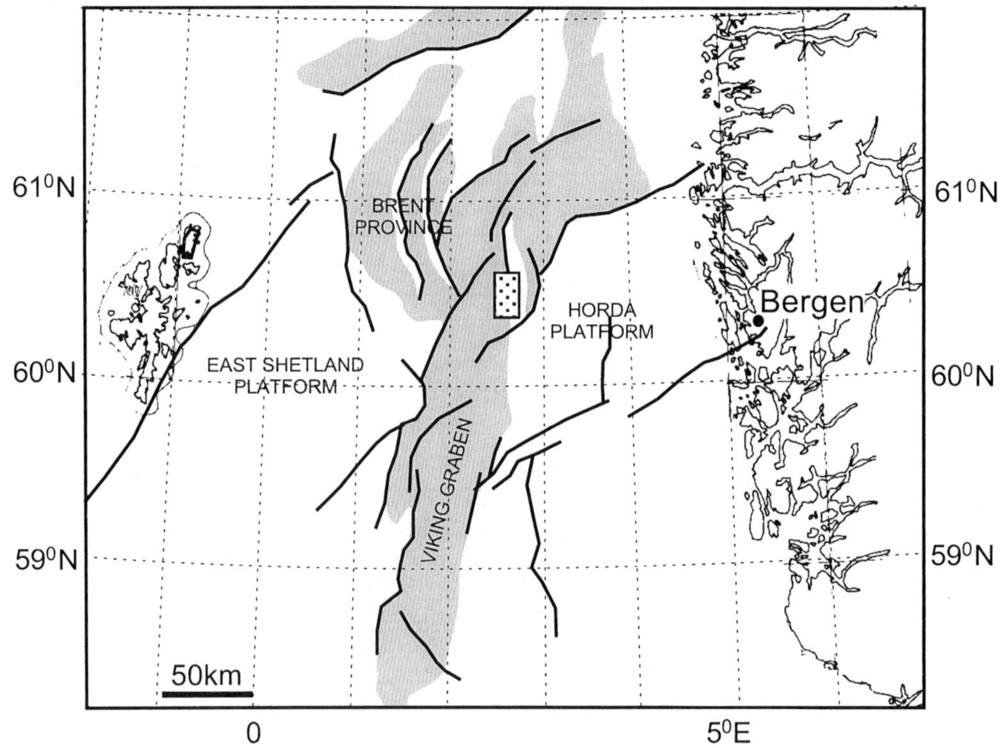

Fig. 1. Map of the northern North Sea showing the major structural features (after Stewart et al., 1992). The box outlines the area of the Tune Field map in Fig. 2 (filled box).

Faulting and potential flow barriers

The Tune Field lies within a 2–4 km wide block bounded to the east and west by large W-dipping N- to NNE-trending normal faults (Figs. 2 and 5). The eastern boundary fault comprises two left stepping segments separated by a region of complex minor normal faulting (Figs. 2 and 6). The zone of minor faulting is interpreted to be a relay zone which transfers displacement from the large fault segment in the north (throw up to 400 m) to the equally large segment to the south. The relay zone is characterised by a high density of variably oriented, relatively low throw (<100 m) faults. Given the low throws in the relay zone we expect that the magnitude of the pressure step across the eastern boundary fault is controlled by the relay zone and our study concentrates on this area.

The 128 bar pressure difference across the eastern segmented fault system is likely to be supported by the largest continuous fault trace connecting the two large relay bounding fault segments (Fig. 6). The Allan diagram for this fault (Fig. 7) shows that, in the region of lowest throw (ca 50 m) on this fault, the Tarbert reservoir interval is not offset across the fault. The throw is however sufficient to offset the high permeability Tarbert 2e–g and b–c intervals resulting in juxtaposition of the 2e–g on the downthrown side against 2b–c on the upthrown side, at throws of 50–90 m. These areas of juxtaposed high permeability reservoir sands are expected to be the hydrodynamically ‘weakest’ points on the eastern boundary fault and to control the magnitude of the pressure difference which can be supported between the two distinct pressure blocks. The area of juxtaposition of these reservoir units occurs 50–100 m above the oil–water contact, so that there are likely to be gas-bearing sands on the downthrown side juxtaposed against water-bearing sands on the upthrown side.

Flow modelling

Dissipation rate of excess pressure through a low permeability fault

To determine whether the observed pressure distribution can be modelled statically we have calculated the time required for the dissipation of a pressure pulse imposed on one side of a fault. The calculation is based on the diffusion equation formulated to describe the pressure in a medium as a function of distance and time, based on the condition (at time = 0) of an instantaneous increase in pressure on one side of a low diffusivity region (e.g. Deming, 1994, 1995; He and Corrigan, 1995; cf. Carslaw and Jaeger, 1959, Section 3.3).

Formulated for the problem of pressure dissipation across a low permeability fault, the important rela-

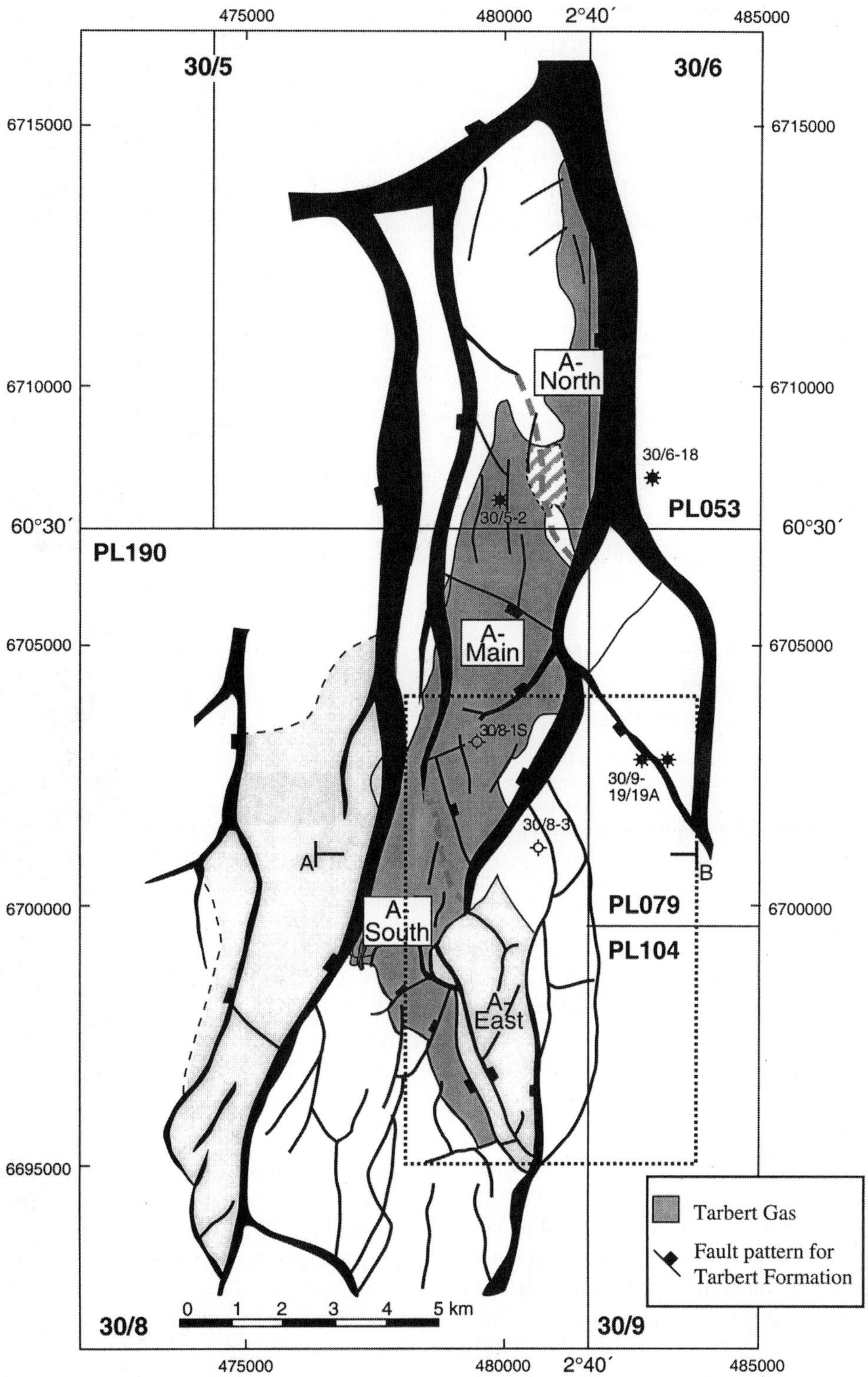

Fig. 2. Map of the Tune Field (for location see Fig. 1). The grey fill is the area of the gas accumulation. *A–B* indicates the location of the seismic line in Fig. 5. The map area in Fig. 6, which is the same as the model area in Fig. 9, is outlined.

tionships are that the time to dissipation (t) varies as:

$$t \propto \frac{1}{k}, \quad t \propto T^2 \text{ and } t \propto \alpha$$

where k is the permeability, T is the thickness and α is the compressibility of the fault rock. Although the thickness of the fault is therefore the most influential variable, it is also the one for which the best (order of magnitude) estimate may be made (see Childs et al., 1997; Walsh et al., 1998; Manzocchi et al., 1999, and references therein). For a 50 m displacement fault, we can say with some confidence that:

$$0.1 \text{ m} < T < 1 \text{ m}$$

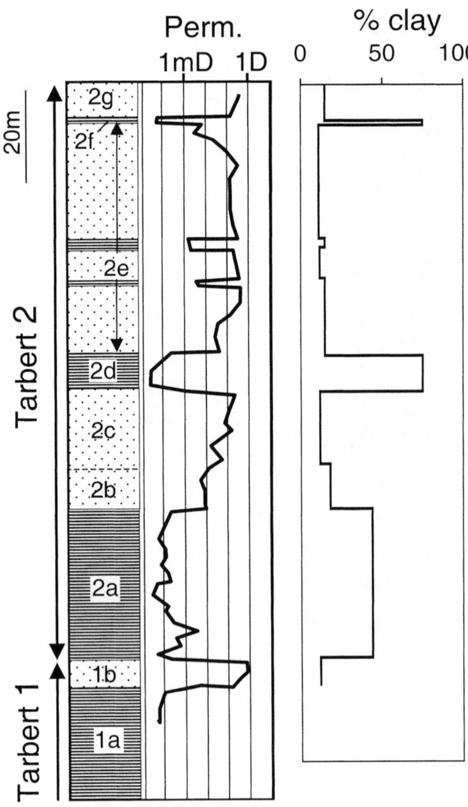

Fig. 3. The Middle Jurassic Tarbert reservoir sequence in the Tune Field. The log to the right of the sequence shows average core plug permeability determinations. The % clay curve on the right shows the values input to each of the units incorporated in the flow model.

The permeability (k) and compressibility (α) are more difficult to estimate for fault rocks, the most detrimental of which are likely to be shale-gouge fault rocks. Deming (1994, 1995) and He and Corrigan (1995), discuss the compressibility of shales, and reach the consensus that 10^{-8} Pa^{-1} is a likely upper limit, and that 10^{-9} Pa^{-1} is a 'most likely' compressibility for shale. Similarly, shale permeabilities lower than 10^{-11} D are considered unlikely, and the lowest fault rock permeabilities reported in the literature are ca 10^{-9} D (Chu et al., 1981; Morrow et al., 1984). For faults in sandstone, fault rock permeabilities are higher than for shaley rocks and compressibilities are likely to be lower, so that shaley fault rocks represent the slowest dissipation case.

Fig. 8 shows the time needed to dissipate 90% of the excess pressure through the fault as a function of the fault thickness, and the fault rock compressibility and permeability, using the slightly different assumptions used by He and Corrigan (1995) and Deming (1994). Using the optimally sealing set of properties implies that less than 10^5 years are needed to dissipate 90% of an excess pressure pulse through a fault. The values used for minimum permeability and maximum compressibility are based on shales, not shale gouges. However, without further data to better constrain the fault gouge properties, it seems unlikely

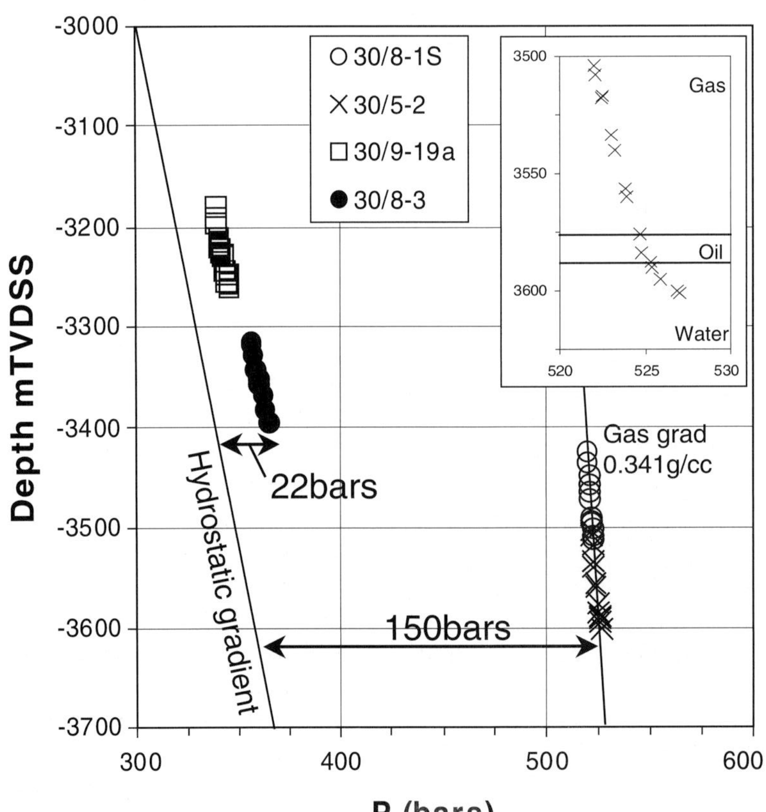

Fig. 4. RFT data for the wells in the Tune area. The inset figure is an enlargement showing the data for the 30/5-2 well.

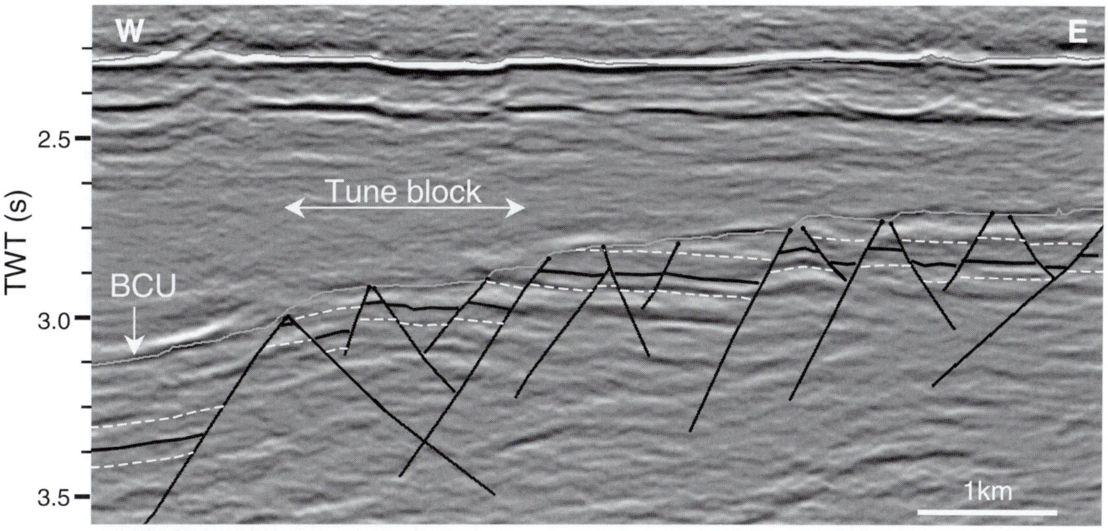

Fig. 5. Part of a seismic line across the Tune Field. The bold solid line is the Tarbert 2a horizon. The top and base of the reservoir section (white dashed lines) and the Base Cretaceous Unconformity (BCU) are shown. The location of the seismic line is shown in Fig. 2.

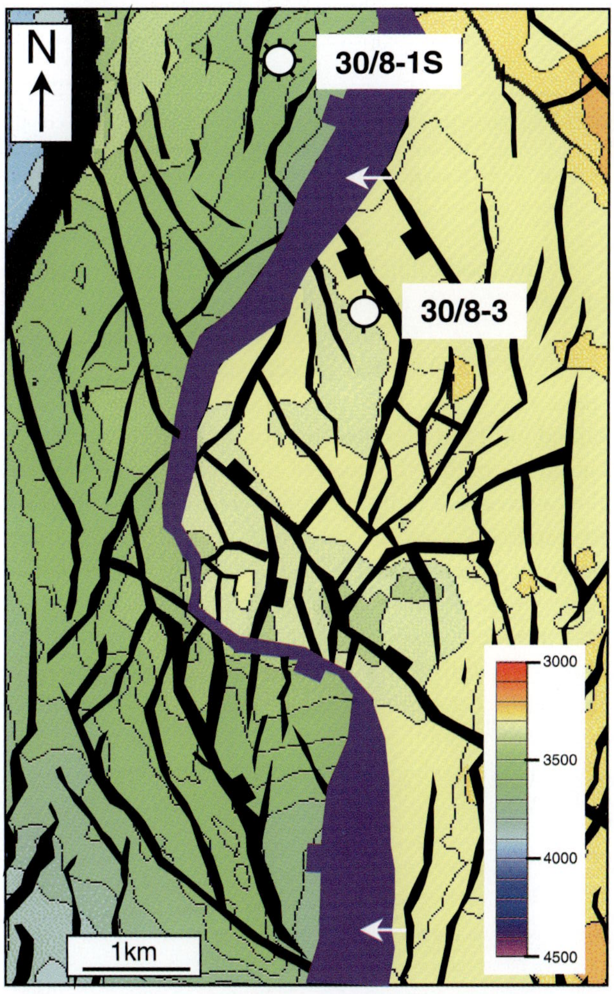

Fig. 6. Detailed fault map of the southern part of the Tune Field. The locations of the 30/8-1S and 30/8-3 wells are shown. The fault highlighted in blue is the trace of the largest continuous fault across the relay zone and is the most likely barrier to flow between the high pressure cell on the left and the low pressure cell on the right. The portion of the fault shown in the Allan diagram (Fig. 7) is between the arrows. The map area is shown in Fig. 2.

that static diffusion across low permeability faults is a slow enough process to support excess pressures over geologically significant times. The area has therefore been modelled hydrodynamically as described below.

Hydrodynamic model construction

A finite difference, Eclipse flow simulation model of a 6 km × 10 km area centred on the relay zone has been generated for the purpose of conducting single-phase (water) flow modelling (Fig. 9). The model contains 13 layers of 100 m square grid cells covering a 130 m thick interval of the Tarbert 1 and 2 intervals (Fig. 3). The model incorporates fault map geometries determined from detailed seismic mapping, and layer porosities, permeabilities and net : gross determined from wells within the area. The model also incorporates percentage clay determined from well logs calibrated to XRD and thin section point count analyses (Fig. 3). The faults are vertical and the fault map pattern is a discretised version of that shown in Fig. 6.

Transmissibility multipliers are attached to the faulted grid cell faces based on empirical estimates of fault rock permeability and thickness according to the method described in Manzocchi et al. (1999). Combining empirical estimates of fault rock thickness, derived from fault displacement, and fault rock permeability, derived from displacement (D) and the clay fraction in the sequence which has slipped past a point on a fault (shale gouge ratio or SGR), gives an estimate of the permeability at a point on a fault surface. By combining fault permeability and thickness with the permeabilities, sizes and net : gross ratios of the adjacent grid blocks in the flow simulator, a fault

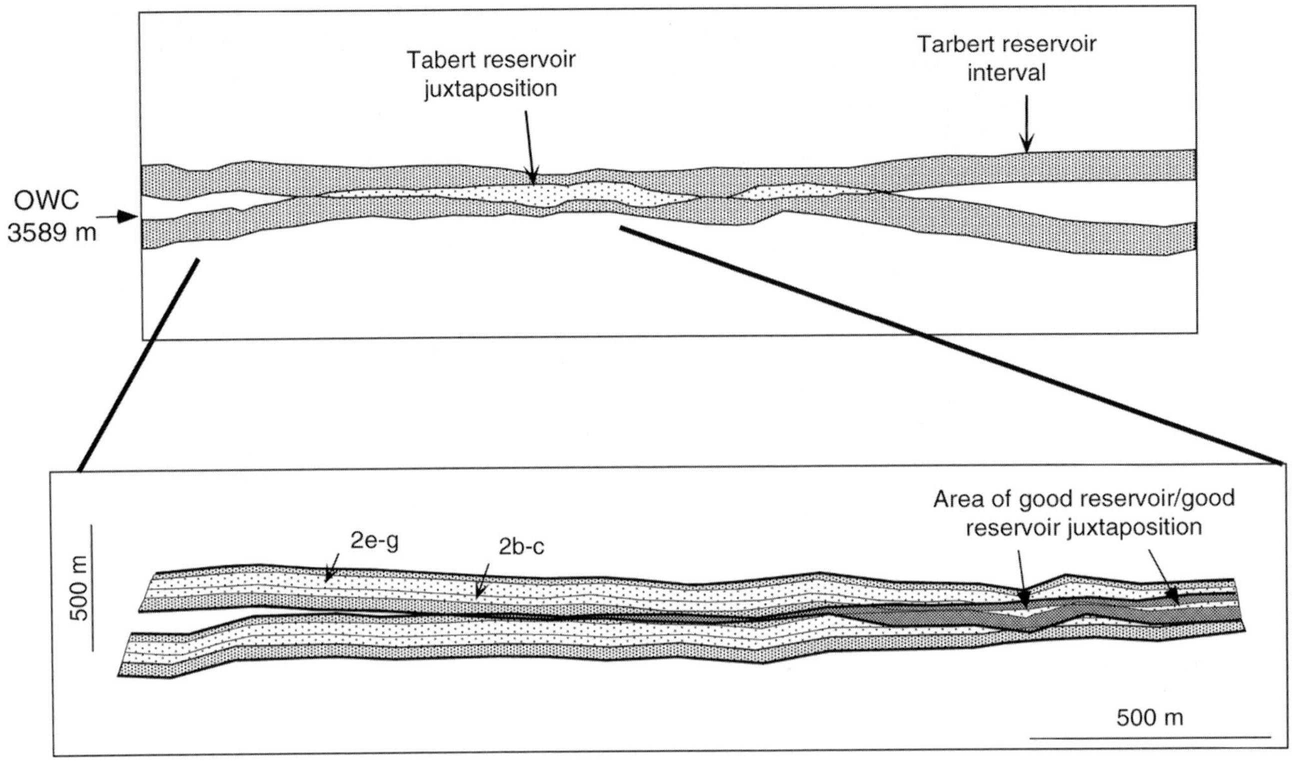

Fig. 7. Allan diagram for the fault coloured blue in Fig. 6 viewed from the west. This shows that at the area of minimum throw (ca 50 m) on the fault, the reservoir sequence is juxtaposed across the fault (coarse stipple). In the enlarged inset, high permeability reservoir rocks 2(e–g) and 2(b–c) are shown (coarse stipple). Areas where high permeability reservoir intervals are juxtaposed are shown (arrows).

transmissibility multiplier is assigned to connections between faulted grid blocks. The advantage of this method of assigning fault transmissibilities is that the values have a geological significance and the flow modelling results can therefore be interpreted in geological terms. In an attempt to match the present-day pressure distribution within the modelled area, five transmissibility multiplier realisations were applied by varying the SGR to fault rock permeability relationship. The five realisations included a datum case, where the faults all have a transmissibility multiplier of 1, i.e. no fault properties are assigned, and four transmissibility multiplier cases derived from the relationships shown in Table 1. These relationships are shown graphically in Fig. 10, superimposed on laboratory fault rock permeability data.

In the modelling, water is injected at the location of well 30/8-1S and produced from the 30/8-3 well (Fig. 9) at variable flow rates which maintain constant pressures equal to the present-day pore pressure in these two wells. For each model run, flow is maintained until the pressure distribution within the model stabilises so that the rates of injection, production and flow across the fault are equal. The time for the flow to stabilise varies for the different model realisations between 1 year and 10 years. To determine the sensitivity of model results to the location of the injector and producer wells, a second boundary condition is employed with water injected into a notional well in the southwest corner of the model and produced from a well in the southeast corner (Fig. 9). The rates of injection and production are constant and equal to those

TABLE 1

Formulae used to calculate fault permeabilities for each of the four fault property cases, and the flow rates at which each model run stabilised

Case	Fault thickness (m)	Fault permeability (mD)	Flow rate (m^3/day)
Datum	–	–	2073.17
1	$T = D/170$	$\log k = 0.4 - 4\text{SGR} - (0.25 \log D)(1 - \text{SGR})^5$	2000.00
2	$T = D/170$	$\log k = -2 - \text{SGR} - (0.3 \log D)$	402.36
3	$T = D/170$	$\log k = -1 - 4\text{SGR} - (0.25 \log D)(1 - \text{SGR})^5$	1014.36
4	$T = D/170$	$\log k = -2 - \text{SGR} - (0.3 \log D)$	10.30

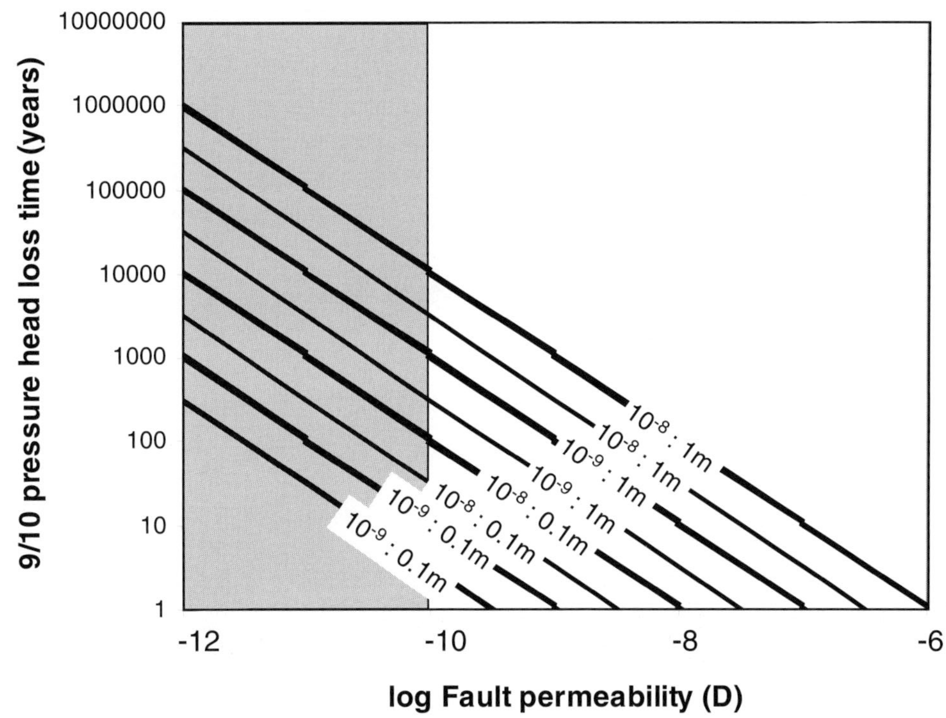

Fig. 8. Graph showing the time required to dissipate 90% of a pressure pulse imposed on one side of a fault versus fault permeability calculated for a fluid viscosity of 1 cp and a matrix permeability of 10 mD. Curves were calculated from equations given in He and Corrigan (1995), shown with solid lines, and Deming (1994) shown with grey lines. Lines are labelled with compressibility and fault rock thickness.

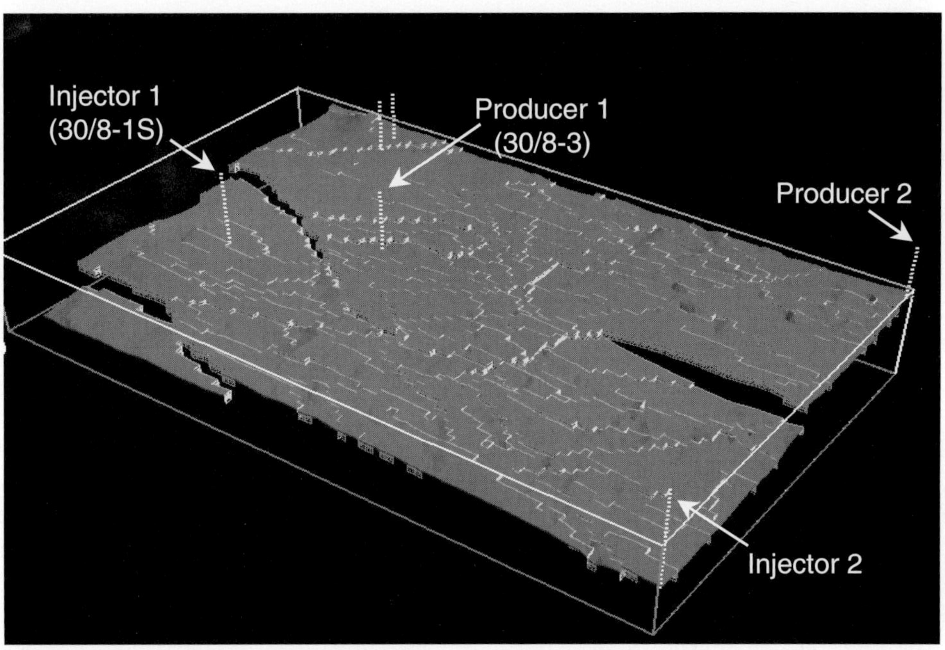

Fig. 9. 3D view of the Eclipse model viewed from the southwest. The area covered by the model is shown in Fig. 2.

determined from the initial flow boundary conditions. The resultant pressure distribution for fault property four is very similar to that obtained by injection and production into the wells, indicating that at low fault permeabilities, the pressure distribution and across-fault flow rate are largely insensitive to the boundary conditions.

Results

The motivation behind the flow modelling is to match the pressure distribution within the modelled area. The pressure in well 30/8-1S has the same pressure as well 30/5-2 (immediately outside the model area to the north) and 30/8-3 has approximately the same pressure as 30/9-19, so that the pressures over

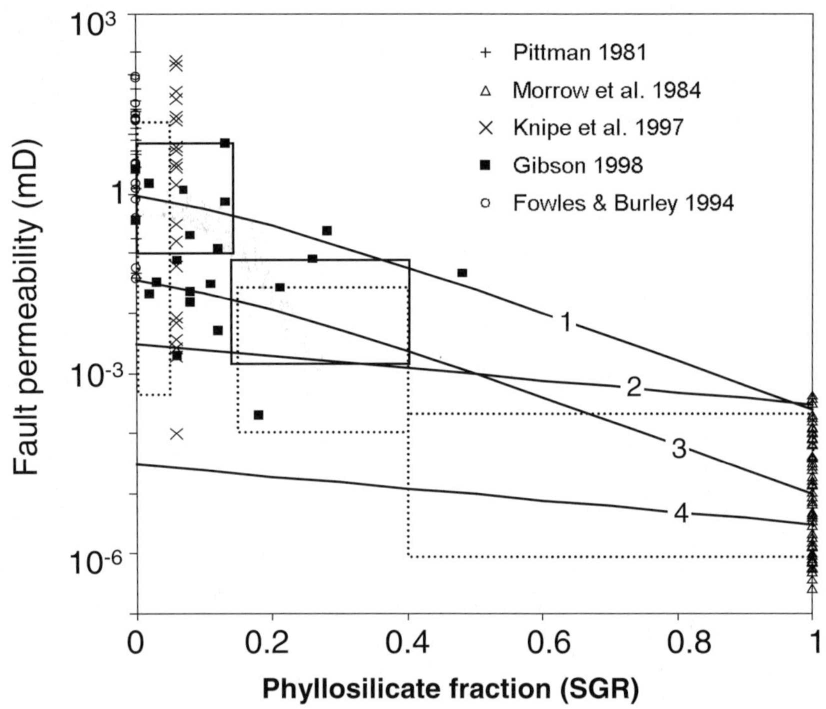

Fig. 10. Graph of fault permeability versus percentage phyllosilicate showing the four relationships (Cases 1 to 4, see Table 1) used to define the four fault transmissibility multiplier realisations for $D = 50$ m. The rectangles are data ranges from Fisher and Knipe, 1998 (dotted lines) and Ottesen Ellevset et al., 1998 (solid lines).

large areas on either side of the fault appear to be constant. Our assessment of the modelling results is based on the accuracy with which the realisations match this pressure distribution.

The pressure distribution for the four flow modelling realisations is shown in Fig. 11. While each realisation, by definition, satisfies the condition that the pressures at the injector and producer wells are equal to the measured pore pressure, only Case 4 provides approximately uniform pressures within the individual fault blocks. The high permeability realisations (i.e. no fault properties and Case 1) provide an approximately uniform pressure gradient between the two wells (Fig. 11). We therefore conclude that extremely low across-fault transmissibilities (Case 4) are required to match the known pressure distribution.

For Case 4 there is not a strong correlation between SGR and fault permeability so that permeabilities are approximately constant and lie within the range of published measurements for pure clays (Fig. 10). SGR values in the areas of juxtaposition of the high permeability reservoir units are between 0.18 and 0.25. At these SGR values the permeabilities used in Case 4 are 1 to 2 orders of magnitude lower than the lower range of the published permeability data, and 3 to 4 orders of magnitude lower than the average of the published data (Fig. 10). Possible causes of a discrepancy between the assigned transmissibility multipliers and the published SGR/fault rock permeability data are discussed later.

The flow rates required to maintain the pressure drop between the wells for each of the fault property cases are shown in Table 1. The range of across-fault flow rates is between 10 and 1000 m^3 per day, but the actual value must be close to the lower limit of this range as discrete pressure blocks are not achieved in the higher permeability cases. Therefore the minimum flow rate required to sustain the observed pressure distribution is 10 m^3/day.

Two-phase effects

The low fault transmissibility multipliers required to maintain the discrete pressure cells in Tune could be due to two-phase effects not considered in our single-phase flow modelling. The area of juxtaposition of the Tarbert reservoir interval on the largest continuous fault separating the two pressure cells lies ca 50–100 m above the Tune oil–water contact (Fig. 7). The reservoir rocks on the downthrown side of this fault will, due to the presence of hydrocarbons, have lower effective permeabilities to water than those incorporated in the single-phase flow modelling. Decreasing the effective permeability of the downthrown reservoir rocks allows the transmissibility multipliers attached to the faults to be increased while maintaining the observed pressure distribution. The amount by which the transmissibility multipliers can be increased has been estimated.

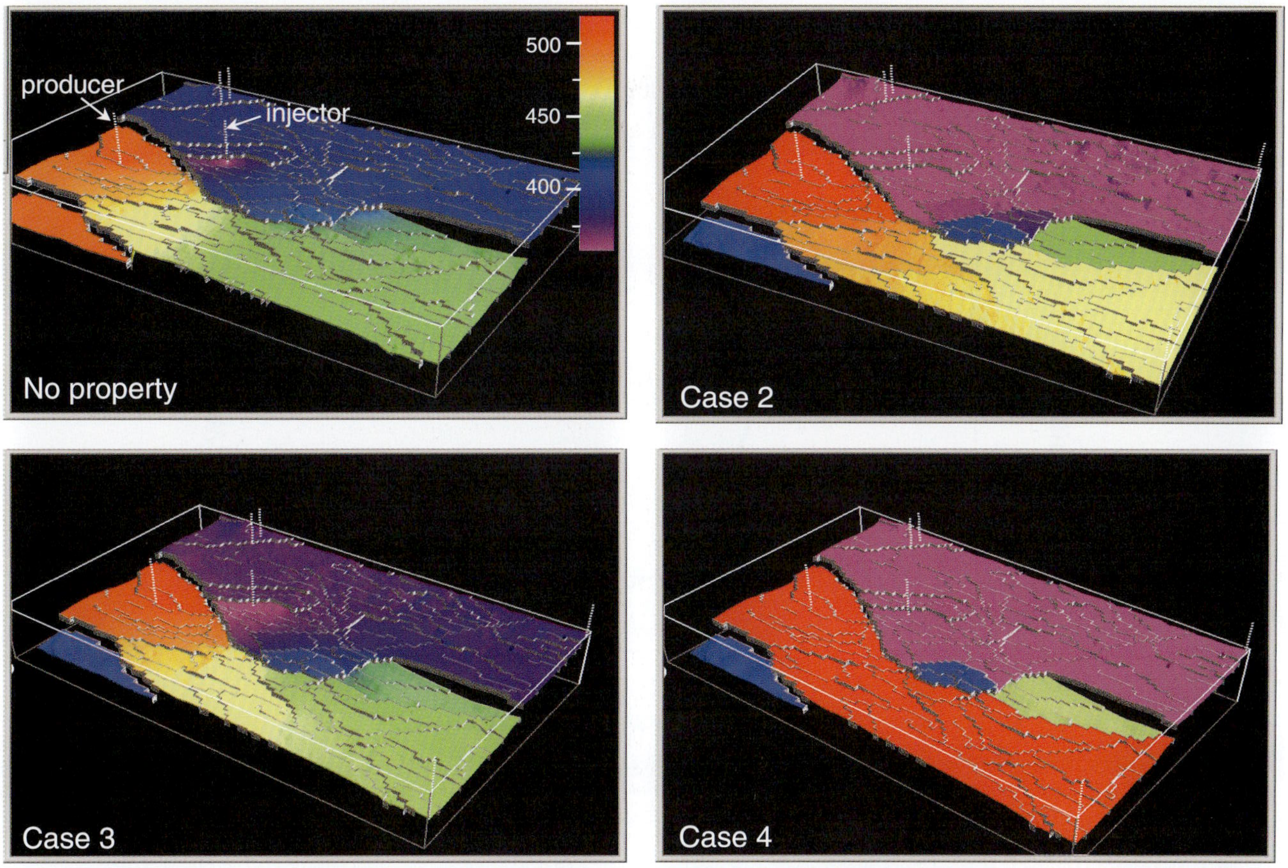

Fig. 11. Final pore pressure distribution (in bars) for the datum, i.e. no fault property case, and three model runs including fault properties (for details, see Table 1). Pressure distribution for Case 1 (not shown) is almost identical to the no property case.

The variation in water saturation (S_w) with height above the OWC (H) was calculated from the 'J' Leverett curve for the Tune reservoir as follows:

$$S_w = 0.34932 \left[\frac{H}{146.6} \sqrt{\frac{k}{\phi}} \right]^{1/-3.04}$$

for permeability (k) and porosity (ϕ). The 'J' curve is based on well log saturations and special core analysis and is applied in the field reservoir simulator with an irreducible water saturation of 0.1 saturation units. The results of special core analysis using reservoir brine and gas on high permeability cores are shown in Fig. 12. Special core analysis indicates that the capillary pressure endpoint for 100–200 mD Tune reservoir rocks occurs at capillary pressures greater than 9 bar. The capillary pressure due to 50–100 m of gas column on the downthrown side of the boundary fault is <6 bar (Tune gas density = 0.341 g/cm^3) so that a continuous and movable water phase will exist at all levels in the accumulation from the OWC to the area of juxtaposition of the reservoir interval.

Relative permeability curves for the high permeability reservoir rocks within the Tarbert Formation in the Tune area are not available. In the absence of these data we have calculated relative permeability to water using the equation (Archer and Wall, 1986)

$$K_{rw} = S_w^4 \times \sqrt{\frac{S_w - S_{wi}}{1 - S_{wi}}}$$

based on an irreducible water saturation (S_{wi}) of 0.1. The calculated effective water permeability for a 100 mD sandstone is 0.029 mD at 50 m above the OWC in Tune and 0.009 mD 100 m above the contact (Fig. 13). In estimating the effective water permeability from the base of the hydrocarbon accumulation to the area of across-fault juxtaposition of the Tarbert reservoir, it is the harmonic mean of the permeabilities over this interval which is important. The harmonic mean is 0.074 mD at 50 m and 0.024 mD at 100 m above the OWC (Fig. 13).

The significance of the reduced effective water permeability on across-fault transmissibilities is illustrated in Fig. 14. This plot shows curves of the effective water permeability of the gas column over the vertical interval (V, Fig. 14 inset) between the oil–water contact and the area of across-fault juxtaposition, against the fault permeability required to maintain the observed pressure distribution. The curves assume that a fault permeability of 0.00001 mD and thickness of 0.3 m is required at a reservoir permeability of 100 mD in the single-phase case, as

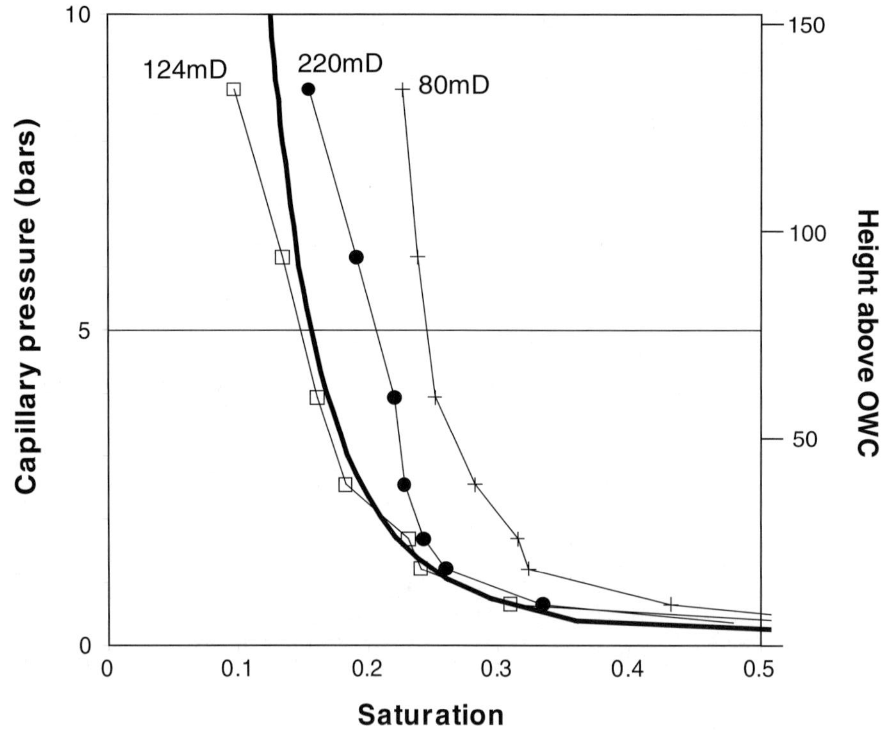

Fig. 12. The results of special core analysis on reservoir sandstones with permeabilities between 80 and 220 mD. The curve shows the variation in saturation with capillary pressure for a 100 mD and 20% porosity sandstone as calculated from the 'J' curve for the Tune reservoir. The axis on the right of the graph is equivalent height above the base of the hydrocarbon accumulation.

indicated by the results of our flow modelling. The curves are drawn for different lateral distances (L) from the fault to the oil–water contact. For $V = 10$ m, the required fault permeability is close to that re-

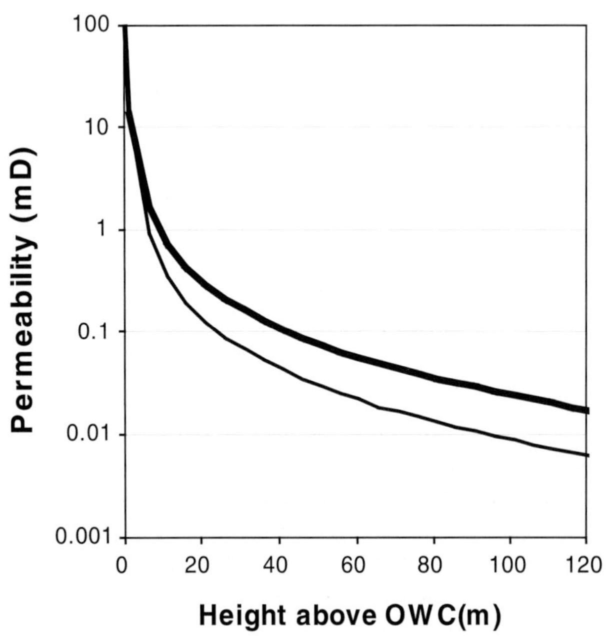

Fig. 13. Graph of calculated effective water permeability of a 100 mD sandstone against height above the base of the hydrocarbon column in Tune (narrow line). The bold line is the harmonic average of effective permeability estimated over the vertical interval down to the oil–water contact.

quired in the absence of gas. However, for V between 50 m and 100 m, the fault permeability required to maintain the pressure distribution rises rapidly, and in certain cases exponentially depending on the magnitude of L. In the area of across-fault juxtaposition of the Tarbert reservoir, the seismic data are not high quality and there are uncertainties in depth conversion and therefore in estimates of both V and L; these values lie somewhere between 50–100 m and 50–500 m, respectively. If $V = 50$ m then fault permeabilities can be raised by only a factor of 1.022 (flow path i, Fig. 14). Where $V = 100$ m and L is 500 m, then the fault permeabilities can be raised by a factor of 1.485 to 3.964, depending on whether flow follows path ii or iii (Fig. 14, table). If L is 1000 m, a value exceeding that indicated by the seismic interpretation, then the fault permeabilities could be raised by several orders of magnitude (flow path iv) into the range expected from published fault permeability data, at gas column heights of ~80 m (Fig. 14). The significance of this case is discussed later. Our calculations therefore indicate that the fault permeabilities required to match the observed pressure distribution can be increased by at most a factor of ~4 (flow path iii, Fig. 14) due to the presence of the gas column.

In the extreme case where the relative permeability to water within the gas column is zero, the gas column acts as a complete seal to the flow of water both within the downthrown reservoir and across the fault.

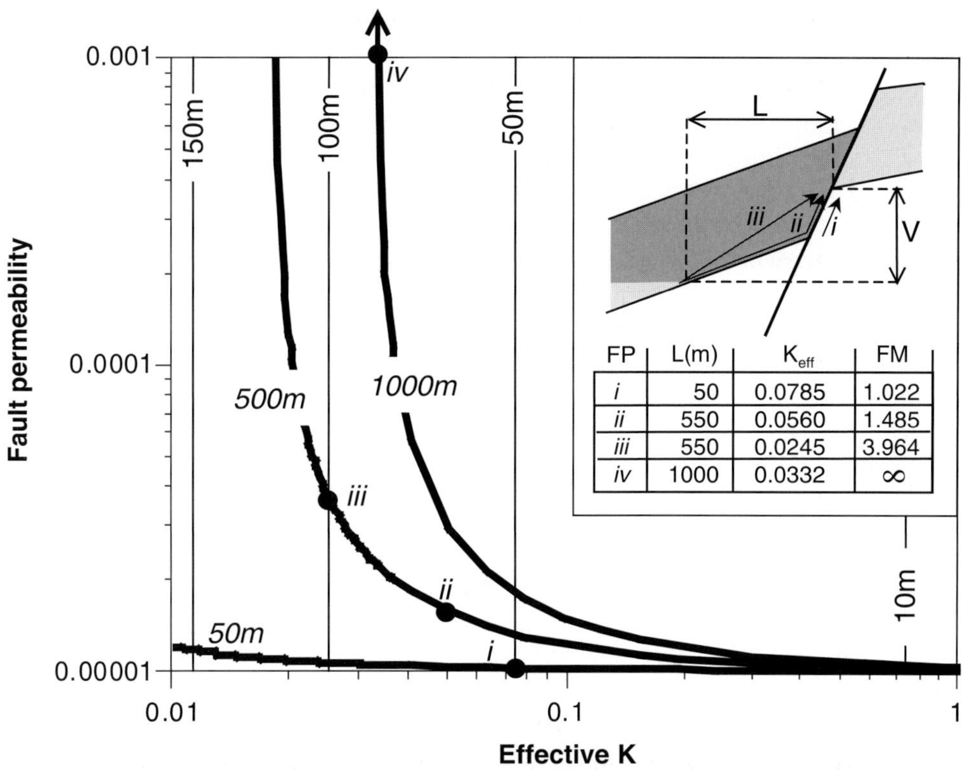

FP	L(m)	K_{eff}	FM
i	50	0.0785	1.022
ii	550	0.0560	1.485
iii	550	0.0245	3.964
iv	1000	0.0332	∞

Fig. 14. Graph of effective water permeability of reservoir rocks versus the fault permeability required to match the pressure distribution in the model area (see text). The vertical lines are contours of the effective water permeability of the gas accumulation at different heights above the oil–water contact. The curves are for different lateral distances (L) from the oil–water contact to the area of across-fault reservoir juxtaposition. Shown as large dots are flow paths i to iv indicating the minimum possible effective permeabilities for each curve; flow paths i to iii are illustrated in the inset figure. The distance (V) in the inset is the vertical interval between the oil–water contact and the area of across-fault reservoir juxtaposition. Flow path i is a straight path through the gas column for $V = 50$ m. Flow paths ii and iii are for $V = 100$ m. Flow path ii follows the highest permeability route through the gas, i.e. along the base of the reservoir interval, and flow path iii is a straight line from the oil–water contact to the area of reservoir juxtaposition. Flow path iv is directly through the gas column for $V = 100$ m but a lateral distance (L) of 1000 m. The inset table gives the flow path label (*FP*), flow path length (L), the effective water permeability along the flow path (K_{eff}) and the corresponding fault property multiplier (*FM*) for flow paths i to iv. The fault property multiplier is the factor by which the fault permeability can be increased due to the decrease in effective permeability of the reservoir.

This implies however that the pressure difference across the fault is maintained by capillary forces at the reservoir/fault interface. The capillary pressure in such a situation is the sum of the buoyancy pressure due to the gas column (3–6 bar) and the pressure difference across the fault in the water phase (128 bar) so that the capillary threshold pressure of the fault gouge must be greater than 130 bar, i.e. equivalent to a trapped gas column of ~2 km. While fault rock capillary threshold pressures of this magnitude have been measured in the laboratory, it is unlikely that such high values occur over significant fault surface areas. A 130 bar threshold pressure is, for example, over 15 times higher than estimated static fault seal capacities in the Oseberg South area (Fristad et al., 1997) immediately to the east of Tune. Furthermore, the special core analysis data (Fig. 12) indicate that a finite relative permeability to water exists on the downthrown side of the fault at heights of >100 m above the base of the hydrocarbon column. Bjørkum et al. (1998) argue that in a water-wet reservoir, a continuous water phase will always exist so that there will be a finite permeability to water at all levels within a hydrocarbon accumulation. In this case the degree of overpressure does not contribute to capillary pressure which is therefore equivalent to the buoyancy pressure of the trapped hydrocarbon. We conclude therefore that the observed pressure difference is not due to the capillary properties of the reservoir and fault rocks, and is not static, but is governed by hydrodynamic flow, perhaps coupled with relatively minor two-phase effects in the gas column.

Although seismic data indicate that it is not the case, it is interesting to consider the situation where the vertical and lateral distances from the OWC to the area of Tarbert juxtaposition are greater than 100 m and 1000 m, respectively. In this situation the observed pressure difference could be attributed to reduced transmissibility due to two-phase effects (Fig. 14, flow path iv) and fault permeabilities would be as high as those predicted from experimental measurement. The presence of the gas column would have a significant effect on overpressure distribution in providing a thick zone of low effective water permeability capable of supporting a large pressure gradient in a hydrodynamic regime. In this situation the stability of the gas column depends only on the capillary pressures at the interface between the gas-filled reservoir and the water-wet fault rock. In the area studied here this capillary pressure is between 3 and 6 bar and well within known fault rock threshold pressures.

While the effects of overpressuring on seal capacity have been widely discussed in the literature (e.g. Clayton and Hay, 1994; Heum, 1996; Bjørkum et al., 1998; Clayton, 1999) there are, to our knowledge, no descriptions of the effects of hydrocarbon accumulation on overpressure distribution. Multiphase effects on overpressure distribution has been previously discussed by Snowdon (1995). The effect considered, however, is only the reduced total permeability, i.e. the summed effective permeability of each of the phases, which for a three-phase system may be a factor of five lower than the intrinsic permeability. In the situation postulated here, the gas phase, which occupies the majority of the pore space, would be capillary sealed and static, so that only the water phase would flow with an effective permeability over three orders of magnitude less than the intrinsic permeability. An implication of this discussion is that thin caprocks may provide capillary seal to a hydrocarbon accumulation which can significantly impede the upward flow of water through the accumulation and give rise to pressure differences which would not be predicted from the seal characteristics.

Flow rate estimation

The pressure distribution within the study area cannot be modelled statically and the boundary between the two overpressured cells cannot be considered as a 'no-flow' boundary. Flow modelling indicates that the high degree of overpressuring within the Tune block and the large pressure drop to the adjacent block must be maintained by a process which generates pore fluids within the Tune fault block at a rate of at least 10 m^3/day. These high flow rates are required irrespective of whether the low transmissibilities are due to extremely low across-fault transmissibilities or to two-phase effects. Many of the processes invoked as sources of overpressure in 'perfectly sealed' rock volumes, e.g. diagenetic and aquathermal expansion, are therefore not applicable here. We describe below three likely processes which may contribute to the high flow rates and we have made a first order estimate of their likely contribution.

Cracking of oil to gas

The thermal cracking of oil to gas can result in a large pore fluid volume expansion particularly at the depth of burial of the Tune Field. The fraction of oil converted to gas at time t is given by:

$$F = 1 - \exp[-\phi(t)]$$

where

$$\phi(t) = \frac{A}{GS}\left[\frac{T(t)\exp(-E/RT(t))}{2+E/RT(t)} - \frac{T_i\exp(-E/RT_i)}{2+E/RT_i}\right]$$

G = geothermal gradient, 30°C/km); S = sedimentation rate, 0.067 km/Ma); $A = 1 \times 10^{28}$; $R = 1.98$ cal/mol K; T_i = 297 K; E = activation energy, 52,000 cal/mol (Berg and Gangi, 1999; Carcione and Gangi, 2000).

One volume of oil cracks to 534.3 volumes of gas at standard temperature and pressure (Barker, 1990). For an initial oil volume of 6×10^6 m^3, the present oil volume in Tune Field, the volume of gas generated during burial has been calculated and converted to in situ pressure and temperature conditions. The rate of increase in pore fluid volume due to gas generation for a subsidence rate of 150 m/my is shown in Fig. 15a. The peak rate of gas generation is 0.00225 m^3/day and occurs at a burial depth of ca 3.6 km, the present-day burial depth of the Tune reservoir. This rate of volume increase is more than three orders of magnitude lower than the minimum flow rate required by flow modelling. Cracking of oil to gas is therefore thought to make a negligible contribution to the flow rates in the study area.

Basin/disequilibrium compaction

Detailed basin modelling to evaluate the rates of mechanical compaction, porosity loss and fluid expulsion in the Tune area are beyond the scope of this paper. Rather, we have calculated the rates of pore fluid expulsion from a compacting sand/shale sequence from standard porosity–depth functions (Sclater and Christie, 1980). The calculations assume that all of the fluids expelled from the compacting sequence are channelled into the Tarbert reservoir section. Fig. 15b shows the rate of fluid expulsion versus drainage area from a compacting sequence of variable thickness and under different rates of subsidence. An average subsidence rate for the last 4 my within the Tune area is ca 150 m/my (black lines in Fig. 15b). The map area of the Tune fault block is ca 125 km^2 so that the rate of fluid expulsion is between 0.2 and 2 m^3/day for a compacting sequence of 0.1 to 1 km. The highest expected fluid expulsion rate is 20%

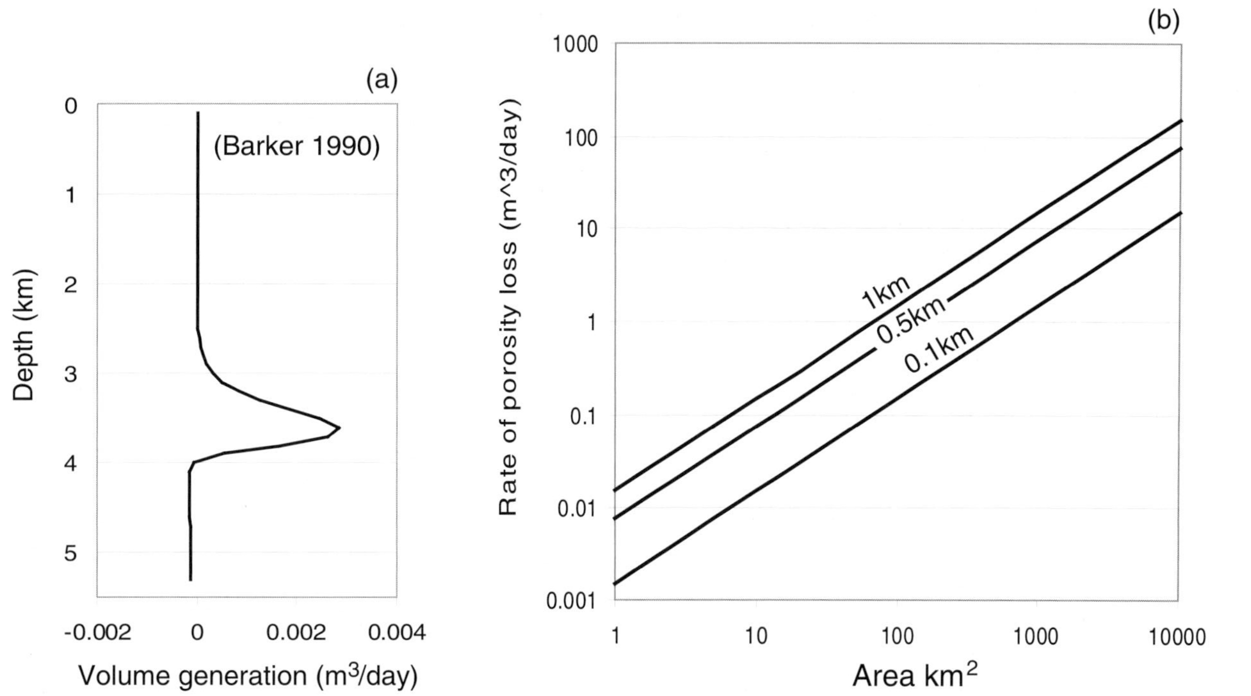

Fig. 15. (a) Rate of volume generation due to cracking of oil to gas against depth, calculated for a geothermal gradient of 30°C/km and sedimentation rate of 0.15 km/Ma. (b) Curves of porosity loss in a compacting sequence buried from 3 to 3.5 km against drainage area. The curves are for compacting sequences of 0.1, 0.5 and 1 km and a subsidence rate of 0.15 km/Ma.

of the 10 m^3/day minimum flow rate required from the conclusions of our single-phase flow modelling. However, if the Tune block is in hydraulic communication with the deeper parts of the Viking Graben to the west then this drainage area and therefore fluid expulsion rate could be greatly increased.

Gas generation

Gas generation is believed to be a major source of overpressure in the Viking Graben (Buhrig, 1989). Basin modelling carried out by Norsk Hydro indicates that peak gas migration into the Tune Field occurred 1–2 Ma but probably initiated 2–3 my earlier. The present-day volume of gas within the Tune Field is 45×10^9 m^3. This volume, averaged over a 2 my period, yields an average rate of volume generation of 63 m^3/day. Although the present-day rate of migration into Tune is probably significantly less than this average value, the present-day rate of gas accumulation could be of the order of the 10 m^3/day derived from the single-phase flow models. We conclude that the migration of gas into the Tune Field is the most likely source of high flow rates within the area.

Discussion

Our calculations indicate that the two-phase effects of the gas column on the downthrown side of the fault separating the distinct pressure cells can be ignored and the observed pressure distribution is therefore due to low across-fault transmissibilities. Low across-fault transmissibilities indicate that fault rock permeabilities are very low and/or the thickness of fault rock is higher than would be predicted from published data. We consider that there are two possible causes of low across-fault transmissibilities in the Tune area. The first possibility is that the 8 m thick mudstone unit towards the centre of the Tarbert reservoir interval (Tarbert 2d, Fig. 3) provides a continuous shale/clay smear over those areas of the fault where there is juxtaposition of the high permeability Tarbert 2b–c and 2e–g reservoir intervals. The depth of burial at the time of formation of the faults within the Tarbert Formation was low (<200 m) which would promote the formation of shale smears (Lehner and Pilaar, 1997). Pure clay/shale smears would be expected to have permeabilities in the range 10^{-7} to 10^{-9} D and are low enough to explain our model results (Fig. 10). An alternative explanation is that the relay zone between the two main strands of the fault separating the pressure cells is characterised by large numbers of subseismic faults. Fault population studies, carried out from seismically resolvable faults, indicate that the density of subseismic faults within the relay zone may be 8 times higher than outside for faults with throws of >1 m and 50 times higher for faults >1 cm. If these faults have been subjected to quartz cementation, the resulting low permeabilities combined with anomalously high fault rock thicknesses could provide low fault transmissibilities. Evidence of pressure solution and precipitation of quartz cements has been found in the 30/5-2 well and is expected to occur at the ca 120ºC Tune reservoir temperature. However, given the shallow depth of formation of the faults it is unlikely that there has been significant cataclasis of quartz grains within the faults which would otherwise further enhance quartz cementation.

A detailed study of across-fault fluid pressures was conducted in the Oseberg Syd area immediately to the east of the Tune Field (Fristad et al., 1997). Here the largest across-fault pressure difference encountered was 9 bar and the authors found that considerable fault seal is expected at SGR > 0.18. For the sequence within the Tune area, SGR values at the areas of reservoir juxtaposition are everywhere >0.18 for faults with throws greater than 50 m, and our result is broadly in agreement with those of Fristad et al. (1997). However, their analysis was directed primarily at deriving static pressure gradients due to hydrocarbon accumulation. We believe that the large pressure gradient across the fault(s) studied here is of dynamic origin and is irrelevant to the estimation of the potential of the fault to seal hydrocarbons. It is possible that the ca 3–6 bar capillary pressure due to the presence of 50–100 m of gas below the point of reservoir juxtaposition across the fault represents the capillary seal capacity of the fault.

The high pressure gradient across the fault(s) in the Tune area requires that the Tune block is dynamically charged by incoming pore fluid, either water from compacting sediments, or gas generation. While a large pressure difference can theoretically be achieved by having either very low fault transmissibilities or very high fluid flow rates, the uniformity of pressure within the high and low pressure blocks precludes the second option. Our modelling therefore places an approximate upper limit on fault transmissibility and constrains the flow rates required to honour both the pressure step across the fault, and the internal pressure distribution within the two blocks. The pressure distribution within the study area could be matched by even lower fault transmissibilities with correspondingly lower across-fault flow rates. However, fault rock property data suggest that the fault permeability to thickness ratio (the critical factor) cannot realistically be much lower than our preferred model (Case 4). While we cannot altogether dismiss two-phase effects, our considerations show that the results from the single-phase modelling described, are sufficient to characterise the nature of the pressure distributions in the Tune area, and to crudely quantify both the fault properties and hydrodynamic flow rates.

Conclusions

(1) A pressure step of 128 bar occurs across a fault with a throw of ~50 m which juxtaposes high permeability reservoir sandstones.

(2) Flow modelling indicates that relatively high fluid flow rates and extremely low transmissibilities are required to maintain the pressure gradient for geologically significant periods.

(3) The most likely cause of the high flow rate is gas generation.

(4) Low transmissibilities are most likely due to fault rock properties, possibly exacerbated by a reduction in effective reservoir permeability due to a gas accumulation on the downthrown side of the fault.

(5) Low across-fault transmissibilities may result from extremely low fault rock permeabilities (clay smear) or anomalously wide quartz cemented fault zones.

Acknowledgements

Thanks to the staff at the Norsk Hydro Research Centre, Bergen for provision of data and particularly to Hans Helle for useful discussion throughout the project. This work was funded by the EC Joule III Programme (Contract No. JOF3-CT97-0036). We are grateful to A.G. Milnes for his helpful review of the manuscript.

References

Archer, J.S. and Wall, C.G., 1986. Petroleum Engineering. Principles and Practice. Graham and Trotman, London.

Barker, C., 1990. Calculated volume and pressure changes during the thermal cracking of oil to gas in reservoirs. Am. Assoc. Pet. Geol. Bull., 74: 1254–1261.

Berg, R.R. and Gangi, A.F., 1999. Primary migration by oil-generation microfracturing in low-permeability source rocks: application to the Austin Chalk, Texas. Am. Assoc. Pet. Geol. Bull., 83: 727–756.

Bjørkum, P.A., Walderhaug, O. and Nadeau, P.H., 1998. Physical constraints on hydrocarbon leakage and trapping revisited. Pet. Geosci., 4: 237–239.

Buhrig, C., 1989. Geopressured Jurassic reservoirs in the Viking Graben: modelling and geological significance. Mar. Pet. Geol., 6: 31–48.

Carcione, J.M. and Gangi, A., 2000. Gas generation and overpressure: effects on seismic properties. Geophysics, 65: 1769–1779.

Carslaw, H.S. and Jaeger, J.C., 1959. Conduction of Heat in Solids. Oxford University Press.

Childs, C., Walsh, J.J. and Watterson, J., 1997. Complexity in fault zone structure and implications for fault seal prediction. In: P. Møller-Pedersen and A.G. Koestler (Editors), Hydrocarbon Seals: Importance for Exploration and Production. Norwegian Petroleum Society (NPF), Special Publication 7. Elsevier, Amsterdam, pp. 61–72.

Chu, C.L., Wang, C.-Y. and Lin, W., 1981. Permeability and frictional properties of San Andreas Fault gouges. Geophys. Res. Lett., 8: 565–568.

Clayton, C.J., 1999. Discussion: physical constraints on hydrocarbon leakage and trapping revisited. Pet. Geosci., 5: 99–101.

Clayton, C. and Hay, S.J., 1994. Gas migration mechanisms from accumulation to surface. Bull. Geol. Soc. Den., 41: 12–23.

Deming, D., 1994. Factors necessary to define a pressure seal. Am. Assoc. Pet. Geol. Bull., 78: 1005–1009.

Deming, D., 1995. Factors necessary to define a pressure seal — Reply. Am. Assoc. Pet. Geol. Bull., 79: 1079–1081.

Fisher, Q.J. and Knipe, R.J., 1998. Fault sealing processes in siliclastic sediments. In: G. Jones, Q.J. Fisher and R.J. Knipe (Editors), Faulting, Fault Sealing and Fluid Flow in Hydrocarbon Reservoirs. Geol. Soc. London Spec. Publ., 147: 117–134.

Fowles, J. and Burley, S., 1994. Textural and permeability characteristics of faulted, high porosity sandstones. Mar. Pet. Geol., 11 (5): 608–623.

Fristad, T., Groth, A., Yielding, G. and Freeman, B. 1997. Quantitative fault seal prediction — a case study from Oseberg Syd. In: P. Møller-Pedersen and A.G. Koestler (Editors), Hydrocarbon Seals: Importance for Exploration and Production. Norwegian Petroleum Society (NPF), Special Publication 7. Elsevier, Amsterdam, pp. 107–124.

Gibson, R.G., 1998. Physical character and fluid-flow properties of sandstone-derived fault zones. In: M.P. Coward, T.S. Daltaban and H. Johnson (Editors), Structural Geology in Reservoir Characterisation. Geol. Soc. London Spec. Publ., 127: 83–97.

He, Z. and Corrigan, J., 1995. Factors necessary to define a pressure seal — Discussion. Am. Assoc. Pet. Geol. Bull., 79: 1075–1078.

Heum, O.R., 1996. A fluid dynamic classification of hydrocarbon entrapment. Pet. Geosci., 2: 145–158.

Knipe, R.J., Fisher, Q.J., Jones, G., Clenell, M.R., Farmer, A.B., Harrison, A., Kidd, B., McAllister, E., Porter, J.R. and White, E.A., 1997. Fault seal analysis: successful methodologies, application and future directions. In: P. Møller-Pedersen and A.G. Koestler (Editors), Hydrocarbon Seals: Importance for Exploration and Production. Norwegian Petroleum Society (NPF), Special Publication 7. Elsevier, Amsterdam, pp. 15–40.

Lehner, F.K. and Pilaar, W.F., 1997. On a mechanism of clay smear emplacement in synsedimentary normal faults. In: P. Møller-Pedersen and A.G. Koestler (Editors), Hydrocarbon Seals: Importance for Exploration and Production. Norwegian Petroleum Society (NPF), Special Publication 7. Elsevier, Amsterdam, pp. 113–123.

Manzocchi, T., Walsh, J.J., Nell, P.A.R. and Yielding, G., 1999. Fault transmissibility multipliers for flow simulation models. Pet. Geosci., 5: 53–63.

Morrow, C.A., Shi, L.Q. and Byerlee, J.D., 1984. Permeability of fault gouge under confining pressure and shear stress. J. Geophys. Res., 89: 3193–3200.

Ottesen Ellevset, S., Knipe, R.J., Olsen, T.S., Fisher, Q.J. and Jones, G., 1998. Fault controlled communication in the Sleipner Vest Field, Norwegian Continental Shelf; detailed, quantitative input for reservoir simulation and well planning. In: G. Jones, Q.J. Fisher and R.J. Knipe (Editors), Faulting, Fault Sealing and Fluid Flow in Hydrocarbon Reservoirs. Geol. Soc. London Spec. Publ., 147: 283–297.

Pittman, E.D., 1981. Effect of fault-related granulation on porosity and permeability of quartz sandstones, Simpson Group (Ordovician), Oklahoma. Am. Assoc. Pet. Geol. Bull., 65 (11): 2381–2387.

Sclater, J.G. and Christie, P.A.F., 1980. Continental stretching: an explanation of the post-mid-Cretaceous subsidence of the central North Sea Basin. J. Geophys. Res., 85: 3711–3739.

Snowdon, L.R., 1995. Multiphase systems, overpressure and migration. Bull. Can. Pet. Geol., 43: 315–319.

Stewart, I.J., Rattey, R.P. and Vann, I.R., 1992. Structural style and the habitat of hydrocarbons in the North Sea. In: R.M. Larsen, H. Brekke, B.T. Larsen and E. Talleraas (Editors), Structural and Tectonic Modelling and its Application to Petroleum Geology. Norwegian Petroleum Society (NPF), Special Publication 1. Elsevier, Amsterdam, pp. 197–220.

Walsh, J.J., Watterson, J., Heath, A.E. and Childs, C., 1998. Representation and scaling of faults in fluid flow models. Pet. Geosci., 4: 241–251.

C. CHILDS — *Fault Analysis Group, Department of Geology, University College Dublin, Dublin, Ireland*
E-mail: fault@fag.ucd.ie

T. MANZOCCHI — *Fault Analysis Group, Department of Geology, University College Dublin, Dublin, Ireland*

P.A.R. NELL — *Fault Analysis Group, Department of Earth Sciences, University of Liverpool, Liverpool L69 3BX, UK*

J.J. WALSH — *Fault Analysis Group, Department of Geology, University College, Dublin, Dublin, Ireland*

J.A. STRAND — *Fault Analysis Group, Department of Geology, University College Dublin, Dublin, Ireland*

A.E. HEATH — *Fault Analysis Group, Department of Geology, University College Dublin, Dublin, Ireland*

T.H. LYGREN — *Norsk Hydro, Oseberg Exploration, Bergen, Norway*

Fault reactivation, leakage potential, and hydrocarbon column heights in the northern North Sea

David Wiprut and Mark D. Zoback

To investigate the question of how faults affect the migration of fluids in petroleum reservoirs, we evaluated the state of stress and pore pressure acting on the major faults in four oil and gas fields in the northern North Sea. Many of the faults bound hydrocarbon reservoirs. Our goal was to test the hypothesis that faults that are being reactivated in the current stress field are permeable and thus tend to leak, whereas those that are not (i.e. faults that are inactive in the current stress field) are likely to seal. To address this question, we utilize a detailed analysis of the magnitude and orientation of all three principal stresses in a number of wells in each field. These data, along with information on pore pressure, allowed us to resolve the shear and effective normal stress acting on distinct ~100 m × 100 m elements of individual fault planes. By comparing the stress state resolved on each fault element to expected stress at failure (using a Coulomb failure criterion) we created color-shaded maps showing the proximity to fault slip (and hence leakage) along each fault. Fault reactivation and hydrocarbon leakage in this area appears to be caused by three factors: (1) locally elevated pore pressure due to buoyant hydrocarbons in reservoirs abutting the faults, (2) fault orientations that are nearly optimally oriented for frictional slip in the present-day stress field, and (3) a relatively recent perturbation of the compressional stress caused by postglacial rebound. We demonstrate that the combination of these three factors may have recently induced fault slippage and gas leakage along sections of previously sealing reservoir-bounding faults in some fields, whereas in others, the stress and pore pressure are not sufficient to cause fault reactivation. We show that only in cases where reservoir-bounding faults are not potentially active, the pore-pressure difference across faults can become quite high. Hence, the leakage potential of reservoir-bounding faults appears to exert an important influence on potential hydrocarbon column heights.

Introduction

The question of how faults affect the migration of fluid in petroleum reservoirs is complicated, as some faults contribute dramatically to formation permeability (Dholakia et al., 1998) and allow hydrocarbon migration between different reservoir units (Finkbeiner et al., 2001), yet others provide effective barriers separating distinct reservoir compartments (Hunt, 1990). The sealing potential of a fault can be related to the juxtaposed lithologies across the fault and the presence or absence of seals resulting from the structure and content of the fault zone (Weber et al., 1978; Downey, 1984; Allan, 1989; Nybakken, 1991; Knipe, 1992; Berg and Avery, 1995; Fristad et al., 1997). However, the process by which a previously sealing fault begins to leak is unclear.

In this paper we consider the effect of fault reactivation on fault seal and fluid flow in the context of in-situ stress and pore pressure. We test the hypothesis that faults that are critically stressed in the current stress field (i.e. capable of slipping) are permeable, whereas those that are not critically stressed are not permeable. A number of permeability studies in fractured and faulted crystalline rock appear to confirm this hypothesis (Barton et al., 1995, 1998; Hickman et al., 1998; Townend and Zoback, 2000). Studies in hydrocarbon reservoirs in sedimentary basins in the Santa Maria Basin (Finkbeiner et al., 1997), the Gulf of Mexico (Finkbeiner et al., 2001), the Timor Sea (Castillo et al., 2000), and on a single partially leaking fault in the northern North Sea (Wiprut and Zoback, 2000b) appear to confirm that critically stressed faults are responsible for promoting hydrocarbon leakage and migration.

In this study we expand upon the work presented by Wiprut and Zoback (2000b) in the Visund field. The point of departure from our previous work is that we evaluate here the leakage potential of seismically mapped faults throughout the Visund field as well as three other fields in the northern North Sea (Fig. 1). We also address the effect of critically stressed faults and water-phase pore pressure on the potential height of hydrocarbon columns.

Hydrocarbon Seal Quantification edited by A.G. Koestler and R. Hunsdale. NPF Special Publication 11, pp. 203–219, Published by Elsevier Science B.V., Amsterdam.

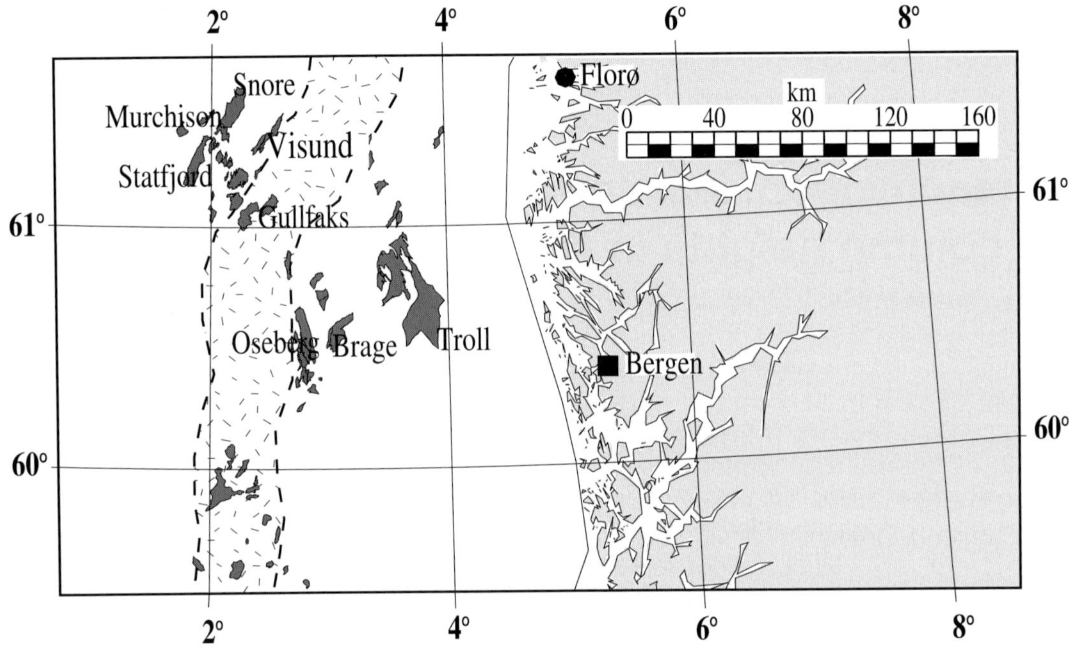

Fig. 1. Map of the northern North Sea showing the west coast of Norway and major offshore oil and gas discoveries. A rough outline of the Viking graben is indicated by the hachured area. Map is modified from the Norwegian Petroleum Directorate, 1997. (http://npd.no/Webdesk/netblast/pages/index.html)

Methodology

We focus here on the influence of excess pressures resulting from buoyant hydrocarbon columns on the sealing capacity of reservoir-bounding faults. Fig. 2 shows conceptually how we apply the hypothesis that buoyant hydrocarbons can increase the pore pressure and trigger fault reactivation. As hydrocarbons accumulate in a permeable reservoir bounded by a sealing fault, the pore pressure at the fault–reservoir interface increases because the pore-pressure gradient in the hydrocarbon column is considerably less than the hydrostatic gradient owing to its low density. As the height of the hydrocarbon column increases, at some point the pore pressure will be sufficient to induce fault slip, providing a mechanism to increase fault

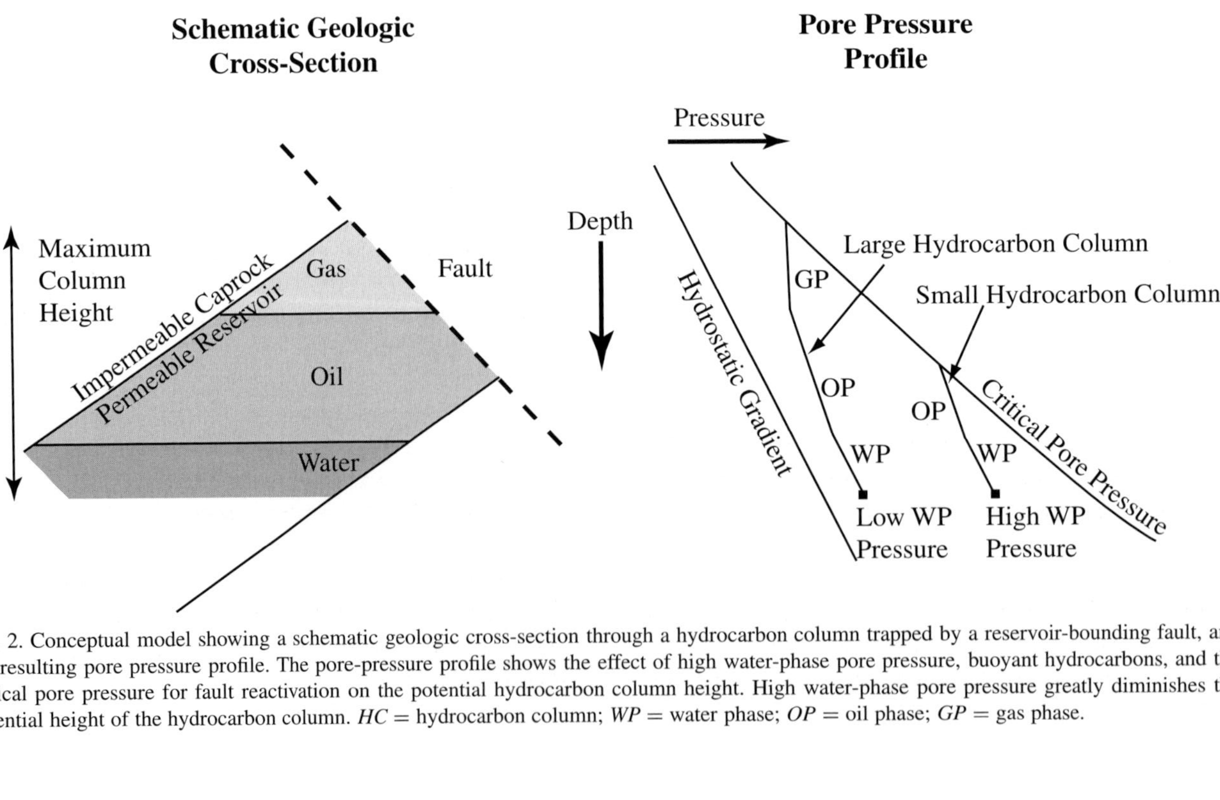

Fig. 2. Conceptual model showing a schematic geologic cross-section through a hydrocarbon column trapped by a reservoir-bounding fault, and the resulting pore pressure profile. The pore-pressure profile shows the effect of high water-phase pore pressure, buoyant hydrocarbons, and the critical pore pressure for fault reactivation on the potential hydrocarbon column height. High water-phase pore pressure greatly diminishes the potential height of the hydrocarbon column. *HC* = hydrocarbon column; *WP* = water phase; *OP* = oil phase; *GP* = gas phase.

permeability and allow leakage from the reservoir. Fig. 2 also demonstrates the combined effect of high water-phase pore pressure and leakage along reactivated faults on the height of the hydrocarbon column. Because high water-phase pore pressure brings the fault closer to the critical pore pressure for failure and leakage, the potential hydrocarbon column height is diminished. Finkbeiner et al. (2001) showed that only small hydrocarbon columns could be trapped against reservoir-bounding faults near frictional failure, whereas larger columns could be trapped against faults of different orientation, or with lower water-phase pore pressures, initially further from failure.

To evaluate the hypothesis that critically stressed faults or portions of faults that are critically stressed are permeable and are the cause of localized leakage, we resolve the stress orientations and stress magnitudes we determine in each field onto distinct ~100 m × 100 m triangular elements on individual fault planes to calculate the shear and normal stress on each part of the fault. We use Coulomb frictional failure to determine which fault element is expected to slip. Coulomb frictional failure is defined in Eq. 1:

$$\tau = \mu \sigma_n \qquad (1)$$

where τ is the shear stress, σ_n is the effective normal stress ($\sigma_n = S_n - P_p$), and μ is the coefficient of sliding friction (Jaeger and Cook, 1979). We solve Eq. 1 to determine the pore pressure at which a fault element will begin to slip (Eq. 2), and refer to this pore pressure as the critical pore pressure.

$$P_p^{crit} = S_n - \tau/\mu \qquad (2)$$

In order to calculate the shear and normal stress we determine the orientation of the unit normal to the fault element in a coordinate system defined by the stress field. Fig. 3A shows a fault element defined in the stress coordinate system. S_1, S_2, and S_3 are the principal stresses, a, b, and c are the vertex points of the fault element, $\hat{n}$ is the unit normal to the fault element, and t is the traction acting on the surface of the fault element. The unit normal to the fault element is defined by the cross product in Eq. 3,

$$\hat{n} = \frac{f x g}{|f||g|} \qquad (3)$$

where f and g are any two vectors defined by the points a, b, and c. The traction acting on the fault plane is the product of the stress tensor and unit normal vector (Eq. 4).

$$t = S\hat{n} \qquad (4)$$

Because the stresses do not vary significantly between the study wells in individual fields, we define one stress tensor for each field using a single one-dimensional model that varies with depth. The stress tensor is defined in Eq. 5,

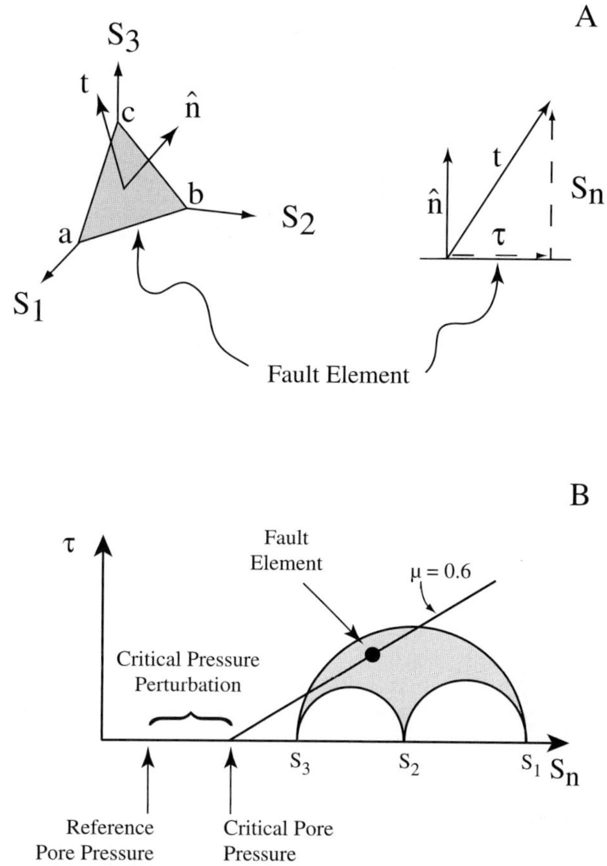

Fig. 3. (A) Orientation of a fault element in a coordinate system defined by the stress field. (B) Mohr-Coulomb plot showing determination of the critical pore pressure and comparison to the reference pore pressure. See text for an explanation of both diagrams.

$$S = \begin{bmatrix} S_1 & 0 & 0 \\ 0 & S_2 & 0 \\ 0 & 0 & S_3 \end{bmatrix} = \begin{bmatrix} S_{Hmax} & 0 & 0 \\ 0 & S_v & 0 \\ 0 & 0 & S_{Hmax} \end{bmatrix} \qquad (5)$$

where S_{Hmax} is the maximum horizontal stress, S_v is the vertical stress, and S_{Hmin} is the minimum horizontal stress. We obtain the stress magnitudes from an analysis of drilling-induced tensile fractures and breakouts (Wiprut and Zoback, 2000a). Taking the dot product of the unit normal vector and the traction vector gives the magnitude of the normal stress (Eq. 6). The magnitude of the shear stress is determined simply by the Pythagorean theorem (Eq. 7, Fig. 3A).

$$S_n = \hat{n} \times t \qquad (6)$$

$$\tau^2 = t^2 - S_n^2 \qquad (7)$$

We calculate the critical pore pressure at which the fault element will slip using Eqs. 2, 6 and 7, and by utilizing a coefficient of sliding friction of 0.6 (Byerlee, 1978; Townend and Zoback, 2000). This coefficient is a reasonable lower bound that is nearly independent of the internal fault structure, the rock type, the depth, or the stresses resolved on the fault surface (Byerlee, 1978). Fig. 3B shows a graphical representation of the preceding calculation. A fault element is plotted as a point within the 3-D Mohr circle according to the shear and normal stress resolved on the fault element. The slope of the Coulomb frictional failure line passing through the fault element point uniquely defines the critical pore pressure where the failure line intersects the normal-stress axis. The critical pore pressure is compared to a reference pore pressure line drawn through the data, where the pore-pressure data are combined across the entire field into a single one-dimensional model that varies with depth. The difference between the critical pore pressure and the reference pore pressure is called the critical pressure perturbation. This value shows how close the fault element is to slipping given the reference pore pressure determined for the field, and hence is a measure of the leakage potential.

The Visund field

The Visund field is located offshore Norway in the easternmost major fault block of the Tampen Spur (Færseth et al., 1995) along the western edge of the Viking graben. The reservoir is divided into several oil and gas compartments, some of which are separated by the A-Central fault (Fig. 4). Hydrocarbon columns were detected in the Brent group, which is the primary reservoir, as well as in the Statfjord and Amundsen formations. As shown in Fig. 4A, low seismic reflectivity along the southern part of the A-Central fault at the top Brent reservoir horizon is interpreted to be the result of gas leakage from the reservoir. The data in this region are of very high quality and there are no changes in lithology that might account for the change in seismic reflectivity. Fig. 4A also shows the mean orientation of the maximum horizontal stress determined in five wells in and near the Visund field from observations of drilling-induced tensile wall fractures (Moos and Zoback, 1990; Brudy and Zoback, 1993, 1999). Drilling-induced tensile wall fractures have been shown to be reliable indicators of the direction of the maximum horizontal stress (Brudy et al., 1997; Wiprut and Zoback, 2000a).

Fig. 4B shows a contour map of the top Brent reservoir horizon (red lines), with the faults, lateral extent of gas leakage (dashed line, see Fig. 4A), and outline of the map area shown in Fig. 4A (blue rectangle) superimposed on the structural contours. Exploration wells that yielded stress and pore-pressure data are shown with black circles. The Brent reservoir consists of a ridge running northeast–southwest with a saddle crossing perpendicular to the ridge between wells B and C. Comparison of the maps in Fig. 4A,B shows that the ridge is trapping gas along most of its length except for the portion of the ridge defined by the dashed low-reflectivity area. In the lower part of Fig. 4B, the southern boundary of the Brent reservoir plunges steeply into the Viking graben. This is the result of a large northeast–southwest trending graben-bounding fault that intersects the southern end of the A-Central fault. The effect of the graben-bounding fault can be seen in Fig. 4A as well, where there is a sharp transition from high to low reflectivity in the southern portion of the map.

Fig. 4C shows a schematic cross-section running approximately east–west through well D and the A-Central fault. The A-Central fault developed during the Jurassic as a normal fault with an ~60° dip (Færseth et al., 1995) and as much as 300 m of normal throw (L. Arnesen, pers. commun.). Since that time, the fault appears to have rotated and now dips between 30° and 45°. As a result, the A-Central fault is well oriented for being reactivated in a reverse sense in the current stress field. The other major faults in Visund generally dip 20° to 40° to the east, with some smaller antithetic faults dipping to the west.

Fig. 5 shows two views of the A-Central fault as determined from three-dimensional seismic reflection data. In the upper part of Fig. 5, a simplified map view of the fault is shown along with the orientation of the maximum horizontal stress in the three wells closest to the fault. The shaded area shows the lateral extent of gas leakage (simplified from Fig. 4A). In the lower part of Fig. 5, a perspective view of the approximately east-dipping fault surface is shown. A dark circle on the fault plane indicates the point where well D penetrates the A-Central fault. The fault plane is colored to indicate the leakage potential based on the orientation of the fault, the stress, and the pore pressure.

Fig. 6 shows a summary of the in-situ stress and pore-pressure data in the Visund field over the depth range of principal interest for the A-Central fault. The pore-pressure data are direct measurements made in the reservoir. The vertical stress was derived by using the average overburden gradient across the field. We calculated the overburden in each well by integrating density logs. The data for the minimum horizontal stress were derived from analysis of carefully conducted leak-off tests (LOTs). The magnitude of the maximum horizontal principal stress was determined from analysis of drilling-induced tensile fractures (following Zoback et al., 1993; Brudy et al., 1997). Determinations of stress magnitude and orientation in Visund are described in detail by Wiprut and Zoback (2000a).

The evidence for gas leakage in the immediate vicinity of the A-Central fault points to the fault as the possible conduit by which hydrocarbons are

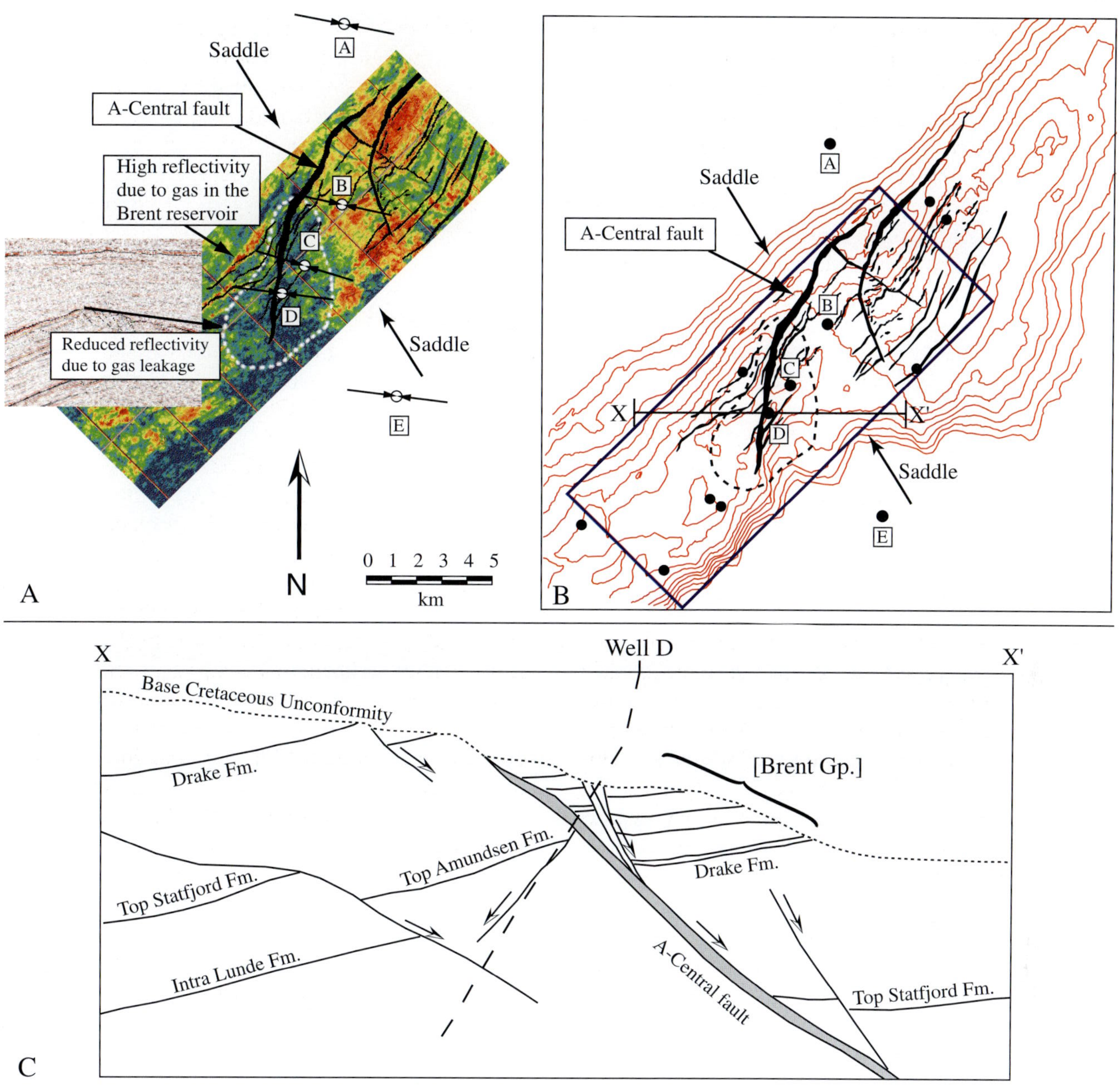

Fig. 4. (A) Map view of the Visund field showing seismic reflectivity of the reservoir horizon as well as the mean orientation of the maximum horizontal stress in five wells (A–E) (after Wiprut and Zoback, 2000b). (B) Contour map of the top Brent reservoir horizon. The saddle defines a local structural low along a ridge running from the northeast to the southwest. The area shown in part A is outlined in blue. (C) East–west cross-section along X–X′ (shown in part B) through the Visund field. Cap rock is defined by the short-dashed line at the base Cretaceous unconformity. The trajectory of well D through the A-Central fault is shown with long dashes.

escaping from the reservoir. We utilize the stress and pore-pressure trends shown in Fig. 6 to create the map of leakage potential shown in Fig. 5. The color shows the difference between the critical pore pressure we calculate and the reference pore-pressure line shown in Fig. 6. This difference is the critical pressure perturbation (defined previously). Hot colors indicate that a small increase in pore pressure ($<\sim 7$ MPa) is enough to bring the fault to failure. Cool colors indicate that the pore pressure must rise significantly (>20 MPa) before those parts of the fault will begin to slip in the current stress field. Note that the largest part of the fault that is most likely to slip (indicated by the white outline) is located along the same part of the fault where leakage seems to be occurring. Note also that this portion of the fault is coincident with a change in the fault plane strike. Thus, there appears to be a qualitative correlation between the critically stressed fault criterion and the places along the fault where leakage appears to be occurring.

Well D was deviated to penetrate the A-Central fault at 2933 m true vertical depth (Fig. 5; horizontal dashed line, Fig. 6 inset), whereas wells B and C were drilled vertically. Because well D penetrates the

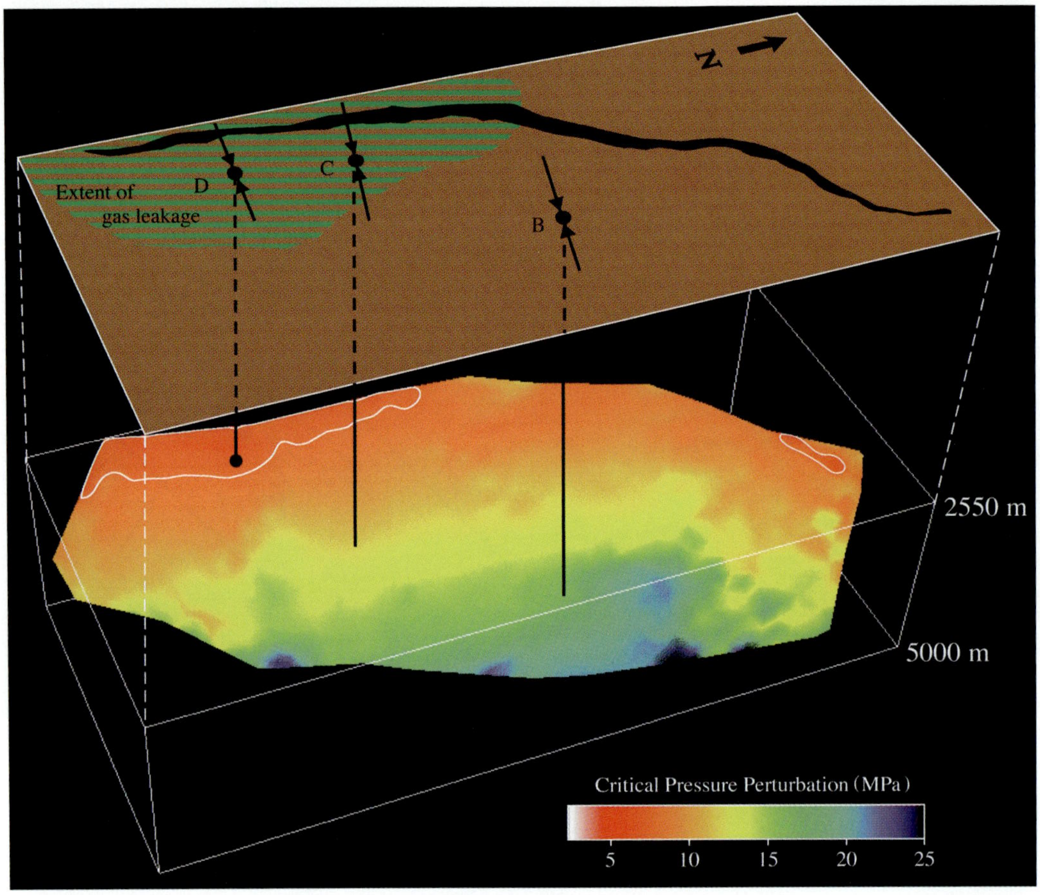

Fig. 5. Map view and perspective view of the A-Central fault as determined from a three-dimensional seismic reflection survey. Map view shows the region of gas leakage inferred from reduced seismic reflectivity (Fig. 4A). Perspective view is colored to show excess pore pressure (critical pressure perturbation) needed to induce fault slip in the current stress field. The white contours indicate portions of the fault that require an excess pore pressure less than approximately 7 MPa above the reference pore pressure. The largest part of the fault that is most likely to slip corresponds to that which appears to be leaking.

fault in this area, we can evaluate the correlation between the gas leakage and our prediction of leakage more quantitatively. Pore pressures in Visund are significantly above hydrostatic throughout the reservoir (Fig. 6). The inset of Fig. 6 shows a detailed view of the pore-pressure measurements in the three wells closest to the A-Central fault. The steep pressure gradient in well D is the result of light oil rather than free gas. A free gas cap was not detected in well D or well C, consistent with the reduced seismic reflectivity shown in Fig. 4A. As shown in the inset of Fig. 6, the pressure below the fault (indicated by the position of the dashed horizontal line) is within $\sim$1 MPa of the theoretical critical pore pressure for fault slippage (the thick dashed line). This value is several megapascals above the reference pore pressure, just as predicted in Fig. 5.

Above the fault, pore pressures are significantly reduced, indicating that there is poor pressure communication across the fault (Fig. 6, inset). The A-Central fault is connected at its southern end with the graben-bounding fault described previously, preventing hydrocarbons from migrating around the southern end of the fault from the footwall to the hangingwall. Geochemical analysis of gas from both sides of the fault indicates that the hydrocarbons are derived from different sources (i.e. no fluid flow across the fault). Hydrocarbons are filling the reservoir on the eastern side of the A-Central fault from the east, and are filling the reservoir on the western side of the A-Central fault from the west (A. Wilhelms, pers. commun.).

It is interesting to note that although the pore pressure in the footwall appears to have caused the A-Central to slip and leak, both the footwall and hangingwall show reduced seismic reflectivity. Increased permeability resulting from fault slip seems to influence pore-pressure compartments on both sides of the fault. In this case, as with cases reported by Hickman et al. (1998) and Finkbeiner et al. (2001), fault slip appears to have principally promoted fault-parallel flow.

The pore pressures shown in the inset of Fig. 6 indicate that wells B, C and D are in approximately the same pressure compartment in the hangingwall, yet well B does not penetrate an area of reduced reflectivity. This is the result of the saddle shown in Fig. 4B.

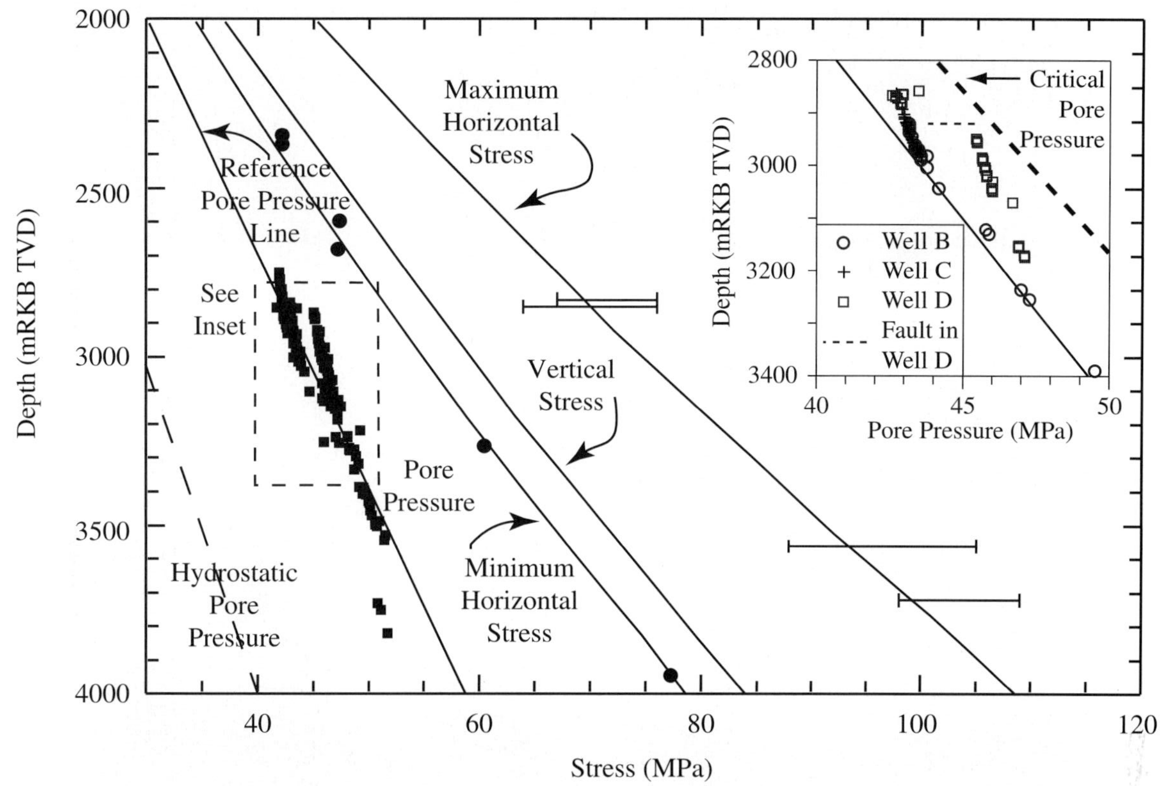

Fig. 6. In-situ stress and pore-pressure data obtained from wells throughout the Visund field. Best-fit lines to data are shown. Inset shows pore-pressure measurements in three wells drilled close to the A-Central fault.

The local structural low provided by the saddle effectively separates the hydrocarbon column in well B from wells C and D. There is an approximately 22-m difference in oil–water contacts between wells B and C; and there is an approximately 18-m gas column in well B that is absent in well C. Assuming the hydrocarbon columns were approximately the same before the fault leaked, the missing gas column in well C nearly accounts for the difference in oil–water contacts between the two wells.

Fig. 7 shows a perspective view, looking down and toward the north, of all the major faults in the Visund field with colors indicating the potential for hydrocarbon leakage. The perspective view in this figure creates distortions, therefore the scales are approximate. The five wells that provided data for the maximum horizontal stress are labeled, and other wells that provided pore-pressure data are shown as white circles. The faults are colored to indicate the likelihood of leakage along the surfaces. The leakage map indicates the potential for hydrocarbon leakage along any fault, and does not imply that any fault with red colors is currently leaking. A reservoir must abut the fault in the proper place, there must be hydrocarbons present to leak and the pore pressure must be high enough to reactivate the fault in order for the leakage to take place. The limits of this analysis are discussed in further detail below.

Fields 1, 2 and 3, northern North Sea

Fig. 8 shows the states of stress observed in Fields 1, 2 and 3, which we also studied in the northern North Sea. The general pore-pressure trend in Field 1 follows a nearly hydrostatic gradient until 3500 m, where it increases significantly in wells A, B and C (Fig. 8A). There is a marked pore-pressure difference between wells A and B, drilled into the hangingwall block of a major north–south trending and eastward dipping fault in Field 1, and well C drilled into the footwall block. Pore pressures in wells A and B follow a steep gas gradient toward the top of the hydrocarbon column, whereas the pore pressures in well C appear to primarily mirror the hydrostatic gradient at the same depth. The pore pressure in well C in a reservoir at greater depth follows a hydrocarbon gradient. The pore-pressure difference across the fault between wells B and C at a depth of 3450 m is shown by the arrow and is approximately 15 MPa. We discuss this large pore-pressure difference subsequently.

Pore pressures in both Field 2 and Field 3 are hydrostatic until approximately 3400 m, where there is an increase in pore pressure in both fields (Fig. 8B,C). The reservoir is highly overpressured in Field 3, and in Field 2 there is only moderate overpressure in the reservoir. The pore-pressure trends in both

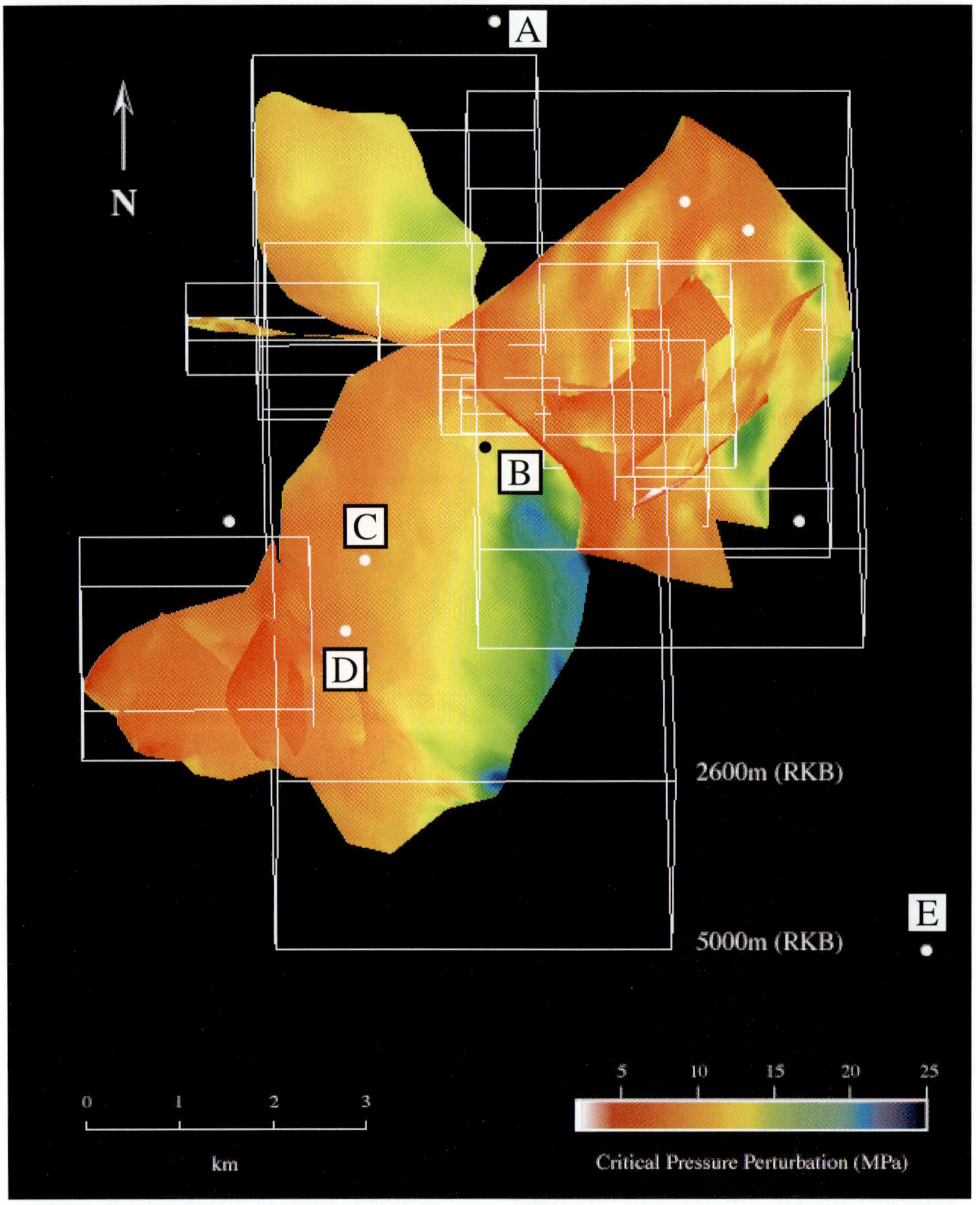

Fig. 7. Perspective view of fault surfaces in the Visund field showing leakage potential. The A-Central fault is shown with depths listed on the bounding box. Perspective view is colored to show excess pore pressure needed to induce fault slip in the current stress field. Hot colors indicate that the fault is close to failure and cool colors indicate that pore pressure must rise significantly (nearly 25 MPa) before fault will be reactivated in current stress field. Note that the scales are approximate, as the perspective view creates distortions.

fields continue to mirror the hydrostatic gradient in the overpressured sections. A number of anomalous pore-pressure measurements in the shallower parts of Field 2 (Fig. 8B) come from approximately five wells scattered throughout the region, and do not reflect the overall pore-pressure trend in any one compartment.

Note that in all three fields the maximum horizontal stress is distinctly larger than the vertical stress, and the minimum horizontal stress is close in magnitude to the vertical stress. This result is consistent with the strike-slip and reverse stress field indicated by earthquake focal-plane mechanisms (at 5 to 30 km depth) in this part of the North Sea (Lindholm et al., 1995).

Fig. 9A shows a map view of Fields 1 and 2 with the faults and mean orientation of the maximum horizontal stress determined in five wells in this area. Other exploration wells that yielded stress and pore-pressure data are shown by black circles. Field 1 is a small discovery approximately 5 km west of Field 2. Reservoirs in Field 1 are quite deep with Brent reservoir sandstones encountered between approximately 3500 and 4100 m. Structural dips are to the east between approximately 1° and 10° in Field 1 and between 2° and 14° in Field 2. Fig. 9B shows a schematic cross-section through well F in Field 2 that gives a generalized picture of the structure in this area. Major reservoir-bounding faults in this area

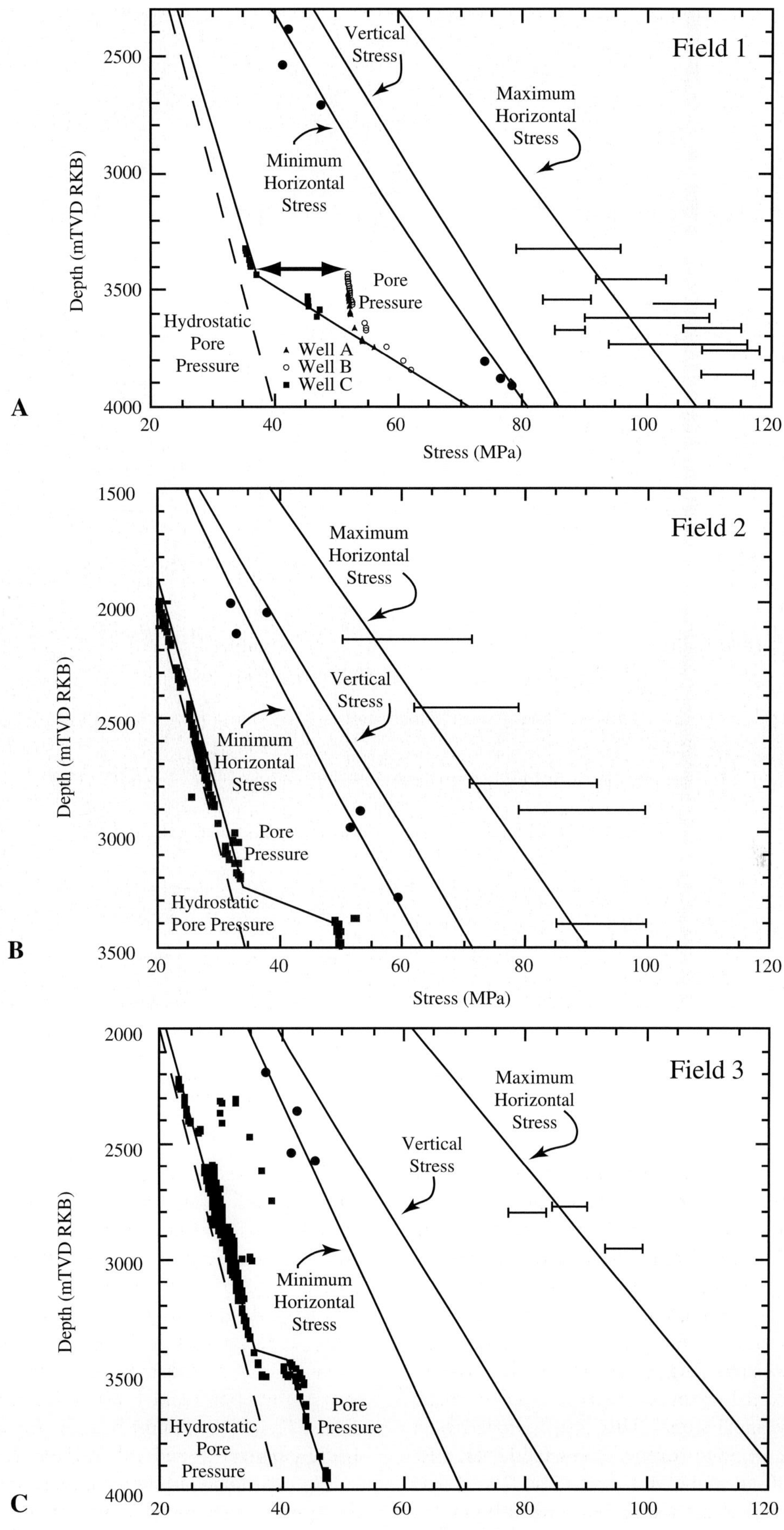

Fig. 8. In-situ stress and pore-pressure data obtained from wells throughout Field 1 (A), Field 2 (B), and Field 3 (C). Best-fit lines to data are shown. See text for explanation of arrow in (A).

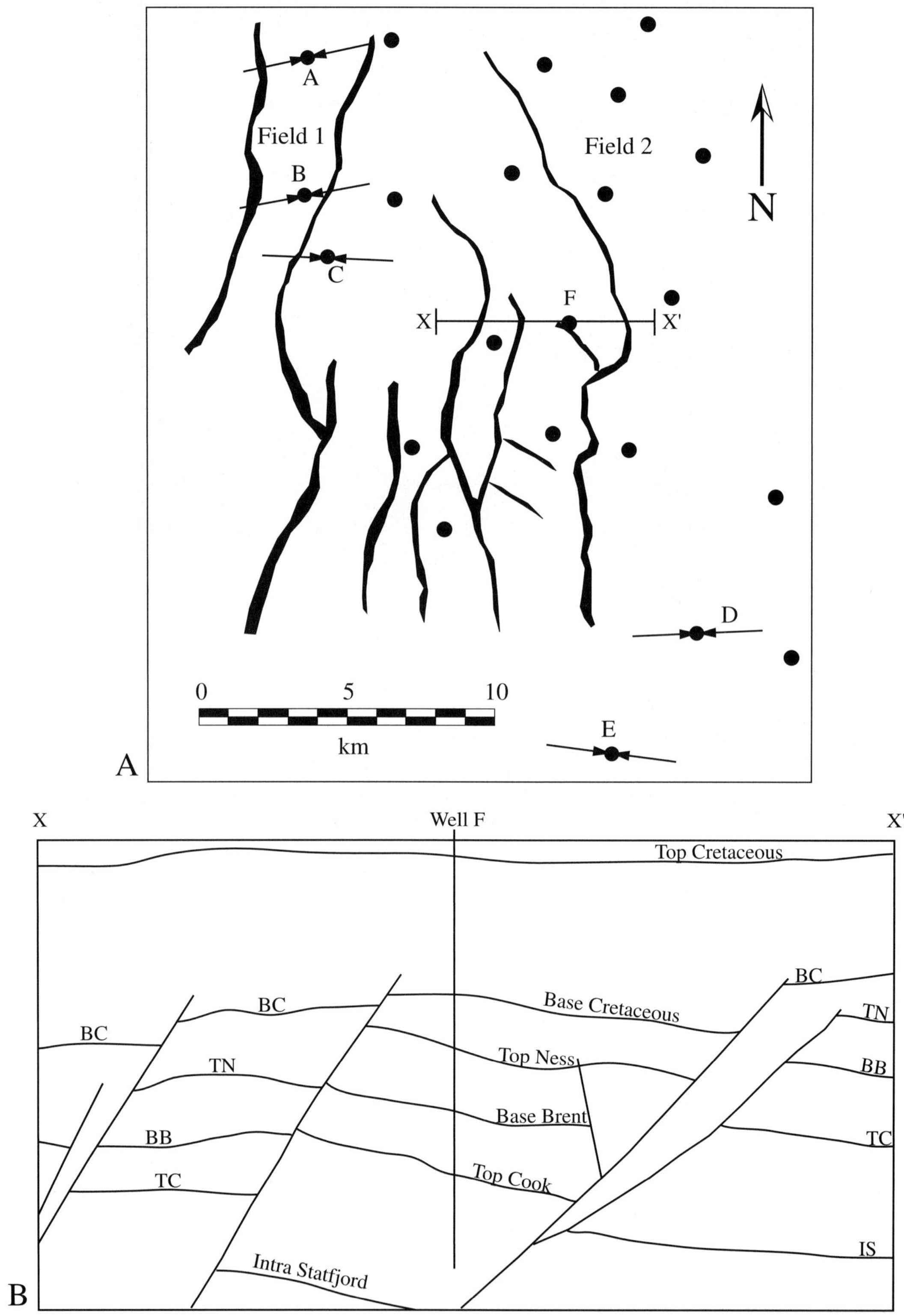

Fig. 9. (A) Generalized map of Field 1 and Field 2 showing the orientation of the maximum horizontal stress (inward pointed arrows), exploration wells (circles), and major faults. (B) East–west cross-section along X–X′ passing through well F in Field 2. Major faults generally dip steeply west toward the Viking graben.

strike approximately north–south, and dip to the west between 40° and 55°. Careful examination of seismic cross-sections in Field 1 and Field 2 revealed no evidence of hydrocarbon leakage, and there is no evidence of hydrocarbon migration at present in these fields. Field 1 and Field 2 are highly compartmentalized by faults.

Fig. 10 shows a perspective view of all the major faults in Fields 1 and 2 with colors indicating the potential for hydrocarbon leakage as in Figs. 5 and 7.

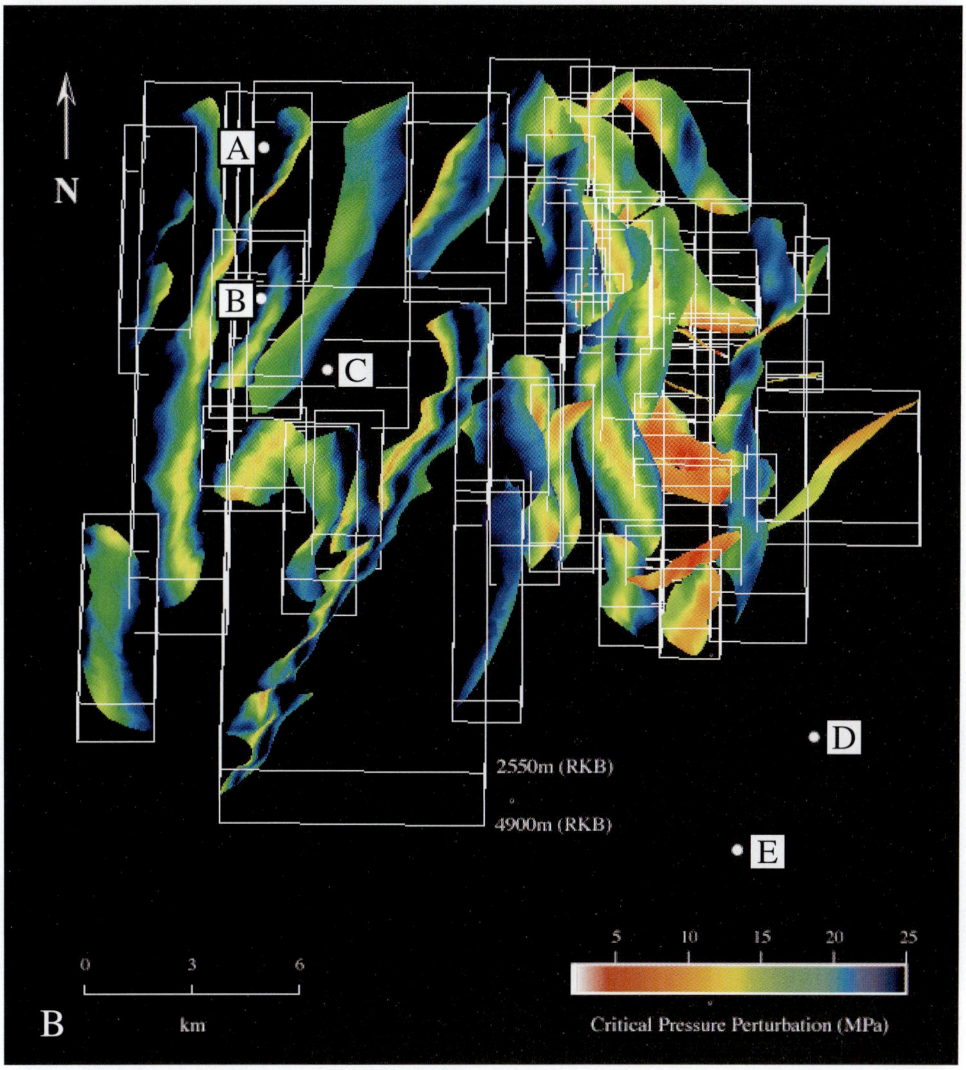

Fig. 10. Fault leakage potential in Field 1 and Field 2. See Fig. 7 for explanation.

The five wells that provided data for the maximum horizontal stress are labeled. Note that most of the faults in Fields 1 and 2 do not show any significant potential for leakage. This is primarily the result of the faults being poorly oriented for frictional failure in the current stress field. This prediction is consistent with the absence of hydrocarbon leakage and migration in these fields. Fig. 10 shows that our analysis predicts there should be no leakage and it also shows that the reservoirs may potentially maintain large pore-pressure differences across compartments. According to our analysis, the major fault to the east of well B in Field 1 can potentially maintain up to approximately 15 to 17 MPa pore-pressure difference across its surface at the weakest points. As noted above, the pore-pressure data in this field show a pressure difference of approximately 15 MPa between the pore-pressure trend used to create Fig. 10 and the hydrocarbon column supported by the major fault east of well B (Fig. 8A, see arrow).

Fig. 11A shows a map view of Field 3 with the faults and mean orientation of the maximum horizontal stress determined in four wells in this area. Other exploration wells that yielded stress and pore-pressure data are shown with black circles. Fig. 11B shows a schematic cross-section through two wells in the field along the line W–W′. The Brent reservoir in Field 3 typically dips between 3° and 10° to the east and southeast in individual fault blocks, but overall becomes shallower to the south-southeast in this region. Reservoir-bounding faults in Field 3 generally strike in two directions, with a northeast–southwest striking set of faults cross-cutting a north–south striking set. The faults typically dip between 50° and 60° throughout the field. Fig. 12 shows three east–west oriented cross-sections cut through wells A and F along the lines X–X′, Y–Y′, and Z–Z′ shown in Fig. 11. Cross-section Y–Y′ indicates that there is some amplitude dimming above the fault east of well A, which is interpreted to be the result of gas

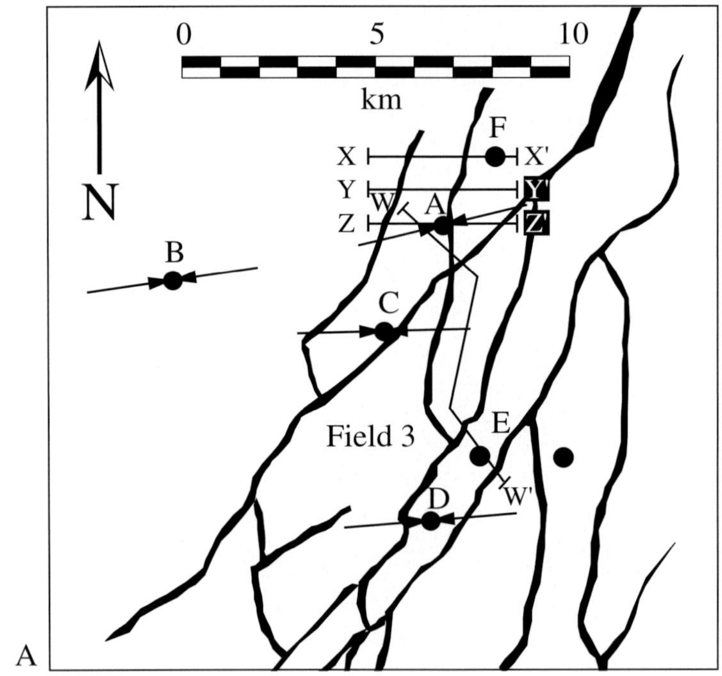

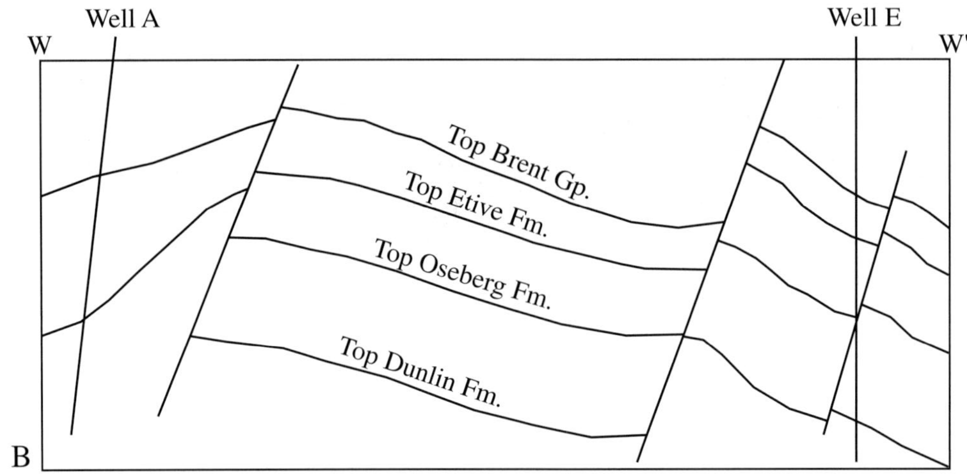

Fig. 11. (A) Generalized map of Field 3 showing the orientation of the maximum horizontal stress (inward pointed arrows), exploration wells (circles), and major faults. Cross-sections X–X′, Y–Y′, and Z–Z′ are shown in Fig. 6. (B) Schematic cross-section along line W–W′ through two wells in Field 3. Structural dips in this area are generally to the east and southeast.

leakage from the reservoir. However, it should be noted that there is evidence of overpressure in the overburden in Field 3 (Fig. 8C), and this overpressure may be responsible for the anomalies seen in the seismic cross-section Y–Y′. It is not known whether these overpressures developed in-situ or if they are the result of gas leakage.

The leakage potential map shown in Fig. 13 indicates that Field 3 is quite likely to leak from many of the more steeply dipping northeast–southwest trending faults, and is less likely to leak from the north–south trending faults. This is a result of the orientation of the faults with respect to the stress field. Faults oriented north–south are nearly perpendicular to the maximum horizontal stress direction in this area, whereas those striking northeast–southwest are well-oriented for strike-slip movement. Note that the amplitude anomaly associated with gas leakage seen on the Y–Y′ cross-section occurs on a portion of the fault that is oriented northeast–southwest.

Discussion

Comparison of Figs. 7, 10 and 13 shows that the shallowly dipping faults in Visund generally have the highest likelihood of leakage. The northeast–southwest trending faults in Field 3 are almost as likely to leak, while the north–south trending faults

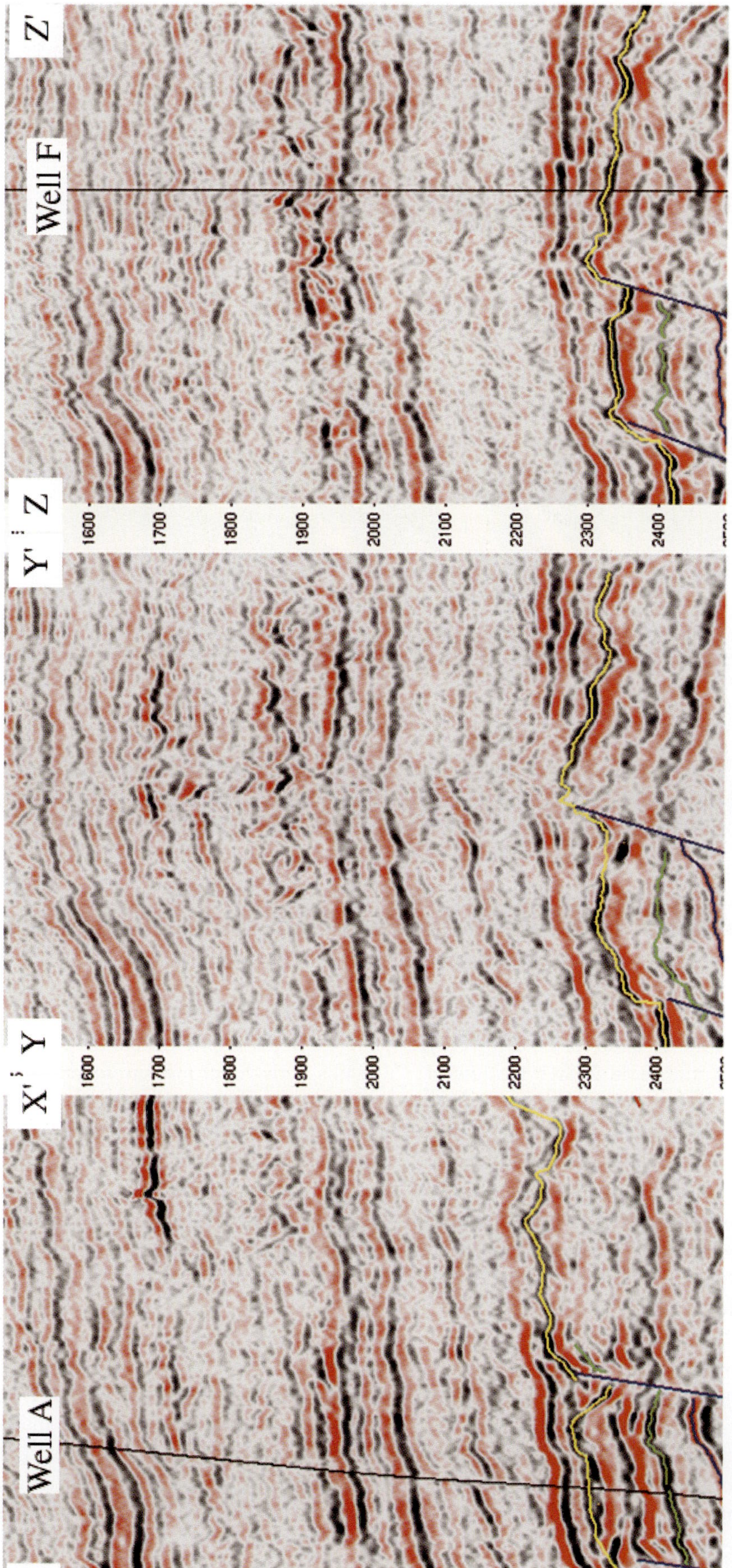

Fig. 12. East–west seismic sections through well A (left panel), and well F (right panel) in Field 3. Middle panel shows leakage along a reservoir-bounding fault between the two wells inferred from reduced seismic amplitudes. Scale is two-way-time in milliseconds.

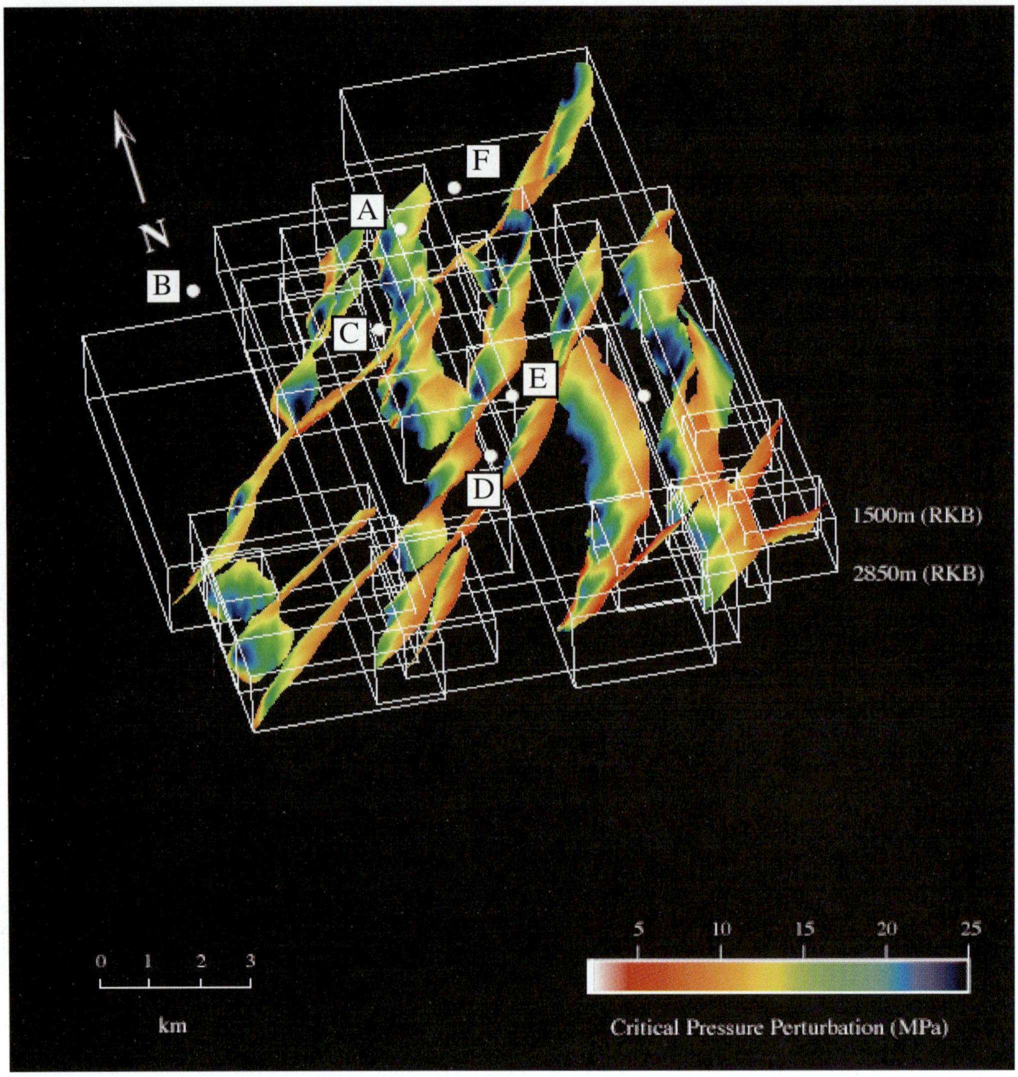

Fig. 13. Fault leakage potential in Field 3. See Fig. 7 for explanation.

are unlikely to leak. The north–south oriented steeply dipping faults in Field 1 and Field 2 have the lowest likelihood of leakage.

The leakage maps indicate the potential for hydrocarbon leakage along any fault, but do not imply that any fault with red colors is currently leaking. As noted previously, a reservoir must abut the fault in the proper place, there must be hydrocarbons present to leak and the pore pressure must be high enough to reactivate the fault in order for the leakage to take place. In order to create leakage potential maps in all of the fields, the stress data must be extrapolated throughout each field. We combine the stress and pore-pressure data across each field into a single one-dimensional model that varies with depth (Fig. 8), and extrapolate this model across the entire field. Therefore, there is some uncertainty about the leakage analysis in areas far removed from the wells with stress data.

The leakage analysis in Field 2 is not as reliable as in the other fields examined in this study because of the need to extrapolate the stress data far from the wells with direct measurements. However, the results are still consistent with the observation that the faults are not conducting hydrocarbons into the caprock in this field. A few smaller east–west trending faults appear to be capable of slipping in Field 2. Closer examination of these faults indicates that they are not large enough to separate pore-pressure compartments, and may be much smaller than indicated.

We noted above evidence of hydrocarbon leakage in Field 3 between wells A and F (Fig. 12). Fig. 13 shows that the lowest critical pressure perturbation along the fault between wells A and F is approximately 10 MPa. The reservoir intersects the fault at 2540 m; however, the shallowest pore-pressure measurement in this reservoir is at 2600 m. A 60-m column of gas above 2600 m increases the pore pressure approximately 3 MPa above the background pore pressure. This results in a pore pressure acting at the fault that is 7 MPa below the pore pressure we calculate is needed to activate the fault in the current stress field. Two factors may contribute to the discrepancy between the observed gas leakage and the predicted pore pressure needed to cause the leakage. First, the maximum horizontal stress (S_{Hmax}) used at this depth may underestimate the actual stress. Because the upper bound of S_{Hmax} was poorly constrained in well A, the lower bound was used as the best estimate of the maximum horizontal stress. Second, the pore pressure on the footwall side of the fault may be higher than in the hangingwall. However, no pore-pressure data were available for well F.

Geochemical studies of hydrocarbons from Field 3 and a field to the south indicate that there is a southward migration of hydrocarbons. The southward migration of hydrocarbons might be expected due to the overall trend of the Brent reservoir becoming shallower to the south-southeast, but the individual fault blocks dip to the east and southeast, which would encourage migration in a westerly and northwesterly direction. Our models predict that the faults in Field 3 have a high likelihood of leaking. We speculate that in order for hydrocarbons to migrate to the south and southeast, the reservoir-bounding faults might be conducting hydrocarbons along strike. However, we have no direct observations indicating that hydrocarbons are being conducted along strike in these faults.

Pore-pressure-induced faulting and leakage may be a dynamic mechanism that acts to control the maximum overpressure in reservoirs bounded by faults, and may explain the observation that throughout the world most economically recoverable hydrocarbons occur in reservoirs with pore-pressure gradients less than 17.4 MPa/km (Law and Spencer, 1998). In Visund the hydrocarbon column heights were smaller than expected (R. Færseth, pers. commun.). Because the water-phase pore pressure was so high in Visund, and the faults so close to slipping and leaking, we believe that this prevented larger hydrocarbon columns from remaining in the reservoir. Field 1 and Field 2 have a very low likelihood of leakage and low water-phase pore pressures. Hence, these fields appear to be capable of maintaining large columns of hydrocarbon in the deeper reservoirs where the faults are unlikely to slip and leak. We show that Field 3 is slightly less likely to leak than Visund, and the pore pressures are predominantly hydrostatic. Hydrocarbon column heights in Field 3 may be limited because the faults may conduct hydrocarbons to the south.

The relationship between overpressure and fault leakage observed in the study fields appears to be present throughout the northern North Sea. Hermanrud et al. (1997) have deduced from seismic chimneys and hydrocarbon shows in caprocks that most overpressured hydrocarbon-bearing structures in the northern North Sea are currently leaking.

The timing of the compression in the northern North Sea has implications for the timing of hydrocarbon leakage and migration. Long-term compression in the North Sea has been inferred from inversion structures observed offshore Norway (Rohrman et al., 1995; Vågnes et al., 1998), and bordering other parts of the northeast Atlantic Margin (Doré and Lundin, 1996). These studies generally indicate that compression may have started with ridge push from Meso–Cenozoic time and extended into the Neogene. However, Grollimund (2000) has shown that glacial loading may have reduced the compressive stresses and stopped active faulting in the northern North Sea while the glacial ice sheet was present. A number of investigators have suggested that the current compressional stress observed in this region may be related to, or enhanced by lithospheric flexure associated with the subsequent deglaciation in the Pleistocene (Stephansson, 1988; Klemann and Wolf, 1998; Grollimund and Zoback, 2000). If this interpretation is correct, the existence of the current compressional stress in this area is a geologically recent ($\sim$10,000–15,000 years old) phenomenon.

As noted by many workers, juxtaposition of lithologies, fault structure and content may be responsible for sealing faults in hydrocarbon reservoirs. However, it appears that the stress and pore pressure acting on a fault surface determine whether a fault will begin to leak after it has sealed, regardless of the fault structure and content. Faults may creep and re-seal once they have slipped and leaked, essentially behaving as pressure-release valves (e.g. see Sibson, 1992; Finkbeiner et al., 2001). Once the pore pressure rises to the critical level (e.g. as a result of hydrocarbons migrating into the field or reservoir compaction), the fault could then slip again and release more hydrocarbons. Hydrofracture is unlikely to play a role in hydrocarbon leakage, as the pore pressure must rise to the level of the minimum principal stress in order to fracture the formation. Well-oriented faults will begin slipping before the pore pressure can rise to such a level. One notable exception to this is severely overpressured formations in which very small pressure perturbations can induce either fault slip or hydrofracture.

Conclusions

Stress, pore pressure, and fault orientation appear to be important factors in controlling hydrocarbon leakage and migration in the northern North Sea. Faults that are critically stressed in the current stress field (i.e. capable of slipping) tend to leak, whereas those that are not critically stressed are more likely to be sealing. Fault reactivation and hydrocarbon leakage in this area appear to be caused by three factors: (1) locally elevated pore pressure due to buoyant hydrocarbons abutting faults, (2) fault orientations that are nearly optimally oriented for frictional slip in the present-day stress field, and (3) a recent perturbation of the compressional stress associated with postglacial rebound. The combination of these three factors may have recently induced fault slippage and gas leakage along sections of previously sealing reservoir-bounding faults in some fields, whereas in others, the stress and pore pressure are not sufficient to cause fault reactivation. In cases where reservoir-bounding faults are not potentially active, the pore-pressure difference across faults can become quite high. Hence, the leakage potential of reservoir-bounding faults appears to exert an important influence on potential hydrocarbon column heights.

Acknowledgements

We thank Norsk Hydro for providing the data and financial support for this project. We thank Bjorn Larsen for suggesting that this work be initiated at Norsk Hydro. We also thank Linn Arnesen, Nils Kageson-Loe, Roald Færseth, Arnd Wilhelms, and Paul Gillespie at Norsk Hydro for their efforts to provide us with data and for helpful discussions.

References

Allan, U.S., 1989. Model for hydrocarbon migration and entrapment within faulted structures. Bull. Am. Assoc. Pet. Geol., 73: 803–811.

Barton, C.A., Zoback, M.D. and Moos, D., 1995. Fluid flow along potentially active faults in crystalline rock. Geology, 23: 683–686.

Barton, C.A., Hickman, S.H., Morin, R., Zoback, M.D. and Benoit, D., 1998. Reservoir-scale fracture permeability in the Dixie Valley, Nevada, geothermal field. In: Proceedings of SPE/ISRM Rock Mechanics in Petroleum Engineering, Trondheim, 2. Society of Petroleum Engineers, pp. 315–322.

Berg, R.R. and Avery, A.H., 1995. Sealing properties of Tertiary growth faults, Texas Gulf Coast. Am. Assoc. Pet. Geol. Bull., 79: 375–393.

Brudy, M. and Zoback, M.D., 1993. Compressive and tensile failure of boreholes arbitrarily-inclined to principal stress axes: application to the KTB boreholes, Germany. Int. J. Rock Mech. Min. Sci. Geomech. Abstr., 30: 1035–1038.

Brudy, M. and Zoback, M.D., 1999. Drilling-induced tensile wall-fractures: implications for determination of in-situ stress orientation and magnitude. Int. J. Rock Mech. Min. Sci., 36: 191–215.

Brudy, M., Zoback, M.D., Fuchs, K., Rummel, F. and Baumgärtner, J., 1997. Estimation of the complete stress tensor to 8 km depth in the KTB scientific drill holes: Implications for crustal strength. J. Geophys. Res., 102: 18453–18475.

Byerlee, J.D., 1978. Friction of rocks. Pure Appl. Geophys., 116: 615–629.

Castillo, D.A., Bishop, D.J., Donaldson, I., Kuek, D., de Ruig, M., Trupp, M. and Shuster, M.W., 2000. Trap integrity in the Laminaria High–Nancar Trough region, Timor Sea: prediction of fault seal failure using well-constrained stress tensors and fault surfaces interpreted from 3D seismic. Aust. Pet. Produc. Explor. Assoc. J., 40, part 1: 151–173.

Dholakia, S.K., Aydin, A., Pollard, D.D. and Zoback, M.D., 1998. Fault-controlled hydrocarbon pathways in the Monterey Formation, California. Am. Assoc. Pet. Geol. Bull., 82 (8): 1551–1574.

Doré, A.G. and Lundin, E.R., 1996. Cenozoic compressional structures on the NE Atlantic margin: nature, origin and potential significance for hydrocarbon exploration. Pet. Geosci., 2: 299–311.

Downey, M.W., 1984. Evaluating seals for hydrocarbon accumulations. Am. Assoc. Pet. Geol. Bull., 68: 1752–1763.

Finkbeiner, T., Barton, C. and Zoback, M.D., 1997. Relationships among in-situ stress, fractures and faults, and fluid flow; Monterey Formation, Santa Maria Basin, California. Am. Assoc. Pet. Geol. Bull., 81 (12): 1975–1999.

Finkbeiner, T., Zoback, M.D., Flemings, P. and Stump, B., 2001. Stress, pore pressure, and dynamically constrained hydrocarbon columns in the South Eugene Island 330 Field, northern Gulf of Mexico. Am. Assoc. Pet. Geol. Bull., 85(6): 1007–1031.

Færseth, R.B., Sjøblom, T.S., Steel, R.J., Liljedahl, T., Sauar, B.E. and Tjelland, T., 1995. Tectonic controls on Bathonian–Volgian syn-rift successions on the Visund fault block, northern North Sea. In: R.J. Steel, V. Felt, E.P. Johannessen and C. Mathieu (Editors), Sequence Stratigraphy of the Northwest European Margin. Norwegian Petroleum Society (NPF), Special Publication 5. Elsevier, Amsterdam, pp. 325–346.

Fristad, T., Groth, A., Yielding, G. and Freeman, B., 1997. Quantitative fault seal prediction — a case study from Oseberg Syd. In: P. Møller-Pedersen and A.G. Koestler (Editors), Hydrocarbon Seals: Importance for Exploration and Production. Norwegian Petroleum Society (NPF), Special Publication 7. Elsevier, Amsterdam, pp. 107–125.

Grollimund, B., 2000. Impact of Deglaciation on Stress and Implications for Seismicity and Hydrocarbon Exploration. Ph.D. dissertation, Stanford University.

Grollimund, B. and Zoback, M.D., 2000. Post glacial lithospheric flexure and induced stresses and pore pressure changes in the northern North Sea. Tectonophysics, 327: 61–81.

Hermanrud, C., Norgård Bolås, H.M., Mari, H., Fichler, C., Rornes, A. and Heggland, R., 1997. Studies of hydrocarbon migration; an important discipline in hydrocarbon exploration. Am. Assoc. Pet. Geol. Bull., 81: 1383.

Hickman, S.H., Zoback, M.D. and Benoit, R., 1998. Tectonic controls on fault-zone permeability in a geothermal reservoir at Dixie Valley, Nevada. In: Proceedings of SPE/ISRM Rock Mechanics in Petroleum Engineering, Trondheim, 1. Society of Petroleum Engineers, pp. 79–86.

Hunt, J.M., 1990. Generation and migration of petroleum from abnormally pressured fluid compartments. Am. Assoc. Pet. Geol. Bull., 74: 1–12.

Jaeger, J.C. and Cook, N.G.W., 1979. Fundamentals of Rock Mechanics (3rd ed.). Chapman and Hall, New York, 593 pp.

Klemann, V. and Wolf, D., 1998. Modeling of stresses in the Fennoscandian lithosphere induced by Pleistocene glaciations. Tectonophysics, 294: 291–303.

Knipe, R.J., 1992. Faulting processes and fault seal. In: R.M. Larsen, H. Brekke, B.T. Larsen and E. Talleraas (Editors), Structural and Tectonic Modelling and its Application to Petroleum Geology. Norwegian Petroleum Society (NPF), Special Publication 1. Elsevier, Amsterdam, pp. 325–342.

Law, B.E. and Spencer, C.W., 1998. Abnormal pressure in hydrocarbon environments. Am. Assoc. Pet. Geol. Mem., 70: 1–11.

Lindholm, C.D., Bungum, H., Villagran, M. and Hicks, E., 1995. Crustal stress and tectonics in Norwegian regions determined from earthquake focal mechanisms. In: Proceedings of the Workshop on Rock Stresses in the North Sea, Trondheim, Feb. 13–14. SINTEF Rock and Mineral Engineering, pp. 77–91.

Moos, D. and Zoback, M.D., 1990. Utilization of observations of well bore failure to constrain the orientation and magnitude of crustal stresses: Application to continental, Deep Sea Drilling Project, and Ocean Drilling Program boreholes. J. Geophys. Res., 95: 9305–9325.

Nybakken, S., 1991. Sealing fault traps — an exploration concept in a mature petroleum province: Tampen Spur, northern North Sea. First Break, 9: 209–222.

Rohrman, M., van der Beek, P., Andriessen, P. and Cloetingh, S., 1995. Meso-Cenozoic morphotectonic evolution of southern Norway: Neogene domal uplift inferred from apatite fission track thermochronology. Tectonics, 14 (3): 704–718.

Sibson, R.H., 1992. Implications of fault-valve behaviour for rupture nucleation and recurrence. In: T. Mikumo, K. Aki, M. Ohnaka, L.J. Ruff and P.K.P. Spudich (Editors), Earthqake Source Physics and Earthquake Precursors. Tectonophysics, 211: 283–293.

Stephansson, O., 1988. Ridge push and glacial rebound as rock stress generators in Fennoscandia. Bull. Geol. Inst. Univ. Uppsala, N.S., 14: 39–48.

Townend, J. and Zoback, M.D., 2000. How faulting keeps the crust strong. Geology, 28: 399–402.

Vågnes, E., Gabrielsen, R.H. and Haremo, P., 1998. Late Cretaceous–Cenozoic intraplate contractional deformation at the Norwegian continental shelf: timing, magnitude and regional implications. Tectonophysics, 300: 29–46.

Weber, K.J., Mandl, G., Pilaar, W.F., Lehner, F. and Precious, R.G., 1978. The role of faults in hydrocarbon migration and trapping in Nigerian growth fault structures. Offshore Technology Conference, 10, Paper OTC 3356, pp. 2643–2653.

Wiprut, D.J. and Zoback, M.D., 2000a. Constraining the full stress tensor in the Visund field, Norwegian North Sea: application to wellbore stability and sand production. Int. J. Rock Mech. Min. Sci., 37: 317–336.

Wiprut, D. and Zoback, M.D., 2000b. Fault reactivation and fluid flow along a previously dormant normal fault in the northern North Sea. Geology, 28 (7): 595–598.

Zoback, M.D., Apel, R., Baumgärtner, J., Brudy, M., Emmermann, R., Engeser, B., Fuchs, K., Kessel, W., Rischmüller, H., Rummel, F. and Vernik, L., 1993. Upper crustal strength inferred from stress measurements to 6 km depth in the KTB borehole. Nature, 365: 633–635.

D. WIPRUT *Department of Geophysics, 397 Panama Mall, Stanford University, Stanford, CA 94305-2215, USA*
Present address: GeoMechanics International Inc., Parmelia House Level 1, 191 St. George's Terrace, Perth, WA 6000, Australia. E-mail: wiprut@geomi.com

M.D. ZOBACK *Department of Geophysics, 397 Panama Mall, Stanford University, Stanford, CA 94305-2215, USA*

Leakage from overpressured hydrocarbon reservoirs at Haltenbanken and in the northern North Sea

Christian Hermanrud and Hege Marit Nordgård Bolås

Most of the deeply buried reservoirs at Haltenbanken and in the Northern North Sea offshore Norway are significantly overpressured. In the North Sea, several of these overpressured Jurassic structures contain hydrocarbons, and few of the failures are thought to be results of cap rock leakage. To the contrary, none of the eight wells which penetrated overpressured Jurassic reservoir rocks at Haltenbanken prior to 1996 hit hydrocarbons, and cap rock leakage seems to be the main explanation for these results. Later drilling in this area, aiming at deeper targets, has however resulted in three hydrocarbon discoveries in overpressured reservoirs at Haltenbanken.

Analysis of pore pressures and leak off pressures in the two areas demonstrate that the high pore pressures at Haltenbanken follow a well-defined gradient with depth. The pressures are however significantly lower than the highest pore pressures which are encountered in the North Sea at similar depths. The leak off pressures and overburden weights are broadly similar between the two areas at any given depth. It is argued that the maximum pore pressures at Haltenbanken are controlled by leakage through fracturing. As a consequence, the stress state at Haltenbanken at the time of leakage was such that rock failure occurred. With this assumption, the maximum compressive stress at the time of failure can be calculated. Such calculations demonstrate that the maximum stress at the time of fracturing at Haltenbanken was significantly higher than the weight of the overburden, especially at shallow burial depths. Such high (paleo) horizontal stresses apparently did not exist in the North Sea.

The proposed high horizontal paleostress at Haltenbanken is suggested to be a result of flexuring due to repeated glaciations and deglaciations during the Quaternary. As the ice rested on the shelf edge west of the Haltenbanken at times, and was floating west of this shelf edge, a large contrast in iceload over a short distance resulted in the downflexing of the crust beneath the ice. As there is no shelf edge in the North Sea, the ice tapered off much more gradually here, and the crustal flexuring was spread over a larger area resulting in a more gentle bending and less amplification of the horizontal stress.

As a result of this work, it is suggested that the probability of cap rock leakage from overpressured reservoirs is comparatively high in areas which are situated close to the shelf edge. This risk of leakage is decreasing with depth.

Introduction

Both Haltenbanken and the northern North Sea are situated on the Norwegian continental shelf (Fig. 1). The geology of both areas has been extensively described; see Koch and Heum (1995) and Jacobsen and Spencer (1987) for further information. The geological setting in the two areas share several similar features, the most prominent being the following.

The main reservoirs consist of pre-rift (Middle Jurassic) and syn-rift (Upper Jurassic) sandstones. Rifting of Late Jurassic to Early Cretaceous age formed the structural traps which contain most of the proven hydrocarbons, and organic-rich marine shales of Late Jurassic age are the main source rocks. The areas subsided gently throughout early Tertiary times, and more rapidly in the Pliocene and Pleistocene. The source rocks are presently in the oil window in large parts of the study area, and the late subsidence has resulted in continuous hydrocarbon supply to most structural traps, a supply which is probably still ongoing. The reservoirs are typically in the 2–5 km depth range.

Mainland Norway was significantly uplifted in the late Cenozoic, and areas close to the Norwegian coastline are not presently at their maximum burial depths. The Haltenbanken and North Sea also experienced repeated periods of glaciation and deglaciation during the Quaternary.

The main differences between the two areas are the following: (1) sandstone units of reservoir quality of Late and Middle Cretaceous age are present at Haltenbanken (the Lange and Lysing formations); (2) the Pliocene to Pleistocene subsidence at Haltenbanken is about 20–30% larger than in the northern North Sea; and (3) Quaternary erosion formed the Norwegian Trench in the North Sea, an up to 300 m deep trench which parallels the Norwegian coastline.

The exploration history of Haltenbanken was summarized by Koch and Heum (1995). At that time,

Hydrocarbon Seal Quantification edited by A.G. Koestler and R. Hunsdale. NPF Special Publication 11, pp. 221–231, Published by Elsevier Science B.V., Amsterdam.

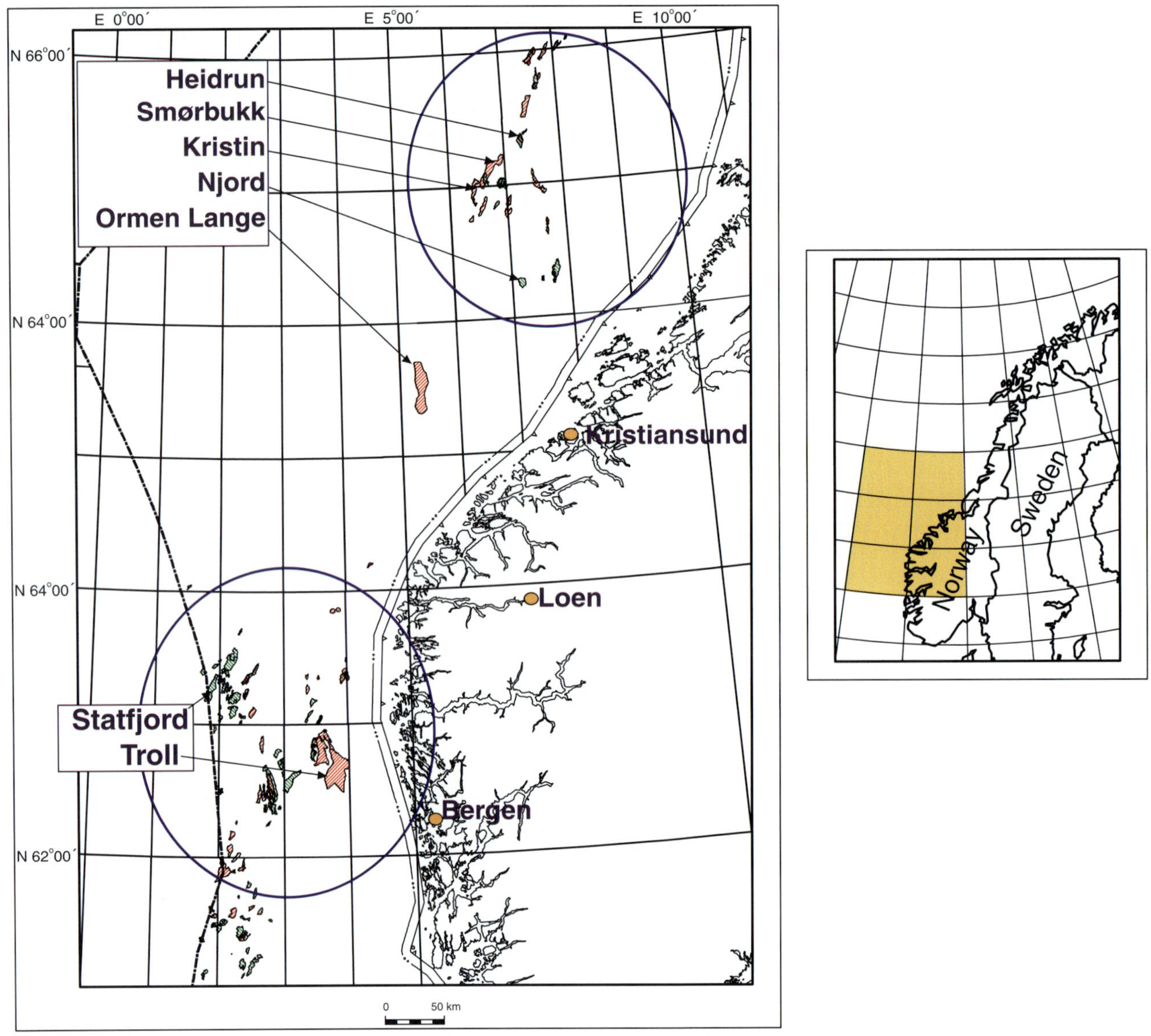

Fig. 1. Location of the Haltenbanken and northern North Sea areas.

it was realized that the fluid pressures in the area fall into three classes: an overpressured region to the west, an area with close to hydrostatic fluid pressures to the east, and an intermediate pressure region which separates the two in the southern part of Haltenbanken (Fig. 2). All eight exploration wells which had been drilled in the overpressured region had been failures. Later drilling in this region has however resulted in three discoveries (Kristin, Erlend and Ragnfrid), and significant exploration potential still exists in this region.

Knag et al. (1995) discussed the exploration history of the northern North Sea. Leakage between different Jurassic sandstone units is often seen in this area. Cap rock leakage from the uppermost Jurassic reservoirs does occur, but it is less common than in the overpressured region of Haltenbanken. No sound explanation of the different leakage frequencies between the two areas appears to have been given. Neither has the fact that the three deepest wells at Haltenbanken (Kristin, Erlend and Ragnfrid) were successful, contrary to all earlier wells, been explained.

The work which is presented here was aimed at explaining these observations. This was done by first investigating the (paleo-) stress fields and thus the conditions for leakage through fracturing, and then by seeking a geological explanation for the observations. As an introduction to the discussion of the results, some important aspects of hydrocarbon leakage which are often overlooked will be addressed.

Volumetric constraints on hydrocarbon seals

Permeable reservoir rocks belong to fluid pressure compartments (Buhrig, 1989). The fluid over-

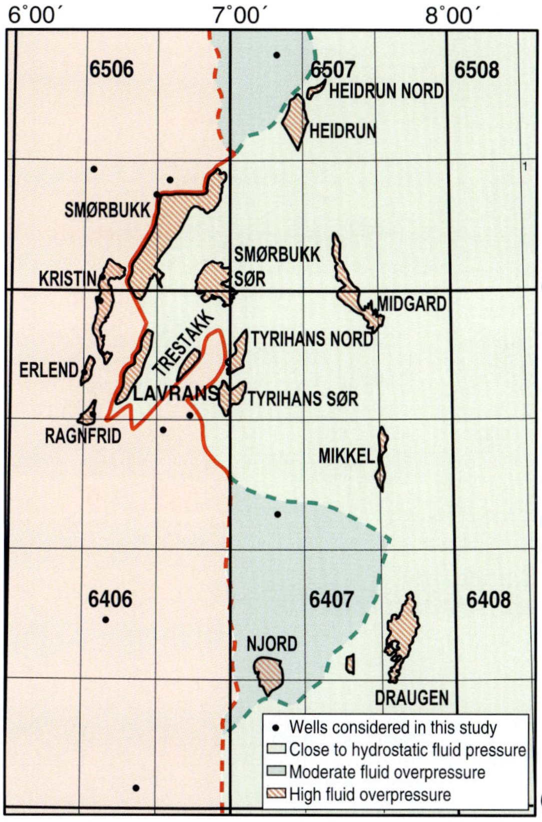

Fig. 2. Different pressure regimes at Haltenbanken. The position of the division line between the pressure regimes is fairly well constrained in the 65°15′ to 65°45′ area, and is less well constrained north and south of this area.

pressure in the water zone is constant within such compartments, and fluids can flow freely within each compartment. Each compartment is separated from its neighboring compartments by impermeable or semi-permeable boundaries: faults, fractures or sealing rocks. When such compartments subside, fluid must leave. This is because (a) the porosity in the compartment is being reduced during burial, (b) most compartments receive fluids from deeper compartments (where the porosity is also being reduced), and (c) because of hydrocarbon supply — the hydrocarbons must occupy a volume, and corresponding volumes of water (or hydrocarbons) must leave to create a space for these new fluids. Fluid overpressuring can only compensate for a minor part of the excess volumes, due to the low compressibility of formation water (Hermanrud et al., 1998). Exceptions to this relationship may exist (e.g. in uplifted areas and areas where all the basin's porosity reduction is due to mechanical compaction and not a result of diagenesis), but the 'all compartments leak' concept has been valid both in the Haltenbanken and in the North Sea in the time period that was of interest in the study which is reported here.

A consequence of this concept is that the important

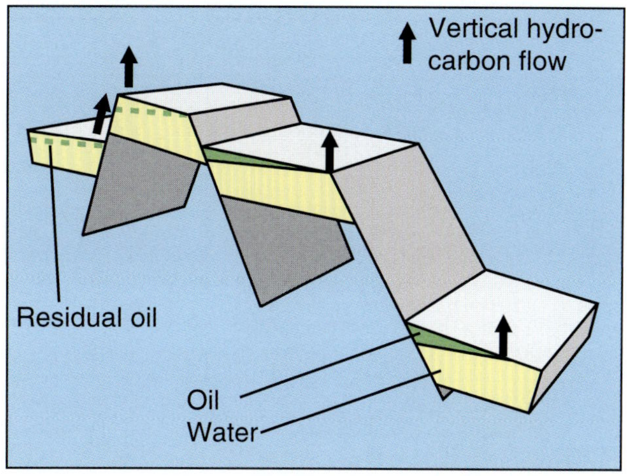

Fig. 3. Leaking pressure compartments. All compartments leak during subsidence, but leakage only restricts any in-place hydrocarbons if it takes place above the hydrocarbon–water contact.

issue may not be whether a pressure compartment leaks or not, but (a) how it leaks (which processes are rate-limiting), (b) where it leaks, (c) and whether such leakage can be identified from seismic data. The last two topics are addressed by Nordgård Bolås and Hermanrud (2002) and Teige et al. (2002).

As visualized in Fig. 3, a pressure compartment can leak through numerous locations, e.g. from its apex (shallowest point), along a fault which intersects the reservoir in a downdip position, through the topseal, side seal or base seal. The leakage can take place through pore networks or through faults or fractures; it can be vertical or lateral; and it can take place from the hydrocarbon-filled or the (downflanks) water-filled zone of the pressure compartment. The prospectivity of an undrilled reservoir is intimately linked to the fluid escape characteristics from the compartment to which it belongs.

Stress and fluid pressure observations

Fig. 4 shows the pore pressures, leak off pressures and overburden pressures for the overpressured structures at Haltenbanken. The fluid pressures were taken from RFT measurements. The measurements include all overpressured (6406/2-3T3, 6406/2-6, 6406/2-7, 6406/3-1, 6406/6-1, 6406/8-1, 6406/11-1S, 6406/12-1S, 6506/11-1, 6506/11-3, 6506/12-4), and moderately pressured (6507/7-1, 6407/4-1) Jurassic reservoirs in the study area, and also all RFT measurements in Cretaceous rocks in the same area. Each data point represents the shallowest measurement on the structure (only one pressure measurement per structure), and only pressure values from the shallowest of the Jurassic formations were selected. The Njord field was not included, for reasons to be discussed later.

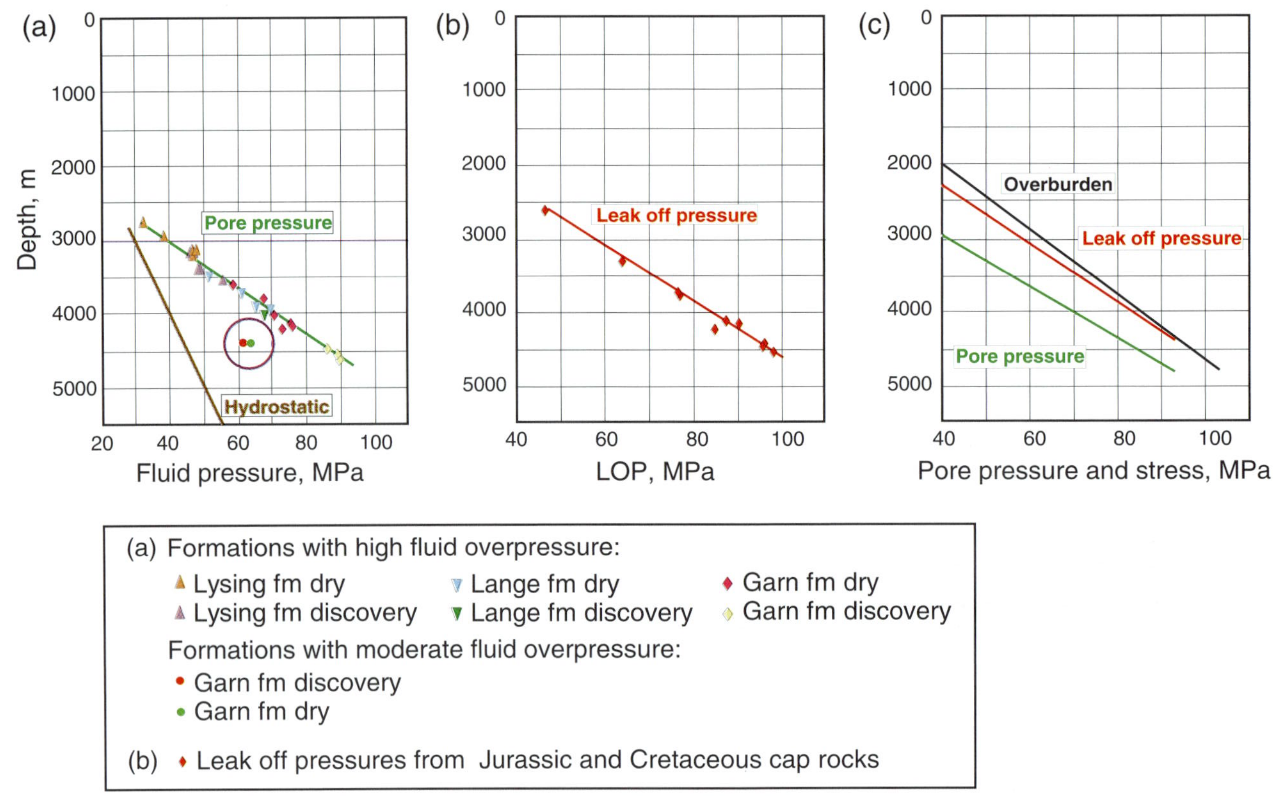

Fig. 4. Haltenbanken well data: (a) pore pressures for overpressured and intermediately pressured wells, (b) leak off pressures from the same wells, (c) pore pressure, leak off pressure and overburden gradients for the overpressured wells combined. The three deepest data represent the Kristin, Erlend and Lavrans discoveries. The circled data represent the intermediately pressured wells 6407/4-1 and 6507/7-1.

The pressures in the two moderately pressured wells may actually be higher than suggested from Fig. 4, as the reservoir quality in both wells was badly influenced by diagenetic illite, and no reliable RFT measurements could be taken. Well 6507/7-1 did not encounter hydrocarbons, and hydrocarbon shows were virtually absent in the well, probably due to lack of charging. Well 6407/4-1 encountered gas, but the discovery is non-economical due to the poor reservoir quality.

The leak off pressure (LOP) values, which are measurements of the maximum drilling mud pressure a well can tolerate before it fractures, are commonly taken as approximations of the least compressive stress (see Nordgård Bolås and Hermanrud, 2002). One LOP measurement for each overpressured well is included in Fig. 4, with the exception of well 6506/12-4, where the drilling report states distrust in the obtained LOP value. The LOP tests were in most cases taken immediately above or just in the uppermost part of the reservoir. The overburden curve was taken from Svare (1995). The approach of using one single (average) overburden curve was selected because of the uncertainty attached to deriving overburden curves from density logs (Svare, 1995; Nordgård Bolås and Hermanrud, 2002).

Fig. 5 shows corresponding data from the northern North Sea. One measurement for each overpressured Jurassic structural trap was recorded. This figure includes data from all overpressured structures where reliable pre-production data could be retrieved, but omits data where several structures are contained within the same pressure compartment. Most of these wells penetrated hydrocarbon-bearing reservoirs (Gullfaks, Kvitebjørn, Vigdis, Visund, 30/2-1, 30/3-1, 30/4-2, 30/7-7, 30/7-8, 34/10-23, 34/10-35, 35/10-2), but some were water-bearing. Of these, well 30/4-1 may have failed because of leakage, although no shows were reported from the well, while the remaining wells (25/1-10, 35/4-1 and 35/10-1) contained ample shows of hydrocarbons, and probably failed because of leakage. The only remaining Jurassic target where failure is attributed to leakage in the study area is well 30/11-4. This well encountered close to hydrostatic fluid pressures, and was for this reason not included in Fig. 5. It is stressed that post-mortem classification of why individual wells have failed is subjective and prone to error, especially for older wells. Several other wells have failed in the study area, and it is entirely possible that some of these failed because of cap rock leakage. However, none of these wells were significantly overpressured.

Throughout this study, a well is classified as a discovery also when the structural trap is not filled

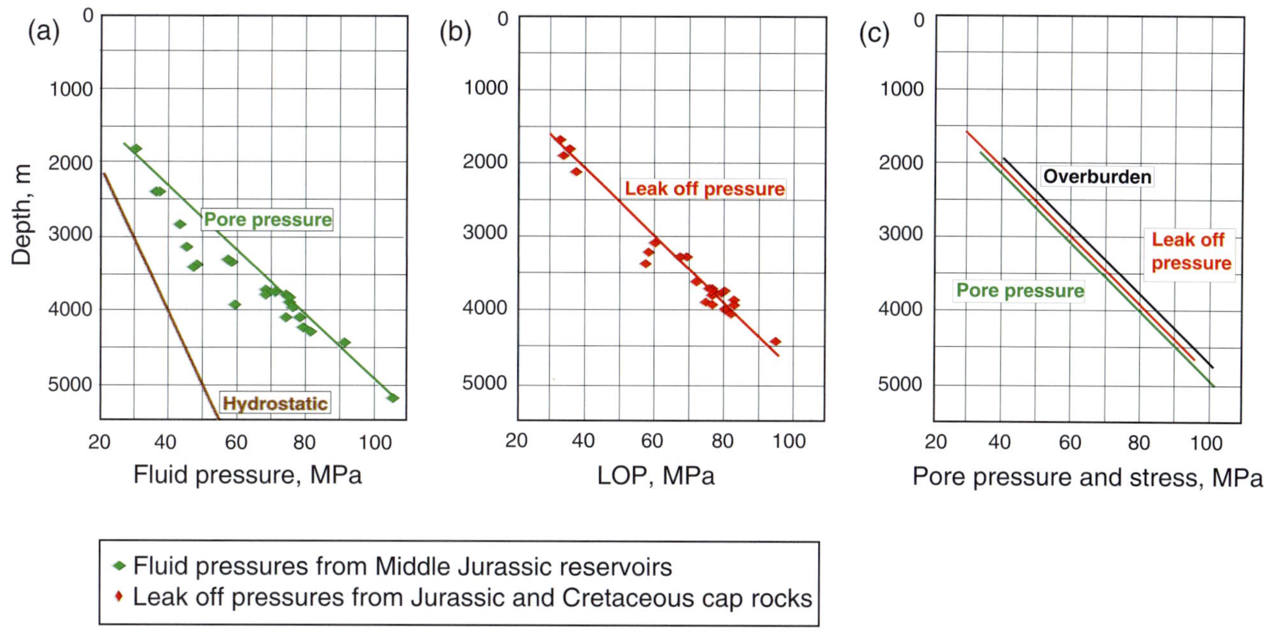

Fig. 5. Northern North Sea well data: (a) pore pressures from overpressured wells, (b) leak off pressures from the same wells, (c) pore pressure, leak off pressure and overburden gradients for the overpressured wells combined.

to structural spill point, and when the lack of filling may be due to cap rock leakage. The uncertainty attached to spillpoint definition in general, and for subcommercial discoveries covered with sparse data in particular, led to this somewhat inappropriate well classification. This classification does however not influence the conclusions of this study.

The LOP data from the North Sea were taken from the same wells as those where fluid pressure data were recorded, based on the same criteria as for the Haltenbanken data. As the overburden rocks at Haltenbanken and the North Sea are quite similar, and because of the uncertainty attached to overburden curves derived from integration of density logs in general, it was decided to apply the Haltenbanken curve also for the North Sea analysis. Averaging of overburden gradients from several North Sea wells were attempted as a control measure, and the resulting curves proved to be virtually identical to the Haltenbanken curve produced by Svare (1995).

The pressure data from Haltenbanken show a remarkable consistent linear increase with depth. The same observation applies to the LOP data from the same area. However, the pore pressure vs. depth gradient is significantly below the overburden and LOP gradients. These results differ distinctively from those of the northern North Sea, where the pore pressures of the structures with the highest overpressures show significantly more scatter, but where straight lines through the highest pressure and leak off values with depth are seen to be very close to each other, and also very close to the overburden weight.

Why do the LOP and pore pressures differ between the areas?

The significant differences in LOP and pore pressures vs. depth between the two areas is hardly coincidental. The suggested explanations which resulted from this work are the following.

(1) The pore pressures at Haltenbanken were limited by vertical cap rock leakage. This leakage took place through faults or fractures.

(2) The stress regime at the time of leakage at Haltenbanken was such that failure was initiated, and pore pressures could not get any higher than their present value. The state of stress can be computed under these conditions. Such computations suggest a horizontal stress significantly in excess of the vertical stress at the time of leakage.

(3) High horizontal paleostress resulted from flexuring initiated by advance and withdrawal of ice over the shelf edge during the Quaternary, where an ice edge of significant thickness was developed. This stress resulted in a large stress anisotropy, which may have triggered leakage from overpressured fields. No shelf edge existed in the North Sea, and so the flexuring due to glaciation was less pronounced in this area.

The next three subsections address each of these statements separately.

Mode of leakage at Haltenbanken

The well-defined pore pressure vs. depth relationship at Haltenbanken is hardly accidental, and points to a common mechanism which has resulted in these pore pressures. As the pore pressure/depth relationship is not parallel to the hydrostatic line, lateral communication in the area is poor. It is therefore inferred that the excess fluids which must leave the pressure compartments as porosity is reduced during burial, leave vertically through the cap rocks. (If the fluids were expelled horizontally, then most of the structural traps would be hydrocarbon-bearing. These traps have received ample amounts of hydrocarbons through time, and residual hydrocarbons (shows) and hydrocarbon inclusions testify to previous hydrocarbon filling of most of these structures (Teige et al., 2002).)

The vertical fluid discharge could take place either through the pore network, or through faults or fractures. However, leakage through the pore network appears unlikely, notably for the following two reasons.

First, hydrocarbons would have to overcome the capillary entry pressure of the cap rocks to leak from the traps. The capillary trapping capacity of a cap rock is related to the diameter of the largest connected pore throats through the cap rock. However, the existence of a hydrocarbon column of about 600 m at the Smørbukk field testifies that the pore throats are sufficiently small for sealing purposes.

One may speculate whether high fluid pressures could amplify the driving force (buoyancy), which must overcome the capillary entry pressure to initiate flow of hydrocarbons in the pore throats. This may however not necessarily be the case, especially when the hydrocarbon columns are small, as pressure communication in the pendular water (the water between the grains and the hydrocarbons) may be transmitted into the cap rock. Such a transfer would reduce the pressure difference between the oil phase in the reservoir and the water phase in the cap rock, which is the capillary threshold pressure. Anyhow, high fluid pressures alone do not explain leakage at Haltenbanken, as hydrocarbons have been discovered in the Kristin, Erlend and Lavrans fields, where the fluid pressures are even higher than in the leaky and somewhat shallower structures at Haltenbanken.

In addition, pore network leakage would favor downflanks leakage of water rather than leakage of hydrocarbons from the apex of the pressure compartment. Hydrocarbons would have to migrate through a largely water-wet cap rock. The relative permeability of hydrocarbons under such conditions would be significantly less than that of water, although this difference may be less for very finely grained rocks (Okui and Waples, 1993).

The stress stage at Haltenbanken at the time of leakage

If pore fluids leaked from the overpressured structures at Haltenbanken as a result of reopening of fractures or renewed faulting, then the faults/fractures must have been critically stressed at the time of leakage: the stress state was such that the conditions for rock failure was met. In a Mohr diagram, this is equivalent to stating that Mohr's circle touched the failure envelope (Fig. 6).

As shown in Fig. 6, the Mohr–Coulomb fracture criteria (Jaeger and Cook, 1979) are reached when

$$\frac{S_1 - P}{S_3 - P} = \left(\sqrt{\mu_2 - 1} + \mu\right)^2 \approx 3 \qquad (1)$$

where P is pore pressure, S_1 and S_3 are the largest and least principal stresses, respectively, and μ is the

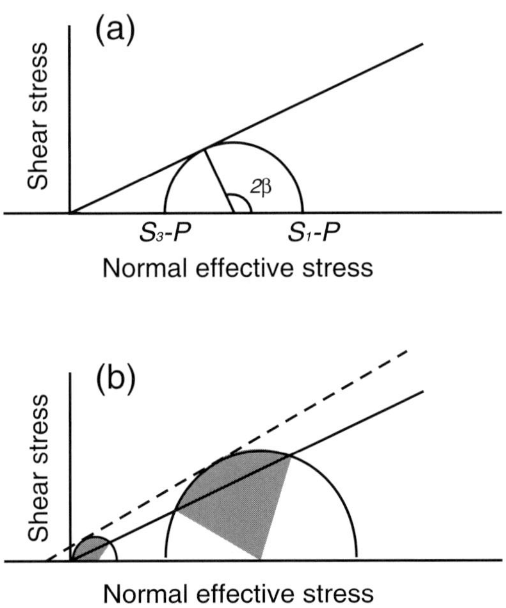

Fig. 6. (a) Mohr's circle with linear failure envelope, describing a stress state which indicates slippage along a fracture which has the most favorable orientation for slippage (i.e. the normal to the fault plane makes an angle (with the S_1 direction). (b) Mohr's circle with one failure envelope for reopening of preexisting fractures (solid line), and one for creation of new fractures (dotted line). The assumption that reactivation of a preexisting fracture which is most favorably oriented for slippage requires less stress than creation of a new fracture with the same strike, implies that the 'reactivation envelope' (solid) is below the 'new fracture envelope' (dotted). The circle to the left, which illustrates a close to isotropic stress state, has a larger portion of its circumference between the two failure envelopes than the bigger circle, which describes a more anisotropic stress state. The portion of Mohr's circle which lies between the two failure envelopes corresponds to a set of angles between the normal of the fault plane and the S_1 direction where slippage along preexisting fractures will be favored relative to initiation of new fractures (as stated implicitly by Barton et al., 1995). This set of angles (shaded area) is seen to increase as the stress state becomes more isotropic (the circle moves to the left). Thus, a close to isotropic stress state will favor reopening of old fractures rather than initiation of new ones.

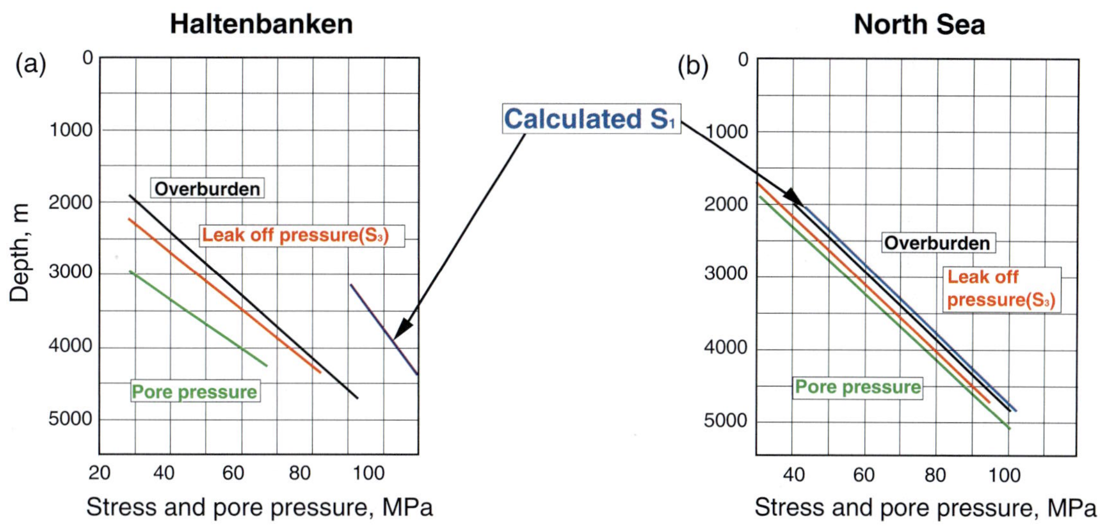

Fig. 7. Calculated maximum stress S_1 (blue line) for the Haltenbanken and North Sea areas at the time of leakage.

coefficient of friction. On the assumptions that the pore pressures were at their maximum value before fracturing was initiated, that the LOP can be taken as a measure of the least compressive stress, and with the commonly used value $\mu = 0.6$, the maximum principal stress S_1 is the only unknown parameter. It follows from this equation that the difference between the S_1 and the pore pressure should be approximately three times the difference between the LOP and the pore pressure at the time and depth of failure.

Fracturing may also result through creation of new fractures (as opposed to reopening of old ones). The conditions for such creation of new fractures (at the apex of the structure, where the effective stress is the least) can also be described by a failure envelope in a Mohr diagram. This failure envelope will be different from that which is critical for reactivation of old fractures. It is presently unclear how the difference between these two fracture envelopes should be quantified. However, at small stress anisotropy at fracturing conditions (or equivalently: at high pore pressure), the probability for failure along preexisting fractures will be favored, as long as opening of new fractures requires more stress than reopening a preexisting fracture which strikes in the most favorable orientation for slippage (Fig. 6). This observation suggests that the initiation of new fractures at the apex of a structural trap would be expected to take place more easily at Haltenbanken than in the northern North Sea, and also that the probability of such initiation of new fractures at the apex will diminish with increased burial depth at Haltenbanken.

The computed S_1 for the Haltenbanken and North Sea areas is shown in Fig. 7. The calculations for the North Sea point to a stress state which is close to isotropic. Indeed, if the maximum pore pressures here were limited by fracturing, then the S_1 at the time of fracturing may well have been equal to the overburden. To the contrary, a significant stress anisotropy apparently existed at Haltenbanken at the time when fracturing adjusted the pore pressures to their present values. This anisotropy seems to have been reducing with depth.

The fact that the calculated S_1 is significantly (25–35 MPa) in excess of the overburden, suggests that a significant horizontal stress may have existed at the time of leakage. The significant Plio/Pleistocene subsidence (1 km or more) with its resulting pulse of hydrocarbon generation suggests that leakage took place during the last few million years, pointing to processes related to Quaternary glaciations or deglaciations as the most likely causes of hydrocarbon leakage. The challenge is thus to identify a geological process which (a) promotes hydrocarbon leakage, (b) operates with different intensity at Haltenbanken and in the northern North Sea, (c) increases the horizontal stress of the sediments, and (d) is related to Quaternary glaciation or deglaciation events.

Glacial flexuring — a suggested mechanism for hydrocarbon leakage at Haltenbanken

The ice sheets are known to have covered the entire North Sea and Great Britain at their maximum extent. At Haltenbanken, they extended all the way to the shelf edge, as is evidenced by glacial erosional events at the seabed. The global water level was significantly (at least 100 m) lower at the time of maximum glaciation than it is today (Fairbanks, 1999). The maximum thickness of the ice sheets at the shelf edge are not well constrained. In general, ice which flows over shelf areas have low reliefs, and flowing ice does not reach thicknesses much in excess of that of an ice sheet floating on water (that is, if the water depth is

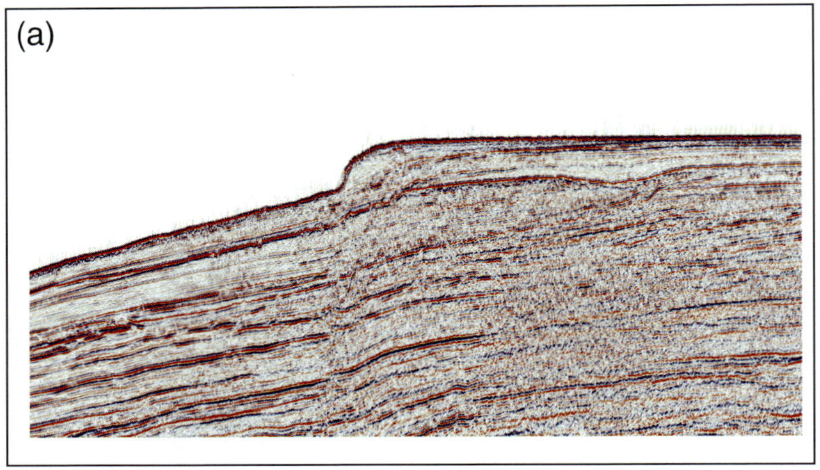

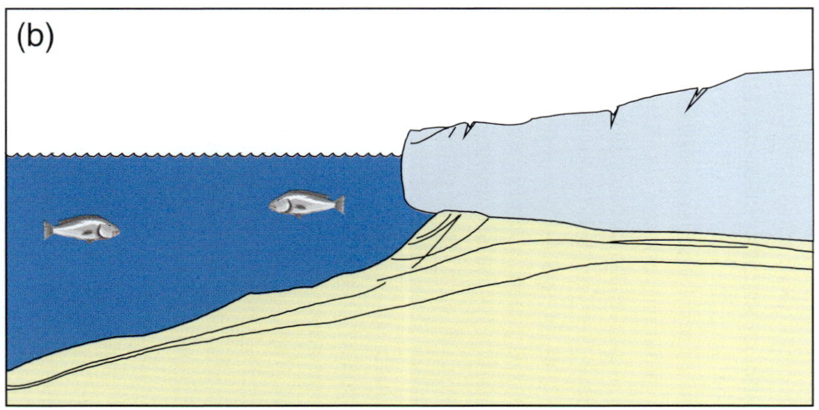

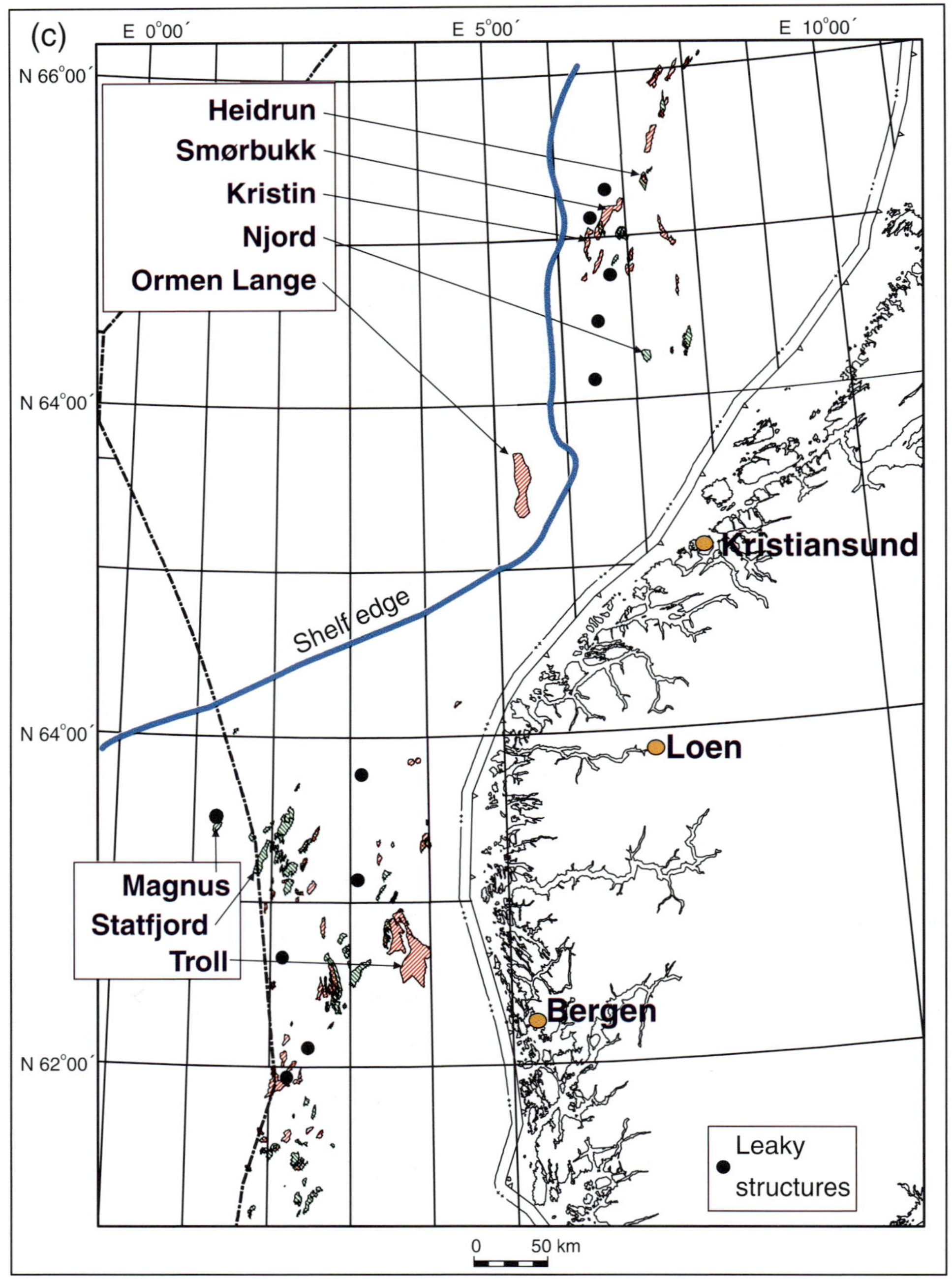

Fig. 8. (a) Seismic line over the shelf edge at approximately 63ºN. (b) Conceptual model of the ice sheet resting at the shelf edge. (c) The position of the shelf edge (blue line) and leaky wells in the study area. The leaky wells are (from north to south): 6506/12-4, 6506/11-1, 6506/3-1, 6406/8-1, 6406/11-1S, 35/4-1, UK 211/12a-14, 35/10-1, 30/4-1 (leaky?), 30/7-6, 30/11-4 and 25/1-10.

300 m, the ice thickness is about 350 m). Under such conditions, the ice does not exert significant vertical stress on the underlying sediments. However, in some areas, the ice will be stagnant, and significant ice domes may develop (as is the case at Berkner Island of western Antarctica today). Fjeldskaar (2000), who based his assessments of maximum ice thickness above sea level on the crust's flexural response to the glaciations, suggest ice thicknesses which tapered off from 1500 to 0 m from mainland Norway towards the shelf edge at 20,000 years B.P.

The shelf edge marks a fairly sharp transition between shallow and deep water areas offshore Mid-Norway. A seismic line over the shelf edge, a conceptual model of the position of the ice sheet, and a map view of its position, are shown in Fig. 8. As is seen from this figure, the shelf edge does not extend to the North Sea — the water here was nowhere deep enough to create such a feature. To the contrary, the shelf edge is present immediately west of Haltenbanken. At the time of maximum glaciation, the ice was resting at the shelf edge at Haltenbanken, and was floating in the deep water areas west of the shelf edge. This situation resulted in increased vertical stress at Haltenbanken (at the times when ice was stagnant and developed ice domes), but in no such increase in the area west of the shelf edge. As a consequence, flexural downbending of the Haltenbanken area took place, with the bend positioned at the shelf edge. The short lateral distance over which the vertical stress varied, probably resulted in a narrower bend (and thus a larger concentration of horizontal stress) than what was probably the case in the North Sea. Here, too, the ice tapered off gradually, but as it was resting on the substratum across the North Sea and covered the British Isles, no abrupt change in vertical load took place along a transect across the North Sea. In fact, the ice thickness maps of Fjeldskaar (2000) point to ice thickness variations in the range of 0–500 m from the Norwegian coastline across the North Sea at the time of maximum glaciation.

Inversion of earthquake focal mechanisms suggests that flexuring due to deglaciation of the Baltic shield significantly influences the present-day stress state both on- and offshore Norway (Hicks et al., 2000). It would be of significant interest to model the implications of these glacial events on the sediments in the two areas. Unfortunately, such modelling cannot be made with high accuracy, because of the imprecise knowledge factors such as maximum ice thickness, the duration of its position at various locations, paleowaterdepth, and viscous and elastic parameters of the various sediments and crust which are appropriate on the time scales of interest (ten-thousands to millions of years).

Grollimund et al. (1998) and Grollimund (2000) modelled the stress perturbations which resulted from emplacement and withdrawal of the latest (Weichselian) ice sheet. They concluded that the maximum horizontal stress could increase to 150% of the vertical load due to these processes in the northern North Sea, decreasing with depth. These numbers fit remarkably well with the results from the S_1 calculations based on Haltenbanken LOP and pore pressure data (150% at 3 km, 135% at 4 km).

Because of the above-mentioned uncertainties, the results of Grollimund et al. (1998) are not here taken as proof of paleostress in the North Sea or Haltenbanken areas. However, the calculations demonstrate that the order of magnitude of the stress variations which can be expected from glaciations are similar to those inferred from the pressure and LOP data. These results are thus taken as supportive, rather than conflicting, evidence for the leakage mechanisms which are suggested as results of this study.

Discussion

Among the leaky structures in the Norwegian sector of the northern North Sea, several (penetrated by wells 35/4-1, 35/10-1 and the underfilled gas discovery 35/10-2) have pore pressures which fall on the Haltenbanken gradient (Fig. 5). Whether this is accidental or not is presently unknown. It would be tempting to speculate that these wells, which are located close to the hinge line of late Cenozoic uplift (Doré and Jensen, 1996), also leaked because of flexuring. These wells, which are closer to the Norwegian coastline than the other leaky wells that have been identified, would probably be influenced by this uplift. Whether this uplift (which culminated during the Plio/Pleistocene) is the cause of this leakage, remains speculative.

As a curiosity, it is noted that the UK well 211/12a-14 close to the Magnus field also has leaked (Hull, 1994). The Magnus oil field area lies in the area in the North Sea which is closest to the shelf edge, and it might for that reason be expected to be leakage-prone. As for the above-mentioned Norwegian leaky wells, the observation of the single leaky 211/12a-14 structure fits with the general picture, but should not be given much weight as supporting evidence.

The Njord field was not included in this study. This field, which is situated at the southern part of Haltenbanken, has been classified as being in the intermediate pressure regime (Koch and Heum, 1995). This field is not filled to its structural spill point, and ample amounts of hydrocarbon shows testify that the field has leaked (Lilleng and Gundesø, 1997). The

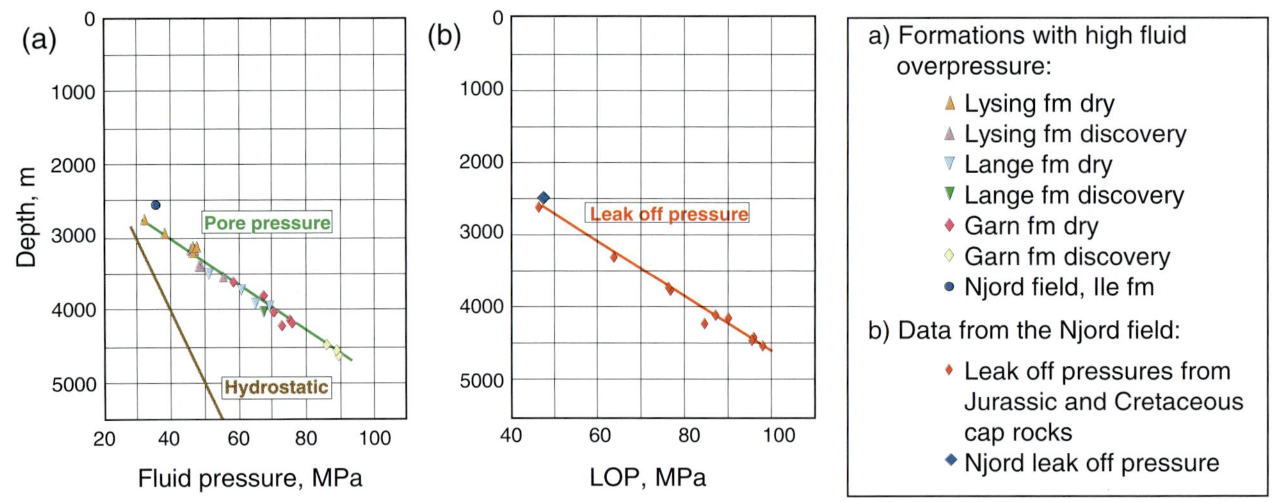

Fig. 9. Pore pressures in overpressured wells at Haltenbanken, including the Njord field. This field, which is further away from the shelf edge than the other overpressured wells at Haltenbanken, has comparatively higher pore and leak off pressures, and has also retained hydrocarbons despite its shallow burial depth.

fluid pressure data from this field are in fact somewhat above those of the regional pressure vs. depth relationship at Haltenbanken (Fig. 9). Its classification as of intermediate pressure apparently reflects that this field is shallower than those of the overpressured region. The facts that Njord (a) has a pore pressure slightly above the regional trend, (b) that it is situated further away from the shelf edge, and (c) that it is leaky, but still contains hydrocarbons, all fit with the general model of hydrocarbon leakage at Haltenbanken resulting from glacial flexuring.

The three deep discoveries at Haltenbanken (Kristin, Erlend and Lavrans) were modelled to be less influenced by glacial flexuring than their shallower overpressured counterparts, and the data also show that the pressure and stress conditions at such large depths at Haltenbanken are closer to those in the North Sea than observed at shallower depths (Fig. 4). Whether all of these fields are filled to spill point, or if they have leaked between the apex and the structural spill point (as the Njord field) remains to be seen. The observation that discoveries are made in the deepest parts of the overpressured region once again fits the concepts of glacial flexuring being the prime cause for leakage in the region, as the stress perturbations which results from such flexuring is decreasing with depth.

As has previously been noted, all pressure compartments in the area have leaked during the Plio/Pleistocene subsidence. Why, then, are hydrocarbons retained in the Kristin, Erlend and Lavrans compartments, while they have been removed from the other overpressured compartments?

Certainly, the mode of leakage must be different in the two situations. In general, one would expect that some compartments would leak from their apex, while others leaked downflanks, e.g. along faults which intersect the reservoirs somewhere downdip. Apparently, the stress increase due to glacial flexuring has promoted leakage from the apex. This could suggest that the induced stress, which is most pronounced at shallow levels, did not primarily reactivate preexisting faults (which extend deeper), but rather promoted fluid escape on local scale where the fracturing criteria were first overcome (at the apex). The calculated small difference between the stress anisotropy which is needed to initiate new fracturing and that which is needed to reopen preexisting faults or fractures at Haltenbanken clearly supports this statement. Where the stress impacts of glacial flexuring were less concentrated (deeper or further away from the shelf edge), a higher pore pressure could be tolerated, reactivation of faults may have taken over as the main mode of fluid leakage, and hydrocarbons may or may not have been retained depending on whether these faults intersected the reservoir at or below the apex.

Summary and conclusions

Leakage of hydrocarbons from overpressured reservoirs is a significant risk factor at Haltenbanken, but has been less severe in the North Sea. Hydrocarbons have lately been discovered also in overpressured reservoirs at Haltenbanken, but only in reservoirs below 4500 m MSL.

The leakage at Haltenbanken is suggested to have taken place as a result of fracturing or faulting, and not as migration through the pore network of the cap rock shales. If this suggestion is correct, then the stress state at the time of leakage can be computed. Such computations demonstrate that the horizontal stress at the time of leakage was significantly higher than the overburden weight.

It is argued that such high horizontal stresses may have resulted from flexuring due to glacial advances and withdrawal during the Quaternary. The flexuring was most pronounced at the shelf edge, where the ice was floating to the west, while resting on the shelf to the east, thus creating a significant difference in ice loading over a short lateral distance. As there is no shelf edge in the North Sea, the glacial flexuring was less pronounced here.

It is further suggested that the glacial flexuring at Haltenbanken resulted in fatal leakage due to formation of new fractures from the apex of the shallower structures. The deeper structures may also have leaked as a result of the flexuring, but reactivation of preexisting faults which did or did not intersect the apex of the structures was the main mode of leakage here.

As a result of this work, it is suggested that the probability of hydrocarbon leakage from overpressured reservoirs offshore Norway is comparatively high in areas which are situated close to the shelf edge. This risk of leakage is reducing with depth in the overpressured region of Haltenbanken.

Acknowledgements

The license partners of Pl 134 and 199, Agip, ElfTotalFina, ExxonMobil, Norsk Hydro and Statoil are thanked for permission to disclose confidential data for this publication. Professor Mark Zoback and students at Stanford University are thanked for guidance and patience during the first author's stay at Stanford, where parts of this work were performed. G.V. Søiland and A.G. Koestler are thanked for constructive reviews of an earlier version of the manuscript.

References

Barton, C.A., Zoback, M.D. and Moos, D., 1995. Fluid flow along potentially active faults in crystalline rock. Geology, 23 (8): 683–686.

Buhrig, C., 1989. Geopressured Jurassic reservoirs in the Viking Graben: modelling and geological significance. Mar. Pet. Geol., 6: 31–48.

Doré, A.G. and Jensen, L.N., 1996. The impact of late Cenozoic uplift and erosion on hydrocarbon exploration: offshore Norway and some other uplifted basins. Global Planet. Change, 12: 415–436.

Fairbanks, R., 1999. A 17,000 year glacio-eustatic sea-level record — influence of glacial melting rates on the Younger Dryas event and deep-ocean circulation. Nature, 342 (6250): 637–642.

Fjeldskaar, W., 2000. An isostatic test of the hypothesis of ice-free mountain areas during the last glaciation. Nor. Geol. Tidsskr., 80: 51–56.

Grollimund, B., 2000. Impact of Deglaciation on Stress and Implications for Seismicity and Hydrocarbon Exploration. Ph.D. thesis, Stanford University, CA.

Grollimund, B., Zoback, M. and Arnesen, L., 1998. Flexurally-induced stresses in the northern North Sea — preliminary comparison of observation and theory. Soc. Pet. Eng. SPE/ISPM 47243, pp. 189–198.

Hermanrud, C., Teige, G.M.G., Vik, E., Wensaas, L. and Nordgård Bolås, H.M., 1998. Overpressures in shales — do we know what they are and why they are there? Bull. Centre Rech. Elf. Explor. Prod., 22: 43–48.

Hicks, E.H., Bungum, H. and Lindholm, C.D., 2000. Stress inversion of earthquake focal mechanism solutions from onshore and offshore Norway. Nor. Geol. Tidsskr., 80: 235–250.

Hull, J., 1994. Prediction of the timing of overpressure development and top-seal failure in 211/12a-14 and 22/7-2RE. Project report, part fulfillment for M.S., Univ. of Aberdeen, TFN 1605.

Jacobsen, B. and Spencer, A.M., 1987. Bibliology of petroleum geology for the Norwegian North Sea (56°–62°N). In: A.M. Spencer, E. Holter, C.J. Campbell, S.H. Hanslien, P.H.H. Nelson, W. Nysæther and A.G. Ormaasen (Editors), Geology of the Norwegian Oil and Gas Fields. Graham and Trotman, London, pp. 429–443.

Jaeger, J.C. and Cook, N.G.W., 1979. Fundamentals of Rock Mechanics, 3rd ed. Chapman and Hall, London.

Knag, G.Ø., South, D. and Spencer, A.M., 1995. Exploration trends in the northern North Sea (60–62°N). In: S. Hanslien (Editor), Petroleum Exploration and Exploitation in Norway. Norwegian Petroleum Society (NPF), Special Publication 4. Elsevier, Amsterdam, pp. 115–134.

Koch, J.-O. and Heum, O.R., 1995. Exploration trends of the Halten Terrace. In: S. Hanslien (Editor), Petroleum Exploration and Exploitation in Norway. Norwegian Petroleum Society (NPF), Special Publication 4. Elsevier, Amsterdam, pp. 235–251.

Lilleng, T. and Gundesø, R., 1997. The Njord field: a dynamic hydrocarbon trap. In: P. Møller-Pedersen and A.G. Koestler (Editors), Hydrocarbon Seals: Importance for Exploration and Production. Norwegian Petroleum Society (NPF), Special Publication 7. Elsevier, Amsterdam, pp. 217–229.

Nordgård Bolås, H.M. and Hermanrud, C., 2002. Rock stress in sedimentary basins — implications for trap integrity, In: A.G. Koestler and R. Hunsdale (Editors), Hydrocarbon Seal Quantification. Norwegian Petroleum Society (NPF), Special Publication 11. Elsevier, Amsterdam, pp. 17–35 (this volume).

Okui, A. and Waples, D.W., 1993. Relative permeabilities and hydrocarbon expulsion from source rocks. In: A.G. Doré, J.H. Augustson, C. Hermanrud, D.J. Stewart and Ø. Sylta (Editors), Basin Modelling: Advances and Applications. Norwegian Petroleum Society (NPF), Special Publication 3. Elsevier, Amsterdam, pp. 293–301.

Svare, E., 1995. Relations between Rock Stresses and Pore-Pressure on the Norwegian Margin, 62°–67° North. A Study Based on Leak-Off Tests and Formation Pressure. M.S. thesis, Norw. Techn. High School, Trondheim.

Teige, G.M.G., Hermanrud, C., Kløvjan, O.S., Eliassen, P.E., Løseth, H., Gading, M., 2002. Evaluation of caprock integrity in the western (high-pressured) Haltenbanken area — a case history based on analyses of seismic signatures in overburden rocks. In: A.G. Koestler and R. Hunsdale (Editors), Hydrocarbon Seal Quantification. Norwegian Petroleum Society (NPF), Special Publication 11. Elsevier, Amsterdam, pp. 233–242 (this volume).

C. HERMANRUD *Statoil's Research Centre, N-7005 Trondheim, Norway*
E-mail: che@statoil.com

H.M. NORDGÅRD BOLÅS *Statoil's Research Centre, N-7005 Trondheim, Norway*
E-mail: hnb@statoil.com

Evaluation of caprock integrity in the western (high-pressured) Haltenbanken area — a case history based on analyses of seismic signatures in overburden rocks

Gunn M.G. Teige, Christian Hermanrud, Oddbjørn S. Kløvjan, Per Emil Eliassen, Helge Løseth and Marita Gading

Seal failure is a significant risk factor in the western, high-pressured part of the Haltenbanken area. Accordingly, an investigation of seismic expressions of hydrocarbon leakage was initiated to aid further exploration in this area. The main objective of this caprock integrity study was to evaluate the caprock of the high-pressured Kristin structure in western Haltenbanken, and more generally to identify seismic expressions of hydrocarbon leakage. This was initially done on a regional 2D dataset, and later on a semi-regional 3D dataset.

The results show a relationship between vertical caprock leakage from Jurassic reservoirs and strong seismic dim-zones on the 2D data. The correlation between dim-zones on the 2D data and dim-zones on the 3D data was very good. However, the strong leakage-related dim-zones on the 2D data were weaker on the 3D data, and consequently an equivalent relationship between caprock leakage and dim-zones on the 3D data could not be established.

The implications of the fact that the relationship between reservoir leakage and strong dim-zones — which was clear on the 2D data — was unclear on the 3D data, is that no reliable tool for identifying hydrocarbon leakage from seismic 3D data exist in the Haltenbanken area. However, analyses from the 2D data have demonstrated that such data can be used as a positive contribution in exploration leakage assessments in this area.

Introduction

A significant number of all exploration failures on the Norwegian Continental Shelf are due to lack of hydrocarbon charge or leakage. As a consequence, an improved understanding about migration and leakage mechanisms and processes in the subsurface is crucial in prospect evaluation.

Hydrocarbon leakage is often expressed on seismic data, although this expression may vary a lot. The different seismic expressions of leakage are dependent on (a) geological setting and stratigraphy, (b) geographical area, (c) leakage mechanism (leakage through fractures versus capillary leakage, fatal versus non-fatal leakage), (d) depth, and (e) seismic data quality (e.g. different processing of seismic data). Such seismic expressions of leakage are often referred to as gas chimneys. In the subsequent discussion, vertically disturbed zones on seismic data are referred to as dim-zones.

Historically within Statoil, seismic anomalies and attributes have been used to detect possible leakage on seismic data. However, since the causes of various seismic anomalies or dim-zones are not fully understood yet, no good criteria or rules appear to have existed to distinguish between leakage from hydrocarbon bearing structures and dry structures. In particular, how to interpret dim-zones in areas with high reservoir pressures — such as the Haltenbanken west area — has been associated with great uncertainties.

This study will show how the investigation of dim-zones from 2D and 3D seismic data has had some influence on the exploration strategy at Haltenbanken, and what we have learned so far from the application of seismic dim-zones as a tool in seal evaluations.

Study area and database

The Haltenbanken area is situated offshore Mid-Norway (Fig. 1). The geological history of the area was dominated by extensive tectonic activity in the Late Jurassic and subsequently by gentle subsidence, although some rifting occurred during the Late Cre-

Hydrocarbon Seal Quantification edited by A.G. Koestler and R. Hunsdale. NPF Special Publication 11, pp. 233–242, Published by Elsevier Science B.V., Amsterdam.

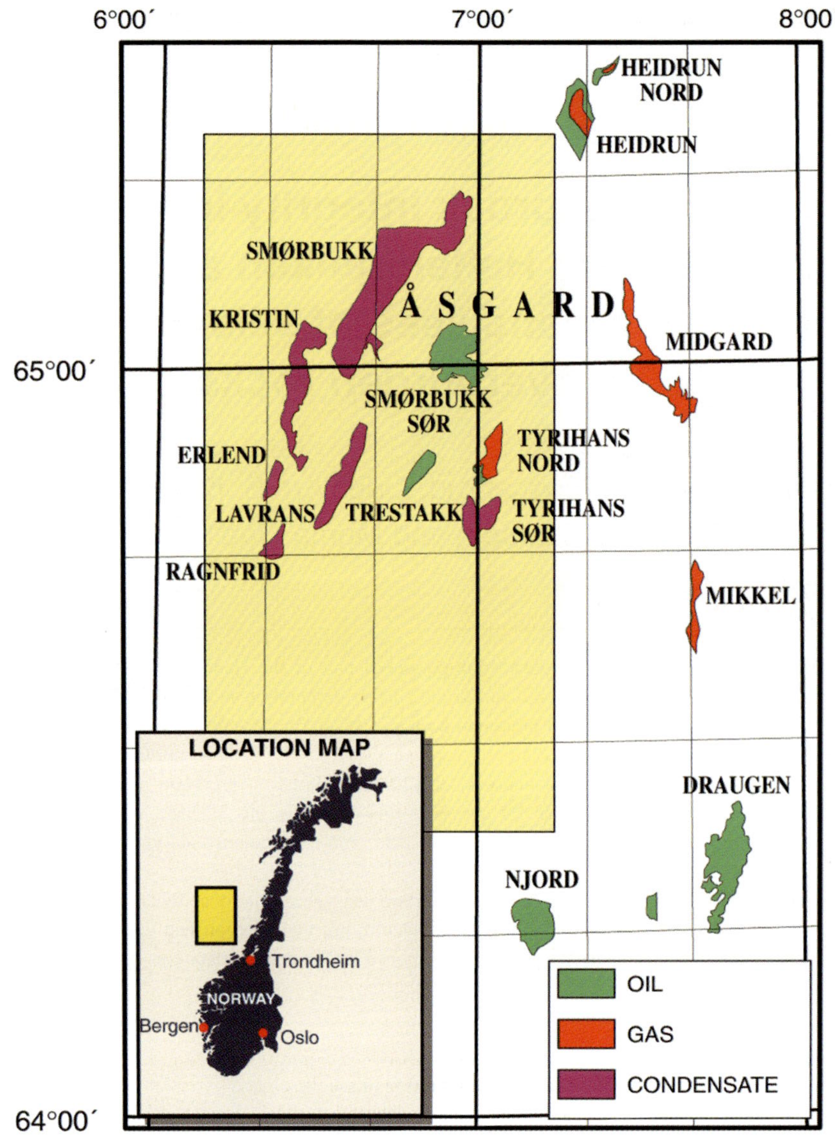

Fig. 1. Haltenbanken, offshore Mid-Norway — location map and study area.

taceous. The main hydrocarbon accumulations have been found in sandstone reservoirs of Middle Jurassic age, although major hydrocarbon accumulations have also been found in sandstone reservoirs of Early and Late Jurassic age. The hydrocarbons have been sourced from coals and shales of Early Jurassic age and from shales of Late Jurassic age (Heum et al., 1986; Karlsen et al., 1995).

Details about the geological history of the Haltenbanken area have been extensively described by several authors, e.g. Hollander (1984), Heum et al. (1986), Ziegler et al. (1986), and Koch and Heum (1995).

The exploration history of Haltenbanken can be divided into two periods: prior to and after the discovery of the Kristin structure in 1996. The study area can be separated into two main pressure regimes: a high-pressured area to the west, and a normal- and intermediate-pressured area to the east (Fig. 2). In the early exploration phase, the eight exploration wells drilled in the high-pressured region (mud weight equivalent greater than 1.6 g/cm^3) turned out to be dry, i.e. water-filled in the Lower and Middle Jurassic sandstone (for detailed pressure information, see Hermanrud and Nordgård Bolås, 2002). Only hydrocarbon shows and traces of hydrocarbons were encountered. In six of these wells, the results were attributed to caprock leakage related to the high fluid pressures in this region (two failures were attributed to lack of charge). Numerous hydrocarbon discoveries in the eastern, normal/intermediate pressured part of Haltenbanken, e.g. Smørbukk, Smørbukk South, Heidrun, Heidrun North, Tyrihans, Trestakk, Njord, Mikkel, and Draugen, demonstrated that leakage was not a major problem in the eastern region. Indeed, of the 55 eastern exploration wells drilled prior to drilling of the Kristin structure in the western region,

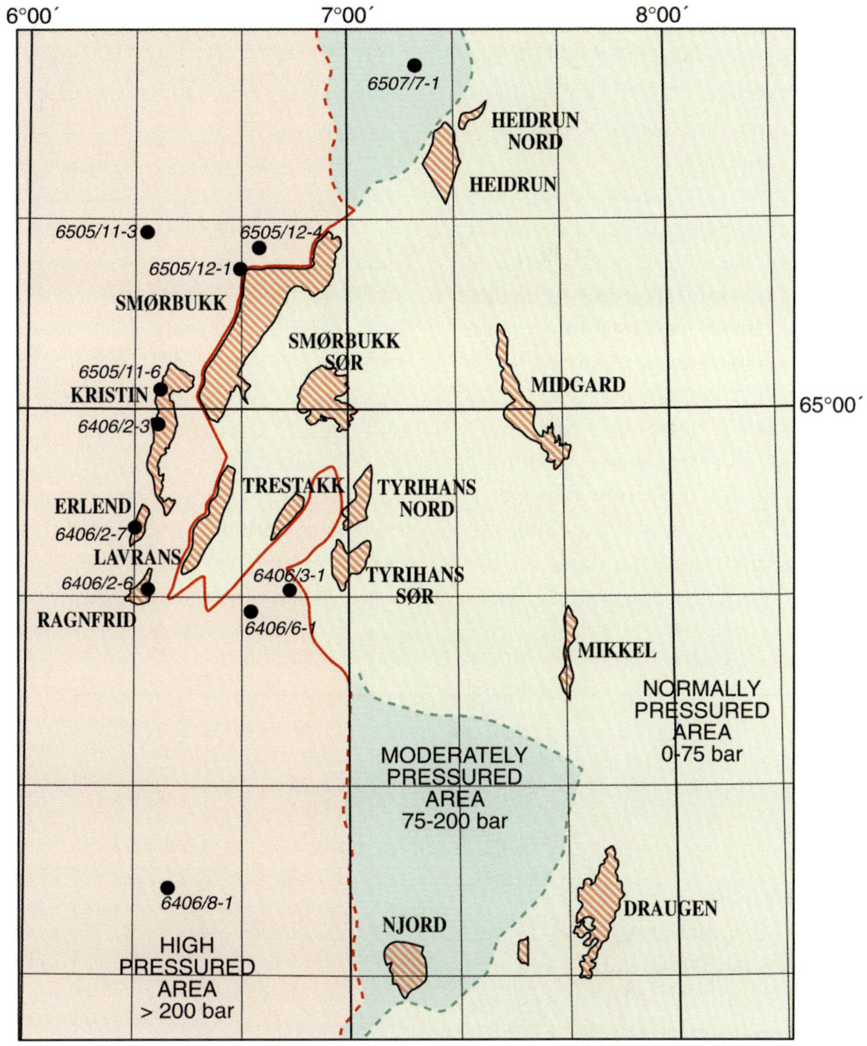

Fig. 2. Pressure regimes and high-pressured wells at the Haltenbanken area. The region is separated into a high-pressured area to the west, and a normal- and intermediate-pressured area to the east. In the early exploration phase the eight exploration wells drilled in the high-pressured region (mud weight equivalent greater than 1.6 g/cm^3) turned out to be water bearing in the Lower and Middle Jurassic sandstone. However, in the eastern part of the Haltenbanken area several hydrocarbon discoveries have been found. The last few years of exploration in the high-pressured area has also resulted in discoveries in this region, i.e. the Kristin, Ragnfrid and Erlend discoveries.

40% failed, and none of the failures were attributed to leakage.

The caprock integrity study in 1995 (prior to drilling of the Kristin structure) was performed on a regional 2D seismic dataset. Subsequent, several older 3D datasets were merged with new 3D seismic data, resulting in a regional 3D cover across large parts of the original caprock leakage study area. Mapping of seismic dim-zones was then performed on these 3D data.

Drilling of the Kristin structure (late 1996) in the high-pressured area, introduced a new period: hydrocarbon discoveries in the high-pressured region of Haltenbanken. The Kristin discovery consisted of both gas and condensate. This success led to renewed interest for hydrocarbon exploration in the high-pressured region, which in turn resulted in other discoveries in this area (i.e. the Ragnfrid and Erlend discoveries).

Evaluation of leakage prior to the discovery of the Kristin structure

The main focus of the 1995 caprock leakage study in the western, high-pressured Haltenbanken area was to evaluate the caprock of the Kristin structure in blocks 6506/11 and 6406/2. Was Kristin in any way different from the eight previously drilled and dry structures in the area?

It has been debated whether hydrocarbon charge was different in the two main pressure areas at Haltenbanken, and whether this could explain the lack of discoveries to the west. However, all six leakage failures in the west encountered hydrocarbon shows/traces, which demonstrated that charge was not the main risk.

The basis for this study was the known relationship between leakage and high fluid pressures on

Haltenbanken (Fig. 2), and a basin-wide dim-zone study done on a regional 2D seismic dataset (Amaliksen, 1989). 3D data coverage existed across parts of the study area at the time, but was not extensive enough to permit regional analysis.

The systematic observations of seismic dim-zones on a 2D grid (4 × 5 km line-spacing, and even coarser in some areas) throughout the whole study area at Haltenbanken classified the confidence and intensity of the dim-zones into six classes. Later modification of Amaliksen's (1989) observations resulted in a seismic dim-zone map with two main classes: strong (previously very strong, strong, good, and moderate) and weak (previously weak and very weak) dim-zone appearances (Fig. 3). By using only two classes of dim-zones, a separation, which may be related to hydrocarbon leakage, emerged from the data. As the main conclusions of our caprock leakage study focus on the different appearances of seismic dim-zones on 2D and 3D data, only the part of the regional 2D map, which was later covered by a merged set of 3D data, is shown in Fig. 3. The appearance of seismic dim-zones related to hydrocarbon discoveries over the whole study area is summarized in Table 1.

One example of typical seismic dim-zones on 2D data is shown in Fig. 4. As is clear, from this figure at least, these dim-zones are relatively well defined and therefore classified as strong. The classification criteria for strong dim-zones, are non-continuous reflectors within the dim-zone and continuous reflectors outside the dim-zone. Within a weak dim-zone, it is possible to interpret the reflectors, although the reflectors are partly non-continuous inside the dim-zone.

Special attention was paid to the analyses of dim-zones related to five of the eight previously drilled, water-filled structures in the high-pressured area, i.e.

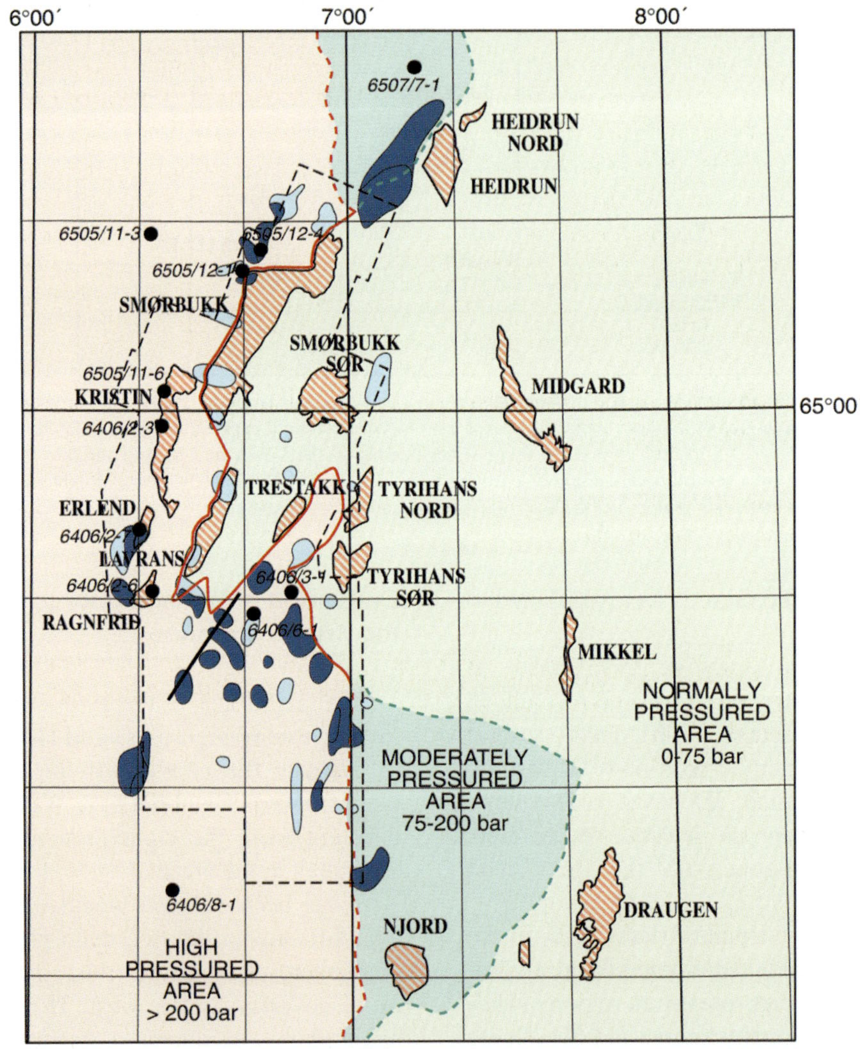

Fig. 3. Dim-zone observations from 2D seismic data at the Haltenbanken area (the figure is modified from a non-published work by Amaliksen, 1989) (the high-pressured area in the west, the normal- and intermediate-pressured area in the east). The blue colours symbolize the two dim-zone classes, i.e. dark blue implying strong dim-zones, and light blue implying weak dim-zones. For the sake of later comparisons, the black dotted outline shows the area of focus and investigation for the subsequent 3D analyses (i.e. where several pre-existing 3D surveys were merged with new data). 2D dim-zones outside this focus area are not presented in this work.

TABLE 1

Summary of dim-zone observations on 2D data at the Haltenbanken area prior to drilling of the Kristin structure.

Well/discoveries	Strong dim-zones (2D)	Weak/no dim-zones (2D)
Dry (due to leakage) high-pressured wells	6506/11-1, 6506/11-3, 6506/12-4, 6406/8-1 and 6406/3-1	
Leaky discoveries	Mikkel, Njord	
Non-leaky discoveries	Midgard, Smørbukk	Heidrun, Heidrun North, Smørbukk South, Trestakk, Tyrihans North, Tyrihans South, Draugen

Note that the classification of dry wells and discoveries was the interpretation at the time the investigations were made (Fig. 3). The high-pressured and dry wells (due to leakage) are all associated with strong dim-zones on the 2D data. In addition, two leaky and two non-leaky discoveries are associated with strong dim-zones (see text for further explanations). The seven non-leaky discoveries are associated with weak or absence of dim-zones on the 2D data. Based on these observations, a relationship between vertical hydrocarbon leakage and strong seismic dim-zones is suggested. See text for further explanations.

6506/11-1, 6506/11-3, 6506/12-4, 6406/8-1, and 6406/3-1. The three remaining high-pressured wells were not included in the caprock integrity study, as two of the exploration failures were interpreted to be dry due to lack of hydrocarbon charge (6406/6-1 and 6507/7-1), and one additional leaky well was drilled outside the study area (6406/11-1S). (In order to include all six leaky wells, a complete new regional 2D study must have been undertaken to ensure consistency throughout the area. Based on the thorough investigations that have been done by Amaliksen (1989), it was decided not to duplicate this work. The subsequent discussion is therefore focused on five of the six leaky wells in the high-pressured part of Haltenbanken.)

The dim-zone observations at Haltenbanken show three major distinct trends (Fig. 3):

(1) More dim-zones are observed in the high-pressured area to the west than in the normal pressured area to the east.

(2) Nearly all of the strong dim-zones are observed in the high-pressured (and leaky) area to the west.

(3) All of the five investigated leaky structures have strong dim-zones associated with them — either in well position or uphill from where the well was drilled. In fact, there seems to be a good correlation between the occurrence of strong dim-zones and exploration failures due to leakage in the high-pressured western area.

A summary of dim-zones from 2D data related to leaky structures in the high-pressured area and hydrocarbon discoveries in the normal- and intermediate-pressured area of Haltenbanken is shown in Table 1. In addition, one leaky discovery in the normal-pressured area (Mikkel) and one leaky discovery in the intermediate-pressured area (Njord) are also associated with strong dim-zones, namely the Mikkel and Njord discoveries (Fig. 3; note that 2D dim-zones outside the focus area (Fig. 3) are not presented in this contribution). Both of these discoveries were interpreted as leaky at the time of investigation, but later evaluations of the Mikkel discovery has concluded that insufficient charge may well explain why this field is not filled to structural spill point. Whether

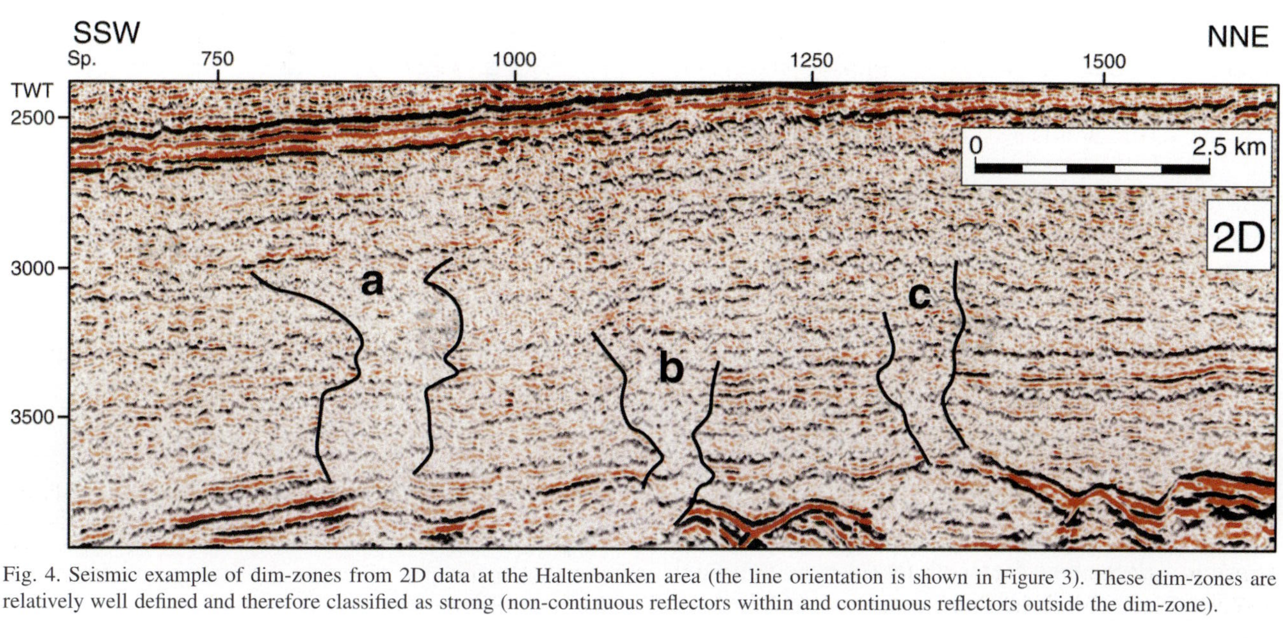

Fig. 4. Seismic example of dim-zones from 2D data at the Haltenbanken area (the line orientation is shown in Figure 3). These dim-zones are relatively well defined and therefore classified as strong (non-continuous reflectors within and continuous reflectors outside the dim-zone).

the Mikkel field does or does not comply with the suggested relationship, i.e. the relationship between strong dim-zones (on 2D data) and vertical hydrocarbon leakage, is presently unclear.

In contrast, seven non-leaky discoveries in the normal-pressured area have no, or weak, dim-zones, i.e. the Heidrun, Heidrun North, Smørbukk South, Trestakk, Tyrihans North, Tyrihans South, and Draugen discoveries (Fig. 3). Two non-leaky discoveries are associated with strong dim-zones, i.e. the Midgard and Smørbukk discoveries (Table 1).

Based on 2D data, Table 1 suggests a relationship between vertical hydrocarbon leakage and strong seismic dim-zones. The likelihood for this happening by an accident, is statistically only 0.18% (i.e. 14 of 16 wells/discoveries are compliant with the model, whereas the Midgard and Smørbukk discoveries are not compliant with the strong dim-zone versus leakage relationship). Actually, the relationship may be even stronger, because the strong dim-zones related to the two non-leaky discoveries may still be consistent with the dim-zone versus leakage relationship; the Midgard dim-zone may be related to leakage along the northern bounding fault of the field, which may control the field's fluid contact, and the Smørbukk dim-zone may actually be due to fluid discharge from west of the field (i.e. the high-pressured area) (and therefore not related to the Smørbukk Field itself) (the likelihood for this happening by accident is statistically 0.02%, i.e. 16 of 16 wells/discoveries are compliant with the model.).

If the Mikkel and Njord discoveries are also considered as not compliant with the strong dim-zone versus leakage relationship (because they are discoveries, and yet they are associated with (uphill) strong dim-zones), then the probability of these observations being random is still no more than 2.8% (i.e. 12 of 16 wells/discoveries are compliant with the model, whereas the Midgard, Smørbukk, Mikkel, and Njord discoveries are not compliant with the model). Note that the term 'non-leaky' is relatively speaking — all pressure compartments leak, the matter is rather 'where, when, and how' the leakage occurs (Nordgård Bolås and Hermanrud, 2002).

The Kristin structure, however, had only a *weak* dim-zone associated with it (Fig. 3). This dim-zone intersected the structure in a downflank position. This feature differentiated the Kristin structure from the other leaky and dry structures in the region, which all had dim-zones intersecting the structures in an upflank and 'fatal position' (Fig. 3). Fatal leakage has occurred when only residual hydrocarbons remain downdip of the 'fatal leakage position'. It is suggested that rock stress is a controlling factor on fatal versus non-fatal leakage (Nordgård Bolås and Hermanrud, 2002).

Among all of these high-pressured structures only the Kristin structure had a *weak* dim-zone (downhill). These observations positively influenced the risk evaluation of the Kristin structure ahead of drilling. This conclusion was consistent with the fact that weak dim-zones were observed over several hydrocarbon discoveries in the eastern, normal-pressured region (Fig. 3; Table 1). The slow leakage processes related to these normal-pressured discoveries are confirmed by hydrocarbons in the caprock (Løseth et al., 2000; Wensaas et al., 2000).

Evaluation of leakage after the discovery of the Kristin Field

The discovery of the Kristin structure in late 1996 by well 6406/2-3 proved the presence of significant volumes of gas and condensate in Jurassic sandstones.

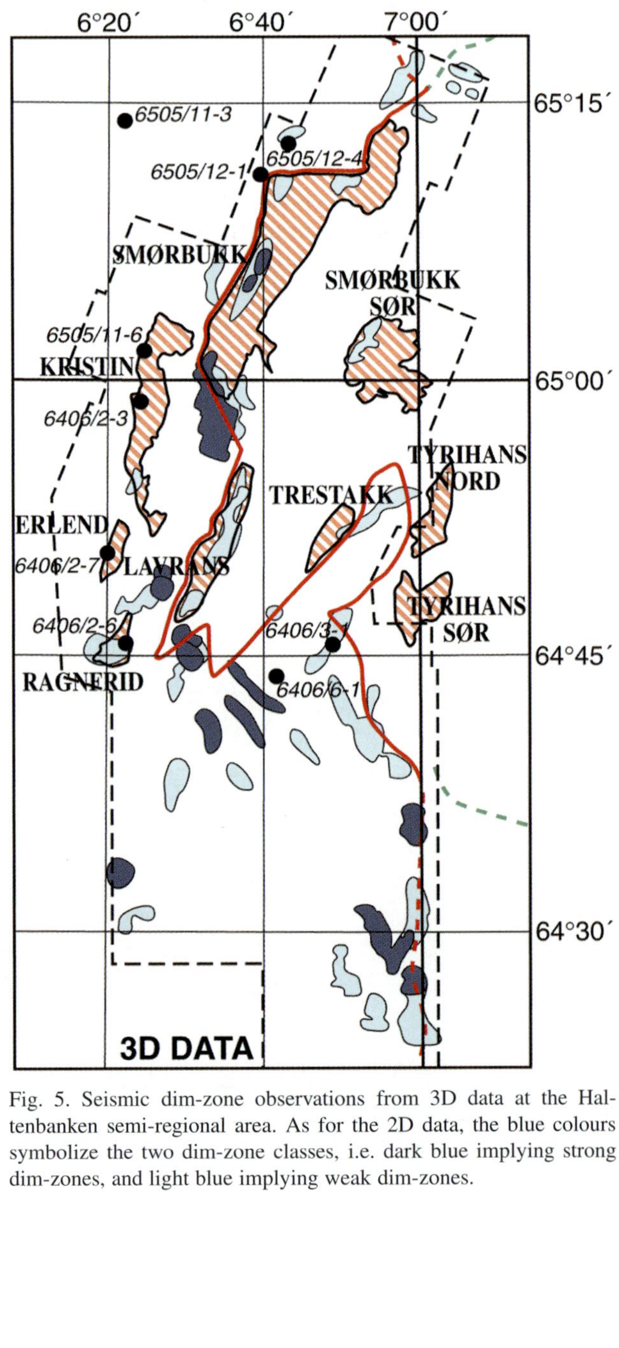

Fig. 5. Seismic dim-zone observations from 3D data at the Haltenbanken semi-regional area. As for the 2D data, the blue colours symbolize the two dim-zone classes, i.e. dark blue implying strong dim-zones, and light blue implying weak dim-zones.

TABLE 2

Summary of seismic dim-zone observations from 2D data and dim-zone observations from 3D data at Haltenbanken

Wells/discoveries	Dim-zones: 2D data	Dim-zones: 3D data
High-pressured area		
6506/11-1	strong	weak
6506/11-3	strong	(outside 3D area)
6506/12-4	strong	weak
6406/8-1	strong	(outside 3D area)
6406/3-1	strong	weak
Kristin	weak	weak, strong
Erlend	strong	none
Ragnfrid	strong	weak
Normal- intermediate-pressured area		
Tyrihans North	weak	weak
Tyrihans South	none	(outside 3D area)
Smørbukk	weak, strong	weak, strong
Smørbukk South	none	weak
Trestakk	none	weak

See Figs. 3 and 5 and text for further explanations.

This success led to renewed interest for hydrocarbon exploration in the high-pressured western area of Haltenbanken. This interest has then resulted in other discoveries in the region (i.e. the Ragnfrid and Erlend discoveries).

At this stage, several pre-existing 3D surveys were merged with new seismic data, resulting in regional 3D coverage across a large part of the study area. Mapping of dim-zones was then carried out using these 3D data. However, the dim-zone observations from the 3D data did not include the two dry wells 6506/11-3 and 6406/8-1 in the high-pressured area (these two wells were not covered by the 3D merge at that time) (Fig. 5). More recently wells were also included in the caprock leakage discussion, namely 6506/11-6 (the northern part of the Kristin Field), 6406/2-6 (the Ragnfrid discovery), and 6406/2-7 (the Erlend discovery).

The dim-zone observations from the 3D data gave the following results (Fig. 5).

(1) The three strong dim-zones appearing on 2D data — associated with the three investigated exploration failures in the high-pressured area (well 6506/11-1, 6506/12-4, 6406/3-1) — were weaker on the 3D data (Table 2).

(2) The Ragnfrid discovery was associated with a weak (less reliable) dim-zone, while the Erlend discovery was not associated with any dim-zone at all. The 3D data also revealed that other discoveries in the normal-pressured eastern part of Haltenbanken, such as Smørbukk South and Trestakk, were now associated with weak (less reliable) dim-zones (Table 2), as opposed to an apparent absence of dim-zones in the 2D data.

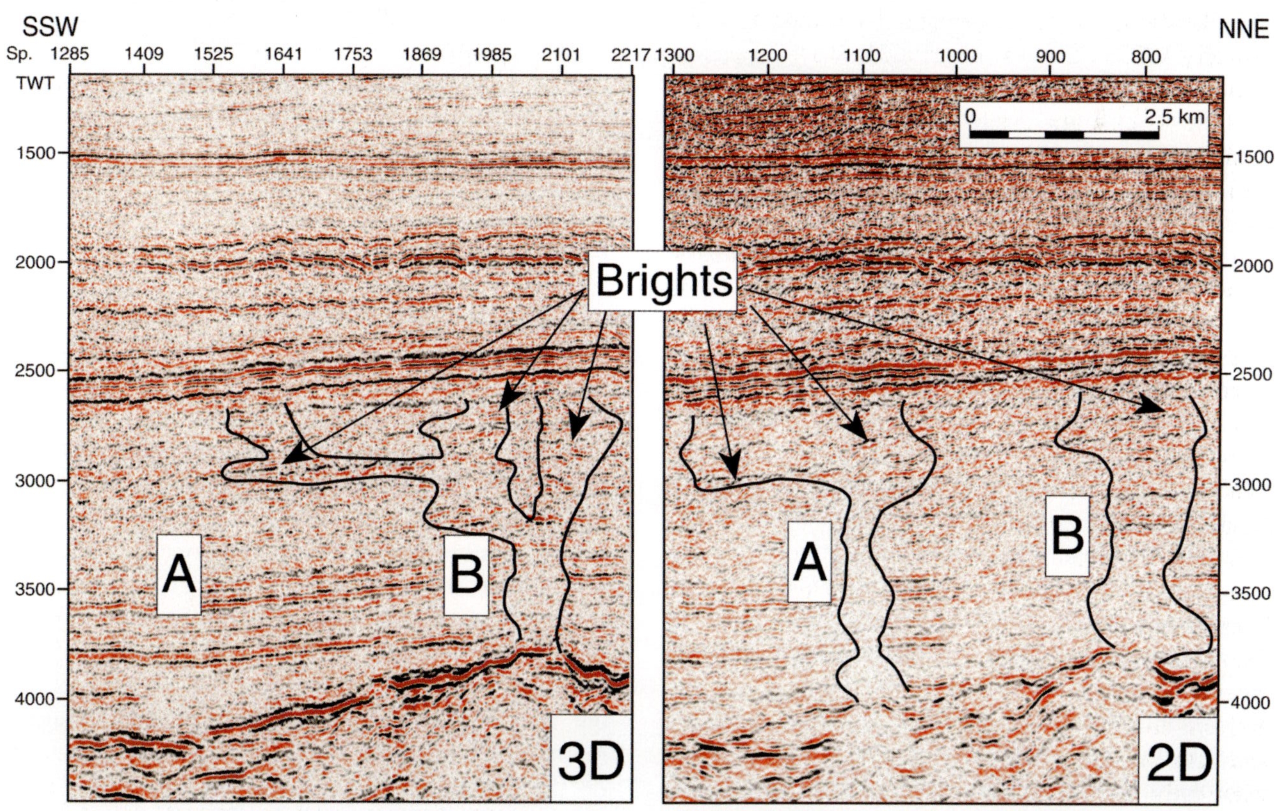

Fig. 6. Seismic example of dim-zone appearances on 2D versus 3D data. Two distinct dim-zones are fairly easy to identify on the seismic 2D line — only one of these dim-zones can be traced on the seismic 3D line, even though this dim-zone seems to be more definable than the equivalent dim-zone on the 2D line.

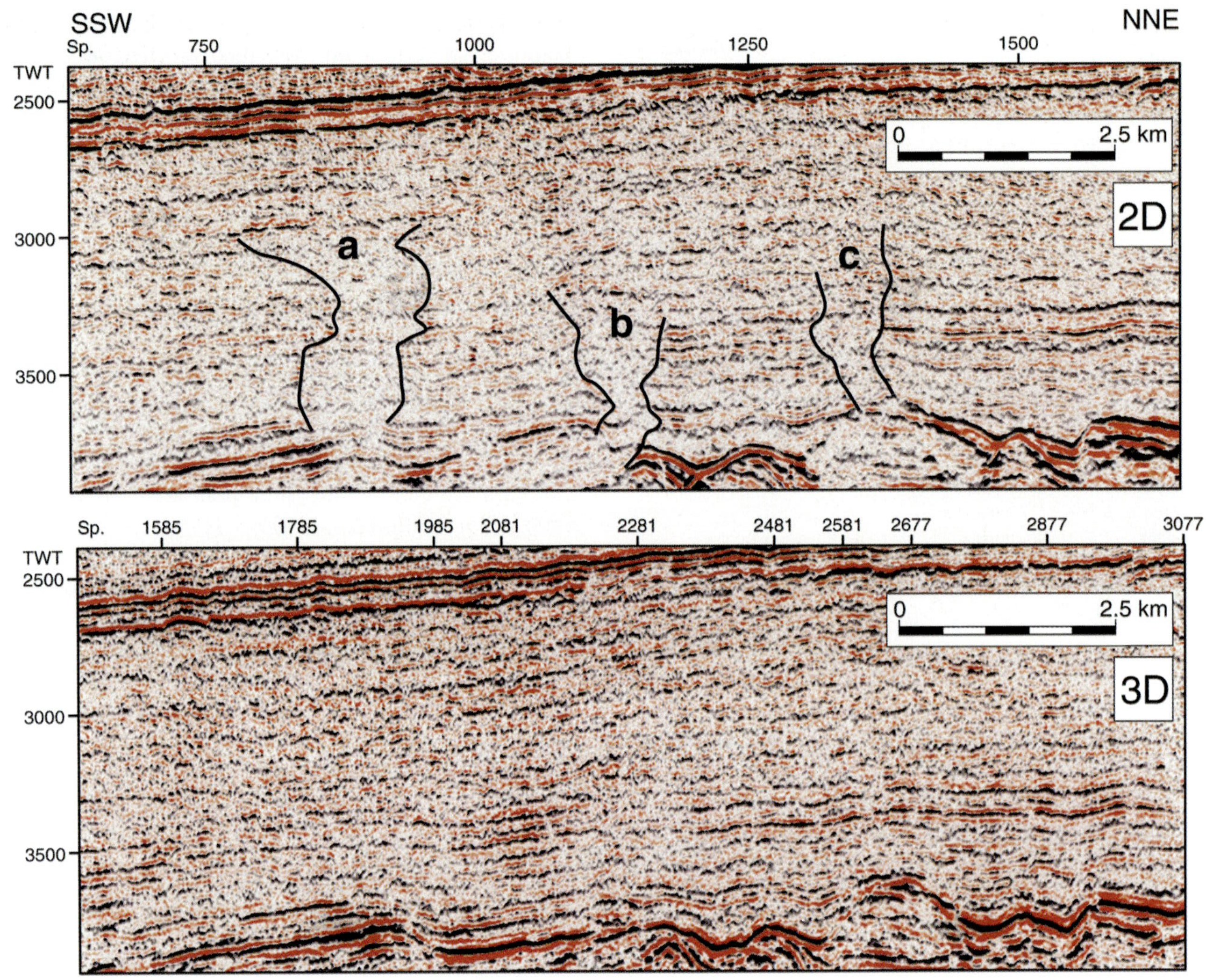

Fig. 7. Seismic example of dim-zone appearances on 2D versus 3D data. Among the three dim-zones that are well defined on the 2D seismic line, only one dim-zone, i.e. zone b, is relatively well defined on the 3D seismic.

(3) The Kristin structure was associated with a strong dim-zone, possibly related to HC movement through an eastern spill point and up through the main Smørbukk Fault to the east. A small and weak (less reliable) dim-zone was observed above the southern part of the structure (Table 2).

Seismic examples of dim-zones from both 2D and 3D data are shown in Fig. 6. Two distinct dim-zones are fairly easy to identify on the 2D seismic line. Only one of these dim-zones can be traced on the 3D seismic line, even though this dim-zone seems to be better defined than the corresponding dim-zone on the 2D line. Fig. 7 shows another example: among the three dim-zones that are well defined on the 2D seismic line, only one dim-zone, i.e. zone b, is relatively well defined on the 3D seismic line. Possible causes for the different dim-zone expressions will be discussed in the next chapter.

Discussion

After the Kristin structure had been drilled, other discoveries on Haltenbanken (both in the high- and normal-pressured area) were added to the discussion: the Tau discovery (normal-pressured, west of the Draugen Field) has a weak dim-zone on the 2D data; the Lavrans discovery (normal-pressured) has a weak dim-zone on the 2D data; the Ragnfrid discovery (high-pressured) has a strong dim-zone in a downhill position on the 2D data; and the Erlend discovery (high-pressured) has a strong dim-zone in a downhill position on the 2D data.

These additional discoveries fit into the apparent relationship between vertical reservoir leakage (exploration failures) and strong seismic dim-zones on the 2D data. However, the Ragnfrid and Erlend dim-zones are not related to fatal leakage, as the dim-zones are associated with hydrocarbons in a downhill position. The only strong dim-zone on the 2D data

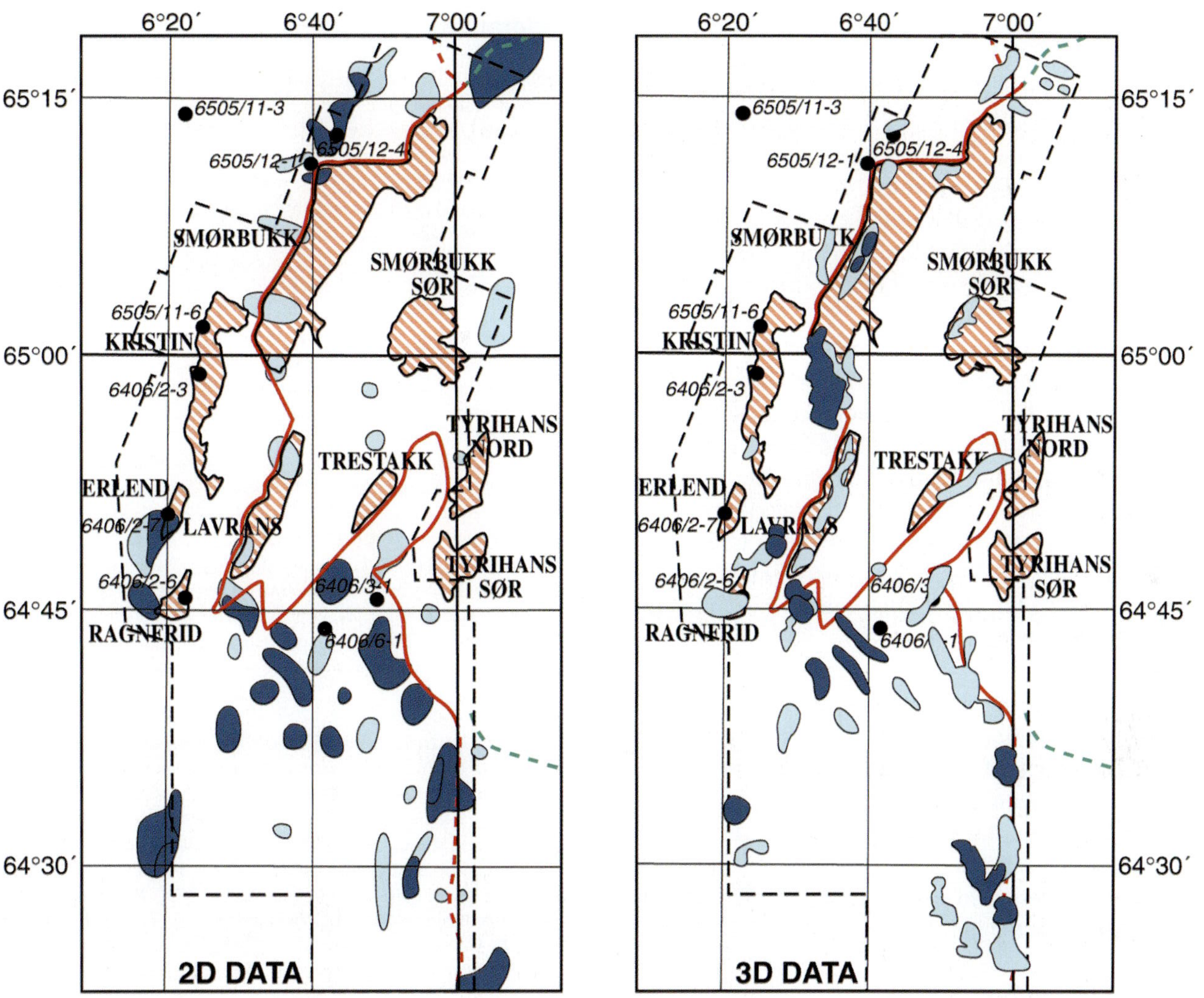

Fig. 8. Dim-zone observations on Haltenbanken: (a) 2D data versus (b) 3D data. The weak dim-zone observations on the 3D data are not only related to strong dim-zones on the 2D data (and thus related to leakage), but also related to other dim-zones on the 2D data. A significant number of weak dim-zones on the 3D data appear in positions where no dim-zones were identified on the 2D data, while others were identified as weak on the 2D data (and thus not related to leakage).

which is not seen on the 3D data is the Erlend dim-zone. In fact, the other strong dim-zones on the 2D data are recognized on the 3D data, either equally strong or weaker (Fig. 8). The correlation between dim-zones on the 2D data and dim-zones on the 3D data is accordingly very good.

However, a relationship between vertical reservoir leakage and dim-zones on the 3D data has not been recognized, because the dim-zones identified on the 2D data are not entirely coincident with the dim-zones on the 3D data. From Fig. 8, it is clear that the weak dim-zone observations on the 3D data are not only related to strong dim-zones on the 2D data (and thus related to leakage), but are also related to other dim-zones on the 2D data. A significant number of weak dim-zones on the 3D data appear in positions where no dim-zones were identified on the 2D data, while others were identified as weak on the 2D data (and thus not related to leakage).

The strong dim-zones on the 2D data appear as weak dim-zones on the 3D data. It was therefore initially also expected that weak dim-zones on 3D data should correlate with leaky structures (only), and therefore be usable in exploration. However, for the reasons just explained, additional weak dim-zones are identified on the 3D data, and these are unrelated to fatal hydrocarbon leakage. Thus, at present, a relationship between dim-zones and fatal hydrocarbon reservoir leakage indicated from 3D data is yet to be determined.

Summary and conclusions

Evaluation of caprock integrity in the Haltenbanken area has indicated the presence of more strong dim-zones on the 2D data in the western high-pressured area than in the normal- and intermediate-pressured area in the east.

All five of the investigated high-pressured exploration failures were associated with strong dim-zone observations on the 1995-2D data. The Kristin structure, however, also located in the western high-pressured Haltenbanken area, had only a weak dim-zone on the 2D data. This separated the Kristin structure from the previously drilled, high-pressured structures and was used successfully to upgrade the prospect ahead of drilling. The result of the drilling in 1996 was a gas/condensate discovery.

Later re-evaluation of the relationship between hydrocarbon reservoir leakage and dim-zones — this time based on 3D data — concluded that no such relationship could be identified from these data. This result was partly due to the fact that the previous distinction between strong (leakage related) and weak (not related to fatal leakage) dim-zones, which was clear from the 2D data, was not evident in the 3D data. As yet, no reliable tool for identifying hydrocarbon leakage from seismic 3D data exists in this area.

Thus, despite its contribution to the exploration success in the region, more work is required before the full potential of dim-zone appearances in seal evaluation has been exploited, at least from seismic 3D data. However, the analyses from seismic 2D data have demonstrated that such data can be used as a positive contribution in exploration and hydrocarbon leakage assessments in this area.

Acknowledgements

The authors would like to thank Knut Gunnar Amaliksen for the thorough work that was done by interpreting seismic dim-zones on 2D data at Haltenbanken (regional) in 1989. Lars Reistad is also thanked for his graphical support. Statoil ASA is acknowledged for granting permission to publish this work.

References

Amaliksen, K.G., 1989. Seismiske 'chimneys' på Haltenbanken (internal Statoil report).

Hermanrud, C. and Nordgård Bolås, H.M., 2002. Leakage from overpressured hydrocarbon reservoirs at Haltenbanken and in the northern North Sea. In: A.G. Koestler and R. Hunsdale (Editors), Hydrocarbon Seal Quantification. Norwegian Petroleum Society (NPF), Special Publication 11. Elsevier, Amsterdam, pp. 221–231 (this volume).

Heum, O.R., Dalland, A. and Meisingset, K., 1986. Habitat of hydrocarbons at Haltenbanken (PVT-modeling as a predictive tool in hydrocarbon exploration). In: A.M. Spencer, C.J. Campbell, S.H. Hanslien, E. Holter, P.H.H. Nelson, E. Nysæther and E.G. Ormaasen (Editors), Habitat of Hydrocarbons on the Norwegian Continental Shelf. Norwegian Petroleum Society, Graham and Trotman, London, pp. 259–274.

Hollander, N.B., 1984. Geohistory and hydrocarbon evolution of the Haltenbanken area. In: A.M. Spencer, S.O. Johnsen, A. Mørk, E. Nysæther, P. Songstad and Å. Spinnanger (Editors), Petroleum Geology of the North European Margin. Graham and Trotman, London, pp. 283–384.

Karlsen, D.A., Nyland, B., Flood, B., Ohm, S.E., Brekke, T., Olsen, S. and Backer-Owe, K., 1995. Petroleum geochemistry of the Haltenbanken, Norwegian continental shelf. The Geochemistry of Reservoirs, Geological Society Special Publication, 86, pp. 203–256.

Koch, J.-O. and Heum, O.R., 1995. Exploration trends of the Halten Terrace. In: S. Hanslien (Editor), 25 Years of Petroleum Exploration in Norway. Norwegian Petroleum Society (NPF), Special Publication 4. Elsevier, Amsterdam, pp. 235–251.

Løseth, H., Gading, M. and Wensaas, L., 2000. Location of leakage points and timing of leakage from seismic data. Abstract to Hydrocarbon Seal Conference, Stavanger, Norway.

Nordgård Bolås, H.M. and Hermanrud, C., 2002. Rock stress in sedimentary basins —implications for trap integrity. In: A.G. Koestler and R. Hunsdale (Editors), Hydrocarbon Seal Quantification. Norwegian Petroleum Society (NPF), Special Publication 11. Elsevier, Amsterdam, pp. 17–35 (this volume).

Wensaas, L., Løseth, H., Arntsen, B., Hermanrud, C. and Hanken, H.M., 2000. Seismic leakage anomalies — Links to well data and field exposures (Abstract to Hydrocarbon Seal Conference, Stavanger, Norway).

Ziegler, W.H., Doery, R. and Scott, J., 1986. Tectonic habitat of Norwegian oil and gas. In: A.M. Spencer, C.J. Campbell, S.H. Hanslien, E. Holter, P.H.H. Nelson, E. Nysæther and E.G. Ormaasen (Editors), Habitat of Hydrocarbons on the Norwegian Continental Shelf. Norwegian Petroleum Society, Graham and Trotman, London, pp. 3–19.

G.M.G. TEIGE *Statoil's Research Centre, N-7005 Trondheim, Norway*
E-mail: gmgt@statoil.com
C. HERMANRUD *Statoil's Research Centre, N-7005 Trondheim, Norway*
O.S. KLØVJAN *Statoil, P.O. Box 40, N-9401 Harstad, Norway*
P.E. ELIASSEN *Statoil ASA, P.O. Box 300, N-4035 Stavanger, Norway*
H. LØSETH *Statoil's Research Centre, N-7005 Trondheim, Norway*
M. GADING *Statoil's Research Centre, N-7005 Trondheim, Norway*

Fault seal analysis in unconsolidated sediments: a field study from Kentucky, USA

Gavin Lewis, Robert J. Knipe and Anren Li

The Eastern Kentucky Coalfield contains Pennsylvanian aged strata deposited into the Central Appalachian Basin, eastern USA. These strata are deltaic to shallow marine facies and represent prograding sediment accumulated into a foreland basin, from an emergent and uplifting Appalachian landmass to the east. The sediments are predominantly unfaulted away from the Appalachian thrust front; however, spectacular examples of metre- to decimetre-scale syn-sedimentary growth faulting are observed. The growth faulting was caused by sedimentary differential loading of underlying mobile shale units. For this reason these fault systems develop soon after deposition while the sediments are semi- to unconsolidated. The fault systems are comprised of listric to planar, down to the basin (regional) and landward (counter-regional) faults as observed in modern deltaic environments such as offshore Texas, Gulf of Mexico and the Niger Delta. Observations made on the structure and architecture within the fault zones identify three critical deformation styles that will affect predicting the fluid flow characteristics of fault systems formed in this environment. These deformation styles are as follows. (1) Faults formed whilst the sediments are unconsolidated show evidence of sands being sheared down the fault plane forming disaggregation bands parallel to the fault plane. (2) Faults propagating through an inter-bedded sequence of sands and shales experience a large strength contrast at the sand–shale interfaces leading to fault plane refraction. Further fault development tends to 'smooth' the fault's surface producing slivers (riders) of sandstone, which are incorporated into the fault zone. (3) Fault motion during soft sediment deformation can lead to sands injected along the fault plane and along planes of weakness in the footwall to increase sand connectivity. Traditional fault seal analysis techniques such as the Shale Gouge Ratio (SGR) assumes the percentage of shale in the fault zone, its petrophysical properties and hence its sealing potential is controlled by the ratio of sand to shale passing a point on the fault. The incorporation or injection of sand into the fault plane is not accounted for. Thus use of the SGR method for assessing intra-reservoir faulting formed by syn-sedimentary processes can overestimate the sealing potential of the system.

Introduction

Extensional faults generated during rifting produce a series of fault rocks that control the flow properties and sealing capacities of faults relative to hydrocarbons. Prediction of the petrophysical properties of the fault rocks is therefore critical to the migration and trapping of hydrocarbons (Jones et al., 1998; Knipe et al., 1998a,b; Møller-Pedersen and Koestler, 1998). Many fault seal studies have been based on the assumption that the fault rock properties (permeabilities and entry pressures) are controlled by the clay content of the fault rocks and that the clay content of fault rocks can be predicted by using algorithms like the Shale Gouge Ratio (SGR) (Yielding et al., 1997). The SGR equates the average clay content of the stratigraphy moved past a point to the clay content of the fault rock. Essentially the clay content of the fault rocks generated is dependent upon the throw of the fault and the distribution of clays in the faulted sequence. Fisher and Knipe (1998) and Knipe et al. (1998a,b) have demonstrated that the geohistory of the fault zone is also important to the petrophysical properties of the fault zone and propose using modifiers such as depth of burial, cementation and reactivation as input parameters for predicting fault properties. At very shallow depths of burial ($<\sim 1$ km) it is important to assess the fault rocks created by the deformation of poorly consolidated sediments.

This paper examines outcrops from the Eastern Kentucky Coalfield (Rice, 1994), where Carboniferous deltaic sediments (Fig. 1) are deformed by growth faults that occurred when the sediments were unconsolidated. These fault exposures allow analysis of the deformation products and fault architectures formed from early faulting processes and provide an opportunity to compare the fault rocks present with those predicted from the application of SGR methods.

Geological setting

The Eastern Kentucky Coalfield is situated within the Appalachian Basin, that extends NE–SW along the

Hydrocarbon Seal Quantification edited by A.G. Koestler and R. Hunsdale. NPF Special Publication 11, pp. 243–253, Published by Elsevier Science B.V., Amsterdam.

Fig. 1. Surface geological map of Kentucky, showing the Eastern Coalfield and the location of the VanCleve outcrop.

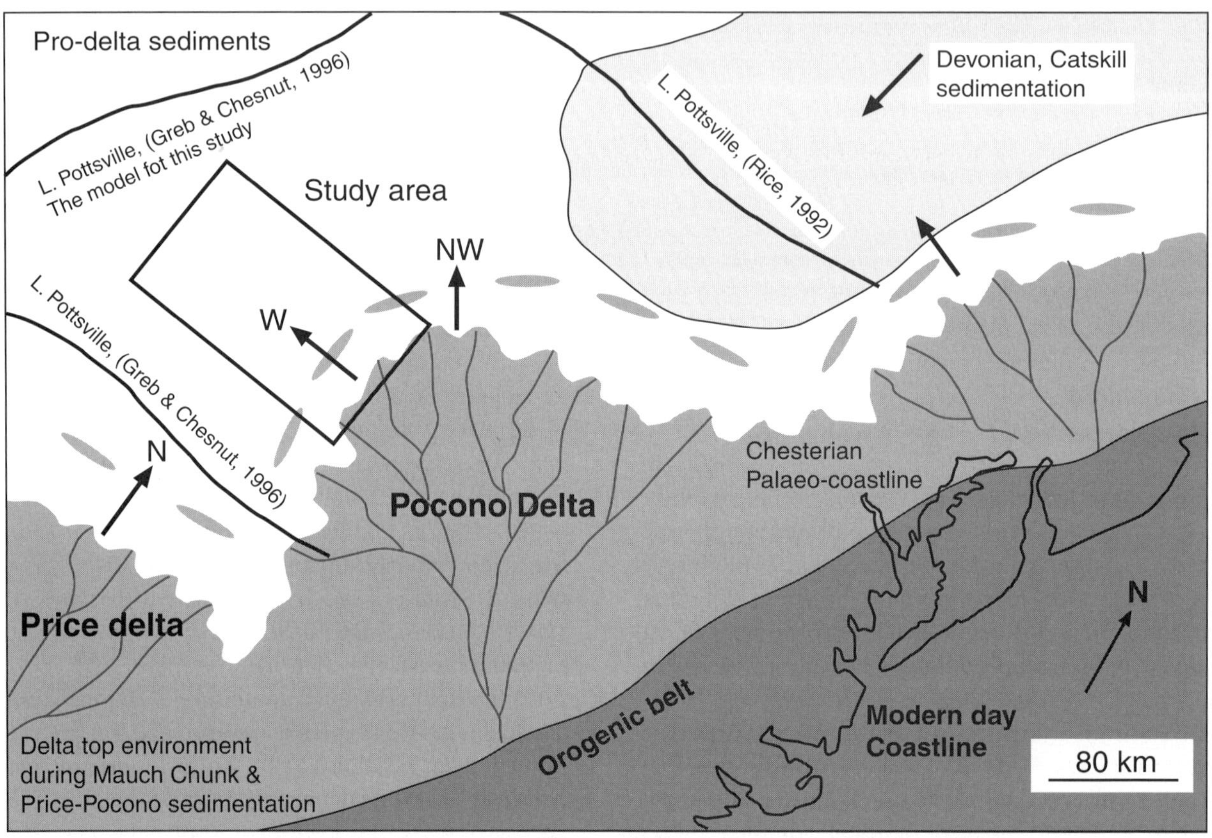

Fig. 2. A model for sedimentation progradation into the Appalachian Basin during the Devonian and Carboniferous.

margin of the North American Craton (Rice, 1994). Following the formation of the proto-Appalachian Basin during the Arcadian orogeny, clastic sediments infilled the basin, with large deltas prograding from the basin margins. Four major deltaic complexes have been identified, the Devonian Catskill delta, the Mississippian Price–Pocono and Mauch Chunk deltas and the Pennsylvanian Pocahontas–Pottsville delta (Fig. 2).

The deltaic sediments studied in Kentucky are from the Pottsville delta system and are derived from the east and southeast and thin in a northwesterly direction onto the Cincinnati Arch. The Pottsville complex (Kentucky) constitutes a prograding system

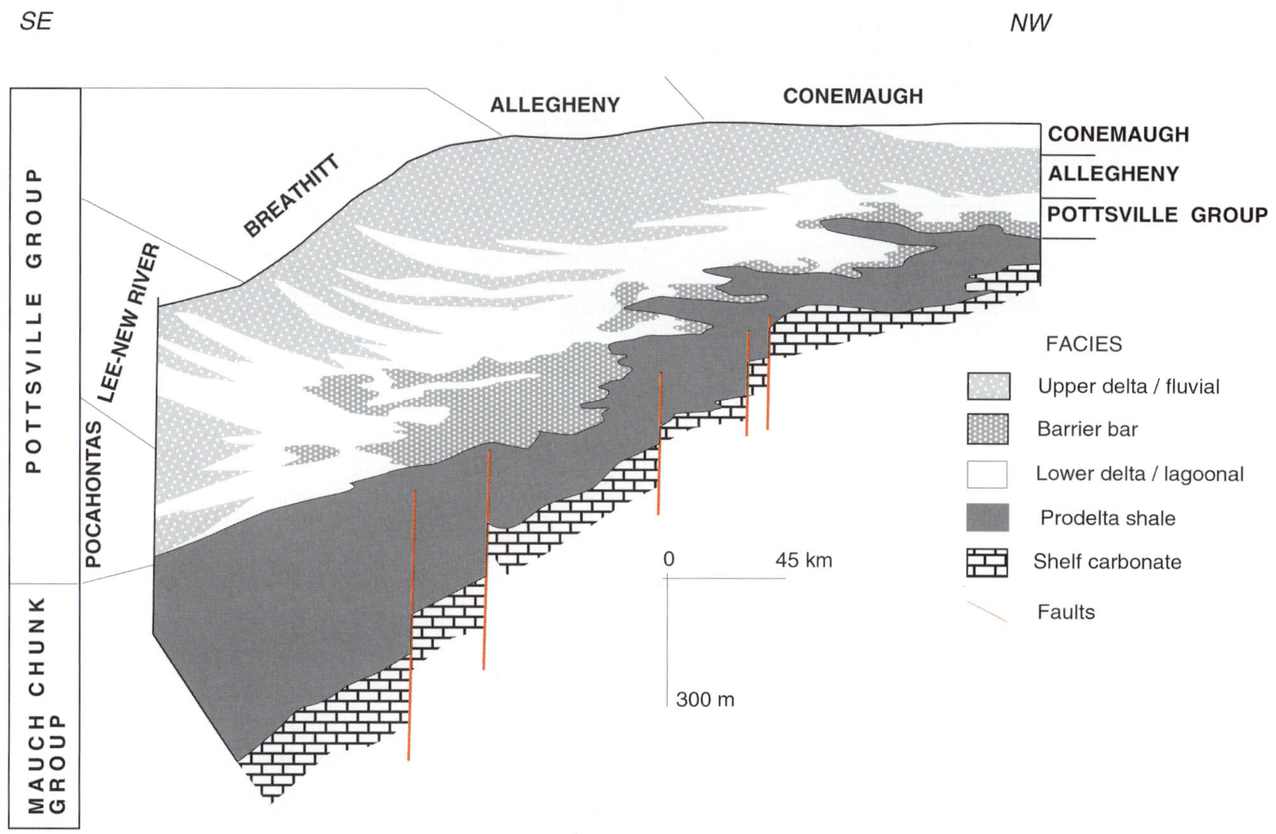

Fig. 3. Cross-section across the study area demonstrating the thickening of units towards the southeast and facies variation within the various stratigraphic units (after J.C. Ferm, written communication).

of offshore barrier island, lagoonal, delta plain and alluvial deposits with coals forming within the lagoonal setting (Horne et al., 1971; Ferm, 1974; Horne et al., 1974; Chestnut, 1992a,b). Curvilinear barrier bars were produced by wave reworking along the delta front, during periodic still stands (Englund et al., 1986). Minor carbonate (>5%) occurs within the bay sequences in both the Pennsylvanian and Mississippian delta sequences (Cobb et al., 1981; Greb et al., 1992). The middle Pennsylvanian strata in Kentucky are subdivided into Lee, Breathitt and Magoffin Formations (Fig. 3). The sediments thicken towards the southeast and the Pine Mountain thrust to a maximum thickness of 1400 m (Rice, 1994).

In eastern Kentucky growth-faulted delta sequences, present in Pennsylvanian strata of the Central Appalachian Basin include the excellent outcrops of syn-sedimentary listric faults described (Brun and Vendeville, 1993; Greb and Weisenfluh, 1996). The fault zones are situated to the west of the Alleghenian–Variscan orogenic front and are unaffected by compressional deformation. The growth faults show expansion of the alluvial to shoreface section and displace mobile delta front shales to detach upon coals at the top of the underlying sequence (Fig. 4). The detailed fault architectures, together with the associated fault rocks and their fluid flow properties are described below. The effect of observed structures on the development of reservoir models and fault seal prediction in unconsolidated sediment is discussed.

Introduction to the outcrops

The analysis of syn-sedimentary deformation in eastern Kentucky is provided by extensive and large roadcuts. The growth-faulted sequences are formed by sedimentary differential loading of marine shales. The displacement of this mobile shale creates accommodation space causing the generation of faulting in the carapace (McClay et al., 1998). The exposures are meso-scale examples of growth fault structures observed on seismic data from the Gulf of Mexico and Niger Delta (e.g. Worrall and Snelson, 1989; Cohen and McClay, 1996). The growth fault exposures are distinguished from slumps by the generation of hanging-wall growth sequences indicating a gradual rather than catastrophic mechanism. The mechanism of differential loading could be either sediment loading driven and/or sediment erosion from the toe area caused by an active fluvial system. Consequently the deformation occurred soon after deposition at shallow depths of burial when the sediments were in an unconsolidated state.

Fig. 4. An example of a growth fault system from VanCleve, eastern Kentucky. The outcrop is composed of a series of basinward dipping listric growth faults which sole onto a coal horizon and displace mobile marine shale. The outcrop is 150 m updip from the outcrop in Fig. 5.

The analysis of fault architectures and sealing was concentrated along a 300 m section of an outcrop near VanCleve (Fig. 5). The exposure is of complex, regional and counter-regional, fault systems located on a toe thrust system caused by earlier updip extension. The discontinuous nature of the fluvial sediment packages renders it difficult to trace marker horizons from the hanging-wall into the footwall. There is a prominent shale detachment horizon, which defines the low limit of deformation and a coal bed present through many fault blocks. The net to gross ratio for the section is high but variable, ranging between 40 and 90%.

Fault zone characteristics

Two critical deformation processes contribute to the finite architecture of the fault zones present. These are (1) sand and shale shearing along faults, where units are deformed to form continuous to discontinuous fault rocks within the fault zone, and (2) sand injections, where mobilized sand enters the fault zone.

Deformation mechanisms

Sand and shale shear

Whilst shale smear occurs along centimetre-scale fault zones (Fig. 6) a common structure found in the larger-scale faults (>1–5 m) is a sand layer parallel to the fault plane (Fig. 6). Faults formed whilst the sediments are unconsolidated show evidence of sands being sheared down the fault plane forming disaggregation zones. This process has been observed from Svalbard, Arctic Norway by Prestholm and Walderhaug (2000). Sand becomes entrained along the fault plane forming a clean sand fault parallel layer, which is available for fluid flow. Such features also form in analogue model experiments where sand layers remain in contact between hanging-wall and footwall via a discrete layer of sand that formed by a shear couple acting on the sand grains during extension. The resultant sand connectivity is analogous to that described in the Kentucky exposures.

The sand shear not only causes entrainment of sand along the fault plane but can also result in enhanced fault drag in the sand layers (Fig. 7). This has the effect of maintaining the connectivity of sands in the fault zone and will thus impact on fault rock predictions using SGR calculations where mixing of units is assumed.

Faults propagating through an inter-bedded sequence of sands and shales experience a strength contrast at the sand–shale interfaces leading to complex fault plane refraction. In unconsolidated sediments sands have lower cohesion and can be weaker than shales. The result is that the original fracture surface is steep through the shale sections and shallow through the sand units. Further fault development tends to 'smooth' the fault's surface by creating 'short-cut' faults and fault-bound slivers (riders) of sandstone which are incorporated into the fault zone (Fig. 8). In SGR analysis sand continuity caused by rider development is not considered.

Sand injection

When sediments are in an unconsolidated state sand can become fluidized during loading or deformation. This can lead to sands being injected along a fault plane either syn- or post-deformation. The example in Fig. 9 shows a sand injection feature that has utilized a weak coal horizon in the footwall of the fault to flow laterally into the footwall and connect with the adjacent fault system. Additional evidence

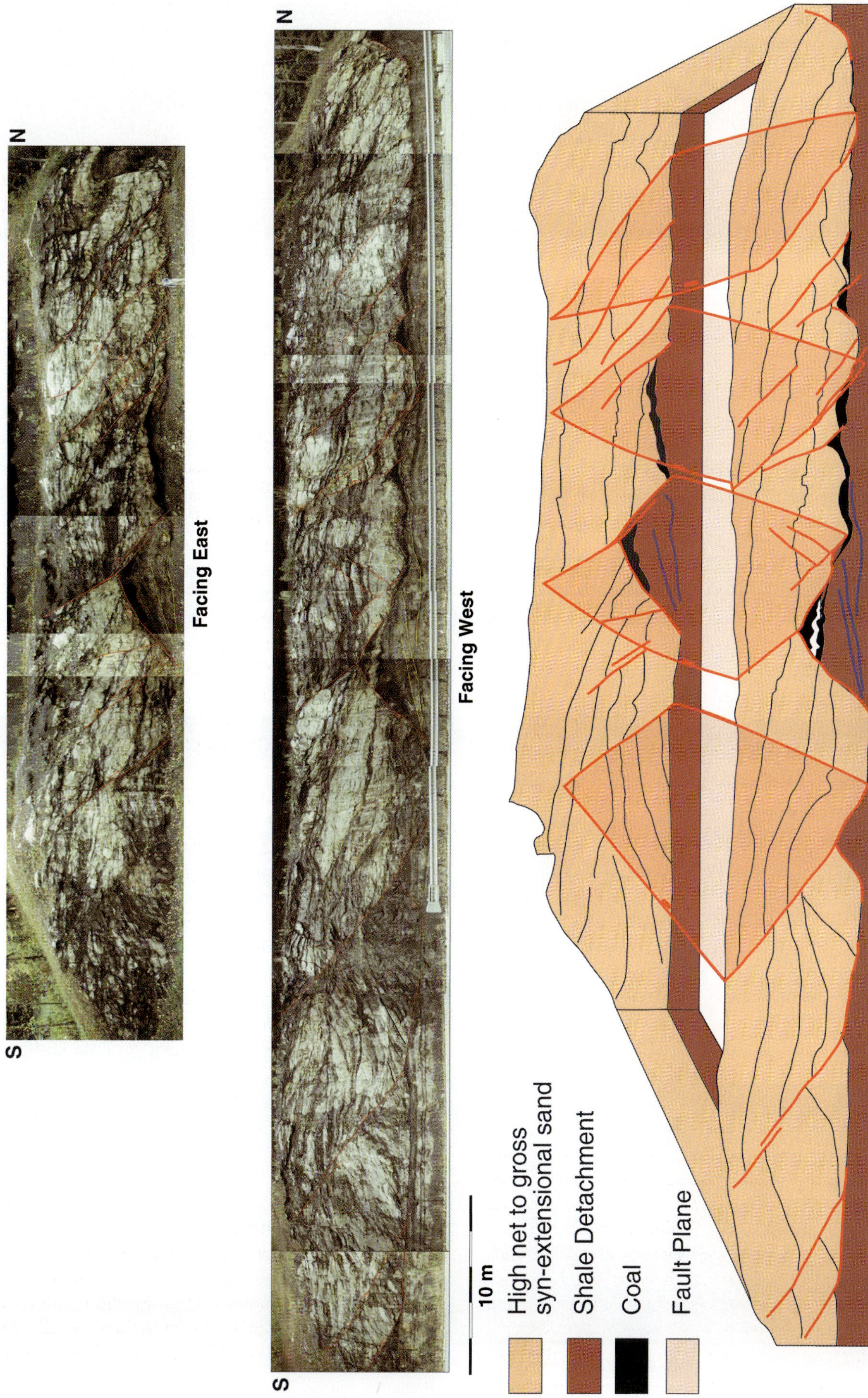

Fig. 5. Outcrop used for the collection of fault seal data. The cliff section has a number of basinward and landward dipping faults. The complex faulting pattern has been controlled by the location of a compression zone formed by extension from faults displayed in Fig. 4.

Fig. 6. A fault plane with a continuous layer of sand parallel to the fault plane. The layer is formed by sand being drawn into the fault plane during sand shear.

Fig. 7. Sand shear can result in enhanced drag on the sand units. In this example the drag of the sand bed is ten times the bed thickness.

Fig. 8. At shallow depths of burial strength differences between sands and shales result in complex fault structures. Sands can deform in a more continuous manner than shales. Variation in slip plane topography in interlayered sequences can result in shortcut faults and the creation of rider blocks and lenses.

Fig. 9. Fluidized sand can be injected along planes of weakness. In this example sand from the hanging-wall of a fault has utilized the fault plane and been injected through the footwall along a coal horizon, increasing the connectivity of the system.

that this has occurred at a shallow burial depth is that post-injection compaction of the coal has caused contortion and buckling of the sand layer.

Fault rock characteristics

Fault rock specimens associated with meso-scale faults where taken from the outcrops and examined in the SEM. Fig. 10, shows a sand-rich fault zone which has been formed by grain boundary sliding and particulate flow. The same process has been documented during deformation within a sand-layered analogue model (Fig. 10). Here the fault forms via the generation of shear zones with the layers remaining in contact despite a large throw on the fault.

The maximum burial depth experienced by the specimens studied is between 3.5 and 4.0 km. This is based upon the quality and rank of the coal beds (Cobb et al., 1981). Because of this post-deformation burial the deformation-induced porosity–permeability (poro-perm) characteristics of the samples have been masked by burial diagenesis. However, analysis of the thin section samples under cathode luminescence discriminates between original grains and the later quartz cement and allows identification of the pre-diagenetic, original fabrics. Comparison of the sand-dominated shear zone and the host rock shows no difference in grain size and similar porosity (Fig. 11a,b). This demonstrates that during the formation of the sand shear zones the grains are rolling into the fault plane without appreciable grain damage, fracturing or size reduction, highlighting that the sediments were unconsolidated at the time of deformation. The fault rocks will have approximately the same poro-perm characteristics as the host lithology and post-deformation diagenetic changes during burial should occur at the same rate (unless grain boundary sliding cleans the grains and helps promote local precipitation which may enhance fault zone cementation).

The relationship between host and fault rock permeability has been described by Prestholm and Walderhaug (2000) in samples from Svalbard. A significant difference in those samples was the growth of diagenetic clay in the fault zone that decreased the permeability. Small-scale fault zones from the Van-Cleve outcrop contain clay; however, there seems to be a net to gross control (Figs. 10 and 11c). Although the outcrop as a whole averages 70–80% sand, the hand specimen (Fig. 10) shows areas of lower net to gross ratio with as little as 50% sand. These ar-

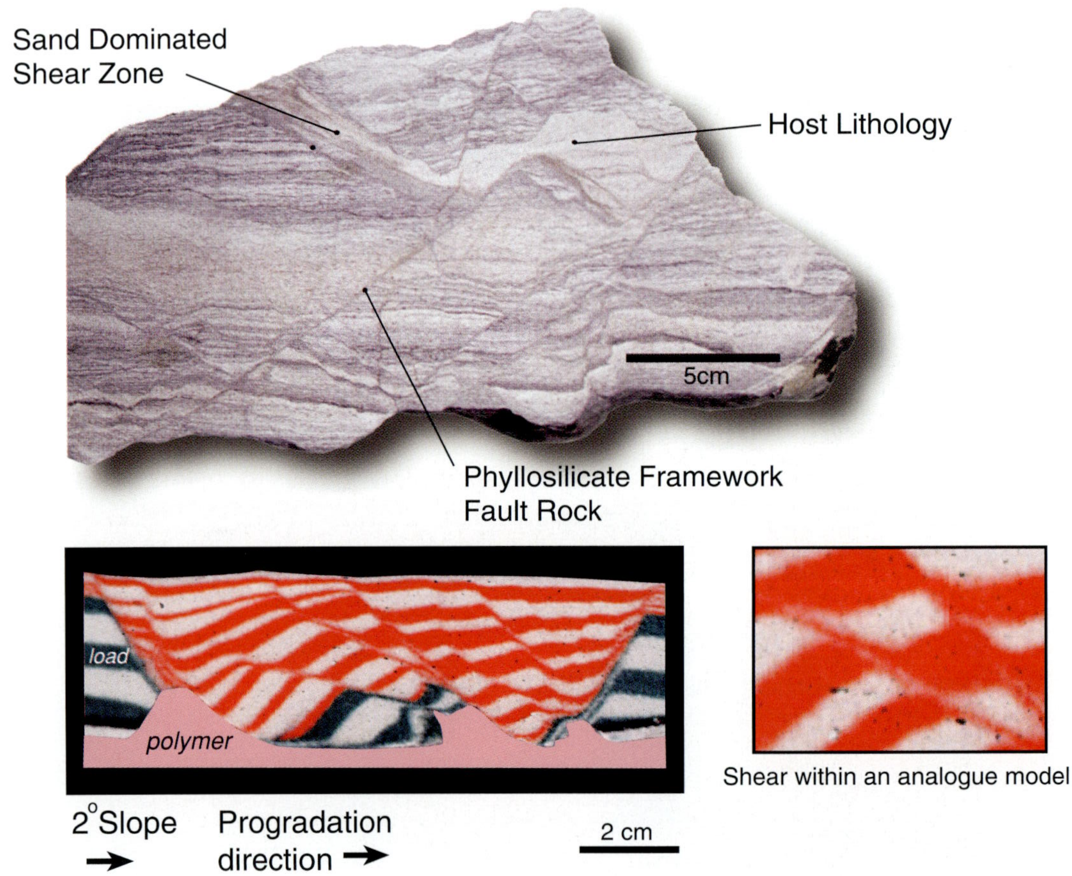

Fig. 10. Hand specimen data from small-scale faults. The specimen shows locations of thin section analysis with the host lithology, sand shear zone and a clay-rich fault. The analogue model example shows fault zones formed by sand shear where the hanging-wall and footwall units remain in communication.

Fig. 11. Microstructural analysis. (a) Cathode luminescence (CL) image of the undeformed host lithology. (b) CL image of deformed sand within a shear zone. (c) A backscattered electron image of a phyllosilicate framework fault rock.

eas have small-scale faults dominated by the local incorporation of clay into the fault zone. The supply of clay is insufficient to form extensive clay smears but transitional phyllosilicate framework fault rocks, with clay contents of ~15–40% (Fisher and Knipe, 1998) are common. The backscattered electron image in Fig. 11c shows discrete clay plates mixing with detrital quartz grains, which will contribute to the early permeability reduction in these fault rocks.

Implications

Outcrop data from eastern Kentucky have illustrated a number of processes, which are critical to understanding the generation of fault rocks formed in unconsolidated strata.

Sand shear (Fig. 6) caused unconsolidated sand to be preferentially incorporated into the fault zone and causes increased sand continuity in fault zones. The characteristics of meso-scale fault zones formed in the poorly consolidated sediments, described above, will be dependent upon both the sand and shale behaviour in the fault zone. The burial history, stress path and consolidation state at the time of faulting will be critical to this behaviour (Jones, 1994). Assuming shallow burial depths (0–300 m) the sand incorporation and continuity in the fault zone is governed by three main variables: the average net to gross ratio of the system, the sand bed thickness, and the sand unit spacing. The environments most prone to sand-rich fault zones will be high net to gross sands with thick beds. The efficiency of the sand incorporation process will be reduced with decreasing net to gross, lower bed thickness and by increased cohesion formed by compaction and diagenesis. Thin, widely spaced sands will have difficulty supplying sufficient sand to form the continuous conduits as observed in the VanCleve outcrop (Fig. 6).

Sand can be injected along fault planes and between hanging-wall and footwall stratigraphies causing increased connectivity between fault blocks (Fig. 8). The evolution of sand and clays/shales properties (porosity, permeability and strength) during compaction and consolidation is complex (Karig and Morgan, 1994; Clennell et al., 1999; Petley, 1999) and strength inversions between the sands and shales are possible at shallow depths. This results in variable 3D fault geometries in interlayered sequences (e.g. undulating slip surfaces) and structures (sheared sediment layers and lenses, Fig. 8). There will be a critical combination of the net to gross, the layer lithologies, thicknesses, spacings and stacking sequences as well as the strength and deformation properties of the individual layers and the effective stress path, which will control the continuity and properties of sand and shale in the fault zones. The shale behaviour is particularly important, as failure in the shales (due to higher cohesion) can promote/accommodate granular flow in sand units, leading to discontinuous (faulted) clay-rich units but more continuous sand layers/lenses along faults. In the outcrops studied here sand units can be continuous within fault zones for up to ten times the undeformed sand bed thickness.

During fault seal analysis the decision to use a methodology, which predicts the fault rock properties, is often taken without consideration of possible sand continuity along the fault. Traditional fault seal analysis techniques such as the Shale Gouge Ratio (SGR) assumes the percentage of shale in the fault zone and hence its sealing potential is based upon the average clay content of the sediments passing a point on the fault. The incorporation or injection of sand into the fault plane is not accounted for. Thus use of the SGR method for assessing intra-reservoir faulting formed by syn-sedimentary processes at shallow depths of burial will overestimate the sealing potential of the system.

Soft sedimentary deformation fault seal has been applied to Chevron's Britannia gas field on the flank of the Witch Ground Graben. Original SGR-based fault seal calculations highlighted areas of the field which should be compartmentalized from the platform area of the field by a complex of growth faults. Later drilling and early production data showed these potentially compartmentalized areas to be in pressure communication with the platform wells as pressure depletion had been recorded. The Britannia reservoir is a series of complex sheet sands (Jones et al., 1999) and pressure drawdown was recorded from unproduced zones. The data suggested that the vertical permeability of the system had been improved and faults were not extensive fluid barriers. The favoured model is for the growth faults to act as fluid conduits due to the incorporation of sand in the fault plane and/or injection caused soon after the deposition of the Britannia reservoir (Porter et al., 2000).

The recognition of fault properties associated with the deformation of poorly consolidated sediments to fault zones is allocable to reservoir studies in many parts of the world in particular, the Cretaceous and Tertiary of Norway, deep-water Nigeria–Angola and Brazil. The example from the UK sector of the North Sea demonstrated a positive impact of faulting on production. However, a negative effect can be that sand-rich fault zones may increase the lateral leakage of exploration prospects. Small faults that cut the reservoir unit and extend into the top seal can become a significant risk if they compromise the top seal.

Acknowledgements

Robert Hunsdale and Alastair Welbon are thanked for useful comments during review which improved earlier versions of the manuscript. Norsk Chevron is thanked for allowing the time for this analysis and for publication of the manuscript. Without the generous support of the AAPG Grants in Aid Program the initial field season from which this research is based would not have been possible. The support of Chuck Kluth of Chevron Overseas Petroleum Inc. and the staff of Rock Deformation Research in Leeds, especially Ned Porter and Tim Needham, is acknowledged. Discussion with Lucy Williams (nee Jones) and Andy Palfrey of Chevron are also acknowledged. Simon Stevens of Wave is thanked for his assistance with the drafting, especially with poster versions of this paper.

References

Brun, J.P. and Vendeville, B., 1993. Linked diapirs and growth fault kinematics: Outcrop analysis and experimental results. AAPG Hedberg Conference, Bath, England, September 13–17 (abstr).

Chestnut Jr., D.R., 1992a. Geologic highway cross sections: Interstate highway 75, Conway (Ky) to Jellico (Te). Map and chart series 3, series XI. Kentucky Geological Survey.

Chestnut Jr., D.R., 1992b. Geologic highway cross sections: Interstate highway 64, Farmers to Cattlesburg (Ky). Map and chart series 4, series XI. Kentucky Geological Survey.

Clennell, M.B., Dewhurst, D.N., Brown, K.M. and Westbrook, G.K., 1999. Permeability anisotropy of consolidated clays. In: A.C. Aplin, A.J. Fleet and J.H. Macquaker (Editors), Muds and Mudstones: Physical and Fluid-Flow Properties. Geological Society, London, Spec. Publ., 158: 79–96.

Cobb, J.C., Chestnut Jr., D.R., Hester, N.C. and Hower, J.C., 1981. Coal and coal bearing rocks of eastern Kentucky. Geological Society of America, Annual Coal Division Field Trip, November 5–8.

Cohen, H.A. and McClay, K.R., 1996. Sedimentation and shale tectonics of the northwestern Niger Delta front. Mar. Pet. Geol., 13: 313–328.

Englund, K.J., Windolph J.F., Jr. and Thomas, R.E., 1986. Origin of thick, low-sulphur coal in the Lower Pennsylvanian Pocahontas Formation, Virginia and West Virginia. Geol. Soc. Am. Spec. Pap., 210: 49–61.

Ferm, J.C., 1974. Carboniferous environmental models in the eastern US and their significance. In: G. Briggs (Editor), Carboniferous of the Southeastern USA. Geol. Soc. Am. Spec. Pap., 148: 79–95.

Fisher, Q.J. and Knipe, R.J., 1998. Fault sealing processes in siliciclastic sediments. In: G. Jones, R.J. Knipe and Q.J. Fisher (Editors), Faulting, Fault Sealing and Fluid Flow in Hydrocarbon Reservoirs. Geol. Soc., London, Spec. Publ., 147: 117–134.

Greb, S.F. and Weisenfluh, G.A., 1996. Palaeoslumps in coal bearing strata of the Breathitt Group (Pennsylvanian), eastern Kentucky coal field, USA. J. Coal Geol., 631.

Greb, S.F., Williams, D.A. and Williamson, A.D., 1992. Geology and stratigraphy of the western Kentucky coalfield. Kentucky Geological Survey, Bull. 2, Series XI.

Horne, J.C., Swinchatt, J.P. and Ferm, J.C., 1971. Lee–Newman barrier shoreline model. In: Carboniferous Depositional Environments in Northeastern Kentucky. Geological Society of Kentucky Guidebook for Annual Spring Field Conference.

Horne, J.C., Ferm, J.C. and Swinchatt, J.P., 1974. Depositional model for the Mississippian–Pennsylvanian boundary in northeast Kentucky. Geol. Soc. Am. Spec. Pap., 148: 97–115.

Jones, G., Knipe, R.J. and Fisher, Q.J. (Editors), 1998. Faulting, Fault Sealing and Fluid Flow in Hydrocarbon Reservoirs. Geological Society, London, Special Publication, 147.

Jones, L.S., Garrett, S.W., Macleod, M., Guy, M., Condon, P.J. and Notman, L., 1999. Britannia Field, UK Central North Sea: modelling heterogeneity in unusual deep water deposits. In: A.J. Fleet and S.A.R. Boldy (Editors), Petroleum Geology of NW Europe. Proceedings of the 5th Conference, Geological Society, London, pp. 1115–1124.

Jones, M.E., 1994. Mechanical principles of sediment deformation. In: A.J. Maltman (Editor), Geological Deformation of Sediments. Chapman and Hall, London, pp. 37–71.

Karig, D. and Morgan, J., 1994. Tectonic Deformation; stress paths and strain histories. In: A.J. Maltman (Editor), Geological Deformation of Sediments. Chapman and Hall, London, pp. 167–204.

Knipe, R.J., Jones, G. and Fisher, Q.J., 1998a. Faulting, fault seal and fluid flow in hydrocarbon reservoir: an introduction. In: G. Jones, Q.J. Fisher and R.J. Knipe (Editors), Faulting, Fault Sealing and Fluid Flow in Hydrocarbon Reservoirs. Geol. Soc., London, Spec. Publ., 147: vii–xxi.

Knipe, R.J., Fisher, Q.J., Jones, G., Clennell, M.B., Farmer, A.B., Kidd, B., McAllister, E., Porter, J.R. and White, E., 1998b. Fault seal prediction methodologies, applications and successes. In: P. Møller-Pedersen and A.G. Koestler (Editors), Hydrocarbon Seals: Importance for Exploration and Production. Norwegian Petroleum Society (NPF), Special Publication 7. Elsevier, Amsterdam, pp. 15–38.

McClay, K.R., Dooley, T. and Lewis, G., 1998. Analogue modeling of progradational delta systems. Geology, 26: 771–774.

Møller-Pedersen, P. and Koestler, A.G. (Editors), 1998. Hydrocarbon Seals — Importance for Exploration and Production. Norwegian Petroleum Society (NPF), Special Publication 7. Elsevier, Amsterdam.

Petley, D.N., 1999. Failure envelopes of mudrocks at high confining pressures. In: A.C. Aplin, A.J. Fleet and J.H. Macquaker (Editors), Muds and Mudstones: Physical and Fluid-Flow Properties. Geol. Soc., London, Spec. Publ., 158: 1–9.

Porter, J.R., Knipe, R.J., Fisher, Q.J., Farmer, A.B., Allin, N.S., Jones, L.S., Palfrey, A.J., Garrett, S.W. and Lewis, G., 2000. Deformation processes in the Britannia Field, UKCS. Pet. Geosci., 6: 241–251.

Prestholm, E. and Walderhaug, O., 2000. Synsedimentary faulting in a Mesozoic deltaic sequence, Svalbard, Arctic Norway — fault geometries, faulting mechanisms and sealing properties. Assoc. Am. Pet. Geol. Bull., 84 (4): 505–522.

Rice, C.L., 1994. Introduction. In: C.L. Rice (Editor), Elements of Pennsylvanian Stratigraphy, Central Appalachian Basin. Geol. Soc. Am. Spec. Pap., 294: 1–5.

Worrall, D.M. and Snelson, S., 1989. Evolution of the northern Gulf of Mexico, with emphasis on Cenozoic growth faulting and the role of salt. In: A.W. Bally and A.R. Palmer (Editors), The Geology of North America — An Overview, A. Geological Society of America, Boulder, CO, pp. 97–138.

Yielding, G., Freeman, B. and Needham, D.T., 1997. Quantitative fault seal prediction. Am. Assoc. Pet. Geol. Bull., 81: 897–917.

G. LEWIS *Norsk Chevron, Karenslyst Alle 2–4, P.O. Box 97 Skøyen, 0212 Oslo, Norway*
E-mail: glew@chevron.com

R.J. KNIPE *Rock Deformation Research Group, Earth Sciences Department, University of Leeds, Leeds LS2 9JT, UK*

A. LI *Rock Deformation Research Group, Earth Sciences Department, University of Leeds, Leeds LS2 9JT, UK*

References index

Subject index